NATO ASI Series

Advanced Science Institutes Series

A series presenting the results of activities sponsored by the NATO Science Committee, which aims at the dissemination of advanced scientific and technological knowledge, with a view to strengthening links between scientific communities.

The Series is published by an international board of publishers in conjunction with the NATO Scientific Affairs Division

A Life Sciences **B Physics**	Plenum Publishing Corporation London and New York
C Mathematical and Physical Sciences **D Behavioural and Social Sciences** **E Applied Sciences**	Kluwer Academic Publishers Dordrecht, Boston and London
F Computer and Systems Sciences **G Ecological Sciences** **H Cell Biology**	Springer-Verlag Berlin Heidelberg New York London Paris Tokyo

The ASI Series Books Published as a Result of Activities of the Special Programme on CELL TO CELL SIGNALS IN PLANTS AND ANIMALS

This book contains the proceedings of a NATO Advanced Research Workshop held within the activities of the NATO Special Programme on Cell to Cell Signals in Plants and Animals, running from 1984 to 1989 under the auspices of the NATO Science Committee.

The books published as a result of the activities of the Special Programme are:

Vol. 1: **Biology and Molecular Biology of Plant-Pathogen Interactions.** Edited by J. A. Bailey. 1986.

Vol. 2: **Glial-Neuronal Communication in Development and Regeneration.** Edited by H. H. Althaus and W. Seifert. 1987.

Vol. 3: **Nicotinic Acetylcholine Receptor: Structure and Function.** Edited by A. Maelicke. 1986.

Vol. 4: **Recognition in Microbe-Plant Symbiotic and Pathogenic Interactions.** Edited by B. Lugtenberg. 1986.

Vol. 5: **Mesenchymal-Epithelial Interactions in Neural Development.** Edited by J. R. Wolff, J. Sievers, and M. Berry. 1987.

Vol. 6: **Molecular Mechanisms of Desensitization to Signal Molecules.** Edited by T. M. Konijn, P. J. M. Van Haastert, H. Van der Starre, H. Van der Wel, and M. D. Houslay. 1987.

Vol. 7: **Gangliosides and Modulation of Neuronal Functions.** Edited by H. Rahmann. 1987.

Vol. 9: **Modification of Cell to Cell Signals During Normal and Pathological Aging.** Edited by S. Govoni and F. Battaini. 1987.

Vol. 10: **Plant Hormone Receptors.** Edited by D. Klämbt. 1987.

Vol. 11: **Host-Parasite Cellular and Molecular Interactions in Protozoal Infections.** Edited by K.-P. Chang and D. Snary. 1987.

Vol. 12: **The Cell Surface in Signal Transduction.** Edited by E. Wagner, H. Greppin, and B. Millet. 1987.

Vol. 19: **Modulation of Synaptic Transmission and Plasticity in Nervous Systems.** Edited by G. Hertting and H.-C. Spatz. 1988.

Series H: Cell Biology Vol. 19

Modulation of Synaptic Transmission and Plasticity in Nervous Systems

Edited by

Georg Hertting

Institut für Pharmakologie, Universität Freiburg
Hermann-Herder-Str. 5, 7800 Freiburg i. Br., FRG

Hanns-Christof Spatz

Institut für Biologie III, Universität Freiburg
Schänzlestr. 1, 7800 Freiburg i. Br., FRG

Springer-Verlag
Berlin Heidelberg New York London Paris Tokyo
Published in cooperation with NATO Scientific Affairs Division

Proceedings of the NATO Advanced Research Workshop on Modulation of Synaptic Transmission and Plasticity in Nervous Systems held in Il Ciocco, Castelvecchio Pascoli, Italy, September 8–13, 1987

ISBN-13:978-3-642-73162-4 e-ISBN-13:978-3-642-73160-0
DOI: 10.1007/978-3-642-73160-0

Library of Congress Cataloging-in-Publication Data. NATO Advanced Research Workshop on Modulation of Synaptic Transmission and Plasticity in Nervous Systems (1987 : Castelvecchio Pascoli, Italy) Modulation of synaptic transmission and plasticity in nervous systems / edited by Georg Hertting, Hanns-Christof Spatz. p. cm.—(NATO ASI series. Series H, Cell biology : vol. 19) "Proceedings of the NATO Advanced Research Workshop on Modulation of Synaptic Transmission and Plasticity Nervous Systems held in Il Ciocco, Castelvecchio Pascoli, Italy, September 8–13, 1987"—T.p. verso. "Published in cooperation with NATO Scientific Affairs Division." Includes index.
ISBN-13:978-3-642-73162-4(U.S.)
1. Neural transmission—Regulation—Congresses. 2. Neuroplasticity—Congresses. I. Hertting, Georg. II. Spatz, Hanns-Christof. III. North Atlantic Treaty Organization. Scientific Affairs Division. IV. Title. V. Series. [DNLM: 1. Neural Transmission—congresses. 2. Neuronal Plasticity—congresses. 3. Neuroregulators—congresses. 4. Synapses—congresses. WL 102.8 N279m 1987] QP364.5.N37 1987 591.1'88—dc 19
DNLM/DLC for Library of Congress 88-12238

Softcover reprint of the hardcover 1st edition 1988

2131/3140-543210

Preface

This volume is based on a workshop
"Modulation of Synaptic Transmission and Plasticity in Nervous Systems"
held in Il Ciocco, Castelvecchio, Pascoli, Italy,
from September 8th to 13th, 1987.

The purpose of the meeting was to bring together scientists working on plasticity in nervous systems on different levels. The contributions can be subgrouped into six fields of research:

1) Presynaptic Modulation of Chemical Neurotransmission
2) Postsynaptic Signal Transduction
3) Modulation of Synaptic Transmission and Plasticity in the Hippocampus
4) Modulation of Neuromuscular Transmission
5) Molecular and Cellular Analysis of Conditioning in Marine Snails
6) Analysis of Learning and Memory in Insects

Understanding how nervous systems and in particular our brain processes and stores information has been a major challenge in science for centuries and will remain for some time to come. Not until recently neurobiologists agreed to seek plasticity of behavior primarily in the modulation of the properties of synapses between nerve cells. This is to be understood within the context provided by a neural circuitry. An important stimulus came from the work on the marine snail Aplysia, where learning processes can be described as a modulation of transmitter release, traced back to a complete chain of molecular events in an identified neuron. Learning became a topic of molecular biology. Three systems appear particularly promising for this approach: insects, in particular Drosophila, marine snails and the mammalian hippocampal tissue.

Our views on neurotransmission have rapidly changed. The new model now emerging has to account for a diversification in different junctions. The list of neurotransmitters is rapidly growing. Many neurons use several transmitters. The nerve impulse pattern may influence not only the amount but also the combination of transmitters released. Multiple receptors coupled to multiple transducers mediate multiple action in effector cells.

Our thinking about the mechanisms of learning and memory in nervous systems is guided by two models:

1) Formation of new junctions between nerve cells or the destruction of junctions.

2) Modulation of the efficiency of already existing synapses.

The complexity and variability of neurotransmission provides the basis for the later model. A great effort, well represented in this volume, is made to elucidate the molecular details of such mechanisms in well defined systems. No unifying picture has yet appeared and may never appear, but the results from various laboratories tie into each other in major aspects.

Georg Hertting Hanns-Christof Spatz

The meeting was made possible by a generous grant from the NATO Scientific Affairs Division.

We gratefully acknowledge further financial support from

E. Merck, Darmstadt, F.R.G.
Biodesign, Freiburg i. Br., F.R.G.
Gry-Pharma, Kirchzarten, F.R.G.
Goedecke AG, Freiburg i. Br., F.R.G.
Knoll AG, Ludwigshafen, F.R.G.
Bayer AG, Leverkusen, F.R.G.
Farmitalia, Freiburg i. Br., F.R.G.
Deutsche Biophysikalische Gesellschaft, F.R.G.

The preparation of this meeting profited greatly from the helpful services of the Institut für Biologie III, Albert-Ludwigs-Universität, Freiburg i. Br.

C.Allgaier, J.Buhmann

A.Ferrús, T.Tully

P. Evans, R.A. North, R. Malenka

H.Gerschenfeld, W.H.Gispen

J. Dudel, K. Starke, E. Muscholl

H.-CH.Spatz, G.Hertting, L.Beani

Table of Contents

PRESYNAPTIC MODULATION OF CHEMICAL NEUROTRANSMISSION: THE BLOOD VESSEL PARADIGM

Klaus Starke
Department of Pharmacology
University of Freiburg
Hermann-Herder-Strasse 5
D-7800 Freiburg i.Br.
Federal Republic of Germany

Various types of modulation of synaptic transmission probably are of the utmost importance in complex neuronal networks. Yet, modulation can already be observed in rather simple systems. An example is the prejunctional modulation of neuroeffector transmission in mammalian blood vessels. Figure 1 shows a postganglionic sympathetic axon varicosity in the tunica adventitia. Action potentials release noradrenaline or the co-transmitters noradrenaline, ATP and neuropeptide Y. The transmitters diffuse to their respective receptors on smooth muscle cells in the tunica media. The release of transmitter per pulse, however, far from being constant, is subject to a manifold modulation by substances that act at receptors located in the varicosity membrane. Prejunctional receptors have been identified in blood vessels for prostaglandins, dopamine, noradrenaline (α- and β-adrenoceptors), neuropeptide Y, adenosine (P_1 receptors), serotonin, GABA, histamine, acetylcholine (muscarine and nicotine receptors), angiotensin, opioids (mu, delta and kappa receptors), atrial natriuretic factor and ACTH. Using, for instance, exogenous opioid delta or kappa agonists in rabbits, one can demonstrate a sequence of events from inhibition of action potential-evoked transmitter release (an effect of unknown cellular mechanism), via a decrease of smooth muscle excitatory junction potential amplitude, via inhibition of neurogenic vasoconstriction in vitro, to a fall in blood pressure in the intact animal (Starke et al. 1987).

NATO ASI Series, Vol. H19
Modulation of Synaptic Transmission and Plasticity
in Nervous Systems. Edited by G. Hertting and H.-C. Spatz

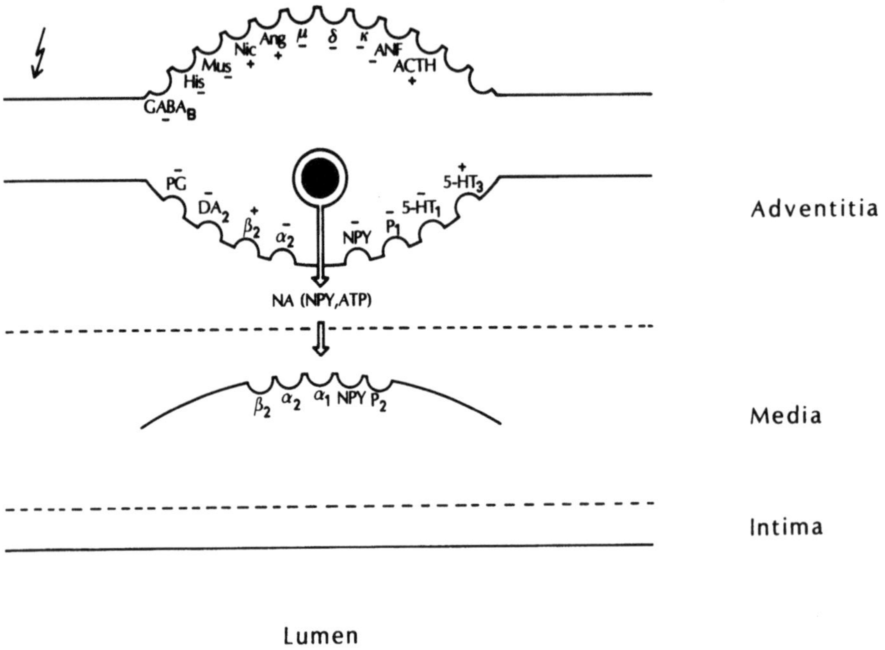

Fig. 1. Vascular sympathetic neuroeffector junction. ANF, atrial natriuretic factor; Ang, angiotensin; DA, dopamine; His, histamine; 5-HT, serotonin; Mus, muscarine; NA, noradrenaline; Nic, nicotine; NPY, neuropeptide Y; PG, prostaglandin.

Although all the prejunctional receptors depicted in Figure 1 have been demonstrated in blood vessels pharmacologically, the question of their physiological role in vivo has rarely been answered. There is no clear evidence for the view that the prejunctional opioid receptors ever receive an endogenous agonist input. The case for a physiological function of other prejunctional sites such as the α-adrenoceptors is more favourable (Starke 1987).

The blood vessel neuromuscular junction is a simple paradigm of neuromodulation. The biochemical, pharmacological and physiological aspects that can be studied here recur in the numerous instances of prejunctional modulation that have been detected in recent years in more complex situations such as the mammalian central nervous system.

REFERENCES

Starke K, Illes P, Ramme D, Ensinger H, Hedler L, Szabo B, von Kügelgen I, Pfeiffer N, Limberger N (1987) Peripheral pre-junctional opioid receptors in cardiovascular control. In: Neuronal Messengers in Vascular Function (Nobin A, Owman C, Arneklo-Nobin B, eds.), pp 247-269. Elsevier, Amsterdam

Starke K (1987) Presynaptic α-autoreceptors. Rev Physiol Biochem Pharmacol 107:73-146

MODULATION OF PURINE RELEASE FROM ELECTRICALLY-STIMULATED CORTICAL SLICES OF THE RAT: INTERACTION WITH THE CHOLINERGIC SYSTEM

F. Pedata and G. Pepeu
Department of Preclinical and Clinical Pharmacology
University of Florence
Viale Morgagni 65
50134 Florence
Italy

The activity and role of adenosine in the central nervous system have been extensively described in several recent reviews (Su et al. 1983; Dunviddie 1985; Snyder 1985) from which it appears that adenosine fulfills many of the criteria which define a neurotransmitter or neuromodulator. Adenosine is present in neuronal somata and axons of discrete brain areas including certain layers of the cerebral cortex (Braas et al. 1986). A release of adenosine from nervous tissue, evoked by electrical stimulation or by depolarizing agents, has been repeatedly demonstrated in both "in vivo" (Sulakhe and Phillis 1975; Lee et al. 1982) and "in vitro" studies (Pull and Mc Ilwain 1972; Hollins and Stone 1980; Mc Donald and White 1985). Adenosine reaches in the extracellular fluid of brain "in vivo" concentrations sufficient to exert physiological effects (Zetterstrom et al. 1982). Adenosine effects are exerted through specific receptors which can be distinguished by their actions on adenylate cyclase (Hamprecht and Van Calker 1980) or by the use of specific receptor agonists (Londos et al. 1977) into A_1 and A_2 subtypes.

The electrophysiological effects of adenosine are characterized by a depressant action on the firing rate of central neurons (Phillis and Kostopulos 1975) and on synaptic trans-

NATO ASI Series, Vol. H19
Modulation of Synaptic Transmission and Plasticity
in Nervous Systems. Edited by G. Hertting and H.-C. Spatz

mission (see Dunwiddie 1985). The effects may be attributable to both pre and postsynaptic actions. The postsynaptic actions of adenosine involve activation of ionic conductance to potassium resulting in hyperpolarization of the membranes (Segal 1982), inhibition of calcium conductance (Henon and McAfee 1983) and decreased sensitivity of postsynaptic receptors to the transmitters (Perkins and Stone 1980).

The best characterized and most frequently described action of adenosine, however, is the presynaptic inhibition of neurotransmitter release presumably caused by a direct effect on the nerve terminals. An example is given by the reduction of the excitatory aminoacids, glutamate and aspartate, released from ippocampal slices (Corradetti et al. 1984) which can be related to the reduction of the $EPSP_s$ evoked by stimulation of the stratum radiatum in the hippocampus (Dunwiddie and Hoffer 1980; Okada and Kuroda 1980).Adenosine also decreases acetylcholine (ACh) release from electrically stimulated cortical slices (Pedata et al. 1983; Fredholm et al.1987) and from hippocampal and cortical synaptosomes depolarized by potassium (Pedata et al. 1986). These effects are antagonized by methylxanthines and mimicked by A_1 receptor agonists.

In behavioural tests, adenosine and its derivatives show marked sedative, hypnotic and anticonvulsant actions (see Dunwiddie 1985) and according to Phillis and Wu (1983) part of the mechanism of action of benzodiazepines could be explained by their effect on adenosine uptake and release.

We reported (Pedata et al. 1984) that methylxanthines at low concentrations stimulate while at high concentrations decrease the release of ACh from electrically stimulated cortical slices. The biphasic effects of methylxanthines are prevented by dipyridamole, an adenosine uptake blocker, and may also depend on changes in adenosine levels in the cortical slices. Furthermore, an interaction seems to exist between the inhibitory effects of cholinergic agonists and adenosine on ACh release (Pedata et al. 1986) and cholinergic agonists also block adenosine inhibition on neuronal firing in hippocampal slices (Worley et al. 1987). In this work, therefore, after defining the characteristics of ^{3}H purine release evoked by different conditions of electrical stimulation, we studied the effect of different concentrations of caffeine and of cholinergic agonists on purine release from rat cortical slices.

MATERIALS AND METHODS

Preparation and electrical stimulation of cortical slices

The experiments were carried out on male Wistar rats of 150-200 g body weight. The cortical slices were prepared and electrically stimulated using the apparatus and the procedure described by Beani et al. (1978). The cortices submerged in Krebs solution were cut into slices 400 µm thick by means of a microtome for fresh tissue. The slice weight was 32 $\pm$ 0.3 mg (n=10). The slices were then kept floating for 30 min in Krebs solution continuously gassed with 95% O_2 and 5% CO_2 at room temperature. This washing procedure was repeated once at room temperature and once at 37°C (Jonzon and Fredholm 1985).

The slices were labelled by incubation for 45 min at 37°C with ^{3}H adenine (specific activity 23 Ci/mmol, Amersham), 0.1 µM final concentration. After labelling, the slices were transferred to Perspex superfusion chambers of 0.3 ml volume and perfused with Krebs solution at 37°C, directly gassed with 95% O_2 and 5% CO_2 at the entrance of the chamber, at a flow rate of 0.8 ml/5 min. After 60 min of washing, a steady state of tritium efflux was reached and 5 min fractions were collected from then on.

After collecting two 5 min fractions, the slices were routinely field-stimulated with rectangular pulses of alternating polarity, with a current strength of 30 mA/cm^2 and a pulse duration of 5 ms at 10 Hz stimulation frequency for 5 min (S_1). A second 5 min stimulation period (S_2), at the same frequency, was carried out 45 min after the first. The drugs were added to the superfusion fluid taking into account the dead space and the perfusion speed reaching the slices 20 min before S_2. A third S_3 stimulation period was carried out in the experiments in which calcium was omitted from the superfusion fluid. At the end of experiments the slices were homogenized in 1 ml perchloric acid 0.4 N and radioactivity in the slices and superfusate fractions was measured by means of a liquid scintillation counter (Packard Tri-Carb 460 C) after addition of 8 ml Instagel with a 36% counting efficiency determined by the external standard method. The fractional release of radioactivity (per 5 min) was calculated as percent of the total content of tritium in the slices at the beginning of each collection period. Drug effects on electrically evoked release were evaluated by calculating the ratio (S_2/S_1) of the release evoked by the two stimulation periods. Drug effects on the basal release were evaluated by calculating the ratio (basal ratio) of the mean of the two 5 min fractional releases before S_2 and S_1 (Jackisch et al. 1984).

<u>Characterization of radioactivity</u>

In two experiments aliquots of the superfusate were directly analyzed using reverse phase HPLC with ultraviolet absorbance detection at 254 nm for analysis of adenosine, inosine, hypoxanthine and adenine, according to the method of Fredholm and Sollevi (1981). ATP, ADP, AMP and hypoxanthine were assayed with the same equipment as described by Schweinberg and Loo

(1980). The eluate was collected in 0.5 or 1.0 min fractions and the radioactivity counted with a 31% counting efficiency determined by the external standard method.

Drugs

The following drugs were used: ethylene Glycol bis (aminoethylether) N,N,N',N'-tetraacetic acid (EGTA) Sigma, physostigmine sulphate Sigma, caffeine Sigma, atropine sulphate BDH, tetrodotoxin (TTX) Calbiochem., oxotremorine sesquifumarate Janssen Chemicals, dipyridamole Boeringer Ingelheim.

Statistical analysis

Student's two tailed t test was used to calculate differences in drug effects. The regression line was calculated by least squares analysis and tested for correlation coefficient (r).

RESULTS

Characteristic of the electrically-evoked ^{3}H purine release

Fig. 1 shows the increase of ^{3}H purine release evoked by 10 Hz stimulation frequency at different stimulus intensities. From the comparison of the S_2/S_1 ratios in normal Krebs solution with those obtained in the presence of TTX, added during the second stimulation period, it appears that 98% of the evoked ^{3}H release is TTX-sensitive at 20 mA/cm^2 stimulus intensity, 86% at 30, 84% at 45, 53% at 60 and only 45% at 100 mA/cm^2. A stimulus intensity of 30 mA/cm^2 was selected for our routine experiments because the ratio between basal and evoked release was higher than at 20 mA/cm^2. Separation of radioactive compounds released at rest and during 10 Hz stimulation at 30 mA/cm^2 stimulus intensity (Table 1) shows that, both at rest and after stimulation, 98% of the radioactivity consisted of

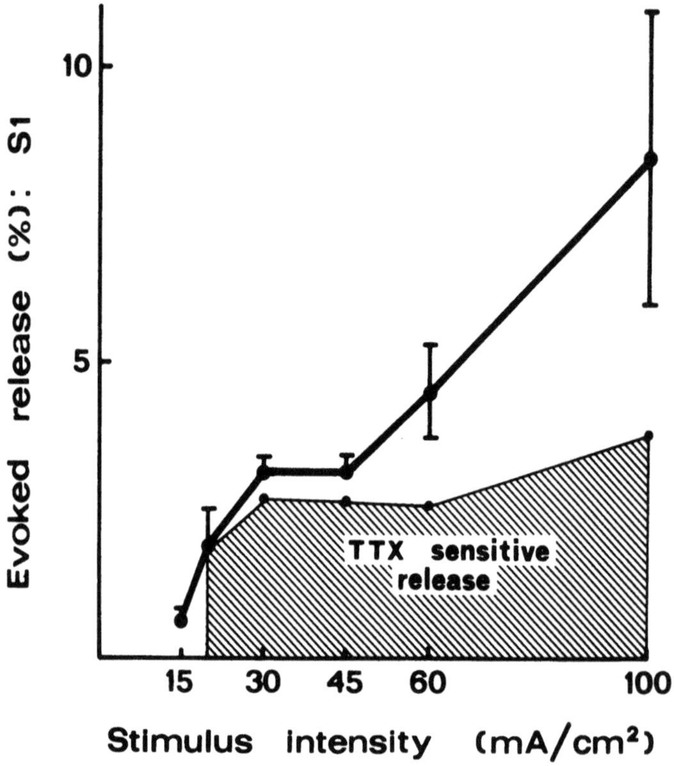

Fig. 1. Tetrodotoxin (TTX) sensitivity of ^{3}H purine release in relationship to the stimulus intensity. Slices were stimulated at 10 Hz for 5 min. Evoked release is expressed as per cent of tritium in the slice evoked by 5 min stimulation. Results are the mean ± S.E. of at least four data for each point. TTX (0.5 µM) was added before a second stimulation period (S_2) and release sensitivity from TTX calculated as per cent variations of S_2/S_1 ratios of control slices.

adenosine and its metabolites inosine and hypoxanthine. Not more than 5% of the radioactivity was recovered in the nucleotides.

The presence of TTX did not significantly change the percent distribution of radioactivity.

A major part of the ^{3}H purine release is calcium dependent (Fig. 2), as shown by the finding that superfusion of the slices with calcium-free Krebs solution containing EGTA 0.5 mM before the second stimulation period S_2 brought about a 69%

TABLE 1. (^3H) COMPONENTS MEASURED IN THE SUPERFUSATE OF THE CORTICAL SLICES

Release	Total counts ± S.E. (dpm/g tissue)	Percentage distribution of (^3H) components ± S.E.			
		Adenosine	Hypoxanthine	Inosine	Nucleotides
Control					
At rest (4)	186.000 ± 49.000	11.7 ± 1.80	36.0 ± 3.91	42.0 ± 5.70	4.0
stimulated (4)	851.000 ± 29.000	13.6 ± 1.91	33.0 ± 4.02	43.0 ± 4.91	3.5 ± 1.50
Tetrodotoxin 0.5 μM					
At rest (2)	160.000 ± 28.000	9.7 ± 2.50	35.4 ± 5.01	43.6 ± 5.31	4.2
stimulated (2)	191.000 ± 60.000	10.9 ± 1.40	46.0 ± 5.20	37.0 ± 5.82	4.7

Release at rest was collected during 5 min before stimulation; stimulated release was collected during 5 min of maximal release (10 min after the beginning of stimulation at 10 Hz) . In parenthesis the number of data from two different experiments.(^3H) nucleotides were not recovered in each separation.

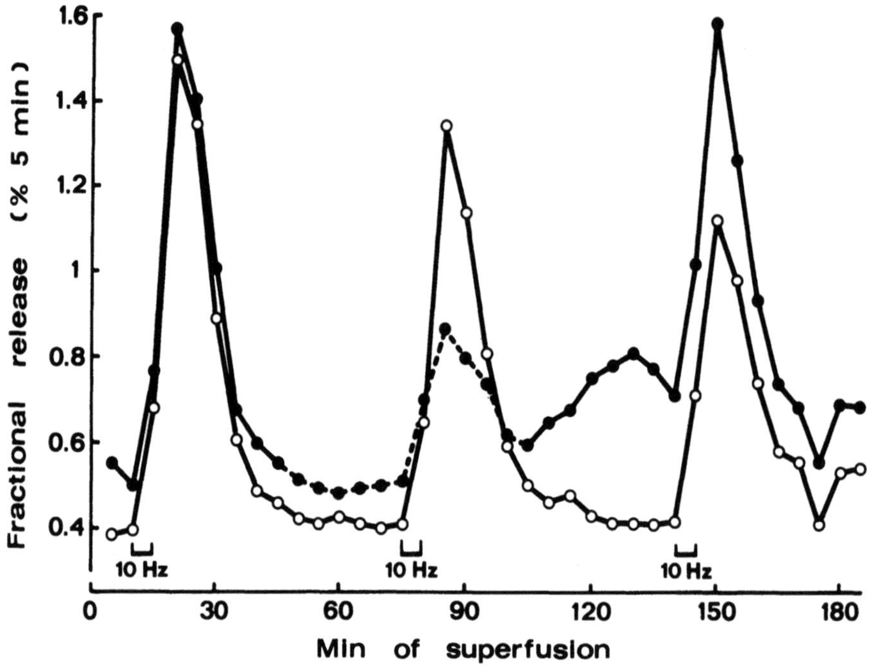

Fig. 2. Ca^{++} dependency of ^3H purine release. The slices were submitted to three consecutive stimulations at 10 Hz as indicated by bars on the abscissa. Each point is the mean of 5 data. S.E. were 3-30% of the mean. ○——○ Control slices; ●——● Ca^{++} deprived slices at the second stimulation period: when indicated by the broken line ●----● Ca^{++} free Krebs, containing EGTA 0.5 mM, reached the slices 30 min before the second stimulation, thereafter the slices were superfused for 30 min in Ca^{++} free Krebs.

reduction of tritium release as calculated by the S_2/S_1 ratio. The ratio was 0.8 $\pm$ 0.07 before and 0.25 $\pm$ 0.06 (n=5; $p < 0.001$) after calcium removal. Perfusion of the slices with normal Krebs solution was followed by an increase in basal efflux and a restoration of the evoked tritium efflux.

The time-course of the evoked ^3H purine release (Fig. 3A) at frequencies ranging from 2.5 to 20 Hz shows that the stimulation period was accompanied by a rapid increase in radioactivity efflux which reached the peak in the following 5 min

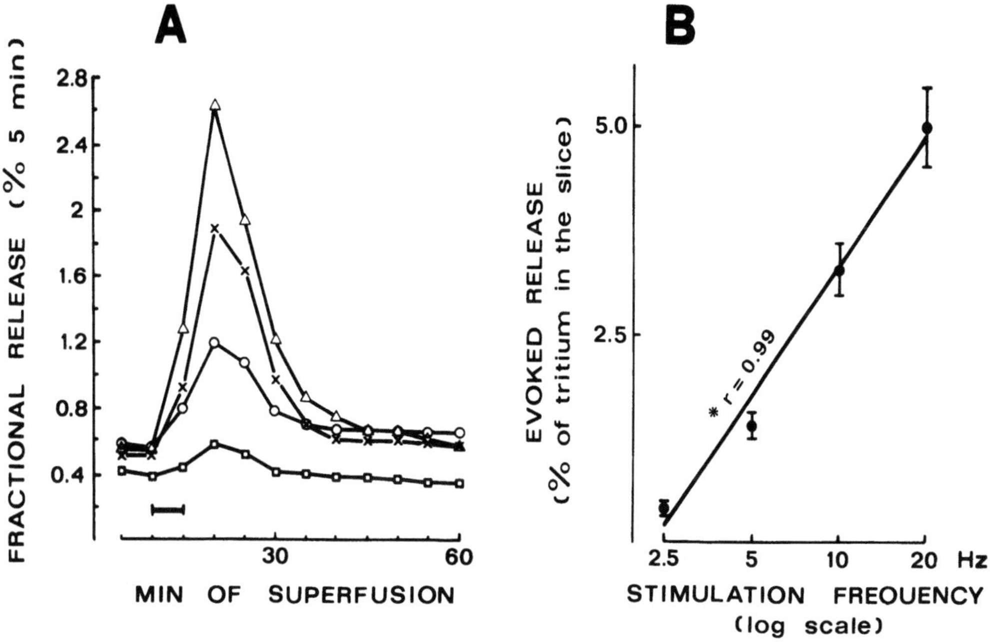

Fig. 3. A) Time-course of ^{3}H purine release from rat cortical slices stimulated, as indicated by the bar on the abscissa, at different stimulation frequencies. □——□ 2.5 Hz; ○——○ 5 Hz; ×——× 10 Hz; △——△ 20 Hz. Each point is the mean of n (number of experiments) = 3 at 2.5 Hz; n=5 at 5 Hz; n=14 at 10 Hz; n=8 at 20 Hz. S.E. were 2-25% of the mean. B) Relationship between the stimulation frequency and the ^{3}H purine evoked release (r= 0.99; p 0.01). Evoked release is expressed as per cent of tritium in the slices evoked by 5 min stimulation.

and returned to resting value in approximately 30 min. No increase in ^{3}H release was detected at 1 Hz stimulation frequency. Fig. 3B illustrates that a linear relationship (r = 0.99, $p < 0.001$) exists between the stimulation frequency and ^{3}H purine release. In Table 2 it is shown that ^{3}H purine release per pulse increases with the stimulation frequency up to 10 Hz but declined at 20 Hz.

Table 2 - ^{3}H purine release per pulse

Stimulation frequency	n	Release/pulse %
2.5 Hz	3	$5.50 \pm 0.10 \times 10^{-4}$
5 Hz	5	$9.30 \pm 0.90 \times 10^{-4}$
10 Hz	14	$11.00 \pm 1.45 \times 10^{-4}$
20 Hz	8	$8.30 \pm 1.00 \times 10^{-4}$

Release is expressed as percent of tritium present in the slice during the 5 min stimulation. Each value is the mean ± S.E., n= number of experiments.

Drug effects on ^{3}H purine release

Muscarinic modulation of ^{3}H purine release - The effects of muscarinic agonist and antagonist on tritium release evoked by a 10 Hz stimulation period are shown in Table 3. Oxotremorine 30 µM significantly enhanced the evoked tritium release.A small effect was already detectable with oxotremorine 5 µM and no futher increase occurred with 10 µM. The effect of oxotremorine showed a 62% reduction in the presence of tetrodotoxin and was antagonized by atropine 1.5 µM. Atropine alone did not significantly affect the evoked tritium release. The effect of oxotremorine was mimicked by physostygmine added to the superfusion fluid at the concentration of 3.8 µM. None of the drugs tested affected the basal ^{3}H efflux as indicated by unchanged basal ratios.

Effect of dipyridamole and caffeine on ^{3}H purine release - Table 4 shows the effect of caffeine in the absence and presence of adenosine uptake inhibitor dipyridamole on ^{3}H efflux evoked by 10 Hz stimulation. Dipyridamole 10 µM did not signi-

Table 3 - Effect of oxotremorine, atropine and physostigmine on ^{3}H purine release from electrically-stimulated rat cortical slices

Drug present at S_2 (concentration in µM)	n	Basal release ratio	Evoked release S_2/S_1	Change (%)
no drug	23	1.15±0.04	0.84±0.02	-
Oxotremorine (5)	3	1.16±0.03	0.91±0.04	+8
Oxotremorine (30)	4	1.12±0.03	1.04±1.12*	+24
Oxotremorine (100)	8	1.21±0.07	1.04±0.07**	+24
Oxotremorine (100) + Atropine (1.5)	4	1.21±0.06	0.84±0.07	-
Atropine (1.5)	4	1.09±0.06	0.75±0.02	-11
Physostigmine (3.8)	4	1.01±0.06	1.02±0.12*	+21
no drug	6	1.27±0.13	0.71±0.13	-
Oxotremorine (100)	6	1.28±0.11	0.92±0.19*	+30
Tetrodotoxin (0.5)	6	1.20±0.10	0.10±0.01***	-86
Oxotremorine (100) +Tetrodotoxin(0.5)	6	1.27±0.06	0.18±0.02♀	-75

The slices were stimulated at 10 Hz for 5 min. Drugs reached the slices 20 min before S_2. The data are the mean ± S.E., n= number of experiments. Statistically significant differences were calculated by Student's two tailed t test from the no drug values.*$p < 0.05$; **$p < 0.01$; ***$p < 0.001$; ♀$p < 0.05$ from tetrodotoxin values.

ficantly affect the electrically stimulated ^{3}H purine release while it tended to increase the basal release as shown by the increased basal release ratio. Caffeine at 50 µM concentration significantly enhanced (30%) the evoked ^{3}H purine release while at 500 µM significantly decreased (20%) the evoked purine release. The effects of both concentrations of caffeine were not significantly modified by the presence of dipyridamole.Caffeine

Table 4 - Effect of dipyridamole and caffeine on ^{3}H purine release from electrically stimulated rat cortical slices

Drug present at S_2 (concentration in μM)	n	Basal release ratio	Evoked release S_2/S_1	Change (%)
no drug	8	1.10 ± 0.08	0.80 ± 0.06	-
Dipyridamole (10)	8	1.43 ± 0.14	0.82 ± 0.15	+2
Caffeine (50)	6	1.15 ± 0.06	1.04 ± 0.11*	+30
Caffeine + (50) Dipyridamole (10)	5	1.34 ± 0.12	1.12 ± 0.12*	+40
Caffeine (500)	6	1.19 ± 0.06	0.64 ± 0.04*	-20
Caffeine + (500) Dipyridamole (10)	4	1.32 ± 0.10	0.59 ± 0.06*	-26

The slices were stimulated at 10 Hz for 5 min. Drugs reached the slices 20 min before S_2. The data are the mean $\pm$ S.E., n= number of experiments. Statistically significant differences were calculated by Student's two tailed t test from the no drug value. $*p < 0.05$.

alone did not affect ^{3}H purine release at rest.

DISCUSSION

Characteristics of ^{3}H purine release and interaction between purinergic and cholinergic system

We have shown that the efflux of radioactivity evoked by electrical stimulation from rat cortical slices preloaded with ^{3}H adenine was frequency dependent, strongly reduced by calcium removal and consisting mainly of adenosine and its metabolites, inosine and hypoxanthine, in agreement with previous data obtained in guinea-pig cortical slices (Pull and McIlwain 1972, 1973) and rat hippocampal slices (Jonzon and Fredholm 1985).

Our experiments demonstrate that at 20 mA/cm^{2} stimulation intensity 98% of ^{3}H purine release is TTX sensitive. At higher stimulation intensities the sensitivity to TTX decreases up to 45% at 100 mA/cm^{2}. This finding indicates that, at high stimulation intensities, only part of the evoked purine release depends on neuronal depolarization and firing brought about by ionic fluxes through voltage sensitive sodium channels blocked by TTX (Narahashi 1974). At high stimulation intensities the release becomes partly unspecific as a consequence of direct depolarization which possibly involves glial cells as well. This could explain why in Pull and McIlwain (1973) experiments only 50% of the radiolabelled purine release is sensitive to TTX. On the other hand, under the stimulation conditions selected for our experiments (30 mA/cm^{2}) 86% of the release was TTX sensitive thus depending mostly on propagated nervous activity.

Although the electrically stimulated ^{3}H purine release shows many features of a transmitter-like release (see Orrego 1979), differences between the evoked release of ^{3}H purine and that of typical transmitter substances emerge from our experiments. First, calcium dependancy was not complete (69%) as also found by Jonzon and Fredholm (1985) in rat hippocampal slices. Conversely, absolute calcium-dependancy was observed in electrically stimulated guinea-pig cortical slices (Pull and McIlwain 1973) and synaptosomes (Kuroda and McIlwain 1974). Second, the electrically evoked efflux of tritium shows a slow time-course, lacking the rapid phasic change which characterize neurotransmitter release under stimulation. The ^{3}H purine efflux begins during depolarization but the peak is reached after stimulation, as also described by Jonzon and Fredholm (1985), and returns to resting level in approximately 30 min. The delayed release could occur via a carrier-mediated process (Jonzon and Fredholm 1985) or could be secondary to the release of excitatory neurotransmitters. Aspartate and glutammate are in fact potent adenosine releasers in the cerebral cortex (Jhamandas and Dumbrille 1980). A third difference concerns the amount of ^{3}H purine released per stimulation pulse. In our experiments the release/ pulse increases from 2.5 to 10 Hz stimulation frequency and declines at 20 Hz. The decline at high frequency could be attributed to exhaustion of the labelled purine pool available for release. A decrease in release/pulse accompanying the decrease in pulse number has been described for various transmitter substances both in central and peripheral nervous system (see Kalsner 1985). Under the same stimulation conditions used to evoke ^{3}H purine release, the release of endogenous ACh per pulse decreases between 0.2 and 20 Hz (Pedata et al. 1983).

A decrease in neurotransmitter/pulse release accompanying the increase in pulse number is attributed to the presence of a negative feedback system by which the increasing concentration of transmitter in the synaptic cleft induces auto-inhibition (Kalsner, 1985). In the case of ^{3}H purine, this autoregulatory mechanism is not operating up to the 10 Hz stimulation rate. On the contrary, a mechanism facilitating release appears to come into play. The stimulatory effect of oxotremirine on evoked purine release demonstrated by our experiments offers an example of facilitation of release. The effect of oxotremorine is presumably mediated through muscarinic receptors since it is blocked by atropine and is mostly dependent on propagated nervous activity as indicated by the reduction of the effect by TTX. Evidence that cholinergic agonists could stimulate purine and nucleotide release was obtained by Curnish et al. (1986) in stimulated frog sympathetic ganglia and by Israel et al. (1980) in Torpedo electric organ prisms. In our experiments the lack of effect of atropine in inhibiting ^{3}H purine release, rules out the possibility that it could depend on muscarinic activation and indicates that ACh exerts only a positive modulatory role on the evoked ^{3}H purine release.

Since it has been shown that adenosine decreases the evoked ACh release (Pedata et al. 1983), a reciprocal modulation of release between the cholinergic and purinergic systems can be envisaged. On the other hand the inhibitory modulation brought about by adenosine on ACh release does not sum with the inhibitory modulation brought about by muscarinic receptors (Pedata et al. 1986). Thus an interference, probably at the level of second messangers (Worley et al. 1987), between muscarinic and adenosine receptors exists.

Effect of caffeine and dipyridamole on ACh and ^{3}H purine release

Our experiments showed that superfusion of the slices in the presence of the adenosine uptake inhibitor dipyridamole facilitates basal release while it had no effect on the evoked ^{3}H purine release. Similar results were obtained in neocortical synaptosomes by Daval and Barberis (1981). A lack of the effect of dipyridamole in increasing the labelled purine release was also reported by Stone et al. (1981) and even a decreased ^{3}H purine overflow evoked by electrical stimulation of hippocampal slices was reported by Jonzon and Fredholm (1985). The possibility exists that dipyridamole exerts a bidirectional block of nucleoside transport across membranes (Jonzon and Fredholm 1985). Furthermore, dipyridamole, increasing extracellular adenosine levels, may dilute a radioactive adenosine pool which sustains the stores responsible by electrical stimulation resulting in an apparent lack of effect on radioactive adenosine extracellular levels. On the other hand, many actions of adenosine are potentiated by dipyridamole (see Wu and Phillis 1984), including the inhibition of ACh release (Pedata et al. 1984). Caffeine at the concentration of 50 µM significantly increases ^{3}H purine release from electrically stimulated cortical slices. The slight stimulatory effect of the adenosine antagonist on ^{3}H purine release suggests the lack of an efficient autoregulatory mechanism of adenosine on its own release. Conversely caffeine at 500 µM concentration significantly decreases ^{3}H purine release evoked by electrical stimulation. Similar biphasic results have been obtained by Stone et al. (1981) on radioactive purine release evoked by ouabain from cortical slices. At high concentrations, the inhibitory effect of caffeine on ^{3}H purine and ACh release (Pedata et al. 1984) may not depend, however, on adenosine receptors but rather on other caffeine actions such as inhibition of phosphodiesterase (Bucher and Sutherland 1962), modification of intracellular calcium mobilization (Johnson and Inesi 1969) and inhibition of 5 nucleotidase (see Stone 1981).

The effects of caffeine on ^{3}H purine release do not however seemingly answer for the byphasic effect (Pedata et al.

1984) of caffeine on ACh release.

Acknowledgement: This investigation was supported by a grant from the University of Florence.

REFERENCES

Beani L, Bianchi C, Giacomelli A, Tamberi F (1978) Noradrenaline inhibition of acetylcholine release from guinea pig brain. Eur J. Pharmacol 48: 179-193

Braas KM, Newby AC, Wilson VS, Snyder SH (1986) Adenosine-containing neurons in the brain localized by immunocytochemistry. J Neurosci 6(7): 1952-1261

Bucher RW, Sutherland EW (1962) Adenosine 3'-5' phosphate in biological material. J Biol Chem 237: 1244-1250

Corradetti R, Lo Conte G, Moroni F, Passani MB, Pepeu G (1984) Adenosine decreases aspartate and glutamate release from rat hippocampal slices. Eur J Pharmacol 104: 19-26

Curnish RR, Bencherif M, Rubio R, Berne RM (1986) Phosphatidylinositol (PI) turnover is inhibited by adenosine (ADO) released during orthodromic stimulation of the frog sympathetic ganglion. Pfluger Archiv Eur J Physiol 407 (Suppl n.1): 541

Daval JL, Barberis C (1981) Release of radiolabelled adenosine derivatives from superfused synaptosomal bed: evidence for the output of adenosine. Biochem Pharmacol 18: 2559-2567

Dunwiddie TV (1985) The physiological role of adenosine in the central nervous system. International review of Neurobiology. Smythies JR, Bradley RJ (eds) 63-139.Academic Press London

Dunwiddie TV, Hoffer BJ (1980) Adenine nucleotides and synaptic transmission in the "in vitro" rat hippocampus. Br J Pharmac 69: 59-68

Fredholm BB,Dunér-Engström M, Fastbom J, Jonzon B, Lindgren E, Nordstedt C, Pedata F, Van der Ploeg I (1987) Interactions between the neuromodulator adenosine and the classic transmitters. In "Topics and perspective in adenosine research" Gerlach E, Becker BF (eds) Springer Verlag Berlin Heidelberg pp 509-520

Fredholm BB, Sollevi A (1981) The release of adenosine and inosine from canine subcutaneous adipose tissue by nerve stimulation and noradrenaline. J Physiol (London) 313: 351-367

Hamprecht B, Van Calker D (1985) Nomenclature of adenosine receptors. TIPS 6: 153-154

Henon BK, McAfee DA (1983) The ionic basis of adenosine receptors action on post-ganglionic neurones in the rat. J Physiol 336: 607-620

Hollins C, Stone TW (1980) Characteristics of the release of adenosine from slices of rat cerebral cortex. J Physiol (London) 303: 73-82

Israel M, Lesbats B, Manaranche R, Meunier FM, Fachon P (1980) Retrograde inhibition of transmitter release by ATP. J Neurochem 34: 923-932

Jackish R, Strittmatter H, Hasakov L, Herting G (1984) Endogenous adenosine as a modulator of hippocampal acetylcholine release. Naunyn-Schmiedeberg's Arch Pharmacol 327: 319-325

Jhamandas K, Dumbrille A (1980) Regional release of ^{3}H adenosine derivatives from rat brain in vivo: effect of excitatory aminoacids, opiate agonists and benzodiazepines. Can J Physiol Pharmacol 58: 1262-1278

Johnson PN, Inesi G (1969) The effect of methylxanthines and local anaesthetics on fragmented sarcoplasmic reticulum. J Pharmac Exp Ther 169: 308-314

Jonzon B, Fredholm BB (1985) Release of purines, noradrenaline and GABA from rat hippocampal slices by field stimulation. J Neurochem 44: 217-244

Kalsner S (1985) Is there feedback regulation of neurotransmitter release by autoreceptors? Biochem Pharmac 34: 4085-4097

Kuroda Y, McIlwain H (1974) Uptake and release of ^{14}C adenine derivatives at beds of mammalian cortical synaptosomes in a superfusion system. J Neurochem 22: 691-699

Lee K, Schubert P, Gribkoff V, Sherman B, Lynch G (1982) A combined in vivo/in vitro study of the presynaptic release of adenosine derivatives in the hippocampus. J Neurochem 23: 309-314

Londos C, Wolff J (1977) Two distinct adenosine-sensitive sites on adenylate cyclase. Proc Natl Acad Sci USA 74: 5482-5486

MacDonald WF, White TD (1985) Nature of extrasynaptosomal accumulation of endogenous adenosine evoked by K^+ and veratridine. J Neurochem 45: 791-797

Narahashi T (1974) Chemicals as tools in the study of excitable membranes. Physiol Rev 54: 813-888

Okada Y, Kuroda Y (1980) Inhibitory action of adenosine and adenosine analogs on neurotransmission in the olfactory cortex slice of guinea-pig structure-activity relationships. Eur J Pharmacol 61: 137-146

Orrego F (1979) Criteria for the identification of central neurotransmitter and their application to studies with some nerve tissue preparation in vitro. Neurosci 4: 1037-1057

Pedata F, Antonelli T, Lambertini L, Beani L, Pepeu G (1983) Effect of adenosine, adenosine triphosphate, adenosine deaminase, dipyridamole and aminophilline on acetylcholine release from electrically-stimulated brain slices. Neuropharmacology 22: 609-614

Pedata F, Pepeu G, Spignoli G (1984) Biphasic effect of methylxanthines on acetylcholine release from electrically-stimulated brain slices. Br J Pharmac 83: 69-73

Pedata F, Giovannelli L, De Sarno P, Pepeu G (1986) Effect of adenosine, adenosine derivatives and caffeine on acetylcholine release from brain synaptosomes: interaction with muscarinic autoregulatory mechanisms. J Neurochem 46: 1593-1598

Perkins MN, Stone TW (1980) 4-aminopyridine blockade of neuronal depressant responses to adenosine triphosphate. Br J Pharmac 70: 425-428

Phillis JW, Kostopulos GK (1975) Adenosine as a putative transmitter in the cerebral cortex. Studies with potentiators and antagonists. Life Sci 17: 1085-1094

Phillis JW, Wu PM (1983) In: Physiology and Pharmacology of Adenosine Derivatives. Daly JW, Kuroda Y, Phillis JW, Shimuzu H, Ui M (eds) Raven New York

Pull I, McIlwain H (1972) Adenine derivatives as neurohumoral agents. The quantities liberated on excitation of superfused cerebral tissues. Biochem J 130: 975-981

Pull I, McIlwain H (1973) Output of (^{14}C)Adenine nucleotides and their derivatives from central tissues. Biochem J 136: 893-901

Schweinsberg PD, Loo TL (1980) Simultaneous analysis of ATP, ADP, AMP and other purines in human erythrocytes by high-performance liquid chromatography. J Chromatogr 181: 103-107

Segal M (1982) Intracellular analysis of a post-synaptic action of adenosine in the rat hippocampus. Eur J Pharmacol 79: 193-199

Snyder SH (1985) Adenosine as a neuromodulator. Ann Rev Neurosci 8: 103-124

Stone TW (1981) Physiological roles for adenosine and adenosine 5'triphosphate in the nervous system. Neurosci 6: 523-555

Stone TW, Hollins C, Lloyd H (1981) Methylxanthines modulate adenosine release from slices of cerebral cortex. Brain Res 207: 421-431

Sulakhe PV, Phillis JW (1975) The release of (^{3}H) adenosine and its derivatives from cat sensorimotor cortex. Life Sci 17: 551-556

Su C (1983) Purinergic neurotransmission and neuromodulation. Ann Rev Pharmacol Toxicol 23: 397-411

Worley PF, Baraban JM, McCarren M, Snyder SM, Alger BE (1987) Cholinergic phosphatidynositol modulation of inhibitory, G protein-linked, neurotransmitter actions: electrophysiological studies in rat hippocampus. Proc Natl Acad Sci 84: 3467-3471

Wu PM, Phillis JW (1984) Uptake by central nervous tissues as a mechanism for the regulation of extracellular adenosine concentrations. Neurochem Int 6: 613-632

Zetterstrom T, Ungerstedt VU, Tossman U, Jonzon B, Fredholm BB (1982) Purine levels in the intact brain. Studies with an implanted perfused hollow fibre. Neurosci Lett 29: 111-115

MODULATION OF GLUTAMATE AND GABA RELEASE BY EXCITATORY AMINO ACID RECEPTOR AGONISTS IN CULTURED CEREBELLAR CELLS

G. Levi, V. Gallo, C. Giovannini and R. Suergiu
Istituto Superiore di Sanità
Laboratorio di Fisiopatologia di Organo e di Sistema
Viale Regina Elena 299
00161 Roma
Italy

INTRODUCTION

The major excitatory input to the cerebellar cortex is represented by the mossy fibers, whose terminals form synaptic contacts with the granule cells dendrites (Palay and Chan-Palay 1974). It has been suggested that some of these fiber terminals use glutamate or aspartate as a transmitter (Kaneko et al. 1987). This proposal would be consistent with the presence of excitatory amino acid receptors on cerebellar granule cells, which was demonstrated both biochemically (Wroblewski et al. 1985; Nicoletti et al. 1986) and electrophysiologically (Cull-Candy and Ogden 1985).

The activation of excitatory amino acid receptors has been often described to modulate the release of neurotransmitter amino acids from various brain tissue preparations (McBean and Roberts 1981; Pearce and Dutton 1982; Ferkany et al. 1982; Ferkany and Coyle 1983; Collins et al. 1983, Potashner and Gerard 1983; Pastuszko et al. 1984; Poli et al. 1985).

We have utilized cerebellar primary cultures which were either 95% enriched in granule cells and contained a small population of GABAergic interneurons (Levi et al. 1984, Aloisi et al. 1985), or 90% enriched in astrocytes (Wilkin et al. 1983) to investigate whether the activation of excitatory amino acid receptors would influence neurotransmitter amino acid release. Our data indicate that at least two subtypes of acidic amino acid receptors coupled to neurotransmitter amino acid release are present in cultured cerebellar interneurons and in a subpopulation of cerebellar astrocytes. The present report is mainly concerned with the neuronal recognition sites. The data on astroglial receptors are discussed in more detail elsewhere (Gallo et al. 1986, 1987a; Gallo et al. 1988).

NATO ASI Series, Vol. H19
Modulation of Synaptic Transmission and Plasticity
in Nervous Systems. Edited by G. Hertting and H.-C. Spatz

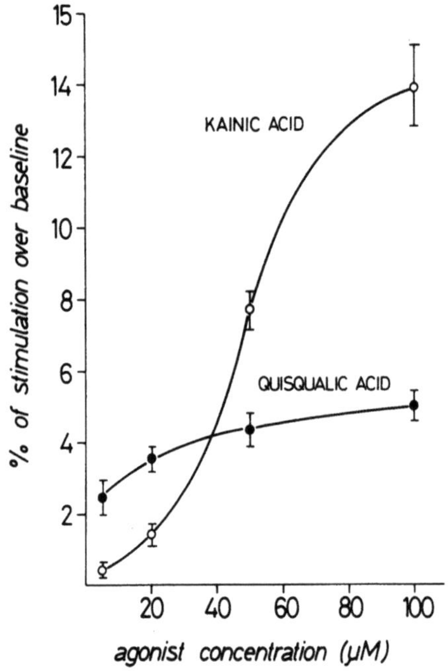

Fig. 1. Concentration dependence of kainate and quisqualate-induced release of ^{3}H-D-aspartate from cerebellar interneuron cultures. Cells were cultured for 11 days, incubated with 0.5 μM ^{3}H-D-aspartate and then subjected to 5 min washes with standard Krebs-Ringer medium, to study ^{3}H-D-aspartate release. After the third wash (min 15) kainic acid and quisqualic acid were present in the release medium at different concentrations for 15 min. The radioactivity recovered in the peak fraction (min 25) was subtracted of that present in the prestimulation fraction (min 10) and expressed as a percentage of the radioactivity present in the cells at the beginning of the release experiment. Bars represent S.E.M. and the curves were obtained from 4 experiments run in duplicate.

EXCITATORY AMINO ACID RECEPTOR SUBTYPES IN CULTURED CEREBELLAR GRANULE CELLS

In all the experiments presented here, granule cell-enriched cultures obtained from 8-day postnatal rat cerebella were used at 10-11 days in vitro. During this culture period the neurons undergo both morphological and functional differentiation (Gallo et al. 1982) and acquire the ability to accumulate the non metabolized glutamate analog ^{3}H-D-aspartate and to

TABLE 1. Effect of kainic acid and quisqualic acid on the release of endogenous glutamate from cerebellar granule cell cultures.

Compound tested	(nmoles/mg prot/5 min) Basal release (prestimulation)	Evoked release (1st fraction)	Evoked release (2nd fraction)	N
Kainic acid (50 μM)	1.6 ± 0.2	4.1 ± 0.2	4.4 ± 0.3	12
Kainic acid (50 μM) + Kynurenic acid (200 μM)	1.2 ± 0.1	2.3 ± 0.1	3.1 ± 0.2	5
Quisqualic acid (50 μM)	1.2 ± 0.1	3.3 ± 0.1	2.5 ± 0.1	4

Granule cell enriched cultures at 10 DIV were incubated with 0.5 μM ^{3}H-D-aspartate and then subjected to 5 min washes with standard Krebs-Ringer medium. A medium containing 50 uM kainic acid (± 200 μM kynurenic acid) or 50 μM quisqualic acid was applied at min 15 for two 5 min fractions. Endogenous glutamate was measured in the medium of the third wash (basal release) and of the fourth and fifth wash (release in the presence of the test compounds: evoked release). The amount of total endogenous glutamate present in the cells at the beginning of the release phase of the experiment was 87.8 ± 3.3 nmoles/mg protein (N = 11). When the values of basal release were subtracted from overall release of glutamate observed in the presence of kainic acid (with or without kynurenic acid), kynurenic acid inhibited by 54% (1st fraction) and by 30% (2nd fraction) the effect of kainic acid. The data shown are from 3-6 different experiments. The data referring to kynurenic acid were obtained in 2 of the experiments in which the effect of kainate was tested. Means ± S.E.M. are presented; N = number of culture dishes examined.

release it in response to depolarization (Levi et al. 1984).

Of the various glutamate agonists and structural analogs tested for their ability to evoke amino acid release, only kainic acid and quisqualic acid caused a significant increase in the release of ^{3}H-D-aspartate previously accumulated by granule cells (Fig. 1), or of endogenous glutamate (Table 1). N-methyl-D-aspartate acid, quinolinic acid and dihydrokainic acid were ineffective (Gallo et al. 1987a and 1987b). The dose response curves shown in Fig. 1 indicate that kainic acid was a more potent releaser at high concentrations (50-100 μM), whereas quisqualic acid was more effective at lower concentrations.

Several observations (Gallo et al. 1987a and 1987b) indicate that the releasing effect of the two glutamate receptor agonists is receptor-mediated, and does not occur through a heteroexchange process (Levi and Rai-

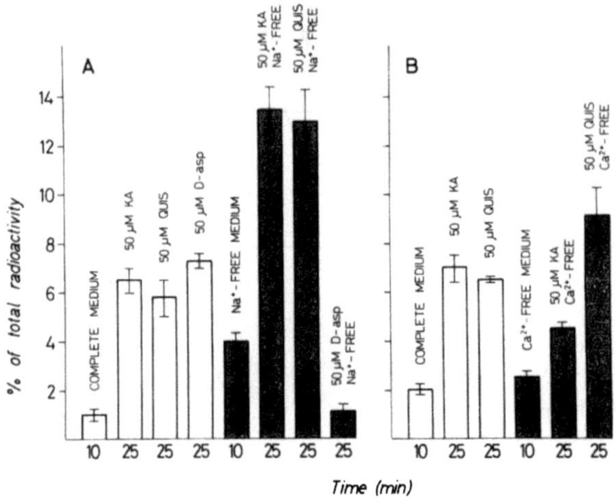

Fig. 2. Ionic dependence of kainic acid and quisqualic acid effects on ^{3}H-D-aspartate release from cerebellar interneuron cultures. Cells were cultured for 11 days, incubated with ^{3}H-D-aspartate (0.5 μM) in Na^+-containing medium for 15 min. ^{3}H-D-aspartate release was studied as described in the legend for Fig. 1, except that, after the second wash, the cells were incubated in a medium in which sucrose replaced NaCl (panel A) or in a medium in which the concentration of $CaCl_2$ was lowered to 0.2 mM and that of $MgSO_4$ raised to 10 mM (panel B). Kainic acid and quisqualic acid were applied at the 15th min for 10 minutes. The radioactivity present in each 5 min release fraction was expressed as a percentage of that present in the cells at the beginning of the release experiment. The enhancement of ^{3}H-D-aspartate release due to the Na^+-free medium (see 5th histogram from the left) was not reproduced in all batches of cultures and was minimal in the set of experiments reported in the last histogram on the right of panel A. Averages of 3 duplicate experiments ± S.E.M. are presented.

teri 1976): i) the release of ^{3}H-D-aspartate induced by kainate and quisqualate was Na^+-independent and partially Ca^{2+}-dependent, whereas that induced by unlabeled D-aspartate (presumably by homoexchange) was totally Na^+-dependent (Fig. 2); ii) the effect of kainate was selectively antagonized by 2,3-cis-piperidine dicarboxylic acid (PDA) and, even more so, by kynurenic acid, two non specific excitatory amino acid receptor antagonists (Fig. 3, see also Table 1 for endogenous glutamate release). Other antagonists (glutamate diethylester, γ-D-glutamylglycine) were without effect, and none of the antagonists tested influenced the quisqualate-induced release (see Fig. 3B, and unpublished observations); iii) ^{3}H-kainic acid was not accumulated by granule cell neurons (not shown). Kainate, however, inhibited ^{3}H-D-aspartate uptake, and its analog dihydrokainate, which has very low affinity for kainate receptors (London and Coyle 1979; Slevin et

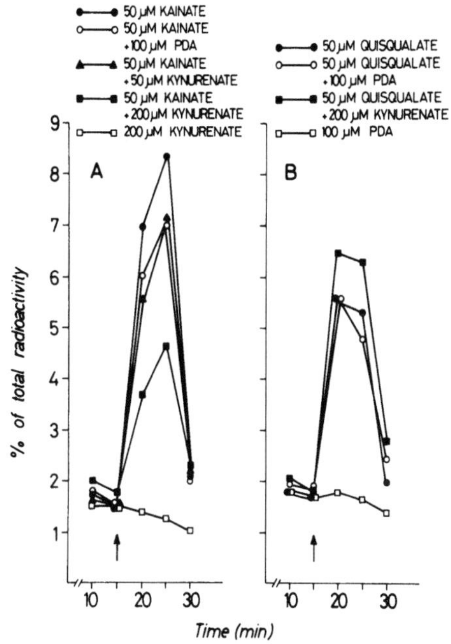

Fig. 3. Effect of excitatory amino acid receptor antagonists on kainate- and quisqualate- induced ^{3}H-D-aspartate release from cerebellar interneuron cultures. Cells cultured for 11 days were labeled with 0.5 µM ^{3}H-D-aspartate for 10 min and then used to study ^{3}H-D-aspartate release (see legend for Fig. 1). Kainic acid, quisqualic acid, kynurenic acid and PDA were present in the release medium from the time indicated by the arrow for 10 min. Results are expressed as in Fig. 2. Averages of 3 duplicate experiments are presented. Differences between kainate-induced ^{3}H-D-aspartate release in the absence or in the presence of kynurenic acid and of PDA were analyzed by a mixed model ANOVA with treatment (kainate vs. kainate + antagonists) as between subjects factor. The antagonistic effects of kynurenic acid and of PDA were statistically significant (see Gallo et al. 1987b).

al. 1983) and does not stimulate ^{3}H-D-aspartate release (Gallo et al. 1987b), had a similar effect on transport. Kainic acid is therefore likely to occupy the glutamate transport sites without being itself transported (see also Johnston et al. 1979; Potashner and Gerard 1983; Pastuszko et al. 1984). The independence between the effects on the transport system and those on release is further supported by the fact that quisqualate elicited ^{3}H-D-aspartate release, but did not have any effect on ^{3}H-D-aspartate uptake (Gallo et al. 1987a and 1987b).

Our results are consistent with the idea that kainate and quisqualate act on two separate receptor subtypes, which could be experimentally distinguished for their different sensitivity to the excitatory amino acid receptor antagonists kynurenic acid and PDA (Fig. 3) and for their different dependence on extracellular Ca^{2+} (Fig. 2). Interestingly, also in the vertebrate retina, kynurenic acid was recently proposed as a potent and selective kainic acid antagonist, and was used to separate kainate-mediated from quisqualate-mediated depolarizing responses (Coleman et al. 1986). The partial Ca^{2+} -dependence of the kainate effect suggests that an increased influx of Ca^{2+} is involved in the releasing action of the neurotoxin. In this respect, it is interesting that kainic acid was able to increase the influx of $^{45}Ca^{2+}$ into cultured cerebellar granule cells (Wroblewski et al. 1985), whereas quisqualic acid was ineffective. Quisqualate, on the other hand, was a more potent activator of phosphatidyl inositol turnover than kainate (Nicoletti et al. 1986) and might, therefore, cause glutamate release by mobilizing Ca^{2+} ions from intracellular stores (Abdel-Latif 1986).

Although the recognition sites of kainic and quisqualic acids appear to be different, they seem to be able to interact with each other. In particular the activation of quisqualate receptors can largely prevent kainate from exerting its biological effect. This is demonstrated by the experiment reported in Fig. 4, which shows that the action of kainic acid and quisqualic acid on ^{3}H-D-aspartate release was much less than additive and that the kainate antagonist kynurenic acid did not antagonize the release of ^{3}H-D-aspartate induced by kainate and quisqualate applied together. A similar type of interaction has already been described in retinal horizontal cells, where quisqualate and glutamate inhibited in a non competitive manner the responses produced by kainic acid (Ishida and Neyton 1985). In a recent patch-clamp study performed in cultured cerebellar Purkinje cells (Cull-Candy and Usowicz 1987), it has been proposed that the three subtypes of glutamate receptors are all linked to a single channel, capable of several sub-conductance states. It is thus possible that also in cerebellar granule cells kainate and quisqualate receptors are coupled to the same channel, and that the occupancy of quisqualate binding sites decreases the affinity of kainate receptors for their ligand. In these conditions, the sub-conductance states of the channel would only be those typical of a quisqualate-induced response.

It seemed important to determine whether the natural excitatory amino acid glutamate would share some of the actions of its agonists on ^{3}H-D-aspartate release from cultured granule cells. When tested in a standard incubation medium, glutamate caused a massive, dose-dependent

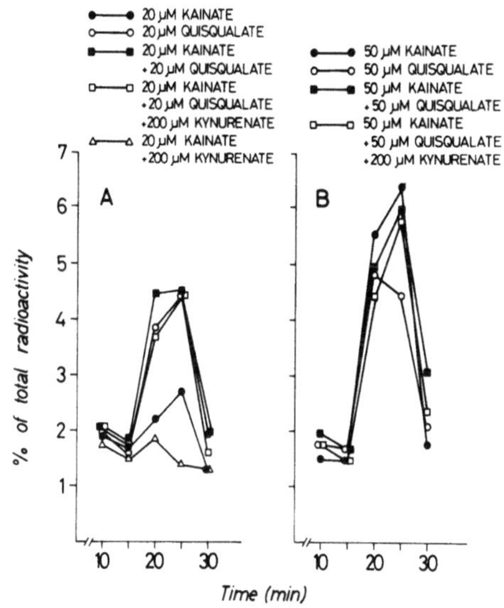

Fig. 4. Functional interaction between kainate and quisqualate receptors in cerebellar interneuron cultures. Cells cultured for 11 days were labeled with 0.5 μM ^{3}H-D-aspartate for 10 min and then used to study ^{3}H-D-aspartate release (see legend for Fig. 1). Kainic acid, quisqualic acid and kynurenic acid were present in the release medium starting from min 15 for 10 min. Results are expressed as in Fig. 2. Averages of 3 experiments, differing less than 20%, are presented.

release of ^{3}H-D-aspartate (Fig. 5). However, similarly to the release induced by unlabeled D-aspartate (Gallo et al. 1987b), the releasing effect of glutamate could be due to heteroexchange (Levi and Raiteri 1976). In fact the two acidic amino acids share the same high affinity transport system (Balcar and Johnston 1972). When the experiment was repeated using a Na^{+}-free medium, a condition previously shown to prevent D-aspartate homoexchange (Fig. 2), glutamate maintained a substantial releasing activity, which might well be receptor-mediated. This hypothesis is corroborated by the additional observation that, in Na^{+}-free conditions, the releasing effect of glutamate was almost totally prevented by kynurenic acid (Fig. 5). If one considers that in our experimental system kynurenic acid antagonized the releasing action of kainate, but not that of quisqualate (Fig. 3), and that N-methyl-D-aspartate was ineffective (Gallo et al. 1987b), one can deduce that the receptor-mediated component of the releasing effect of glutamate is related to the activation of the kainate preferring recognition sites.

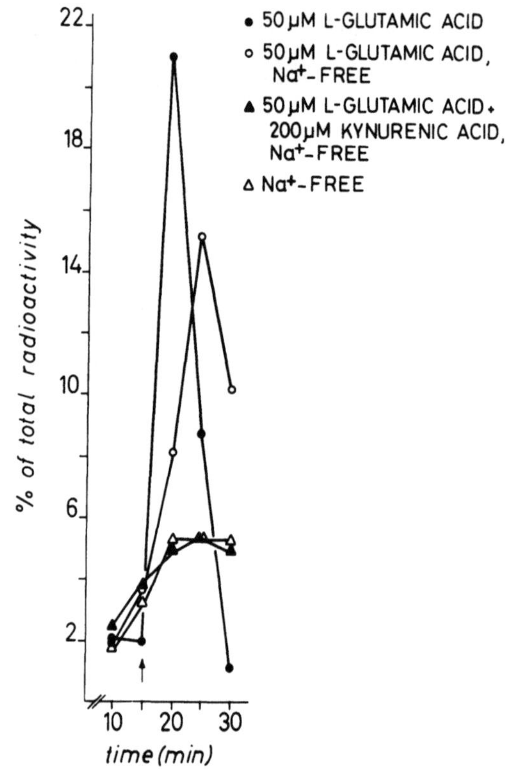

Fig. 5. Effect of L-glutamate on the release of ^{3}H-D-aspartate from cerebellar interneuron cultures. Cells were cultured for 11 days and then incubated with 0.5 μM ^{3}H-D-aspartate in Na^{+}-containing medium for 15 min. The release of ^{3}H-D-aspartate was monitored using a standard medium or a Na^{+}-free medium (see legend for Fig. 2), before and after adding 50 μM L-glutamic acid to the medium. In some cases, 200 μM kynurenic acid was added together with glutamic acid to the Na^{+}-free medium. Averages of 2 different experiments run in duplicate are presented.

EXCITATORY AMINO ACID RECEPTOR SUBTYPES IN CULTURED CEREBELLAR INHIBITORY INTERNEURONS AND ASTROCYTES

Kainate and quisqualate were capable of releasing also ^{3}H-GABA from the cerebellar interneuron-enriched cultures used. In these cultures, ^{3}H-GABA is accumulated by a small population of GABAergic interneurons (about 2% of the total cells; Levi et al. 1984; Aloisi et al. 1985) and by the contaminating astrocytes (about 1% of the total cells; Levi et al. 1983) bearing a type-2 phenotype (Raff et al. 1983; Levi et al. 1986). The characteristics of ^{3}H-GABA evoked release were similar to those described for ^{3}H-D-aspartate release, in terms of Na^{+}-independence, partial Ca^{2+}-de-

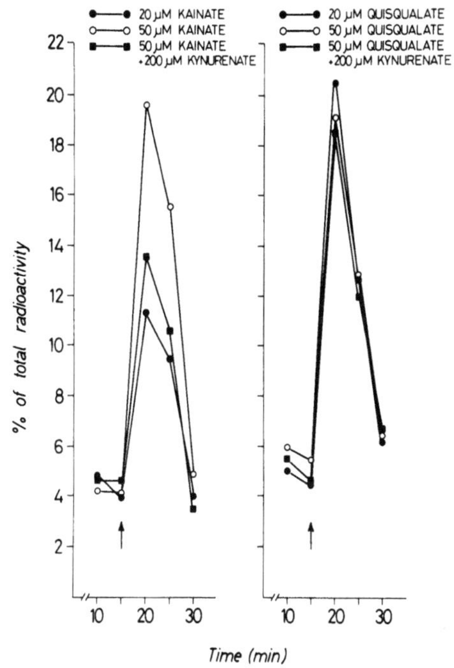

Fig. 6. Effect of kainic acid and quisqualic acid on ^{3}H-GABA release from cerebellar interneuron cultures. Cells cultured for 10 days were incubated for 15 min in a medium containing 10 µM aminooxyacetic acid and 0.5 µM ^{3}H-GABA. The media used for the release phase of the experiments also contained 10 µM aminooxyacetic acid. Kainic acid, quisqualic acid and kynurenic acid were applied starting from the time indicated by the arrow for 10 min. Averages of 3 experiments run in duplicate are shown. Results are expressed as in Fig. 2.

pendence and sensitivity to kynurenic acid (for the latter point, see Fig. 6). These features, and the fact that the release evoked by the two glutamate agonists was much more pronounced than that elicited by unlabeled GABA (not shown), exclude that ^{3}H-GABA was released through a heteroexchange process and suggest the involvement of acidic amino acid receptors.

We have reported elsewhere (Gallo et al. 1986) that kainate and quisqualate release ^{3}H-GABA also from type-2 astrocytes through a receptor-mediated mechanism. Therefore the effects observed in interneuron-enriched cultures might, in principle, be related to the activation of acidic amino acid receptors present on the membrane of the contaminating astrocytes, rather than of GABAergic neurons. However, the bulk of the ^{3}H-GABA evoked release observed in the neuronal cultures seems to originate from GABAergic neurons and not from astrocytes. This discrimination could be done on the basis of the results obtained with neuron-free, astrocyte-enriched cultures

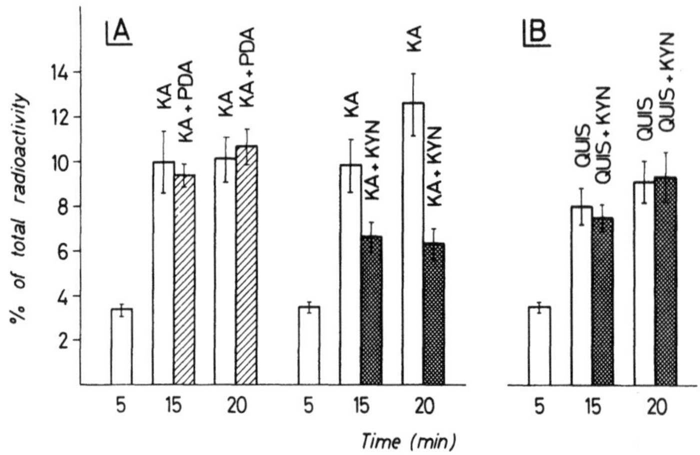

Fig. 7. Effect of kainic acid and quisqualic acid on ^{3}H-GABA release from cerebellar astrocyte cultures. Cells cultured for 5 days were incubated for 15 min in a medium containing 10 µM aminooxyacetic acid and 0.5 µM ^{3}H-GABA and then used to study ^{3}H-GABA release. Kainic acid (50 µM), quisqualic acid (50 µM), kynurenic acid (100 µM) and PDA (100 µM) were applied from min 10 to min 20. The radioactivity recovered in the prestimulation fraction (5 min) and in the two fractions in which the glutamate agonist was present (15 and 20 min) was expressed as a percentage of that present in the cells at the beginning of the release experiments. Averages of 3 duplicate experiments ± S.E.M. are presented. Data from Gallo et al. 1987b).

containing a subpopulation of cells expressing the type-2 astrocyte phenotype. In marked contrast with the behavior of neuronal cultures, in astrocyte-enriched cultures the releasing effects of kainate and quisqualate were totally Na^+-dependent (Gallo et al. 1986) and that of kainate was not antagonized by PDA (Gallo et al. 1987a, see Fig. 7). In other respects (kynurenic acid and Ca^{2+} sensitivity) neurons and type-2 astrocyte receptors did not appear to differ significantly one from the other (Gallo et al. 1987a and 1987b).

CONCLUSIONS

Although excitatory amino acid recognition sites coupled to neurotransmitter amino acid release appear to be present both in neurons and in type-2 astrocytes, the mechanism whereby their activation produces the biological effect may differ profoundly. While in the case of astrocytes the evoked release may be directly related to the opening of Na^+ channels and to the consequent depolarization, in the case of neurons it may be related to the entry of Ca^{2+} ions or to the activation of second messenger systems.

The existence of acidic amino acid receptor subtypes on cerebellar granule cells can be refered to the excitatory innervation that these cells receive from the mossy fibers (Kaneko et al. 1987). In this case, the receptors described in the present study should be considered as post-synaptic. Other authors (Ferkany et al. 1982) have postulated the presence of kainate receptors coupled to glutamate release on the parallel fiber nerve terminals.

Autoradiographic studies have shown that the highest density of kainate binding sites in the cerebellar cortex is in the granule cell and in the molecular layers (Henke et al. 1981; Monaghan and Cotman 1982; Unnerstall and Walmsley 1982). As to quisqualate receptors, the cerebellar cortex showed the highest density in the molecular layer (Halpaim et al. 1984). Our present studies do not help in elucidating the subcellular localization of these receptors, and are compatible with both their post-synaptic or presynaptic localization.

ACKNOWLEDGMENTS

We tank Mr. Mario Patrizio for excellent assistance. This work was partly supported by Grant n. 86.02034.56 of the Italian National Research Council (Target Project "Preventive Medicine and Rehabilitation", Sub-project "Disease of the Nervous System").

REFERENCES

Abdel-latif AA (1986) Calcium-mobilizing receptors, phosphoinositides and the generation of second messengers, Pharmacol Rev 38: 227-272.

Aloisi F, Ciotti MT, Levi G (1985) Characterization of GABAergic neurones in cerebellar primary cultures and selective neurotoxic effects of a serum fraction, J Neurosci 5: 2001-2008.

Balcar VJ, Johnston GAR (1972) The structural specificity of the high affinity uptake of L-glutamate and L-aspartate by rat brain slices, J Neurochem 19: 2657-2666.

Coleman PA, Massey SC, Miller RF (1986) Kynurenic acid distinguishes kainate and quisqualate receptors in the vertebrate retina, Brain Res 381: 172-175.

Cull-Candy SG, Ogden DC (1985) Ion channel activated by L-glutamate and

GABA in cultured cerebellar neurones of the rat, Proc R Soc London B 224: 367-373.

Cull-Candy SG, Usowicz MM (1987) Multiple-conductance levels activated by excitatory amino acids in cerebellar neurons, Nature 325: 525-528.

Ferkany JW, Coyle JT (1982) Kainic acid selectively stimulates the release of endogenous excitatory acidic amino acids, J Pharmacol Exp Ther 225: 399-406.

Ferkany JW, Zaczek R, Coyle JT (1982) Kainic acid stimulates excitatory amino acid neurotransmitter release at presynaptic receptors Nature 298: 757-759.

Gallo V, Giovannini C., Levi G (1988) Expression of excitatory amino acid receptors and neurotransmitter amino acid release in cultured cerebellar neural cells, in Heuter G, ed., "Amino Acid Availability and Brain Function in Health and Disease", Springer-Verlag, in press.

Gallo V, Ciotti MT, Coletti A, Aloisi F, Levi G (1982) Selective release of glutamate from cerebellar granule cells differentiating in culture, Proc Natl Acad Sci USA 79: 7919-7923.

Gallo V, Suergiu R, Levi G (1986) Kainic acid stimulates GABA release from a subpopulation of cerebellar astrocytes, Eur J Pharmacol 133: 319-322.

Gallo V, Suergiu R, Levi G (1987a) Functional evaluation of glutamate receptor subtypes in cultured cerebellar neurones and astrocytes, Eur J Pharmacol 138: 293-297.

Gallo V, Suergiu R, Giovannini C, Levi G (1987b) Glutamate receptor subtypes in cultured cerebellar neurones: modulation of glutamate and GABA release, J Neurochem , in press.

Halpain S, Wieczorek CM, Rainbow TC (1984) Localization of L-glutamate receptors in rat brain by quantitative autoradiography, J Neurosci 4: 2247-2258.

Henke H, Beaudet A, Cuenod M (1981) Autoradiographic localization of specific kainic acid binding sites in pigeon and rat cerebellum, Brain Res 219: 95-105.

Ishida AT, Neyton J (1985) Quisqualate and L-glutamate inhibit retinal horizontal-cell responses to kainate, Proc Natl Acad Sci USA 82: 1837-1841.

Johnston GAR, Kennedy SME, Twitchin B (1979) Action of the neurotoxin kainic acid on high affinity uptake of L-glutamic acid in rat brain slices, J Neurochem 32: 121-127.

Kaneko T, Urade Y, Watanabe Y, Mizuno N (1987) Production, characterization, and immunohistochemical application of monoclonal antobodies to glutaminase purified from rat brain, J Neurosci 7: 302-309

Levi G, Raiteri M (1976) Synaptosomal transport processes, Int Rev Neurobiol 19: 51-74.

Levi G, Wilkin GP, Ciotti MT, Johnstone SR (1983) Enrichment of differentiated stellate astrocytes in cerebellar interneuron cultures as studied by GFAP immunofluorescence and autoradiographic uptake patterns of ^{3}H-D-aspartate and ^{3}H-GABA, Dev Brain Res 10: 227-241.

Levi G, Aloisi F, Ciotti MT, Gallo V (1984) Autoradiographic localization and depolarization-induced release of acidic amino acids in differentiating cerebellar granule cell cultures, Brain Res 290: 77-86

Levi G, Gallo V, Ciotti MT (1986) Bipotential precursors of putative fibrous astrocytes and oligodendrocytes in rat cerebellar cultures express distinct surface features and "neuron-like" γ-aminobutyric acid transport, Proc Natl Acad Sci USA 83: 1504-1508

London ED, Coyle JT (1979) Specific binding of ^{3}H-kainic acid to receptor sites in rat brain, Molec Pharmacol 15, 492-505.

McBean GJ, Roberts PJ (1981) Glutamate-preferring receptors regulate the release of ^{3}H-D-aspartate from rat hippocampal slices, Nature 291: 593-594.
Monaghan DT, Cotman CW (1982) The distribution of ^{3}H-kainic acid binding sites in rat CNS as determined by autoradiography, Brain Res 252: 91-100.
Nicoletti F, Wroblewski JT, Novelli A, Alho H, Guidotti A, Costa E (1986) The activation of inositol phospholipid metabolism as a signal transducing system for excitatory amino acids in primary cultures of cerebellar granule cells, J Neurosci 6: 1905-1911.
Palay SL, Chan-Palay V (1974) Cerebellar cortex. Cytology and organization, Springer-Verlag, Berlin.
Pastuszko A, Wilson DF, Erecinska M (1984) Effects of kainic acid in rat brain synaptosomes: the involvement of calcium, J Neurochem 43: 747-754.
Pearce BR, Dutton GR (1982) L-glutamate increases the spontaneous release of ^{3}H-GABA from cultured cerebellar neurons, Dev Brain Res 3: 492-496.
Poli A, Contestabile A, Migani P, Rossi L, Rondelli C, Virgili M, Bissoli R, Bernabei O (1985) Kainic acid differentially affects the synaptosomal release of endogenous and exogenous amino acidic neurotransmitters, J Neurochem 45: 1677-1686.
Potashner SJ, Gerard D (1983) Kainate-enhanced release of ^{3}H-D-aspartate from cerebral cortex and striatum: reversal by baclofen and pentobarbital, J Neurochem 40, 1548-1557.
Raff MC, Abney ER, Cohen J, Lindsay R, Noble M (1983b) Two types of astrocytes in cultures of developing rat white matter: differences in morphology, surface gangliosides and growth characteristics, J Neurosci 3: 1289-1300
Slevin JT, Collins JF, Coyle JT (1983) Analogue interactions with the brain receptor labeled by ^{3}H-kainic acid, Brain Res 265: 169-172.
Unnerstall JR, Walmsley JK (1983) Autoradiographic localization of high-affinity ^{3}H-kainic acid binding sites in rat forebrain, Eur J Pharmacol 86: 361-371.
Wilkin GP, Levi G, Johnstone S, Riddle PN (1983) Cerebellar astroglial cells in primary culture: Expression of different morphological appearances and differential ability to take up ^{3}H-D-aspartate and ^{3}H-GABA, Dev Brain Res 10: 265-277
Wroblewski JT, Nicoletti F, Costa E (1985) Different coupling of excitatory amino acid receptors with Ca^{2+} channels in primary cultures of cerebellar granule cells, Neuropharmacology 24: 919-921.

SEROTONIN-GLUTAMATE INTERACTIONS IN RAT CEREBELLUM: INVOLVEMENT OF MULTIPLE 5-HT RECEPTOR SUBTYPES

Maurizio Raiteri, Guido Maura, Giambattista Bonanno, Anna Pittaluga and Massimo Ulivi
Istituto di Farmacologia e Farmacognosia, Universitá di Genova, Viale Cembrano 4, 16148 Genova, Italy.

INTRODUCTION

The existence of different 5-HT receptor subtypes in the central nervous system is now accepted. Two different recognition sites, termed 5-HT_1 and 5-HT_2, were originally identified (Peroutka and Snyder 1979). In particular, the 5-HT_1 type is reported to be heterogeneous and has been subdivided into 5-HT_{1A}, 5-HT_{1B} and 5-HT_{1C} (Pedigo et al 1981; Pazos et al 1984; Peroutka 1986). However, the possibility that this multiplicity of binding sites reflects the existence of multiple 5-HT receptors would be investigated in studies in which receptor-mediated functional responses are measured.

Evidence has accumulated to suggest the neurotransmitter role of glutamic acid (GLU) in the cerebellum. In particular, it has been proposed that this excitatory amino acid could be the transmitter of the granule cells and that it is released from the terminals of the parallel fibers (Levi et al 1982; Young et al 1974). Serotonergic afferences from the raphe nuclei (Chan Palay 1976,1977; Shinnar et al 1975) and from the nuclei reticularis gigantocellularis, paragigantocellularis and pontis oralis of the reticular formation reach the cerebellar cortex (Bishop and Ho 1985; Bobillier et al 1976). Some of these fibers terminate with fine varicosities in the granular cell layer and others in the molecular layer (Takeuchi et al 1982).

NATO ASI Series, Vol. H19
Modulation of Synaptic Transmission and Plasticity in Nervous Systems. Edited by G. Hertting and H.-C. Spatz

According to these anatomical observations we have investigated the existence and the pharmacological characteristics of 5-HT receptors modulating GLU release in the rat cerebellum and of the 5-HT autoreceptors regulating 5-HT release from the nerve terminals of the fibers projecting to the cerebellum. To characterize the pharmacological properties and the anatomical distribution of the different receptor subtypes involved, the modulation of ^{3}H-5-HT and endogenous GLU release has been studied using both cerebellar synaptosomes and slices in superfusion.

5-HT AUTORECEPTORS IN THE CEREBELLUM

5-HT can inhibit its own release in rat brain through the activation of autoreceptors sited on 5-HT axon terminals (Cerrito and Raiteri 1979; Gothert and Weinheimer 1979; Maura et al 1986a). 5-HT autoreceptors have so far been identified and characterized essentially in some areas of the rat brain: hypothalamus, cerebral cortex and hippocampus (Baumann and Waldmeier 1981; Martin and Sanders-Bush 1982; Middlemiss 1984; Maura and Raiteri 1984; Maura et al 1986a).

Our results show that 5-HT autoreceptors also exist on 5-HT nerve terminals in the cerebellum of the rat (Fig. 1). Similarly to what was found in cerebral cortex, hypothalamus and hippocampus also in the cerebellum, 5-HT was antagonized by methiothepin (Fig. 1) but not by ketanserin, methysergide, cinanserin or spiperone (data not shown). Moreover, the cerebellar autoreceptor was blocked by (-)propranolol (Fig. 1) which stereoselectively displace ^{3}H-5-HT from the 5-HT$_1$ sites showing equal and high affinity for the 5-HT$_{1A}$ and the 5-HT$_{1B}$ subsites, but low affinity for the 5-HT$_{1C}$ (Hoyer et al 1985) and was insensitive to 8-OH-DPAT (Fig. 2) which selectively display ^{3}H-5-HT from the 5-HT$_{1A}$ sites in the rat frontal cortex (Middlemiss and Fozard 1983). Our results with 8-OH-DPAT and (-)propranolol would allow us to conclude that the 5-HT autoreceptor in rat cerebellum is very similar to that in the cerebral

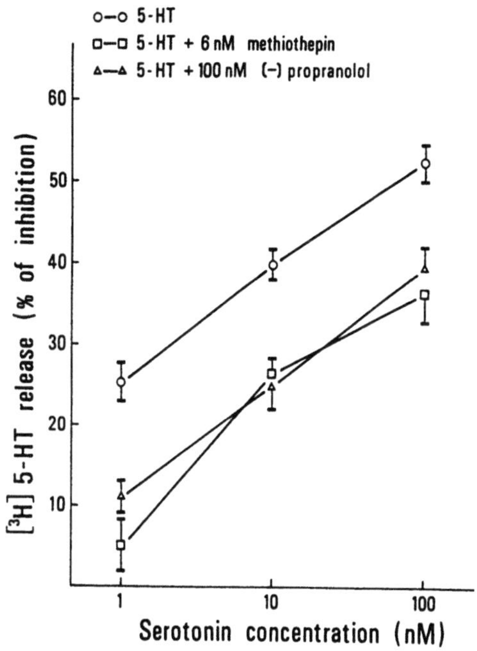

Fig. 1. Concentration-dependent inhibition by 5-HT of the K^+-evoked release of ^{3}H-5-HT: effects of methiothepin and of (-)propranolol. Crude synaptosomes were incubated 15 min at 37° C with ^{3}H-5-HT and superfused as previously described (Bonanno et al 1986). Depolarization was carried out with 15 mM KCl. Exogenous 5-HT was added concomitantly with 15 mM KCl. The antagonists were present starting 8 min before depolarization. Means ± S.E.M. of 6-8 experiments run in triplicate are presented.

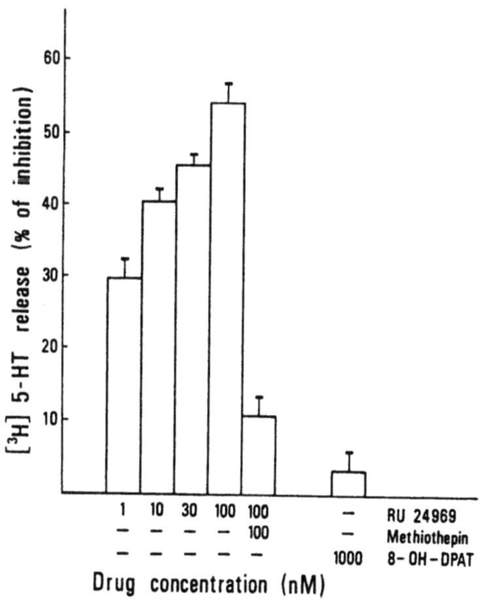

Fig. 2. Effect of RU 24969 and of 8-OH-DPAT on the K^+-evoked release of ^{3}H-5-HT. Experimental details as in the legend to Fig. 1. The data presented are means ± S.E.M. of 6-8 experiments in triplicate.

cortex and hippocampus can be classified as a $5\text{-}HT_{1B}$ receptor subtype.

SEROTONERGIC MODULATION OF GLU RELEASE

a) Synaptosomes

Fig. 3 shows clearly that 5-HT receptors are localized on GLU nerve terminals in rat cerebellum and that the activation of these 5-HT presynaptic heteroreceptors elicits inhibition of the stimulated GLU release.

The presynaptic 5-HT heteroreceptors on GLU terminals are methiothepin-sensitive, but insensitive towards ketanserin, methysergide, cinanserin or spiperone (Table 1). These data and the finding that the $5\text{-}HT_1$ agonist RU 24969 was almost equipotent as an agonist at both auto-

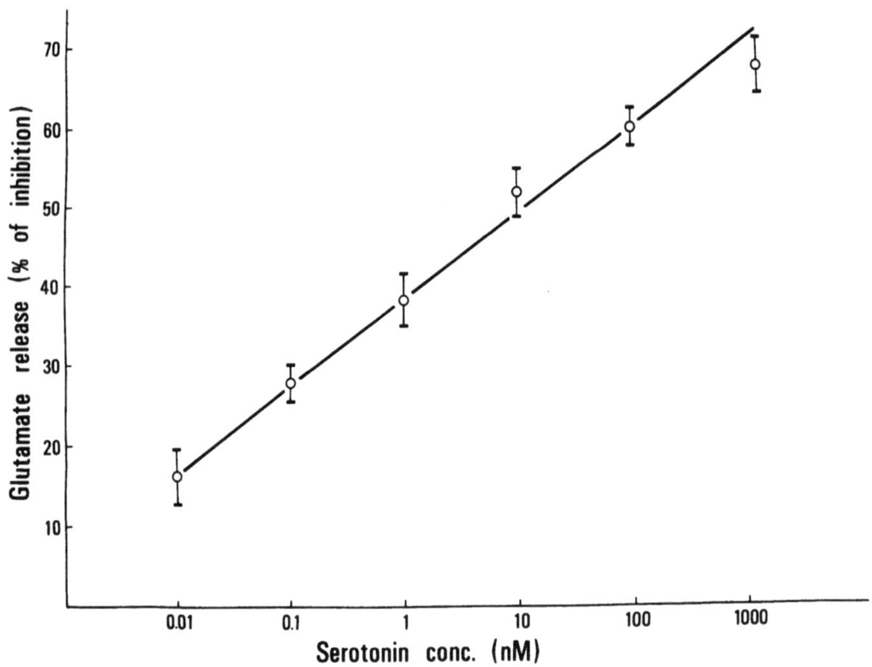

Fig. 3. Concentration-response curve of the inhibition by 5-HT of the K^+-evoked release of endogenous GLU from rat cerebellar synaptosomes. Experimental details as previously reported (Maura et al 1986b). Means ± S.E.M. of 6-8 experiments in triplicate are shown.

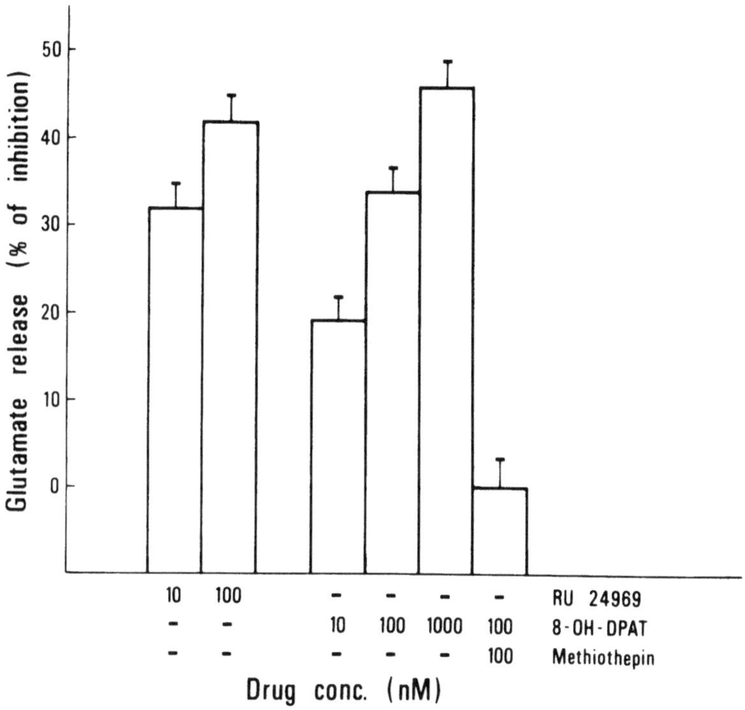

Fig. 4. Inhibition by RU 24969 and 8-OH-DPAT of the K^+-evoked endogenous GLU release from cerebellar synaptosomes. Experimental details as reported in Fig. 3. The data presented are means ± S.E.M. of 5-8 experiments in triplicate.

and heteroreceptors (Figs. 2 and 4) would suggest a strong similarity between the two 5-HT receptors. However, a clear-cut discrimination could be obtained with (-)propranolol and 8-OH-DPAT (Fig. 4, Table 1). The sensitivity to 8-OH-DPAT of the 5-HT heteroreceptors regulating GLU release would therefore legitimate their classification as 5-HT_{1A} receptors. However, it has to be noted that: 1) spiperone, at a concentration (1 μM) adequate to block the 5-HT_{1A} receptors (Middlemiss and Fozard 1983; Sills et al 1984) was ineffective as an antagonist of 5-HT at the heteroreceptors (Table 1); 2) (-)propranolol which, at 1 μM is likely to interact with the 5-HT_1 sites of both A and B subtype

(Tricklebank et al 1985) was ineffective at the heteroreceptors and 3) although RU 24969 was reported to be relatively selective for the 5-HT$_{1B}$ vs. the 5-HT$_{1A}$ binding sites (Sills et al 1984; Doods et al 1985) the data of figures 2 and 4 do not appear to be indicative of such a selectivity. Thus, to classify the heteroreceptors on GLU terminals as 5-HT$_{1A}$ does not seem to be justified easily. On the other hand, the findings that 8-OH-DPAT, effective at the heteroreceptors, had very low affinity for the 5-HT$_{1C}$ binding site (Pazos et al 1984) whereas methysergide, ineffective at the heteroreceptors, had high affinity for that site, would exclude the possibility that the heteroreceptors sited on GLU terminals belong to the 5-HT$_{1C}$ subtype. All together the data would not be incompatible with the hypothesis that the 5-HT heteroreceptors sited on GLU terminals in rat cerebellum represent a new subtype of 5-HT$_1$ receptor (5-HT$_{1D}$).

TABLE 1

Effect of 5-HT antagonists on the inhibition by 5-HT of the K^+-evoked (15 mM) release of endogenous GLU from cerebellum synaptosomes

DRUGS	K^+-evoked release of GLU (% of inhibition)
10 nM 5-HT	50.9 ± 3.12 (16)
10 nM 5-HT + 10 nM methiothepin	10.5 ± 2.01 (5)
10 nM 5-HT + 1000 nM cinanserin	47.0 ± 3.57 (3)
10 nM 5-HT + 1000 nM ketanserin	49.7 ± 2.99 (3)
10 nM 5-HT + 1000 nM methysergide	51.6 ± 4.03 (3)
10 nM 5-HT + 1000 nM (-)propranolol	52.0 ± 4.73 (4)
10 nM 5-HT + 1000 nM spiperone	49.7 ± 3.72 (4)

The data reported represent the percent changes of the K^+-evoked release. For experimental details see Legend to Fig. 3. Each value represents the mean ± S.E.M. of several experiments (number in parenthesis) performed in triplicate.

b) Slices

5-HT elicited a concentration-dependent inhibition of the 35 mM KCl evoked endogenous GLU release from cerebellar slices in superfusion; the effect of 5-HT was mimicked, less potently, by 8-OH-DPAT (Fig. 5, panels a,b). Fig. 5, panel c, also show that the relatively selective $5\text{-}HT_2$ receptor agonist DOI inhibited, also in this case less potently than 5-HT, the release of GLU from cerebellar slices. This inhibition was counteracted by the selective $5\text{-}HT_2$ receptor antagonist ketanserin (Table 3). This result is consistent with the presence in rat cerebellum of a $5\text{-}HT_2$ receptor mediating inhibition of GLU release. Since the effect of 5-HT on GLU release from cerebellar synaptosomes was totally insensitive to drugs

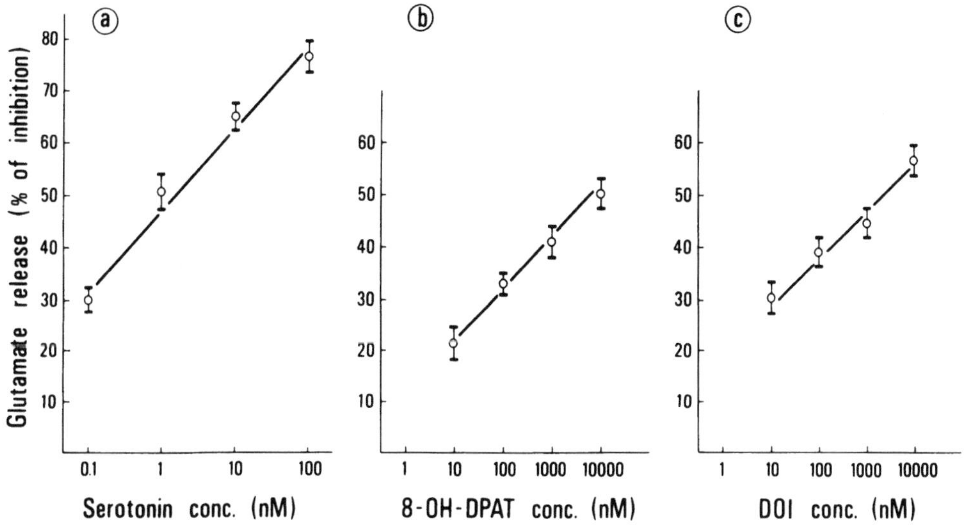

Fig. 5. Concentration-dependent inhibition by 5-HT, 8-OH-DPAT or DOI of the K^+-evoked (35 mM) release of endogenous GLU from rat cerebellar slices. Slices were prepared and superfused as previously described (Maura et al submitted European Journal of Pharmacology). Agonists were added concomitantly with high K^+ (S_2). The data presented are means ± S.E.M. of 3-7 experiments in duplicate.

used to characterize $5\text{-}HT_2$ receptors, the ketanserin-sensitive receptors seen in slices should take part in a mechanism which is lost during the preparation of synaptosomes.

Interestingly, ketanserin could antagonize only part (35-40 %) of the inhibition of GLU release when the agonist used was 5-HT (Table 2). As for the ketanserin-insensitive portion of the effect, one possibility is that it corresponds, at least in part, to the $5\text{-}HT_1$ receptor previously found to be sited on GLU nerve terminals. In fact methiothepin (the mixed $5\text{-}HT_1/5\text{-}HT_2$ receptor antagonist) completely abolished the effect of 5-HT (Table 2) and 8-OH-DPAT (which had been found to activate the $5\text{-}HT_1$ receptors located on GLU terminals) inhibited GLU release also in slices in a way that was insensitive to ketanserin but fully antagonized by methiothepin (Table 3).

As a note of caution, the possibility that 5-HT in slices acts also on $5\text{-}HT_1$ receptors other than those previously characterized in synaptosomes can not be ruled out. However, the results so far obtained allow to conclude that the release of GLU in rat cerebellum can be inhibited by 5-HT through the activation of both $5\text{-}HT_1$ and $5\text{-}HT_2$ receptors.

Using brain slices an increase of the stimulated release caused by a presynaptic receptor antagonist may indicate a disinhibition of the release inhibited by the endogenous transmitter present in the biophase of the receptor. This is often interpreted as an indication that the receptors in question may play a physiological role. Figure 6 shows that blocking only the $5\text{-}HT_2$ receptors with ketanserin the release of GLU increased to a maximum of 23%, reaching a plateau at about 10 nM. The effect of methiothepin which blocks both $5\text{-}HT_1$ and $5\text{-}HT_2$ receptors was much more pronounced reaching at 10 nM a maximum of about 55%. Thus, according to the current view, both the $5\text{-}HT_1$ and the $5\text{-}HT_2$ receptors found to regulate "pharmacologically" the release of endogenous GLU may play a physiological role and the release of GLU could be under tonic inhibition by endogenous 5-HT.

TABLE 2

Effect of methiothepin and ketanserin on the inhibition by 5-HT of the K^+-evoked release of endogenous GLU from cerebellar slices.

DRUGS	K^+-evoked release of GLU (% of inhibition)
10 nM 5-HT	63.8 ± 4.05 (5)
10 nM 5-HT + 100 nM ketanserin	30.6 ± 3.02 (3)
10 nM 5-HT + 300 nM ketanserin	27.4 ± 3.51 (3)
10 nM 5-HT + 1000 nM ketanserin	37.5 ± 4.75 (3)
10 nM 5-HT + 300 nM methiothepin	1.3 ± 2.73 (3)

The data reported represent the percent of the K^+-evoked release. 5-HT was added to S_2 concomitantly with high-K^+. The antagonists were present throughout the experiment. For experimental details see legend to Fig. 5. Each value represents the mean ± S.E.M. of several experiments (number in parenthesis) performed in duplicate.

TABLE 3

Effect of 5-HT antagonists on the inhibition by 8-OH-DPAT and DOI of the K^+-evoked release of GLU from cerebellar slices.

DRUGS	K^+-evoked release of GLU (% of inhibition)
10 nM 8-OH-DPAT	21.4 ± 2.10 (4)
10 nM 8-OH-DPAT + 300 nM ketanserin	19.1 ± 1.50 (4)
10 nM 8-OH-DPAT + 1000 nM ketanserin	21.0 ± 1.22 (4)
10 nM 8-OH-DPAT + 300 nM methiothepin	1.5 ± 2.10 (4)
100 nM DOI	39.2 ± 2.14 (3)
100 nM DOI + 10 nM ketanserin	1.7 ± 2.30 (3)

The data reported represent the percent of the K^+-evoked release. The agonists were added concomitantly with high K^+ at S_2 and the antagonists were present throughout the experiment. For experimental details see Legend to Fig. 5. Each value represents the mean ± S.E.M. of several experiments (number in parenthesis) performed in duplicate.

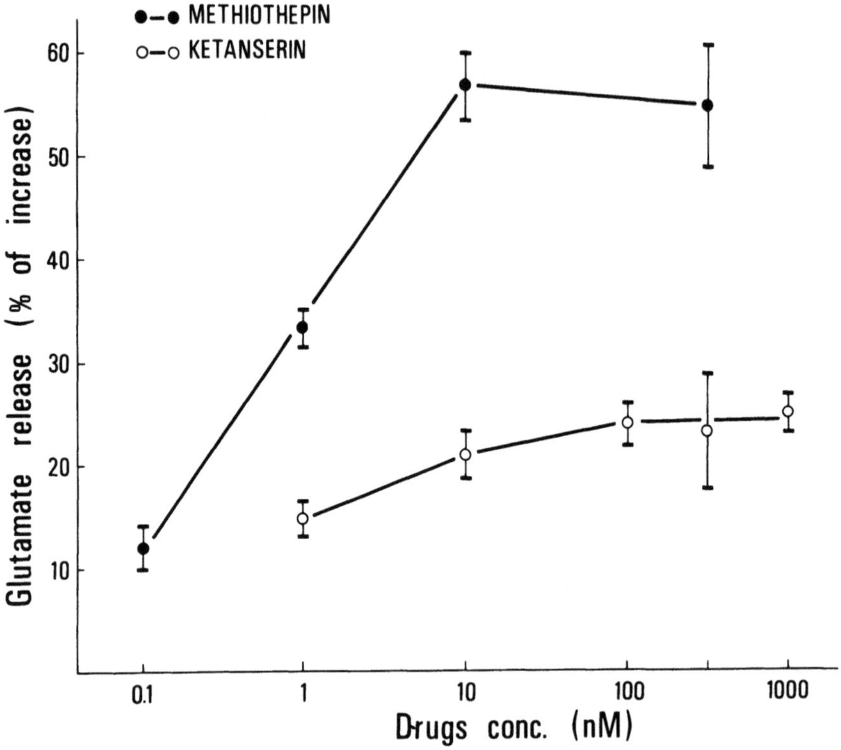

Fig. 6. Effects of ketanserin or methiothepin on the K^+-evoked release of GLU from cerebellar slices. The antagonists were added 8 min before 35 mM KCl (S_2). Experimental details as in the legend to Fig. 5. Means ± S.E.M. of 3-5 experiments in duplicate are presented.

CONCLUSIONS

On the basis of the existing literature and the present results, the interaction between 5-HT neurons and the cerebellar glutamate system appears to be complex and to involve types and subtypes (at least 4) of the 5-HT receptor. The following scheme (see fig. 7) could be tentatively proposed: a) serotonin autoreceptors on serotonergic cell bodies of raphe nuclei are of the 5-HT_{1A} subtype (de Montigny et al 1984; Verge et al 1985); b) the axon terminals of these serotonergic neurons that project to the cerebellar cortex possess autoreceptors of the 5-HT_{1B} subtype (Bonanno

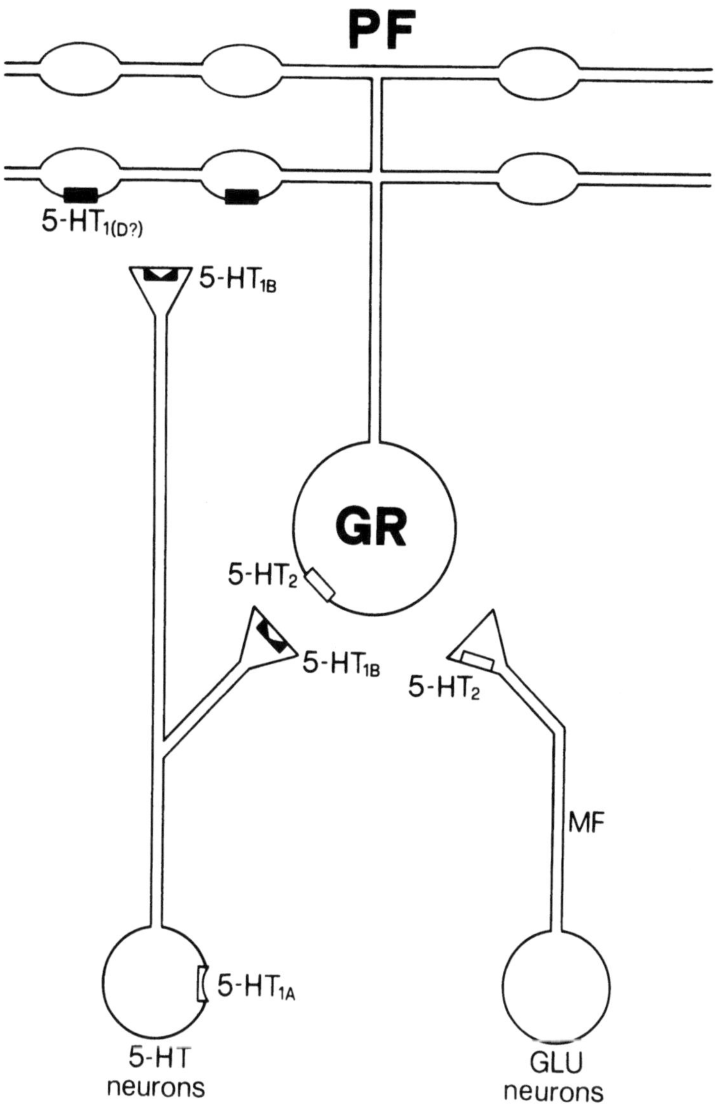

Fig. 7. Schematic model of possible 5-HT and GLU interactions in the rat cerebellum.
5-HT, serotonin; **GLU**, glutamic acid; **GR**, granule cell; **PF**, parallel fibers; **MF**, mossy fibers.
Two alternatives have been considered concerning the 5-HT_2 receptors:
1) serotonin-containing fibers arising from raphe nuclei (Chan Palay 1976,1977; Shinnar et al 1975) or/and nucleus reticularis gigantocellularis, nucleus reticularis paragigantocellularis and nucleus reticularis pontis oralis (Bishop and Ho 1985) release 5-HT on 5-HT_2 receptors sited on granule cells;
2) 5-HT_2 receptors are sited on the terminals of glutamatergic mossy fibers projecting to the cerebellar cortex (Beitz et al 1986).

et al 1986); c) a group of 5-HT axon terminals reach the molecular layer where they release 5-HT onto receptors of the 5-HT_1 type (possibly of a new non-A, non-B, non-C subtype) that are sited on GLU nerve endings (probably the varicosities of the parallel fibers) and mediate inhibition of GLU release; d) a different group of 5-HT axons terminate as classical mossy fibers in the granular cell layer (Chan Palay 1977) and might release 5-HT onto 5-HT_2 receptors localized on the soma and/or dendrites of granule cells. Activation of these receptors also leads to decrease of GLU release, an effect that appears to be lost during the preparation of synaptosomes but is measurable in slices.

It has to be noted that an alternative exists to point d) (see fig. 7). Recent reports have proposed that pontocerebellar fibers project as GLU-containing mossy fibers to synapse on cerebellar granule cells (which are also considered to be glutamatergic) thus constituting a multisynaptic glutamatergic pathway (Beitz et al 1986). In support to this idea, immunogold labeling revealed enrichement of GLU in mossy fiber terminals (Somogyi et al 1986). Interestingly, the endings of mossy fibers from the granular layer of the cerebellar cortex are lost during the classical preparation of synaptosomes because of their sedimentation properties (Israel and Whittaker 1965). Thus, as an alternative to point d), 5-HT_2 receptors might be present on the endings of glutamatergic mossy fibers to regulate GLU release onto granule cells. The technique by Terrian et al (1985) for the isolation of cerebellar glomeruli could be of help to verify the latter possibility.

ACKNOWLEDGEMENTS

This work was supported by grants from the Italian Ministry of Education and from the Italian C.N.R. The authors thank Mrs. Maura Agate for typing the manuscript.

REFERENCES

Baumann PA, Waldmeier PC (1981) Further evidence for negative feedback control of serotonin release in the central nervous system. Naunyn-Schmiedeberg's Arch Pharmacol 317:36-43

Beitz AJ, Larson AA, Monaghan P, Altschuler RA, Mullett MM, Madl JE (1986) Immunohistochemical localization of glutamate, glutaminase and aspartate aminotransferase in neurons of the pontine nuclei of the rat. Neuroscience 17:741-753

Bishop GA, Ho RH (1985) The distribution and origin of serotonin immunoreactivity in the rat cerebellum. Brain Res 331:195-207

Bobillier P, Seguin S, Petitjean F, Salvert D, Touret M, Jouvet M (1976) The raphe nuclei of the cat brain stem: a topographical atlas of their efferent projections as revealed by autoradiography. Brain Res 113:449-486

Bonanno G, Maura G, Raiteri M (1986) Pharmacological characterization of release-regulating serotonin autoreceptors in rat cerebellum. Eur J Pharmacol 126:317-321

Cerrito F, Raiteri M (1979) Serotonin release is modulated by presynaptic autoreceptors. Eur J Pharmacol 57:427-430

Chan-Palay V (1976) Serotonin afferents from raphe nuclei to the cerebellum in mammals. Exp Brain Res Suppl 1:20-25

Chan-Palay V (1977) The indoleamine afferent axons to the cerebellum. In: Cerebellar dentate nucleus: organization, cytology and transmitters. Springer Verlag New York, pp 390-454

De Montigny C, Blier P, Chaput Y (1984) Electrophysiologically-identified serotonin receptors in the rat CNS. Effect of antidepressant treatment. Neuropharmacol 23:1511-1520

Doods HN, Kalkman HO, De Jonge A, Thoolen MJMC, Wilffert B, Timmermans PBMWM, van Zwieten PA (1985) Differential selectivities of RU 24969 and 8-OH-DPAT for the purported 5-HT_{1A} and 5-HT_{1B} binding sites. Correlation between 5-HT_{1A} affinity and hypotensive activity. Eur J Pharmacol 112:363-370

Göthert M, Weinheimer G (1979) Extracellular 5-hydroxytryptamine inhibits 5-hydroxytryptamine release from rat brain cortex slices. Naunyn--Schmiedeberg's Arch Pharmacol 310:93-96

Hoyer D, Engel G, Kalkman HO (1985) Characterization of the 5-HT_{1B} recognition site in rat brain: Binding studies with (-)[^{125}I]iodocyanopindolol. Eur J Pharmacol 118:1-12

Israel M, Whittaker VP (1965) The isolation of mossy fibre endings from the granular layer of the cerebellar cortex. Experientia 21:325-326

Levi G, Gordon RD, Gallo V, Wilkin GP, Balazs R (1982) Putative acidic amino acid transmitters in the cerebellum. I. Depolarization-induced release. Brain Res 239:425-445

Martin LL, Sanders-Bush E (1982) Comparison of the pharmacological characteristics of 5-HT_1 and 5-HT_2 binding sites with those of serotonin autoreceptors which modulate serotonin release. Naunyn-Schmiedeberg's Arch Pharmacol 321:165-170

Maura G, Raiteri M (1984) Functional evidence that chronic drugs induce adaptive changes of central autoreceptors regulating serotonin release. Eur J Pharmacol 97:309-313

Maura G, Roccatagliata E, Raiteri M (1986a) Serotonin autoreceptor in rat hippocampus: pharmacological characterization as a subtype of the $5\text{-}HT_1$ receptor. Naunyn-Schmiedeberg's Arch Pharmacol 334:323-326

Maura G, Ricchetti A, Raiteri M (1986b) Serotonin inhibits the depolarization-evoked release of endogenous glutamate from rat cerebellar nerve endings. Neurosci Lett 67:218-222

Maura G, Roccatagliata E, Ulivi M., Raiteri M (submitted) Serotonin-glutamate interaction in rat cerebellum: involvement of $5\text{-}HT_1$ and $5\text{-}HT_2$ receptors. Eur J Pharmacol

Middlemiss DN (1984) Stereoselective blockade at $[^3H]$5-HT binding sites and at the 5-HT autoreceptor by propranolol. Eur J Pharmacol 101:289-293

Middlemiss DN, Fozard JR (1983) 8-hydroxy-2-(di-n-propylamino)-tetralin discriminates between subtypes of the $5\text{-}HT_1$ recognition site. Eur J Pharmacol 90:151-153

Pazos A, Hoyer D, Palacios JM (1984) The binding of serotonergic ligands to the porcine choroid plexus: characterization of a new type of serotonin recognition site. Eur J Pharmacol 106:539-546

Pedigo NW, Yamamura HJ, Nelson DL (1981) Discrimination of multiple ^3H-5-hydroxytryptamine binding sites by the neuroleptic spiperone in rat brain. J Neurochem 36:220-226

Peroutka SJ (1986) Pharmacological differentiation and characterization of $5\text{-}HT_{1A}$, $5\text{-}HT_{1B}$ and $5\text{-}HT_{1C}$ binding sites in rat frontal cortex. J Neurochem 47:529-540

Peroutka SJ, Snyder SH (1979) Multiple serotonin receptors: differential binding of $[^3H]$5-hydroxytryptamine, $[^3H]$lysergic acid diethyl amide and $[^3H]$spiroperidol. Mol Pharmacol 16:687-699

Shinnar S, Maciewicz RJ, Shofer RJ (1975) A raphe projection to cat cerebellar cortex. Brain Res 97:139-143

Sills MA, Wolfe BB, Frazer A (1984) Determination of selective and nonselective compounds for the $5\text{-}HT_{1A}$ and $5\text{-}HT_{1B}$ receptor subtypes in rat frontal cortex. J Pharmacol Exp Ther 231:480-487

Somogyi P, Halasy K, Somogyi J, Storm-Mathisen J, Ottersen OP (1986) Quantification of immunogold labelling reveals enrichement of glutamate in mossy and parallel fibre terminals in cat cerebellum. Neuroscience 19:1045-1050

Takeuchi Y, Kimura H, Sano Y (1982) Immunohistochemical demonstration of serotonin-containing nerve fibers in the cerebellum. Cell Tissue Res 226:1-12

Terrian DM, Butcher WI, Wu PH, Armstrong DL (1985) Isolation of glomeruli from areas of bovine cerebellum and comparison of $[^3H]$serotonin uptake. Brain Res Bull 14:469-475

Tricklebank MD, Forler C, Fozard JR (1985) The involvement of subtypes of the $5\text{-}HT_1$ receptor and of catecholaminergic system in the behavioural response to 8-hydroxy-2-(di-n-propylamino)tetralin in the rat. Eur J

Pharmacol 106:271-282
Verge D, Daval G, Patey A, Gozlan H, El Mestikawy S, Hamon M (1985) Presynaptic 5-HT autoreceptors on serotonergic cell bodies and/or dendrites but not terminals are of the 5-HT_{1A} subtype. Eur J Pharmacol 113:463-464
Young AB, Moster-granite ML, Herndon RM, Snyder SH (1974) Glutamic acid: selective depletion by viral induced granule cell loss in hamster cerebellum. Brain Res 73: 1-13

MODULATION OF TRANSMITTER RELEASE BY PRESYNAPTIC SEROTONIN RECEPTORS

Manfred Göthert
Institute of Pharmacology and Toxicology
Rheinische Friedrich-Wilhelms-Universität
Reuterstraße 2b
5300 Bonn 1
Fed. Rep. Germany

Introduction

Inhibitory presynaptic serotonin (5-HT) receptors, i.e. 5-HT receptors on the axon terminals of neurones, play a role in the modulation of synaptic transmission both in the central and in the peripheral nervous system. Such receptors which upon stimulation inhibit transmitter release occur either on the 5-HT nerve terminals themselves (presynaptic 5-HT autoreceptors) or on the terminals of non-5-HT neurones (presynaptic heteroreceptors).

In recent years evidence has accumulated that the 5-HT receptors are heterogeneous, and attempts were made to classify them (see Bradley et al. 1986, Göthert and Schlicker 1987). The heterogeneity of 5-HT receptors offers a chance for the development of drugs which may modify transmission in certain synapses, leaving that in others unaffected. Therefore, a current problem is the classification of presynaptic receptors with respect to their pharmacological properties. Closely related to this is the question for species differences. In particular, it is of interest whether the various presynaptic 5-HT receptors are also present in the human nervous system and, if so, to which receptor type or even subtype they belong. In this context, the question for an appropriate animal model relevant for the human nervous system is arising. These problems will be discussed mainly for presynaptic 5-HT autoreceptors in the central nervous system and for presynaptic 5-HT heteroreceptors on peripheral

NATO ASI Series, Vol. H19
Modulation of Synaptic Transmission and Plasticity
in Nervous Systems. Edited by G. Hertting and H.-C. Spatz

noradrenergic nerve terminals. Presynaptic 5-HT heteroreceptors in the central nervous system will be only briefly touched upon in the context of 5-HT receptors on dopaminergic nerve terminals. As a background information for the more detailed discussion, a very brief overview over the current classification of 5-HT receptors will be given.

Classification of 5-HT receptors

The present classification of 5-HT receptors is based on both radioligand binding studies and experiments on isolated tissues, whereas the former attempts to classify them were founded on either binding or functional data and, furthermore were hampered by the nonselective pharmacological tools then available. Two types of 5-HT recognition sites were first identified by Peroutka and Snyder (1979): one site labelled by ^{3}H-5-HT was termed 5-HT_1 and the other one labelled by ^{3}H-spiperone was denoted as 5-HT_2. Taking our present knowledge of 5-HT-mediated effects into account, it may be stated that in most cases the pharmacological properties ascribed to the classical "D" receptors (nomenclature of Gaddum and Picarelli 1957; defined exclusively by functional experiments) resemble those of the 5-HT_2 binding sites and that only a smaller proportion of the "D" receptors are similar to the 5-HT_1 sites. Recently a third class of 5-HT recognition sites could be identified in the rat brain by means of a new radioligand ("^{3}H-GR 65630A"; M.B. Tyers, personal communication) and the site to which this tritiated compound is bound was termed 5-HT_3. Its pharmacological properties are identical to those of the excitatory 5-HT receptors identified by functional experiments on terminals of sympathetic and parasympathetic nerves and on afferent nerve fibres; the effects mediated by these neuronal receptors are analogous (but probably not identical) to those ascribed to the classical "M" receptors of Gaddum and Picarelli (1957).

Since the above-mentioned recognition sites appear to be involved in the effects of 5-HT, the introduction of the term receptor is justified in this context. Accordingly, the 5-HT receptors may be divided into three classes designated $5\text{-}HT_1$, $5\text{-}HT_2$ and $5\text{-}HT_3$. As will be discussed in the following sections, the majority of the inhibitory presynaptic 5-HT auto- and heteroreceptors possess the main characteristics of $5\text{-}HT_1$ receptors. According to the definition (Bradley et al. 1986; Göthert and Schlicker 1987), a given 5-HT receptor may be denoted as $5\text{-}HT_1$ if it is blocked by the nonselective 5-HT receptor antagonist metitepine but not by selective $5\text{-}HT_2$ or $5\text{-}HT_3$ antagonists, and if the agonist 5-carboxamidotryptamine is at least as potent as 5-HT itself; since in the present account only inhibitory presynaptic 5-HT receptors are considered, the involvement of $5\text{-}HT_3$ receptors (which mediate neuronal excitation, see above) is very improbable and, hence, will not be discussed below. By development of new specific radioligands and of drugs with selective actions in certain functional studies three different subtypes of $5\text{-}HT_1$ receptors have been distinguished which are designated $5\text{-}HT_{1A}$, $5\text{-}HT_{1B}$ and $5\text{-}HT_{1C}$. These subtypes have to be considered in the context of presynaptic 5-HT receptors.

Presynaptic 5-HT autoreceptors in the CNS

Basic experimental findings in tissue slices and synaptosomes from rats. In most detail, presynaptic 5-HT autoreceptors were studied in the central nervous system of rats. Superfused slices or synaptosomes from brain areas which only contain 5-HT nerve terminals or from the spinal cord were used for these experiments. The electrically or potassium-evoked release of ^3H-5-HT from slices preincubated with ^3H-5-HT was concentration-dependently inhibited by serotonin. This effect was blocked by metitepine, but not by ketanserin, suggesting that the inhibitory effect of 5-HT is mediated via a $5\text{-}HT_1$ rather than a $5\text{-}HT_2$ receptor. According to this interpretation, metitepine, given alone, increased (or in other words

Table 1. Occurrence of presynaptic 5-HT autoreceptors in the rat CNS

Region of the CNS	References
Cerebral cortex	Göthert and Weinheimer (1979), Göthert (1980)
Hippocampus	Maura et al. (1986)
Striatum	Middlemiss (1984a)
Hypothalamus	Cerrito and Raiteri (1979), Langer and Moret (1982)
Cerebellum	Bonanno et al. (1986)
Medulla oblongata*	Schlicker et al. (1988)
Spinal cord	Monroe and Smith (1985)

* Area of the nucleus tractus solitarii

disinhibited) ^{3}H-5-HT release by preventing endogenous 5-HT from activating the autoreceptors via a short negative feedback loop. Evidence for the assumption that the receptors involved are located presynaptically on the serotoninergic nerve terminals was derived from experiments in synaptosomes, in which ^{3}H-5-HT release was evoked by high K^+. Serotonin inhibited ^{3}H-5-HT release in a manner susceptible to blockade by metitepine. By means of such experiments presynaptic 5-HT autoreceptors could be identified in various regions of the rat CNS (Table 1).

Recently it was shown in cortex slices that the inhibitory effect of serotonin and the facilitatory effect of metitepine on the electrically evoked ^{3}H-5-HT release were not altered in the presence of forskolin and a cAMP phosphodiesterase inhibitor. These findings suggest that the presynaptic 5-HT autoreceptors do not appear to be linked to the adenylate cyclase which is operative in the serotoninergic nerve terminals (Schlicker et al. 1987).

It is also of interest to note that the effects mediated by 5-HT autoreceptors of the cortex, medulla oblongata (region of the nucleus tractus solitarii) and hypothalamus from spontaneously hypertensive rats (5-7, 9-11 and 19-22 weeks old) did not differ from those of age-matched Wistar-Kyoto rats: the 5-HT-induced inhibition of release and the metitepine-induced facilitation were identical in both strains (Schlicker et al. 1988). Since inhibition of 5-HT release in the hypothalamus and nucleus tractus solitarii may be assumed to cause a decrease in blood pressure (see Göthert 1986), it is tempting to speculate that preferential 5-HT autoreceptor agonists may be suitable to lower an increased blood pressure.

Classification of 5-HT autoreceptors in the rat CNS. The receptor type and subtype to which the presynaptic 5-HT autoreceptor belongs was most extensively investigated in the rat brain cortex. Not only 5-HT but also a large number of other 5-HT receptor agonists, including mono- and dihydroxylated or methylated tryptamine derivatives as well as 5-carboxamidotryptamine and RU 24969 (5-methoxy-3(1,2,3,6-tetrahydro-4-pyridinyl)-1H indole succinate) inhibited the electrically evoked ^{3}H-5-HT release. The inhibitory potencies of the 5-HT receptor agonists were compared with their affinities for the various 5-HT binding sites. There was no correlation with 5-HT$_{1C}$ or 5-HT$_2$ sites and correlation was best with 5-HT$_{1B}$ binding sites, suggesting that the autoreceptors belong to this subtype (Engel et al. 1986).

Further support for the 5-HT$_{1B}$ character of the presynaptic 5-HT autoreceptors came from experiments with drugs acting more or less selectively at certain 5-HT receptor subtypes. 8-Hydroxy-2-(di-n-propylamino)tetralin (8-OH-DPAT) and ipsapirone (formerly denoted as TVX Q7821), which are full and partial 5-HT$_{1A}$ receptor agonists, respectively, did not modify ^{3}H-5-HT release (Middlemiss 1984a; Engel et al. 1987).

Furthermore, it was recently described by Neale et al. (1987) that the preferential 5-HT_{1B} receptor agonist, CGS 12066B (7-trifluoromethyl-4(4-methyl-1-piperazinyl)-pyrolo [1,2-a] quinoxaline) acted inhibitory on ^{3}H-5-HT release.

Besides metitepine (see above), also quipazine and metergoline blocked the inhibitory effect of 5-HT in the rat brain, whereas most of the other classical, rather nonselective 5-HT receptor antagonists were ineffective (Engel et al. 1983). In view of the lack of highly potent autoreceptor antagonists, it was an important observation made by Middlemiss (1984b) that propranolol stereoselectively blocked these receptors, a property which is shared by other ß-adrenoceptor antagonists: both pindolol and cyanopindolol produced a shift to the right of the concentration-response curve of 5-HT, the (-)-enantiomer being more potent than the (+)-compound (Engel et al. 1986). Again, the potencies of the 5-HT receptor antagonists to competitively block the autoreceptors were compared with their affinities for 5-HT binding sites. As with the agonists no correlation was found with 5-HT_{1C} and 5-HT_2 sites and correlation was best with 5-HT_{1B} binding sites. These data confirm the 5-HT_{1B} character of the presynaptic 5-HT autoreceptor in the rat brain, and this classification is further strengthened by the ineffectiveness of spiperone which besides 5-HT_2 receptors also blocks 5-HT_{1A} receptors.

Presynaptic 5-HT autoreceptors in the human and porcine brain. Release-inhibiting 5-HT receptors were also identified in human cortical slices prepared from brain tissue which had been removed to gain access to deep-seated tumours: 5-HT inhibited the electrically evoked ^{3}H-5-HT release in a manner sensitive to blockade by metitepine; the latter given alone increased ^{3}H-5-HT release, probably by preventing endogenous 5-HT from activating the receptors (Schlicker et al. 1985). Since in human cortical membranes no 5-HT_{1B} binding sites were found (Hoyer et al. 1986), the presynaptic 5-HT autoreceptors in the human brain may belong to another 5-HT receptor type or subtype.

Since fresh human cortical tissue is hardly obtainable, an appropriate model of the human brain with respect to its presynaptic 5-HT autoreceptors has to be found. The pig brain may be such a model, since it appears to be also devoid of 5-HT_{1B} binding sites, whereas (like in the human and rat brain) 5-HT_{1A}, 5-HT_{1C} and 5-HT_2 binding sites could be identified (Hoyer et al. 1985). Therefore, it was of interest to determine whether presynaptic 5-HT autoreceptors exist in the pig brain cortex as well. In slices of the pig brain cortex, the electrically evoked ^{3}H-5-HT release was inhibited by 5-HT and 5-carboxamidotryptamine. This effect was antagonized by the nonselective antagonist metitepine, whereas ketanserin was ineffective; metitepine given alone increased ^{3}H-5-HT release (Fink, Schlicker and Göthert, unpublished). These data suggest that the release-modulating receptors of the pig brain are of the 5-HT_1 type. They are located presynaptically on the serotoninergic nerve terminals themselves, as suggested by experiments on pig cortical synaptosomes in which 5-HT and 5-methoxytryptamine were found to inhibit ^{3}H-5-HT release; this effect was blocked by metitepine (Fink, Schlicker and Göthert, unpublished). The 5-HT_1 receptor subtype involved in the autoreceptor-mediated feedback inhibition of 5-HT release in the pig brain remains to be determined.

5-HT heteroreceptors on dopaminergic nerve terminals of the rat CNS

There are many open questions concerning 5-HT heteroreceptors on various types of neurones in various brain areas. This will be demonstrated very briefly for only one example, namely the 5-HT receptor modulating 5-HT release in the rat striatum. 5-HT and other nonselective 5-HT receptor agonists at remarkably high concentrations (in the µmolar range) inhibited the K^+- or impulse-evoked ^{3}H-dopamine release, an effect which was antagonized by nonselective 5-HT receptor antagonists, such as methysergide, metitepine and metergoline (Ennis et al. 1981; Westfall and Tittermary 1982). Since neither ketanserin nor compounds which act selectively at 5-HT_1 receptors were used, it can only be stated that a 5-HT receptor appears to be involved, whereas its type and subtype are unknown.

Furthermore, it was uncertain whether this 5-HT receptor is, in fact, located on the dopaminergic nerve terminals. Therefore, we recently carried out experiments on striatal synaptosomes preincubated with ^{3}H-dopamine in order to provide evidence for the presynaptic location of the receptors involved. Both the K^+- and the veratridine-evoked release were inhibited by 5-HT in a manner susceptible to blockade by metitepine, suggesting that the effect is actually mediated by a 5-HT heteroreceptor on the dopaminergic nerve terminals (Fink and Göthert, unpublished).

Presynaptic 5-HT heteroreceptors on peripheral noradrenergic nerve terminals

Classification of inhibitory 5-HT receptors on the sympathetic nerve terminals of the rat vena cava and renal vascular bed. The effects of serotonin receptor agonists and antagonists on the electrically evoked ^{3}H-noradrenaline release were investigated in the rat inferior vena cava preincubated with ^{3}H-noradrenaline (Molderings et al. 1987). 5-HT, 5-carboxamidotryptamine, RU 24969 and four further nonselective 5-HT receptor agonists inhibited ^{3}H-noradrenaline release. Comparison of their potencies in inhibiting release with their affinities for the various 5-HT binding sites revealed that correlation was best with 5-HT_{1B} receptors, whereas there was no correlation with 5-HT_{1A}, 5-HT_{1C} and 5-HT_2 binding sites. The inhibitory effect of 5-HT on ^{3}H-noradrenaline release was competitively antagonized by metitepine, metergoline, quipazine, 6-chloro-2-(1-piperazinyl)pyrazine (MK 212) and propranolol, but not by ketanserin. Correlation of the potencies of these drugs in blocking the autoreceptors with their affinities for the various 5-HT binding sites revealed no clear-cut picture. No significant correlation occurred, but there was at least a tendency towards a positive correlation with 5-HT_{1B}

sites. Since, in addition, 8-OH-DPAT and ipsapirone were ineffective in inhibiting release and since spiperone did not antagonize the effect of 5-HT, it may be concluded that the presynaptic 5-HT receptors on the sympathetic nerves of the rat vena cava are of the 5-HT_{1B} subtype. Hence, in the rat, the same subclass of 5-HT receptors appears to mediate the inhibition of neurotransmitter release from the serotoninergic nerve terminals of the brain cortex and from the noradrenergic nerve terminals of the inferior vena cava.

Furthermore, we found recently that the 5-HT receptor-mediated inhibition of noradrenaline release was more pronounced in the vena cava of spontaneously hypertensive rats than of Wistar-Kyoto rats (Schlicker and Göthert 1987). Accordingly, one may speculate that 5-HT_1 receptor agonists may decrease an elevated blood pressure not only by activating 5-HT autoreceptors of the CNS (see above) but also by inhibiting noradrenaline release in blood vessels.

In contrast to the results in the rat vena cava, the pharmacological properties of the inhibitory 5-HT receptors found by Charlton et al. (1986) in the perfused rat renal vascular bed do not conform to the 5-HT_{1B} character. These receptors mediating inhibition of ^{3}H-noradrenaline release also exhibit the main features of 5-HT_1 receptors, but a subclassification was not possible in view of the failure of propranolol, pindolol and cyanopindolol to act as antagonists.

5-HT receptors on the sympathetic nerve terminals of the human saphenous vein. Inhibitory presynaptic 5-HT_1 heteroreceptors are also involved in the modulation of noradrenaline release in the human saphenous vein. This conclusion was drawn from the findings that the electrically evoked ^{3}H-noradrenaline release was inhibited by 5-HT or high concentrations (µmolar range) of 8-OH-DPAT but not by ipsapirone, and that the 5-HT- or 8-OH-DPAT-induced inhibition of release was abolished by metitepine but not by ketanserin (Göthert et al. 1986). Propranolol, rauwolscine and flupenthixol were also ineffective in this respect. A subclassification of the 5-HT_1 receptors involved is not possible on the basis of these data.

Modulatory effect of 5-HT on noradrenaline release in the pig coronary artery. In search for an appropriate model relevant for human neuronal 5-HT receptors in the periphery, we recently decided to investigate whether the sympathetic nerves of porcine vasculature are also endowed with inhibitory 5-HT receptors. Electrically evoked ^3H-noradrenaline release in the coronary artery was investigated (Molderings and Göthert, unpublished). 5-HT in the nanomolar range acted inhibitory, but surprisingly 5-carboxamidotryptamine was ineffective, as was 4-hydroxytryptamine; ipsapirone slightly increased release. The inhibitory property of 5-HT on noradrenaline release was shared by N,N-dimethyl-5-hydroxytryptamine, 5-aminotryptamine, tryptamine and RU 24969, the maximum effect obtainable with the latter 2 compounds being about 40% lower than that of the other ones. Neither metitepine nor ketanserin nor a 5-HT_3 receptor antagonist produced a shift to the right of the concentration-response curve for 5-HT; yohimbine rather increased the inhibitory effect of 5-HT. Therefore, it is not yet clear whether a receptor (and particularly which class) is actually involved in the inhibitory effect of the indolethylamines on noradrenaline release in the porcine coronary artery. Nevertheless, the data with agonists do not exclude the possibility of a receptor-mediated effect.

Conclusions

For most of the modulatory effects of 5-HT on transmitter release, described in the present report, it has been demonstrated that they are mediated by presynaptic receptors which, on the rule, belong to the 5-HT_1 class. This holds true for the presynaptic 5-HT autoreceptor in the rat, porcine and human brain as well as for the presynaptic 5-HT heteroreceptor on the sympathetic nerves of rat and human blood vessels. Subclassification of the 5-HT_1 receptor involved revealed a less clear picture. In the rat, the 5-HT autoreceptor in the CNS is of the 5-HT_{1B} subtype, but no consistent results were obtained with respect to the pharmacological properties of the presynaptic 5-HT_1 receptors on the sympathetic nerves of different vascular beds. The 5-HT receptor involved in modulation of transmitter release may differ between the species investigated. Clarification of the open questions is a prerequisite for the development of new classes of drugs acting at release-modulating 5-HT receptors. In view of one of the potential indications of such drugs, namely arterial hypertension, it is of interest to note that the receptor-mediated inhibitory effects of 5-HT on 5-HT release in the CNS and on noradrenaline release in blood vessels are not attenuated in spontaneously hypertensive rats.

Summary

Synaptic transmission in the central and peripheral nervous system can be modulated by inhibitory presynaptic serotonin (5-HT) receptors. - Among these, the presynaptic 5-HT autoreceptors in the CNS have been studied in most detail. In slices and synaptosomes prepared from the spinal cord or various brain regions of several species, the electrically or K^+-evoked release of ^3H-5-HT was inhibited by 5-HT or other 5-HT receptor agonists. This effect was blocked by nonselective 5-HT receptor antagonists (e.g. metitepine) or ß-adrenoceptor blockers with antagonistic properties at 5-HT receptors (e.g. pindolol), but not by the 5-HT_2 receptor antagonist ketanserin. Further analysis of the pharmacological properties of the presynaptic 5-HT autoreceptors in the rat brain revealed that they are of the 5-HT_{1B} subtype. In the pig and human brain, the 5-HT autoreceptors may belong to another subtype, since in cortical membranes of these species no 5-HT_{1B} binding sites were identified. - A well-defined presynaptic 5-HT heteroreceptor modulating synaptic transmission in the peripheral nervous system is the presynaptic 5-HT receptor on the sympathetic nerve terminals, which upon activation inhibits noradrenaline release (e.g., in the rat vena cava and the human saphenous vein). In the rat, the presynaptic 5-HT receptors of the vena cava resemble the 5-HT autoreceptors in the brain inasmuch as they also are of the 5-HT_{1B} subtype. The subtype involved in the human sympathetic nerves has not yet been determined.

Acknowledgement

The experimental work carried out in the author's laboratory was supported by a grant of the Deutsche Forschungsgemeinschaft. I thank Mrs. U. Claren for typing the manuscript.

References

Bonanno G, Maura G, Raiteri M (1986) Pharmacological characterization of release-regulating serotonin autoreceptors in rat cerebellum. European J Pharmacol 126:317-321

Bradley PB, Engel G, Feniuk W, Fozard JR, Humphrey PPA, Middlemiss DN, Mylecharane EJ, Richardson BP, Saxena PR (1986) Proposals for the classification and nomenclature of functional receptors for 5-hydroxytryptamine. Neuropharmacology 25:563-576

Cerrito F, Raiteri M (1979) Serotonin release is modulated by presynaptic autoreceptors. European J Pharmacol 57:427-430

Charlton KG, Bond RA, Clarke DE (1986) An inhibitory prejunctional 5-HT_1-like receptor in the isolated perfused rat kidney. Apparent distinction from the 5-HT_{1A}, 5-HT_{1B} and 5-HT_{1C} subtypes. Naunyn-Schmiedeberg's Arch Pharmacol 332: 8-15

Engel G, Göthert M, Müller-Schweinitzer E, Schlicker E, Sistonen L, Stadler PA (1983) Evidence for common pharmacological properties of [^{3}H]5-hydroxytryptamine autoreceptors in CNS and inhibitory presynaptic 5-hydroxytryptamine receptors on sympathetic nerves. Naunyn-Schmiedeberg's Arch Pharmacol 324:116-124

Engel G, Göthert M, Hoyer D, Schlicker E, Hillenbrand K (1986) Identity of inhibitory presynaptic 5-hydroxytryptamine (5-HT) autoreceptors in the rat brain cortex with 5-HT_{1B} binding sites. Naunyn-Schmiedeberg's Arch Pharmacol 332:1-7

Ennis C, Kemp JD, Cox B (1981) Characterisation of inhibitory 5-hydroxytryptamine receptors that modulate dopamine release in the striatum. J Neurochem 36:1515-1520

Gaddum JH, Picarelli ZP (1957) Two kinds of tryptamine receptor. Brit J Pharmacol 12:323-328

Göthert M (1980) Serotonin-receptor-mediated modulation of Ca^{2+}-dependent 5-hydroxytryptamine release from neurones of the rat brain cortex. Naunyn-Schmiedeberg's Arch Pharmacol 314:223-230

Göthert M (1986) Serotonin receptors in the circulatory system. In: Progress in Pharmacology, Vol. 6/2 (Van Zwieten PA, Schönbaum E, eds.), pp. 155-172, Gustav Fischer Verlag, Stuttgart

Göthert M, Schlicker E (1987) Classification of serotonin receptors. J Cardiovasc Pharmacol 10:S3-S7

Göthert M, Weinheimer G (1979) Extracellular 5-hydroxytryptamine inhibits 5-hydroxytryptamine release from rat brain cortex slices. Naunyn-Schmiedeberg's Arch Pharmacol 310:93-96

Göthert M, Kollecker P, Rohm N, Zerkowski HR (1986) Inhibitory presynaptic 5-hydroxytryptamine (5-HT) receptors on the sympathetic nerves of the human saphenous vein. Naunyn-Schmiedeberg's Arch Pharmacol 332:317-323

Hoyer D, Engel G, Kalkman HO (1985) Molecular pharmacology of 5-HT_1 and 5-HT_2 recognition sites in rat and pig brain membranes: radioligand binding studies with $[^3H]$5-HT, $[^3H]$8-OH-DPAT, (-)$[^{125}I]$ iodocyanopindolol, $[^3H]$mesulergine and $[^3H]$ketanserin. European J Pharmacol 118:13-23

Hoyer D, Pazos A, Probst A, Palacios JM (1986) Serotonin receptors in the human brain. I. Characterization and autoradiographic localization of 5-HT_{1A} recognition sites. Apparent absence of 5-HT_{1B} recognition sites. Brain Res 376:85-96

Langer SZ, Moret C (1982) Citalopram antagonizes the stimulation by lysergic acid diethylamide of presynaptic inhibitory serotonin autoreceptors in the rat hypothalamus. J Pharmacol Exp Ther 222:220-226

Maura G, Roccatagliata E, Raiteri M (1986) Serotonin autoreceptor in rat hippocampus: pharmacological characterization as a subtype of the 5-HT_1 receptor. Naunyn-Schmiedeberg's Arch Pharmacol 334:323-326

Middlemiss DN (1984a) 8-Hydroxy-2-(di-n-propylamino)tetralin is devoid of activity at the 5-hydroxytryptamine autoreceptor in rat brain. Implications for the proposed link between the autoreceptor and the $[^3H]$5-HT recognition site. Naunyn-Schmiedeberg's Arch Pharmacol 327:18-22

Middlemiss DN (1984b) Stereoselective blockade at $[^3H]$5-HT binding sites and at the 5-HT autoreceptor by propranolol. European J Pharmacol 101:289-293

Molderings GJ, Fink K, Schlicker E, Göthert M (1987) Inhibition of noradrenaline release via presynaptic 5-HT_{1B} receptors of the rat vena cava. Naunyn-Schmiedeberg's Arch Pharmacol 336:245-250

Monroe PJ, Smith DJ (1985) Demonstration of an autoreceptor modulating the release of $[^3H]$5-hydroxytryptamine from a synaptosomal-rich spinal cord tissue preparation. J Neurochem 45:1886-1894

Neale RF, Fallon SL, Boyar WC, Wasley JWF, Martin LL, Stone GA, Glaeser BS, Sinton CM, Williams M (1987) Biochemical and pharmacological characterization of CGS 12066B, a selective serotonin-1B agonist. European J Pharmacol 136:1-9

Peroutka SJ, Snyder SH (1979) Multiple serotonin receptors: differential binding of [^{3}H] 5-hydroxytryptamine, [^{3}H] lysergic acid diethylamide and [^{3}H] spiroperidol. Mol Pharmacol 16:687-699

Schlicker E, Göthert M (1987) Noradrenaline release and its modulation via presynaptic serotonin receptors in blood vessels of spontaneously hypertensive rats. J Cardiovasc Pharmacol 10 (Suppl. 4), S141-S143

Schlicker E, Brandt F, Classen K, Göthert M (1985) Serotonin release in human cerebral cortex and its modulation via serotonin receptors. Brain Res 331:337-341

Schlicker E, Fink K, Classen K, Göthert M (1987) Facilitation of serotonin (5-HT) release in rat brain cortex by cAMP and probable inhibition of adenylate cyclase in 5-HT nerve terminals by presynaptic α_2-adrenoceptors. Naunyn-Schmiedeberg's Arch Pharmacol 336:251-256

Schlicker E, Classen K, Göthert M (1988) Presynaptic serotonin receptors on central serotoninergic and noradrenergic neurons of normotensive and spontaneously hypertensive rats. J Cardiovasc Pharmacol in press

Westfall TC, Tittermary V (1982) Inhibition of the electrically induced release of [^{3}H] dopamine by serotonin from superfused rat striatal slices. Neurosci Lett 28:205-209

OPIOID PEPTIDES, OPIOID RECEPTORS AND MODULATION OF CATECHOLAMINERGIC NEUROTRANSMISSION PROCESSES IN THE CENTRAL NERVOUS SYSTEM

Arie H. Mulder, Abraham L. Frankhuyzen and Anton N.M. Schoffelmeer

Department of Pharmacology, Free University Medical Faculty,
Van der Boechorststraat 7, 1081 BT Amsterdam, The Netherlands

In the past decade a variety of peptides with opioid activity have been discovered in the CNS, which are derived from three different precursor-molecules and are denoted as the enkephalins, endorphins and dynorphins. These three opioid peptide families appear to be localized in separate neuronal systems and are thought to function as neurotransmitters or neuromodulators (Cuello, 1983; Akil et al., 1984; Khatchaturian et al., 1985). Furthermore, at present at least three pharmacologically different opioid receptor subtypes are known to exist, viz. mu-, kappa- and delta-receptors (Paterson et al., 1983; Martin, 1984; Zukin and Zukin, 1984; Goldstein and James, 1984). However, the physiological implications of multiple opioid peptides and multiple opioid receptors still are largely unknown. Important questions to be answered in this respect are whether different opioid receptors are involved in the actions of different opioid peptides and whether different opioid receptors mediate effects on different physiological (cellular) processes. These questions underly the investigations briefly reviewed in this paper.

The monoaminergic neuronal systems in the CNS are considered to be of major importance for the regulation of mood, mental and motorial activities and vegetative functions, all of which may be influenced by opioid drugs (Iwamoto and Way, 1979; Martin, 1984; Akil et al, 1984). Therefore, numerous studies have explored possible interactions between opioid and monoamine neuronal systems. Obviously, opioids may interfere with monoaminergic neurotransmission at different levels. At the presynaptic level, for example, synthesis and/or release of the neurotransmitter may be modulated. At the postsynaptic level opioids may either mimic or modulate the receptor-mediated effects of the neurotransmitter. Modulation of postsynaptic effects may involve allosteric changes in the affinity of the transmitter receptors concerned or in the effector mechanisms, such as the production of intracellular second messengers. This paper deals with in vitro studies on the

NATO ASI Series, Vol. H19
Modulation of Synaptic Transmission and Plasticity
in Nervous Systems. Edited by G. Hertting and H.-C. Spatz

modulation of the release of noradrenaline (NA), dopamine (DA) and acetylcholine (ACh) and of catecholamine-sensitive adenylate cyclase in rat brain by opioid peptides and other opioid drugs and the receptor subtypes involved.

1. MODULATION OF NEUROTRANSMITTER RELEASE.

Thus far most studies on neurotransmitter release in the brain and its pharmacological modulation, e.g. through presynaptic receptors, have been carried out in vitro, using brain slices or synaptosomes, prelabelled with the radioactive transmitter (Langer, 1981; Starke, 1981; Chesselet, 1984; Mulder et al., 1984a). The experimental procedure generally involves selective labelling of the appropriate nerve terminals by incubating brain slices with the radiolabelled transmitter or its precursor and subsequently superfusion of the labelled slices to determine the spontaneous efflux of radioactivity as well as depolarization-induced calcium-dependent release of the radiolabelled transmitter. The depolarizing stimulus may consist of electrical pulses (e.g. 1 pps., 15-30 mA, 2 ms duration) or an increased K^+-concentration (e.g. 10 - 20 mM), applied during 5-10 min. Drugs are added to the superfusion medium to study their effects on transmitter release.

1.1 NORADRENALINE RELEASE

Numerous studies have shown that activation of mu-opioid receptors results in an inhibition of NA release from slices of various regions of the rat brain (for reviews see Langer, 1981; Starke, 1981; Chesselet, 1984; Mulder et al., 1984a). Recent experiments using synaptosomes have confirmed that these opioid receptors, just like the presynaptic alpha2-adrenoceptors mediating inhibition of NA release (De Langen et al., 1979), are localized on noradrenergic nerve terminals (Mulder et al., 1987; Fig. 1). Delta-opioid receptor activation, e.g. with the highly delta-selective cyclic enkephalin analogue [D-Pen^2, D-Pen^5] enkephalin (DPDPE), had no effect on NA release from rat brain slices or synaptosomes (Mulder et al., 1987; Werling et al., 1987). Also drugs with selectivity for kappa-opioid receptors, e.g. bremazocine or U50,488H, did not affect NA release at the low (nanomolar) concentrations, that have been found to produce maximal effects with regard to the kappa-receptor mediated inhibition of striatal dopamine release (Werling

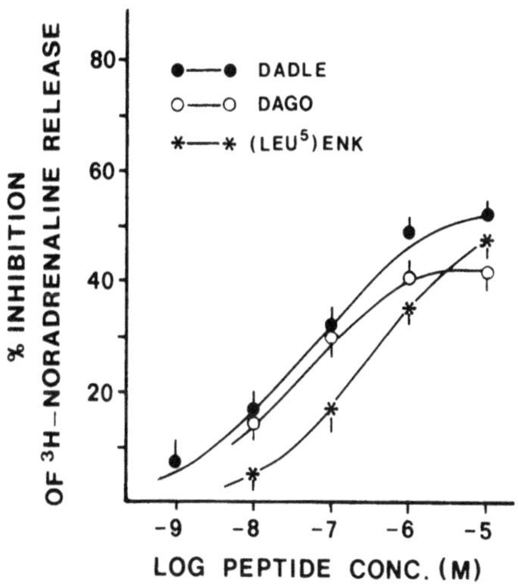

Fig. 1. Inhibitory effect of enkephalins on K^+(10 mM)-induced release of 3H-NA from superfused rat cortex synaptosomes. (adapted from Mulder et al., 1987)

et al., 1987; see also below). Therefore, it appears that, at least in rat brain, NA release is subject to presynaptic inhibition by opioids only if they display affinity for mu-receptors. Thus, activation of mu-opioid receptors mimics the inhibitory effect of activation of presynaptic alpha2-adrenoceptors on NA release.

Nevertheless, although activation of both mu-opioid receptors and alpha2-adrenoceptors results in presynaptic inhibition of NA release, studies in our laboratory have revealed marked differences in the mechanisms involved (Schoffelmeer et al., 1983a,b). In these studies evidence was obtained indicating that, in contrast to mu-opioid receptors, alpha2-adrenoceptors on noradrenergic nerve terminals are negatively coupled to a presynaptic adenylate cyclase system (Schoffelmeer et al., 1985) and may thus mediate a reduction of neurotransmitter release by counteracting the facilitatory action of cyclic AMP on the secretion process. The most decisive evidence in this respect was obtained in a study examining the release of radiolabelled NA from cortex slices induced by the calcium-ionophore A23187 (Schoffelmeer et al., 1986b). In contrast to NA release evoked by depolarizing stimuli, that induced by A23187 appears not to be inhibited by activation of mu-opioid receptors or alpha2-adrenoceptors. However, when adenylate cyclase activity was stimulated with forskolin NA release induced by A23187 became susceptible to inhibition by alpha2-adrenoceptor, but not mu-opioid receptor activation. No presynaptic inhibition of A23187-induced NA release was found in the presence of 8-bromo-cyclic AMP, that increases the intracellular cyclic AMP concentration in a manner bypassing adenylate cyclase, suggesting that activation of alpha2-adrenoceptors does not affect a step in the secretory process beyond the generation of cyclic AMP, but results in an inhibition of adenylate cyclase (Schoffelmeer et al., 1986b).

The mechanisms involved in the presynaptic inhibition of NA release by activation of mu-opioid receptors are less clear thus far, although a critical role of adenylate cyclase can virtually be excluded. We have found recently that NA release induced by veratrine in the absence of extracellular Ca^{2+} was inhibited by activation of mu-opioid receptors (unpublished observations), as shown previously for presynaptic alpha2-adrenoceptors (Schoffelmeer et al., 1983b). This finding precludes the possibility that inhibition of voltage-sensitive calcium-channels is the primary effect involved in the presynaptic inhibition of NA release mediated by these receptors, whereas other data from this laboratory (Schoffelmeer et al., 1984) have made a primary role of hyperpolarization, via opening of K^{+}-channels, unlikely. Therefore, although the data thus far suggest that activation of presynaptic mu-opioid receptors may directly affect the availability of intracellular Ca^{2+} for some step(s) in the secretory process in noradrenergic nerve terminals, the precise mechanisms involved remain to be elucidated.

Recent evidence indicates that the mu-opioid receptors and the alpha2-adrenoceptors mediating presynaptic inhibition of NA release are localized in the same noradrenergic nerve terminals and that they are not functionally independent from each other. This was derived from data, showing that activation of alpha2-adrenoceptors prevented the inhibitory effect of mu-opioid receptor activation on NA release and that this phenomenon could not be explained by assuming a common pathway by which activation of the respective receptors is translated into inhibition of release (Schoffelmeer et al., 1986a). It appears, therefore, that the decline in presynaptic mu-opioid receptor efficacy following activation of presynaptic alpha2-adrenoceptors is due to some sort of local interaction between these receptors within the membrane of noradrenergic nerve terminals. Similar findings have been reported by Limberger et al. (1986) with regard to interactions between alpha2-adrenoceptors and kappa-opioid receptors involved in presynaptic inhibition of NA release in the rabbit brain.

1.2 DOPAMINE AND ACETYLCHOLINE RELEASE

Activation of DA receptors has been shown to inhibit DA release as well as ACh release from striatal slices; pharmacologically these receptors appear to belong to the D-2 subtype of DA receptors (Stoof et al., 1980; Starke et al., 1983; Stoof and Kebabian, 1984). The receptors mediating presynaptic inhibition of DA release are generally referred to as DA autoreceptors, whereas the (presynaptic) inhibition of ACh release caused by D-2 receptor activation is considered to reflect one of the postsynaptic actions of DA in the neostriatum (Lehmann and Langer, 1983; Starke et al., 1983; Stoof and Kebabian, 1984).

An interesting discovery, in view of opioid receptor and opioid peptide multiplicity, was made when we studied the effects of enkephalin and dynorphin on the release of DA and ACh from rat striatal slices (Mulder et al., 1984b). In the neostriatum both dynorphinergic and enkephalinergic neurons are found with a striking abundance of the latter (Cuello, 1983; Khatchaturian et al., 1985) and in the early 1980's numerous studies had already suggested interactions between opioidergic and dopaminergic systems (for review see Iwamoto and Way, 1979; Schmauss and Emrich, 1985). We examined whether dynorphin and enkephalin would perhaps differentially modulate the release of DA and ACh in the neostriatum and found this indeed to be the case. Thus, dynorphin and other agonists with kappa-opioid receptor selectivity appear to inhibit DA release, with little effect on ACh release (Fig. 2). On the contrary, Leu- and Met-enkephalin and various enkephalin analogues cause a strong inhibition of ACh release, with little effect on DA release. Further data, particularly those obtained with the highly delta-selective enkephalin analogue DPDPE, strongly suggest that opioid-induced inhibition of striatal ACh release involves delta-receptors (Table 1). Morphine nor the enkephalin analogue DAGO, both of which are selective mu-receptor agonists, affect the release of DA nor that of ACh. On the other hand, as noted above mu-, but not delta- or kappa-opioid receptor activation appears to inhibit NA release from rat cortical or hippocampal slices as well as synaptosomes (Mulder et al., 1984a, 1987).

A striking discrepancy between our findings and those of Lubetzki et al. (1982) and Petit et al. (1986), regarding the opioid-induced modulation of DA release from rat striatal slices should be mentioned here. Whereas we have found consistently that, in contrast to activation of kappa-receptors, delta-receptor activation does not affect the release of 3H-DA previously

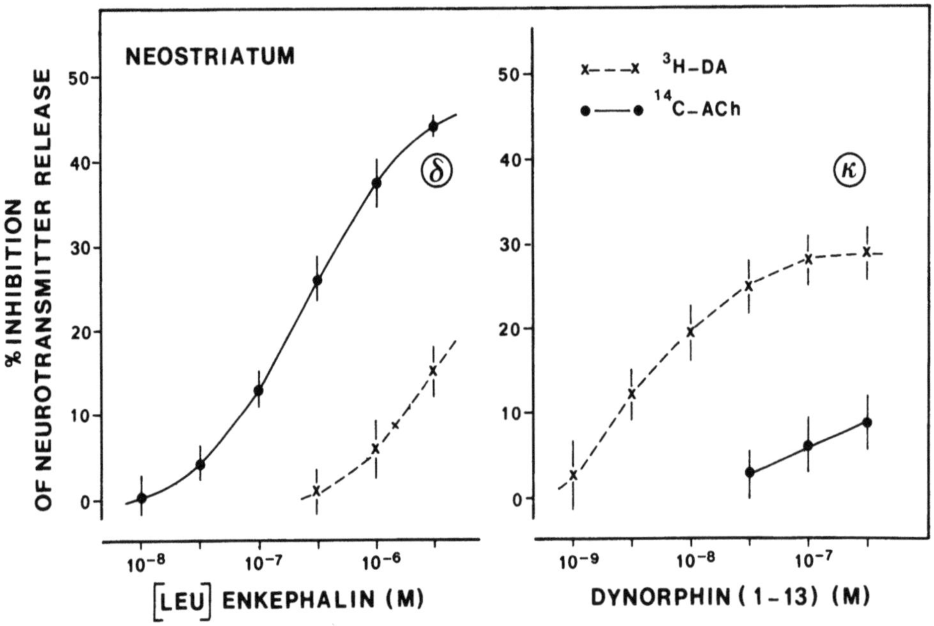

Fig. 2. Differential inhibitory effects of [Leu]enkephalin and dynorphin(1-13) on the K^+(15 mM)-induced release of 3H-DA and 14C-ACh from rat neostriatal slices. (adapted from Mulder et al., 1984b)

Table 1. Inhibition of electrically-evoked neurotransmitter release from brain slices by some opioid peptides and morphine.

	inhibitory effect on:					
	3H-DA release (neostriatum)	(EC50)	14C-ACh release (neostriatum)	(EC50)	3H-NA release (cortex)	(EC50)
[Leu]enkephalin	±	-	++	50 nM	++	300 nM
DPDPE	±	-	++	50 nM	-	-
Dynorphin(1-13)	++	5 nM	±	-	++	300 nM
Morphine	-	-	-	-	+	300 nM
Opioid receptor involved	κ		δ		μ	

* Percentage inhibition of release caused by a concentration of 1 μM:
- < 10%   ± 10-20%   + 40-50%   ++ > 50%

taken up by striatal slices, they reported an increase of the release of newly-synthesized radiolabelled DA in response to delta-receptor activation. The reason for this discrepancy, which may somehow be related to the rather pronounced differences in methodology, remains to be established.

Taken together our data indicate that opioid drugs including various opioid peptides, acting on different receptors, may indeed selectively inhibit the release of distinct neurotransmitters. Thus, in the neostriatum, enkephalin and dynorphin, released from opioid-peptidergic neurons may differentially regulate DA and ACh release. It is of interest to note that dynorphin and other kappa-opioid receptor agonists mimic the effect of DA autoreceptor activation, whereas enkephalin and other delta-opioid receptor agonists mimic the effect of postsynaptic DA receptor activation on ACh release. It is also noteworthy that enkephalin inhibits ACh release in the rat striatum by activation of delta-opioid receptors, whereas it inhibits NA release in the rat cortex/hippocampus by activating mu-opioid receptors (Table 1). Similarly, dynorphin inhibits striatal DA release by activation of kappa-opioid receptors, but cortical/hippocampal NA release by activating mu-opioid receptors.

1.3 REGIONAL AND SPECIES DIFFERENCES

Recent data from our laboratory indicate that a particular opioid receptor subtype is not necessarily involved in presynaptic modulation of transmitter release from a particular class of neurons. These data resulted from a study comparing the effects of opioid drugs on the release of both radiolabelled NA and ACh from slices of rat amygdala to those on NA release and ACh release from cortical and striatal slices, respectively. Both in cortex and amygdala NA release appeared to be inhibited exclusively by activation of mu-opioid receptors. However, whereas ACh release in the striatum is inhibited only by activation of delta-opioid receptors, ACh release in the amygdala was found to be inhibited only upon activation of mu-opioid receptors, just like NA release in this brain region (Table 2).

Finally, it should be emphasized, that all of our studies were carried out using rat brain, but that striking species differences appear to exist with regard to the type of opioid receptor involved in the modulation of release of a particular neurotransmitter. Thus, it has been shown that in the rabbit NA release from cortical or hippocampal slices is inhibited by activation of kappa- rather than mu- or delta-opioid receptors (Jackisch et al., 1986a;

Table 2. Different opioid receptor subtypes (delta and mu) mediate inhibition of ACh release from rat striatum and amygdala slices by opioid drugs.

Agonist	14C-ACh release striatum % inhibition at 1 μM	pD2	14C-ACh release amygdala % inhibition at 1 μM	pD2
DADLE (δ/μ)	65 + 2	7.8	76 + 3	7.4
Morphine (μ)	no effect	-	64 + 4	6.7
DPDPE (δ)	73 + 3	7.1	12 + 4	-

Limberger et al., 1986). Similarly, opioid drugs have been demonstrated to inhibit the release of ACh from rabbit hippocampal slices through activation of kappa- rather than mu- or delta-opioid receptors (Jackisch et al., 1986b). Recently, Werling et al. (1987) examined opioid-induced modulation of NA release from slices of various guinea-pig brain regions and found that agonists with selectivity for mu-, delta- and kappa-opioid receptors, respectively, were all active in inhibiting NA release. It thus appears that the guinea-pig brain deviates from that of rat and rabbit in that NA release is not selectively inhibited by activation of a distinct opioid receptor subtype.

Obviously, it will be of interest to further explore these species differences, e.g. to see whether in rabbit and guinea-pig neostriatum DA and ACh release are modulated by opioid drugs in a selective manner through receptors different from those involved in the modulation of NA and ACh release in other brain regions, as has been demonstrated in the rat.

2 MODULATION OF ADENYLATE CYCLASE ACTIVITY.

One of the possible postsynaptic actions of monoamine neurotransmitters in the CNS is modulation of adenylate cyclase activity, resulting in changes in the production of the second messenger cyclic AMP. For instance, cyclic AMP production may be increased by NA through activation of beta-adrenoceptors or decreased through activation of alpha2-adrenoceptors (Daly, 1977; Fain and Garcia-Sainz, 1980). Also DA may either stimulate or inhibit adenylate cyclase activity through activation of D-1 and D-2 receptors, respectively (Stoof and Kebabian, 1981, 1984).

Inhibition of adenylate cyclase activity in brain tissue by morphine and other opioid drugs has been demonstrated in several studies over the past 15 years (West and Miller, 1983) and, obviously, opioid receptor activation might interfere with the stimulatory effects of some monoamine neurotransmitters on this enzyme. Inhibition of DA-sensitive adenylate cyclase activity in neostriatal homogenates by opioid drugs was reported round about 1980 by Neff and coworkers. These investigators suggested that delta-opioid receptors are coupled to DA-sensitive adenylate cyclase in the neostriatum (Neff et al., 1981; Gentleman et al., 1983).

Using striatal slices we have further explored the potential interactions between opioid and dopaminergic neuronal systems at the postsynaptic level with particular attention for the opioid receptor subtypes involved and the consequences of D-2 receptor blockade (Schoffelmeer et al., 1986c). As noted above, in the neostriatum adenylate cyclase is subject to a dual regulation by DA, involving both stimulatory D-1 and inhibitory D-2 receptors. Thus, blockade of D-2 receptors by neuroleptic drugs will remove the inhibitory constraint mediated by these receptors and result in an increased cyclic AMP formation by striatal slices exposed to DA (Stoof and Kebabian, 1981). The experimental method used in our laboratory involves superfusion of the slices with a physiological medium and measurement of the efflux of cyclic AMP (which has been shown to mirror its intracellular production, see Stoof and Kebabian, 1981), using a radioimmuno assay. The effects of (opioid) drugs, added to the superfusion medium, are determined, both on the basal level of cyclic AMP efflux (averaging 1 fmole / mg of tissue / min) and on the stimulated cyclic AMP efflux, e.g. following DA-receptor activation.

Exposure of striatal slices to 30 µM DA, i.e. upon simultaneous activation of D-1 and D-2 receptors, approximately doubles the efflux of cyclic AMP. Both the delta/mu-agonist [D-Ala^2,D-Leu^5]enkephalin (DADLE) and the mu-selective agonist morphine appear to reduce the increase in cyclic AMP efflux, with respective EC50 values of about 100 and 800 nM (Fig. 3). Blockade of D-2 receptors by 10 µM (-)sulpiride strongly enhances the response of the slices to 30 µM DA, resulting in a 4-6 fold increase of the cyclic AMP efflux. Also under these conditions, corresponding to selective D-1 receptor activation, both DADLE and morphine were found to inhibit the increase in cyclic AMP efflux. However, whereas D-2 receptor blockade did not affect the EC50 of morphine, it greatly reduced that of DADLE (from 100 nM to about 3 nM), implying the involvement of different receptor subtypes (Fig. 3). Further data, obtained with the highly selective delta-opioid

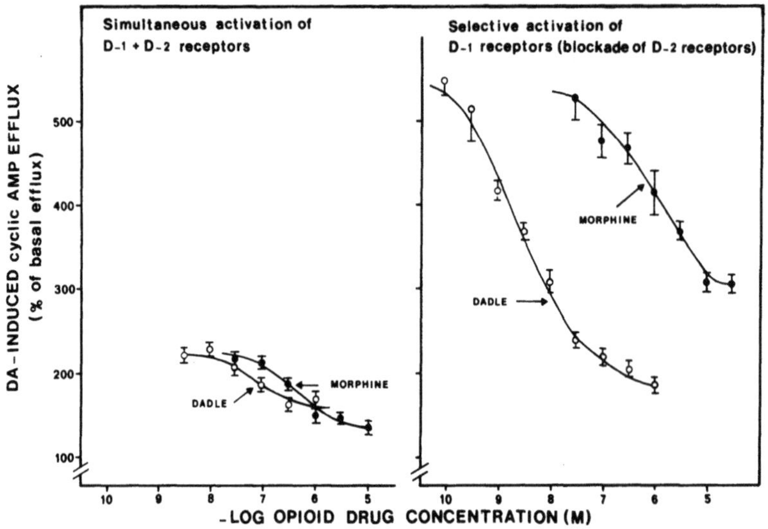

Fig. 3. Effects of the delta/mu-opioid receptor agonist DADLE and the mu-opioid receptor agonist morphine on the efflux of cyclic AMP from superfused rat neostriatal slices exposed to either 30 µM DA alone (i.e. simultaneous D-1 + D-2 receptor activation) or to 30 µM DA + 10 µM (-)sulpiride (i.e. selective D-1 receptor activation after D-2 receptor blockade).
(adapted from Schoffelmeer et al., 1986c)

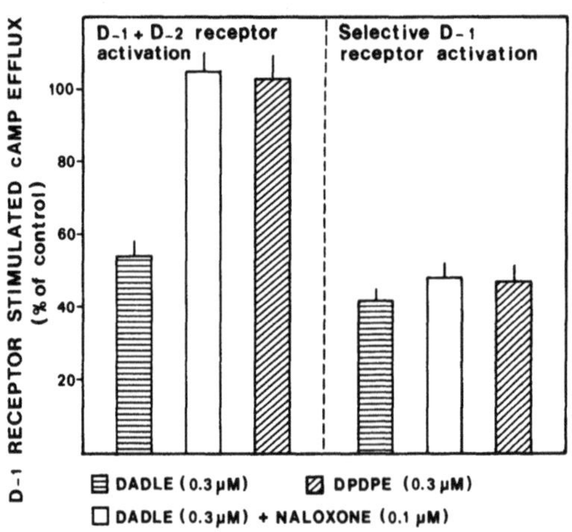

Fig. 4. Effects of the delta/mu-opioid receptor agonists DADLE and the highly selective delta-opioid receptor agonist DPDPE on the efflux of cyclic AMP from superfused rat neostriatal slices, that were exposed to either 10 µM SKF 38393 + 10 µM LY 141865 (i.e. simultaneous D-1 + D-2 receptor activation) or 10 µM SKF 38393 alone (i.e. selective D-1 receptor activation). Note that naloxone does not antagonize the inhibitory effect of DADLE in the latter case. (adapted from Schoffelmeer et al., 1986c)

receptor agonist DPDPE and the delta-selective antagonist ICI 174864, indicate that upon blockade of D-2 receptors both mu- and delta-opioid receptors are involved in the opioid-induced inhibition of DA-stimulated cyclic AMP production. On the other hand, upon simultaneous activation of D-1 and D-2 receptors the opioid receptors mediating inhibition of DA-stimulated cyclic AMP production appear to display a pronounced mu-selectivity (Schoffelmeer et al., 1986c, 1987).

Similar data have been obtained in experiments using the selective D-1 and D-2 dopamine receptor agonists SKF 38393 and LY 141865, respectively, instead of DA in the absence or presence of (-)sulpiride (Schoffelmeer et al., 1986c). For example, DPDPE only inhibited the increased cyclic AMP efflux in response to SKF 38393; simultaneous activation of D-2 receptors by LY 141865 fully prevented the expression of the delta-opioid receptor mediated inhibitory effect of DPDPE. Further, the inhibitory effect of DADLE was only antagonized by a relatively low concentration of naloxone after simultaneous D-1 and D-2 receptor activation (Fig. 4).

There are various other neurotransmitter substances, including monoamines other than DA, that cause an increase in the production of cyclic AMP and the question arises whether inhibition of transmitter-stimulated adenylate cyclase by mu- and delta-opioid receptor activation might be a ubiquitous phenomenon. This does not seem to be the case, however, since it has been shown that the increased cyclic AMP production caused by activation of beta-adrenoceptors and A2-adenosine receptors in striatal slices was not affected by opioid receptor activation (Kelly and Nahorski, 1986; Schoffelmeer et al., 1987). Further, the increase in cyclic AMP production in response to vasoactive intestinal peptide (VIP) was inhibited by mu-opioid receptor activation only (Schoffelmeer et al., 1987), although such an inhibition could not be demonstrated by others (Kelly and Nahorski, 1986). The scarce data available thus far suggest, therefore, that the coupling of both mu- and delta-opioid receptors to DA-sensitive adenylate cyclase may be a rather exceptional phenomenon.

Recent findings indicate that the mu- and delta-opioid receptors concerned are localized on the same cells, possibly as a receptor complex, and interact functionally with each other via allosteric mechanisms (Schoffelmeer et al., 1987). The possible existence of an opioid receptor complex in rat brain with non-competitively interacting mu- and delta-binding sites has previously been suggested on the basis of ligand-receptor binding and receptor-autoradiographic studies (Bowen et al., 1981; Rothman et al., 1985) as well as pharmacological studies in vivo (Holaday and Tortella, 1984)). Our studies (Schoffelmeer et al., 1987)) have provided functional evidence in vitro for this hypothesis and suggest that in rat neostriatum closely associated mu- and delta-opioid receptor subtypes are coupled in an inhibitory fashion to DA-sensitive adenylate cyclase, possibly sharing an inhibitory guanine regulatory protein. However, the pharmacological nature of this functional opioid receptor complex appears to be regulated somehow by activation of D-2 dopamine receptors, which prevents delta-opioid receptor mediated inhibition of DA-sensitive adenylate cyclase without affecting mu-opioid receptor efficacy. Thus, functional coupling of delta-receptors to the enzyme appears to occur only after blockade of D-2 dopamine receptors. In this context it should be recalled that the neostriatum is very rich in enkephalinergic neurons and that the enkephalins display a considerably higher affinity for delta- than for mu-opioid receptors (Paterson et al., 1983). Therefore, blockade of D-2 receptors, thought to be the primary action of neuroleptic drugs (Seeman, 1980; Snyder, 1984), would result in a substantially enhanced sensitivity of D-1 receptor stimulated adenylate cyclase to inhibition by enkephalins. Conceivably, this might be of importance for the antipsychotic action of neuroleptic drugs, particularly since others have shown that neuroleptic treatment enhances enkephalin biosynthesis in the neostriatum (Hong et al., 1985) and induces behavioural supersensitivity to opioids (Stinus et al., 1985).

References

Akil H, Watson SJ, Young E, Lewis ME, Khachaturian H, Walker JM (1984). Endogenous opioids: biology and function. Ann. Rev. Neurosci. 7: 223-255.

Bowen WD, Gentleman S, Herkenham M, Pert CB (1981). Interconverting mu and delta forms of the opiate receptor in rat striatal patches. Proc. Natl. Acad. Sci. USA 78: 4818-4822.

Chesselet M-F (1984). Presynaptic regulation of neurotransmitter release in the brain: facts and hypothesis. Neuroscience 12: 347-375.

Cuello AC (1983). Central distribution of opioid peptides. Br. Med. Bull. 39: 11-16.

Daly JW (1977). The formation, degradation, and function of cyclic nucleotides in the nervous system. Int. Rev. Neurobiol. 20: 105-168.

De Langen CDJ, Hogenboom F, Mulder AH (1979). Presynaptic noradrenergic alpha-receptors and modulation of 3H-noradrenaline release from rat brain synaptosomes. Eur. J. Pharmacol. 60: 79-89.

Fain JN, García-Sáinz JA (1980). Role of phosphatidylinositol turnover in alpha1 and of adenylate cyclase inhibition in alpha2 effects of catecholamines. Life Sci. 26: 1183-1194.

Gentleman S, Parenti M, Neff NH, Pert CB (1983). Inhibition of dopamine-activated adenylate cyclase and dopamine binding by opiate receptors in rat striatum. Cell. Mol. Neurobiol. 3: 17-26.
Goldstein A, James IF (1984). Multiple opioid receptors. Criteria for identification and classification. TIPS 5: 503-505.
Herkenham M, Pert CB (1981). Mosaic distribution of opiate receptors parafascicular projections and acetylcholinesterase in rat striatum. Nature 291: 415-418.
Holaday JW, Tortella FC (1984). Multiple opioid receptors: possible physiological function of the mu- and delta-binding sites in vivo. In: Central and peripheral endorphins: Basic and clinical aspects. Eds.: Muller EE, Genazzani AR. Raven Press, New York: pp. 237-249.
Hong JS, Yoshikawa K, Kanamatsu T, Sabol SL (1985). Modulation of striatal enkephalinergic neurons by antipsychotic drugs. Federation Proc. 44: 2535-2539.
Iwamoto ET, Leong Way E (1979). Opiate actions and catecholamines. In: Neurochemical mechanisms of opiates and endorphins. Eds.: Loh HH, Ross DH. Adv. Biochem. Psychopharmacol. 20: 357-407.
Jackisch R, Geppert M, Illes P (1986a). Characterization of opioid receptors modulating noradrenaline release in the hippocampus of the rabbit. J. Neurochem. 46: 1802-1810.
Jackisch R, Geppert M, Brenner AS, Illes P (1986b). Presynaptic opioid receptors modulating acetylcholine release in the hippocampus of the rabbit. Naunyn-Schmiedeberg's Arch. Pharmacol. 332: 156-162.
Kelly E, Nahorski SR (1986). Specific inhibition of dopamine D-1- mediated cyclic AMP formation by dopamine-D2, muscarinic cholinergic, and opiate receptor stimulation in rat striatal slices. J. Neurochem. 47: 1512-1516.
Khachaturian H, Lewis ME, Schafer MK-H, Watson SJ (1985). Anatomy of the CNS opioid systems. TINS 7: 111-119.
Langer SZ (1981). Presynaptic regulation of the release of catecholamines. Pharmacol. Rev. 32: 337-362.
Lehmann J, Langer SZ (1983). The striatal cholinergic interneuron: synaptic target of dopaminergic terminals?. Neuroscience 10: 1105-1120.
Limberger N, Spath L, Holting Th, Starke K (1986). Mutual interaction between presynaptic alpha-adrenoceptors and opioid kappa-receptors at the noradrenergic axons of rabbit brain cortex. Naunyn-Schmiedeberg's Arch. Pharmacol. 334: 166-171.
Lubetzki C, Chesselet MF, Glowinski J (1982). Modulation of dopamine release in rat striatal slices by delta opiate agonists. J. Pharmacol. Exp. Ther. 222: 435-440.
Martin WR (1984). Pharmacology of opioids. Pharmacol. Rev. 35: 283-323.
Mulder AH, Frankhuyzen AL, Stoof JC, Wemer J, Schoffelmeer ANM (1984a). Catecholamine receptors, opiate receptors and presynaptic modulation of neurotransmitter release in the brain. In: Catecholamines: Neuropharmacology and central nervous system. Theoretical Aspects. Eds.: Usdin E, Carlsson A, Dahlstrom A, Engel J. Alan Liss, New York: pp. 47-58.
Mulder AH, Wardeh G, Hogenboom F, Frankhuyzen AL (1984b). Kappa- and delta-opioid receptor agonists differentially inhibit striatal dopamine and acetylcholine release. Nature 308: 278-280.
Mulder AH, Hogenboom F, Wardeh G, Schoffelmeer ANM (1987). Morphine and enkephalins potently inhibit [3H]noradrenaline release from rat brain cortex synaptosomes: further evidence for a presynaptic localization of mu-opioid receptors. J. Neurochem. 48: 1043-1047.

Neff NH, Parenti M, Gentleman S, Olianas MC (1981). Modulation of dopamine receptors by opiates. In: Apomorphine and other dopaminomimetics, Vol. 1: Basic Pharmacology. Eds.: Gessa GL, Corsini GU. Raven Press, New York: pp. 193-200.

Paterson SJ, Robson IE, Kosterlitz HW (1983). Classification of opioid receptors. Br. Med. Bull. 39: 31-36.

Petit F, Hamon M, Fournie-Zaluski M-C, Roques BP, Glowinski J (1986). Further evidence for a role of delta-opiate receptors in the presynaptic regulation of newly synthesized dopamine release. Eur. J. Pharmacol. 126: 1-9.

Rothman RB, Bowen WD, Herkenham M, Jacobson AE, Rice KC, Pert CB (1985). A quantitative study of [3H] [D-Ala2-D-Leu5]-enkephalin binding to rat brain membranes. Evidence that oxomorphine is a noncompetative inhibitor of the lower affinity delta-binding site. Mol. Pharmacol. 27: 399-408.

Schmauss C, Emrich HM (1985). Dopamine and the action of opiates: a re-evaluation of the dopamine hypothesis of schizophrenia; with special consideration of the role of endogenous opioids in the pathogenesis of schizophrenia. Biol. Psychiatry 20: 1211-1231.

Schoffelmeer ANM, Mulder AH (1983a). Differential control of Ca2+ - dependent [3H]noradrenaline release from rat brain slices through presynaptic opiate receptors and alpha-adrenoreceptors. Eur. J. Pharmacol. 87: 449-458.

Schoffelmeer ANM, Mulder AH (1983b). 3H-Noradrenaline release from rat neocortical slices in the absence of extracellular Ca2+ and its presynaptic alpha2-adrenergic modulation. A study on the possible role of cyclic AMP. Naunyn-Schmiedeberg's Arch. Pharmacol. 323: 188-192.

Schoffelmeer ANM, Mulder AH (1984). Presynaptic opiate receptor- and alpha2-adrenoceptor-mediated inhibition of noradrenaline release in the rat brain: role of hyperpolarization?. Eur. J. Pharmacol. 105: 129-135.

Schoffelmeer ANM, Hogenboom F, Mulder AH (1985). Evidence for a presynaptic adenylate cyclase system facilitating [3H]norepinephrine release from rat brain neocortex slices and synaptosomes. J. Neurosci. 5: 2685-2689.

Schoffelmeer ANM, Putters J, Mulder AH (1986a). Activation of presynaptic alpha2-adrenoceptors attenuates the inhibitory effect of mu-opioid receptor agonists on noradrenaline release from brain slice. Naunyn-Schmiedeberg's Arch. Pharmacol. 333: 377-380.

Schoffelmeer ANM, Wierenga EA, Mulder AH (1986b). Role of adenylate cyclase in presynaptic alpha2-adrenoceptor- and mu-opioid receptor-mediated inhibition of [3H]noradrenaline release from rat brain cortex slices. J. Neurochem. 46: 1711-1717.

Schoffelmeer ANM, Hansen HA, Stoof JC, Mulder AH (1986c). Blockade of D-2 dopamine receptors strongly enhances the potency of enkephalins to inhibit dopamine-sensitive adenylate cyclase in rat neostriatum: involvement in delta- and mu-opioid receptors. J. Neurosci. 6: 2235-2239.

Schoffelmeer ANM, Hogenboom F, Mulder AH (1987). Inhibition of dopamine-sensitive adenylate cyclase by opioids: possible involvement of physically associated mu- and delta-opioid receptors. Naunyn-Schmiedeberg's Arch. Pharmacol. 335: 278-284.

Seeman P (1980). Brain dopamine receptors. Pharmacol. Rev. 32: 229-313.

Snyder SH (1984). Drug and neurotransmitter receptors in the brain. Science 224: 22-31.

Starke K (1981). Presynaptic receptors. Ann. Rev. Pharmacol. Toxicol. 21: 7-30.

Starke K, Spath L, Lang JD, Adelung C (1983). Further functional in vitro comparison of pre- and postsynaptic dopamine receptors in the rabbit caudate nucleus. Naunyn Schmiedeberg's Arch. Pharmacol. 323: 298-306.

Stinus L, Winnock M, Kelley AE (1985). Chronic neuroleptic treatment and mesolimbic dopamine denervation induce behavioral supersensitivity to opiates. Psychopharmacology 85: 323-328.
Stoof JC, Horn AS, Mulder AH, (1980). Simultaneous demonstration of the activation of presynaptic dopamine autoreceptors and postsynaptic dopamine receptors in vitro by N,N-dipropyl-5,6-ADTN. Brain. Res. 196: 276-281.
Stoof JC, Kebabian JW (1981). Opposing roles for D-1 and D-2 dopamine receptors in efflux of cyclic AMP from rat neostriatum. Nature 294: 366-368.
Stoof JC, Kebabian JW (1984). Two dopamine receptors: biochemistry, physiology and pharmacology. Life Sci. 35: 2281-2296.
Werling LL, Brown SR, Cox BM (1987). Opioid receptor regulation of the release of norepinephrine in brain. Neuropharmacol. 26: 987-996.
West RE, Miller RJ (1983). Opiates, second messengers and cell response. Br. Med. Bull. 39: 53-58.
Zukin RS, Zukin SR (1984). The case for multiple opiate receptors. TINS 7: 160-164.

NORADRENERGIC MODULATION OF ACETYLCHOLINE RELEASE

L.Beani,C.Bianchi,S.Tanganelli,T.Antonelli,G.Spalluto,C.Tomasini
Department of Pharmacology,University of Ferrara,via Fossato di Mortara 23, Ferrara, Italy.

INTRODUCTION

Many reports have demonstrated the inhibitory modulation directly exerted through α_2 receptors by noradrenaline(NE) on the cholinergic cells of animal species, including humans, both at the periphery and in the CNS (Beani et al.,1978; Vizi,1979,1983).

It is generally believed that the noradrenergic signal affects membrane regulatory G-proteins which, in turn, influence the phosphorylation-dephosphorylation balance and the opening of ion channels.

From the biochemical point of view, inhibition of adenylate cyclases has been reported.

From the electrophysiological point of view, membrane hyperpolarization, increased membrane conductance, reduced calcium influx and increased potassium outward currents have been described (Aghajanian and Wang,1987; Slack,1986).

The end-effects induced by the α_2-mediated inhibition are above all represented by reduced cellular excitability and lesser efficiency of the electrosecretory coupling. These changes in the cholinergic neurons are not always or immediately demonstrable in the brain of the whole animal. This depends on the differences in the overall, functional and behavioural responses elicited by the noradrenergic signal. For instance, during the electrical stimulation of the central noradrenergic neurones some species, such as guinea-pigs, display sedation (Bianchi et al.,1979), whereas others, such as monkeys, show excitation (Redmond et al.,1976). Clearly, the cholinergic inhibition can be better seen in the former case than in the latter, it being

NATO ASI Series, Vol. H19
Modulation of Synaptic Transmission and Plasticity
in Nervous Systems. Edited by G. Hertting and H.-C. Spatz

well known that neurodepression is associated with (or facilitates) the reduced activity of the cholinergic cortical input.

The different reading of the noradrenergic message probably depends on the different prevalence of α_1,α_2 or ß receptors in various brain areas, on the different receptor location onto distinct neuronal pools, and on their different coupling with membrane transducing mechanisms. In this respect, it is worth mentioning that brain adeny - late cyclases are mainly coupled with the ß receptors in rats and with the α-receptors in guinea-pigs (Daly,1977). Consequently, great care must be taken in comparing the findings obtained using different laboratory animals. In fact, neuronal systems concomitantly affected by NE can, in some instances, shadow or counteract the direct inhibition elicited by NE on cholinergic cells. These problems are obviously minimized or overcome in simplified models, such as brain slices or isolated neurons.

My address will summarize the work carried out by our group in the last years, using the guinea-pig as "reference species". This choice was made taking into account the extensive work performed with different guinea-pig isolated preparations, like gut and heart, to study the interactions between the noradrenergic and cholinergic system at the periphery.

The presentation will be divided into three parts:

1- the ex vivo and in vitro observations concerning the direct noradrenergic modulation of cells releasing acetylcholine (ACh);
2- the data demonstrating the GABA-cooperation in NE-induced inhibition;
3- the recent, partly unpublished results, stressing the dramatic changes induced by chronic morphine treatment in the noradrenergic modulation of the cholinergic system. Examples will be presented to show that the reading is now accomplished through α_1 receptors instead of α_2 receptors, and that the end-effect on the cholinergic elements is reversed from inhibition into excitation.

The methods employed were :

A- Freely moving guinea-pigs equipped with epidural cups, in order to study the release of transmitters from their cortical surface (Beani et al.,1968);

B- Brain slices superfused in small chambers provided with electrodes in order to apply field stimulation at various frequencies and measure the released transmitters (Beani et al.,1978)

1-THE DIRECT NORADRENERGIC INHIBITION

As regards the first point, the most prominent findings in the whole animal are the following:

a) NE 0.9μmol injected i.c.v. or clonidine 7-18 nmol/kg administered i.p. cause sedation, E.Co.G. synchronization and reduce the cortical release of ACh to 70-60 percent of normal levels (Beani et al.,1978;Moroni et al.,1983). This effect can be accounted for by the reduced traffic of pulses in the cholinergic axons and nerve endings in the neocortex underlying the cup. In fact, the normal release of ACh is due, for the most part, to sodium-dependent phenomena, since it is drastically reduced to about 15% by the local application of tetrodotoxin (Antonelli et al.,1986). Consequently, the adrenergic drugs appear to dampen the firing rate of the subcortical cholinergic neurons projecting to the cortex, so that no inference can be drawn about the existence and extent of presynaptic inhibition of ACh release in the whole animal.

b) Even the electrical stimulation of locus coeruleus cells, (carried out with stimulation frequencies similar to the firing rate, evoked by sensory activation of this nucleus) is able to reduce the cortical outflow of ACh, as NE and clonidine do. Thus, the inhibition of the cholinergic structures by NE is not merely a pharmacological curiosity, but can be induced by the nearly physiological activation of NE neurons (Bianchi et al.,1979).

c) In accord with the idea of a tonic control on ACh cells exerted by endogenous NE, is the observation that the α-blocking agents phentholamine,yohimbine and idazoxan, not only prevent the inhibition by NE and clonidine and by locus coeruleus stimulation, but per se enhance the basal release of ACh by freely moving animals (Beani et al.,1978;Moroni et al.,1983;Beani et al.,1987). In addition, it was found that prazosin too prevents the inhibition by the adrenergic signal (Moroni et al,1983). This unexpected finding will be reconsidered later.

In summary, these results stress the physiological background of NE modulation in the overall function of the cholinergic system in the forebrain.

Concurrently with these investigations, experiments were carried out to prove, as already performed at the periphery, the existence of presynaptic modulation by NE on the ACh nerve endings. To this end the model of the cortical slices submitted to the electrical field stimulation was chosen.

This approach was considered convenient, in the light of the following arguments:

1) the cerebral cortex is, above all, provided with ACh nerve endings, while the presence of intrinsic ACh neurons does not consistently contribute to the whole amount of cholinergic markers present in this area (Fibiger,1982).

2) the kind of stimulation applied to the slices (to evoke an extra-release of ACh above the "leakage" of transmitter) is crucial to evidence NE effect. In fact, the electrophysiological studies carried out to clarify the mechanism of α_2-mediated inhibition have demonstrated that NE shortens the duration of the active state of the neuronal membrane, by accelerating the recovery of calcium-dependent potassium conductances (Slack,1986). Thus, one can infer that the elementary ionic processes supporting ACh secretion (calcium influx during membrane depolarization) are cut-off by α_2-receptor stimula-

tion.If this is the case, it is essential to reproduce with proper stimuli the physiological condition of sequential cycles of brisk sodium-dependent depolarization interrupted by the recovery of potassium outward currents. Such alternate periods of rest and activity can be easily achieved either by electrical stimulation of the axons or by drugs able to interfere with the active mechanisms ensuring membrane polarization, such as the inhibitor of Na^+-K^+ pump ouabaine. This drug appears to evoke ACh release through nerve-conducted activity because its effect is highly tetrodotoxin-sensitive (Vizi,1979), like the electrically-evoked release.

Taking into account these two methodological considerations, the following data merit mentioning:

i) NE and other α_2 agonists are able to inhibit in a concentration-dependent manner the extra-release of endogenous ACh from cortical slices submitted to field stimulation at 2-10 Hz (Beani et al.,1978). The expected inverse relationship between stimulus rate and inhibition cannot be easily evidenced in this model, since the noise, due to the ACh release at rest (which is tetrodotoxin-insensitive), is so high that it impedes the application of stimulation rates lower than 1 Hz. Such inhibition has also been demonstrated for oubaine-induced ACh release (Vizi,1979). Conversely, our and Vizi's group could not detect any inhibitory effects by NE or other adrenergic agonists when ACh release was evoked by potassium, an ion which causes a steady reduction of membrane potential, although lower potassium concentrations can display a tetrodotoxin-sensitive component (Bianchi et al.,1982 ;Bonanno et al.,1987).

ii) Moreover, the α_2 antagonists not only prevent the inhibition induced by the agonists, but per se enhance the evoked release by field stimulation or oubain. This suggests the persistence of an endogenous noradrenergic control even in the slices.

iii) The involvement of α_1,ß, dopamine, 5-HT, µ heteroceptors in the noradrenergic control of ACh nerve endings has been excluded, as proved by the ineffectiveness of the corresponding antagonists versus NE inhibition and by the lack of significant inhibitory properties on ACh release by phenylephrine, isoprotenerol,dopamine, serotonin,morphine. Furthermore, any involvement of muscarinic autoceptors in the amine effect has also been ruled out (Beani et al.,1978;Beani et al.,1982).

The presentation of this old data being of common knowledge, may be considered redundant; but it represents the basis of the second and third parts of this presentation.

2 THE GABA MEDIATED-INHIBITION

The participation of the GABA system in the noradrenergic control of ACh cells became evident in previous research carried out to ascertain: a) a possible modulation of ACh by GABA; b) a possible modulation of GABA by NE.

Experiments performed both in freely moving guinea-pigs and in brain slices proved that GABA actually inhibited the function of ACh cells. In brief, GABA facilitatory drugs, like benzodiazepines tested in the unanaesthetized animal, reduced the cortical outflow of ACh. Similar effects were elicited by the i.c.v. injection of GABA and GABA-like compounds, like muscimol and THIP, although their pattern of action was complicated by the parallel interference with the serotonergic system. When the 5-HT signal was excluded, sedation, E.Co.G. synchronization and reduced ACh release were clearly evident after the injection of GABA and allied compounds, as after that of benzodiazepines. These effects appear to depend on the stimulation of $GABA_A$ receptors (Tanganelli et al.,1985).

Again, the in vitro experiments on cortical slices agreed with the ex vivo observations. In fact, GABA was previously shown to inhibit the electrically-evoked ACh release through $GABA_A$ receptors (Bianchi et al.,1982). However, in contrast with the ineffectiveness of NE on the

ACh release at rest, GABA increased this kind of basal "leakage". In addition,GABA facilitated the ACh release evoked by moderate potassium concentrations (Bianchi et al.1982). These last two data suggest that GABA can reduce the membrane potential of the nerve endings thus facilitating the spontaneous transmitter release. The physiological significance of GABA control was supported by the following data:

a) The GABA antagonists bicuculline and picrotoxin not only counteracted GABA inhibition but also increased the stimulus-evoked release of ACh in vitro ;

b) The inhibition of the GABA metabolism, induced in vivo and in vitro by ethanolamine-O-sulphate, exactly reproduced the effects of exogenously applied GABA, the changes in ACh release being counteracted by the classical aminoacid antagonists (Bianchi et al.,1982;Tanganelli et al.,1985).

The positive role of GABA in NE modulation of ACh became an attractive hypothesis when we found that the adrenergic drugs, NE and clonidine,as well as the electrical stimulation of locus coeruleus increased the outflow of GABA from the cerebral cortex of freely moving guinea-pigs, by acting through α_1 receptors (Moroni et al.,1982; Moroni et al.,1983). Similarly, NE added at relatively high concentrations to unstimulated cortical slices increased the outflow of this aminoacid and the amine facilitatory effect was prevented not only by prazosin but, interestingly, by tetrodotoxin. Therefore, the inference could be drawn that the amine activated through α_1 receptors the firing rate of GABA interneurones (Beani et al.,1986) . Could the amount of transmitter released by these cells act with NE in restraining ACh release? This hypothesis was tested in brain slices and it was found to be correct.

In fact, it was demonstrated that : 1) the stimulus-evoked release of ACh, inhibited by NE at concentrations able to release GABA, was associated with the increased leakage of ACh in resting conditions, as happened following GABA application ; 2) a consistent part of these NE effects were prevented by prazosin; c) the GABA reuptake inhibitor DABA, at a concentration insufficient to primarily affect ACh release, was able to shift leftwards the concentration-effect relationship of NE on ACh release, so that the amine potency was multiplied by a factor of about 3. Even this kind of NE inhibition in DABA-treated slices was prevented not only by α_2 blocking agents, but also by prazosin (Beani et al.,1986). These data fit well with the observation that prazosin prevented NE and clonidine inhibition of ACh release in freely moving animals(Moroni et al.,1983).

In conclusion, the central noradrenergic system controls the function of the cholinergic structures inpinging upon the cortex, by acting at subcortical and cortical level. The endogenous amine and allied compounds reduce both the neuronal firing rate and the efficiency of the electrosecretory coupling. This effect is partly direct and partly mediated . The direct effect is operated through α_2 receptors located on the different parts of the ACh cell body (somatodendritic complex and nerve endings).

Concomitantly, the noradrenergic signal moves a subsidiary , "in parallel", mechanism which activates GABA interneurones through α_1 receptor.These cells,by increasing the availability of synaptic GABA, further depress the function of the cholinergic system.

3 THE CROSSING-OVER OF NORADRENERGIC MODULATION DURING MORPHINE TOLERANCE

The last point to be discussed concerns the effects elicited by NE on ACh and GABA during morphine tolerance. These observations underline the plasticity of the interplay between these neuronal systems. In addition, these findings can offer the basis for explaining certain aspects of drug tolerance and abstinence.

It is well known that opiates inhibit not only the GABA and ACh cells

(Antonelli et al.,1986) but also locus coeruleus neurones (Aghajanian et al.,1987). It is also firmly estabished that the prolonged functional inhibition of a neuronal pool is generally associated with compensatory changes in the signalling and receiving synaptic mechanisms. Thus, during chronic morphine treatment, the heteroceptors present on the target cells under noradrenergic control would be expected to be upregulated and to increase in number or in affinity for the agonist. On the other hand, the functional and spatial closeness of α_2 and opiate receptors, clearly demonstrated by many groups in peripheral organs and in the CNS (Limberger et al.,1986; Shoffelmeer et al.,1986; North,1986), opens the question of a possible inbalance in the reading mechanisms of the NE signal. In fact, while the α_1 and ß receptors could be upregulated (according to the law of compensation for the pharmacological " denervation"), the α_2 receptors could be "occluded" by the close maximally activated opiate receptors, which employ the same transducing mechanisms in the neuronal membrane. The contradictory results obtained with the standard binding techniques, together with the impelling evidence of the partial overlapping between the opioid and noradrenergic receptors, induced us to test morphine influence in the noradrenergic control of ACh and GABA neurons with a functional approach , i.e. by studying their release.

The results so far obtained both in freely moving guinea-pigs and in cortical slices have proved that dramatic changes in the reading of the NE message take place in tolerant animals. Tolerance was induced by implanting s.c. morphine pellets at proper intervals and the neurochemical responses to NE and other adrenergic agonists were tested on the seventh day of treatment (Antonelli et al.,1986). The preliminary observations were presented at the recent symposium on the cholinergic mechanisms held at Buxton (Beani et al.,1987). They have now been expanded and submitted for publication. The most relevant points so far estabished are the following:

1) clonidine doses as high as 112 nmol/kg are still able to inhibit ACh release but low doses of 7-18 nmol/kg (which would mainly act onto α_2 receptors) increase the release of ACh from the parietal cortex of tolerant guinea-pigs. A similar inverted response has been found after locus coeruleus stimulation. In these animals, the release of ACh is lower than in normal animals, suggesting a lack of adaptation to morphine inhibition by the subcortical cholinergic cells. However, this functional depression is not the reason for the inverted response, since the inhibition of ACh release by e.g. benzodiazepines is still detectable (Antonelli et al.1986). Another relevant finding concerns the receptors involved in this unexpected stimulatory effect: prazosin is able to antagonize the increase in ACh release by the adrenergic signal, whereas idazoxan is ineffective or even further enhances the inverted response. Therefore, the α_1 receptors have replaced the α_2 receptors.

2) In the meantime, clonidine 7-18 nmol/kg and endogenous NE are now able to inhibit the cortical outflow of GABA, and this effect is prevented by idazoxan but not by prazosin. Therefore, the α_2 receptors are now involved. Only high clonidine doses maintain the normal facilitatory effect on GABA outflow. In sum, the control of the ACh system by NE and low clonidine doses in morphine-tolerant guinea-pigs becomes excitatory, whereas the servo-mechanism involving GABA is obliterated and substituted by an opposite, α_2-mediated inhibitory effect. This is, in our opinion, an interesting example of the crossing-over of NE modulation onto ACh and GABA cells.This phenomenon is manifest one week, but not 24 hours, after starting morphine treatment.Interestingly, the stimulation of locus coeruleus in morphine-dependent animals does not give rise to sedation, but is ineffective or in some cases induces excitation, increased locomotion and exploratory activity. This opens the question of possible behavioural deviations detectable in morphine-tolerant animals during activation

of the noradrenergic neurons by physiological, enviromental or sensory inputs.

The experiments in cortical slices (maintained in the presence of morphine 1 μM in order to avoid the the neurochemical signs of withdrawal in vitro) allowed us to confirm the crossing-over of the reading at least for the ACh structures. Clonidine 1 μM increased by 30%, or more,the electrically-evoked ACh release in slices taken from tolerant animals, at variance with the inhibitory effect displayed in control slices. This facilitation was prevented by prazosin, but not by idazoxan. Similar findings have been obtained with NE 30 μM.It is worthy of note that the α_1 agonist phenylephrine,100 μM, consistently (+30%) enhanced ACh release through α_1 receptors in tolerant slices. Isoproterenol up to 100 μM was found inactive both in normal and in tolerant tissue.

A more simplified model, such as isolated neurons and synaptosomes can solve the problem of whether the" occlusion" of the α_2 receptors induced by morphine is associated with the appearance of "new", (or otherwise silent) 1 receptors on the cholinergic nerve membrane or, alternatively, whether chronic opiate treatment opens new excitatory pathways in the still complex circuitry of the slices, so that the adrenergic drugs can now induce, through α_1 receptors ,the release of endogenous compounds (such as glutamate) able to increase the release of ACh. Experiments are in progress in this area.

CONCLUSIONS

To conclude my address I wish to draw your attention to the "ambiguity" of the noradrenergic modulation on ACh release. As stressed in the introduction, certain species normally display excitatory behavioural responses following locus coeruleus stimulation, while others show sedation. In every case the control of NE on ACh cells seems to be inhibitory, at least when simplified models are chosen. A direct and a GABA-mediated component have been demonstrated. In morphine-tolerant guinea-pigs, however, the changes in the behavioural responses to locus coeruleus stimulation are associated with the reversal of NE modulation on ACh and GABA cells, the former being activated through α_1 receptors, the latter being inhibited through α_2 receptors.

These plastic changes require some days to become manifest. Possibly, the reversal of NE modulation represents a mechanism compensatory to chronic morphine neurodepression. Such a mechanism develops even in those parts of the cholinergic cells (i.e. nerve endings) which are not under the direct control of receptors . As a consequence of this functional rearrangement, the α_2 inhibitory signals, normally directed to the "activating" cholinergic system are substituted by α_1 excitatory signals. Conversely the α_1 excitatory message normally directed to the "dampening" GABA system , is substituted by an α_2 inhibitory input. This fact may offer a neurochemical basis to explain the latent hyperexcitability shown by the CNS of morphine-tolerant animals and the dramatic signs of opiate withdrawal, which must be controlled by high clonidine doses, the only ones able to reproduce the physiological modulation exerted by NE on ACh and GABA cells.

REFERENCES

Aghajanian GK, Wang YY (1987) Common α_2 and opiate effector mechanisms in the locus coeruleus : intracellular studies in brain slices. Neuropharmac.26:793-799

Antonelli T,Beani L ,Bianchi C , Rando S ,Simonato M , Tanganelli S,(1986) Cortical acetylcholine release is increased and γ-aminobutyric acid outflow is reduced during morphine withdrawal. Br.J.Pharmac. 89:853-860

Beani L , Bianchi C , Santinoceto L , Marchetti P,(1968) The cerebral acetylcholine release in conscious rabbits with semipermanently

implanted epidural cups.Int.J.Neuropharmac.7:469-481
Beani L , Bianchi C , Giacomelli A , Tamberi F , (1978) Noradrenaline inhibition of acetylcholine release from guinea-pig brain .Eur.J.Pharmac. 48:179-193
Beani L , Bianchi C , Siniscalchi A , (1982) The effect of naloxone on opioid-induced inhibition and facilitation of acetylcholine release in brain slices. Br.J.Pharmac. 76:393-401
Beani L , Tanganelli S , Antonelli T , Bianchi C ,(1986) Noradrenergic modulation of cortical acetylcholine release is both direct and -aminobutyric acid mediated.J.Pharmac.Exp.Ther.236:230-236
Beani L , Bianchi C , Tanganelli S , Antonelli T , Simonato M , Rando S (1987) Reversal of clonidine effect on acetylcholine release during morphine tolerance. In"Cellular and molecular basis of cholinergic function" Dowdall M.J. and Hawthorne J.N. eds, Ellis Horwood, Chichester (England) pp.250-255
Bianchi C , Spidalieri G , Guandalini P , Tanganelli S , Beani L (1979) Inhibition of acetylcholine outflow from guinea-pig cerebral cortex following locus coeruleus stimulation. Neurosci.Letters 14:97-100
Bianchi C , Tanganelli S , Marzola G , Beani L (1982) GABA-induced changes in acetylcholine release from slices of guinea-pig brain. Naunyn Schmiedeberg's Arch.Pharmac.318:253-258
Bonanno G , Raiteri M , (1987) Interaction between 5-HT uptake inhibition and activation of 5-HT autoreceptors by exogenous agonists in rat cerebral cortex slices and synaptosomes.Naunyn Schmiedeberg's Arch.Pharmacol.335:219-225
Daly J (1977) Cyclic nucleotides in the nervous system. Plenum Press,New York and London
Fibiger HC (1982) The organization and some projections of cholinergic neurons of the mammalian forebrain.Brain Res.Reviews 4:327-388
Limberger N , Spath L , Holting TH , Starke K , (1986) Mutual interaction between presynaptic α_2 adrenoceptors and opioid k receptors at the noradrenergic axons of rabbit brain cortex. Naunyn Schmiedeberg's Arch.Pharmac.334:166-171
Moroni F , Bianchi C , Moneti G , Tanganelli S , Spidalieri G , Guandalini P , Beani L ,(1982) Release of GABA from the guinea-pig neocortex induced by electrical stimulation of the locus coeruleus or by norepinephrine.Brain Res.232:216-221
Moroni F , Tanganelli S , Antonelli T , Carla' V , Bianchi C , Beani L , (1983) Modulation of cortical acetylcholine and γ-aminobutyric acid release in freely moving guinea-pigs: effects of clonidine and other adrenergic drugs. J.Pharmacol.exp.Ther. 227:435-440
North RA (1986) Receptors on individual neurones . Neuroscience 17:889-907
Redmond DE , Huang YM , Snyder DR , Maas JW (1976) Behavioural effects of stimulation of the nucleus locus coeruleus in the stump-tailed monkey macaca arctoides.Brain Res.116:502-510
Schoffelmeer ANM , Putters J , Mulder AH (1986) Activation of

presynaptic α_2-adrenoceptors attenuates the inhibitory effect of μ-opioid receptor agonists on noradrenaline release from brain slices.Naunyn Schmiedeberg's Arch.Pharmacol.333:377-380

Slack BE (1986) Pre- and postsynaptic actions of noradrenaline and clonidine on mienteric neurons.Neuroscience 19:1303-1309

Tanganelli S , Bianchi C , Beani L (1985) The modulation of cortical acetylcholine release by GABA,GABA-like drugs and benzodiazepines in freely moving guinea-pigs.Neuropharmac.24:291-299

Vizi ES (1979) Presynaptic modulation of neurochemical transmission. Prog.Neurobiol. 12:181-290

Vizi ES (1983) Release-modulating adrenoceptors.In "Adrenoceptors and catecholamine action.Part B,Kunos G ed.,John Wiley & Sons inc.,pp 65-107

CONTROL OF ACETYLCHOLINE RELEASE AND OF INTESTINAL MOTILITY BY SUBTYPES OF MUSCARINE RECEPTORS

H. Kilbinger, H. Schwörer, A. Stein, and K.-D. Süß
Pharmakologisches Institut
Universität Mainz
Obere Zahlbacher Str. 67
D 6500 Mainz
F.R.G.

Two types of neuronal muscarine receptors have been investigated in the myenteric plexus preparation of the guinea-pig small intestine:

1. Presynaptic receptors activation of which inhibits the depolarization-evoked release of acetylcholine. Pirenzepine and dicyclomine have low affinities to the release-inhibitory receptors (pA_2 values 6.9 and 7.6) which suggests that the presynaptic receptors (similar to the smooth muscle receptors) belong to the M2 subtype. The inhibition of the electrically-evoked acetylcholine release by muscarine (0.01 - 1 µmol/l) was not affected by forskolin (1µmol/l). This indicates that cyclic AMP is not crucially involved in the muscarinic inhibition of acetylcholine release.

2. Somatodendritic receptors whose activation causes an increase in spontaneous acetylcholine release. These receptors have a high affinity to pirenzepine (pA_2 8.5) and dicyclomine (pA_2 9.3), and thus belong to the M1 subtype. M1 receptors are probably involved in the regulation of intestinal motility. Blockade by low concentrations (< 1 nmol/l) of either pirenzepine or dicyclomine causes an increase in intestinal peristalsis. This increase is antagonized by bicuculline (10 nmol/l). The results suggest that the activation of ganglionic M1 receptors by endogenous acetylcholine leads to an increased release of γ-aminobutyric acid which, in turn, causes inhibition of intestinal motility.

NATO ASI Series, Vol. H19
Modulation of Synaptic Transmission and Plasticity
in Nervous Systems. Edited by G. Hertting and H.-C. Spatz

INTRODUCTION

There is now considerable evidence for the existence of more than one type of muscarine receptors. Receptors exhibiting a high affinity to the antagonist pirenzepine have been called M1 receptors, and those with a low affinity M2 receptors (for review see Birdsall and Hulme, 1985). Both subclasses have been detected on guinea-pig myenteric neurones in functional (Kilbinger, 1984; Kilbinger and Nafziger, 1985; North et al., 1985) and autoradiographic (Buckley, 1985) studies. Activation of the somadendritic M1 receptors leads to a slow membrane depolarization (North et al., 1985) and an increase in spontaneous acetylcholine release (Kilbinger, 1984), whereas the activation of the presynaptically located M2 receptors causes an inhibition of the depolarization-evoked acetylcholine release (Kilbinger et al., 1984). Pirenzepine is so far the only antagonist which clearly discriminates between M1 and M2 receptors. Here, we report that dicyclomine is another competitive antagonist which shows a selective effect on M1 receptors, and which facilitates intestinal peristalsis via blockade of myenteric M1 receptors. The present article also examines the mechanisms behind the neuronal muscarine receptors. Inhibition of cyclic AMP formation may be involved in the α2-autoreceptor mediated inhibition of noradrenaline release (see Starke, 1987). However, it is not known whether a similar interaction exists at the level of the muscarine autoreceptor. We have therefore studied the interaction between muscarine and the adenylate cyclase activator, forskolin, on both spontaneous and electrically-evoked release of acetylcholine.

METHODS

1. Release of [^{3}H]acetylcholine

Experiments were performed essentially as described by Kilbinger and Nafziger (1985). Longitudinal muscle myenteric plexus preparations of the guinea-pig small intestine were incubated with [^{3}H]choline (1.25 μCi/ml) and subsequently superfused (2 ml/min) with Tyrode solution (composition in mM: NaCl 137; KCl 2.7; $CaCl_2$ 1.8; $MgCl_2$ 1.0; NaH_2PO_4 0.4; $NaHCO_3$ 11.9; glucose 5.6; choline chloride 0.001). For reasons discussed previously (Kilbinger, 1984) naloxone (0.1 µM) was present in the superfusate from the washout period on. After 60 min washout the superfusate was collected in 3 min fractions, and tritium was determined by liquid scintillation spectrometry. For electrical stimulation square wave pulses of 1 ms duration were applied at frequencies of 0.1 or 1 Hz. The outflow of tritium evoked by electrical stimulation or by drugs was calculated by subtracting the estimated spontaneous outflow from the total outflow of tritium. It was expressed in per cent of the tritium content of the tissue at the onset of each stimulation.

2. Contraction experiments

Isometric contractions of the longitudinal muscle were studied in separate experiments. Three cumulative concentration-response curves to oxotremorine were obtained at 45 min intervals with no significant change in sensitivity being apparent. Dicyclomine was introduced into the incubation solution in cumulatively increasing concentrations immediately after the first and second curves had been obtained.

3. Recording of intestinal peristalsis

The arrangement for eliciting peristalsis was similar to that described by Holzer and Lembeck (1979). In brief, segments of the proximal part of the small intestine, about 6 cm in length were mounted horizontally in an organ bath which contained Tyrode solution. The oral end was connected to a Mariotte bottle with two inlet tubes which contained prewarmed (37° C) Tyrode solution, the aboral end was connected to a mercury valve which prevented back flow of fluid and which was set to open at an outflow pressure above 500 Pa. Peristalsis was elicited in intervals of 15 min by increasing the intraluminal pressure of the intestinal segment by 500 Pa for 2 min. Changes of the fluid volume expelled are expressed as per cent of the fluid volume expelled during two control peristalsis periods in the absence of muscarine receptor antagonists.

4. Statistics

Given are means ± SEM. Statistical significance of differences was assessed by Student's t-test. The pA_2 values were calculated from Schild plots by linear regression analysis.

5. Drugs

Methyl-[^{3}H]-choline (NEN); (+)-bicuculline; forskolin; tolazoline hydrochloride (all Sigma); dicyclomine hydrochloride (Merrell Dow); naloxone hydrochloride (Goedecke); pirenzepine dihydrochloride monohydrate (Thomae); scopolamine hydrobromide (EGA); oxotremorine sesquifumarate (Aldrich).

RESULTS

1. Discrimination by dicyclomine of M1 and M2 receptors

Strips were stimulated twice (S1, S2) by field stimulation (1 Hz, 3 min), and pilocarpine or oxotremorine were added to the superfusate 30 min before the second stimulation. Pilocarpine (0.1-100 µmol/l) caused a concentration-dependent transient increase in spontaneous outflow, and a decrease in the evoked outflow of tritium. Previous experiments had shown that the increase in tritium outflow by pilocarpine is due to the release of [^{3}H]acetylcholine (Kilbinger, 1984). Oxotremorine (0.01-10 µmol/l) did not affect spontaneous outflow but strongly reduced the evoked outflow of tritium. For interaction experiments dicyclomine was added to the medium 30 min before S1. In the presence of 1-100 nmol/l dicyclomine, the concentration-response curves for the facilitatory effect of pilocarpine on tritium outflow were shifted in parallel to the right. Much higher concentrations of dicyclomine were needed (0.03-1 µmol/l) to antagonize the inhibition by pilocarpine or oxotremorine of the electrically-evoked outflow during S2. Likewise, the longitudinal muscle contractions evoked by oxotremorine were antagonized by dicyclomine concentrations ranging from 0.1-3 µmol/l. In all experiments dicyclomine (up to 3 µmol/l) behaved as a competitive antagonist. pA_2 values were calculated from the parallel shifts of the curves for pilocarpine (increase in spontaneous outflow), and for oxotremorine (inhibition of evoked outflow; contraction) in the presence of dicyclomine. The results are summarized in Table 1. It is evident that dicyclomine like pirenzepine discriminates between M1 and M2 receptors. Dicyclomine is at least 50 times more potent on the excitatory ganglionic receptor than on the release-inhibitory or the smooth muscle receptor.

Table 1. Affinity constants (pA_2 values) of competitive antagonists at subtypes of muscarine receptors in the guinea-pig ileum.

Subtype	Response	Scopolamine[a]	Pirenzepine[a]	Dicyclomine
M1	Increase in spontaneous acetylcholine release	9.5	8.5	9.3
M2	Inhibition of evoked acetylcholine release	9.1	6.9	7.6
M2	Longitudinal muscle contraction	9.4	6.7	7.3

[a] from Kilbinger et al. (1984).

2. Effects of forskolin on the muscarinic modulation of acetylcholine release

In control experiments the effects of forskolin alone on both spontaneous and electrically-evoked outflow of tritium from strips preloaded with [^{3}H]choline were studied. Forskolin increased in a concentration-dependent fashion both spontaneous and evoked outflow of tritium (Table 2). The facilitatory effect of forskolin on spontaneous outflow was prevented in the presence of 300 nmol/l tetrodotoxin. In order to identify the nature of the $[^3H]$radioactivity in the medium the effect of 10 μmol/l forskolin on tritium outflow was investigated in the presence of 10 μmol/l neostigmine plus 0.1 μmol/l scopolamine. Labelled choline was separated from labelled acetylcholine by thin layer chromatography as described earlier (Kilbinger and Nafziger, 1985). It was found that the outflow of tritium caused by forskolin consisted to 93±13% (N=4) of $[^3H]$acetylcholine. The outflow of $[^3H]$choline was not affected by forskolin.

Table 2. Effects of forskolin on spontaneous and electrically-evoked outflow of tritium from strips preloaded with $[^3H]$ choline.

Forskolin (μmol/l)	$[^3H]$outflow evoked by S1 (% of tissue content)	Forskolin (% of tissue content)	S2 (% of S1)	N
–	3.09±0.30	-	0.87±0.05	9
0.1	2.66±0.26	0.11±0.08	0.96±0.05	5
1	2.96±0.16	0.67±0.12	1.17±0.11*	8
10	2.66±0.11	1.24±0.13	1.63±0.07**	4
30	2.99±0.12	1.09±0.21	1.55±0.16**	3

Strips were stimulated twice (S1, S2; 0.1 Hz, 60 pulses), and forskolin was added 54 min before S2. * $p<0.05$, ** $p<0.001$

When forskolin (1 μmol/l) was added to the medium 20 min before S1, the ratio S2/S1 (1.01 ± 0.05, N=6) was not significantly different from the control value in the absence of forskolin (0.87 ± 0.05; N=9). In the presence of forskolin, muscarine still inhibited the electrically-evoked outflow of tritium to the same degree as in control experiments (Figure 1, left panel), and the EC 50 values for muscarine in the absence (63 nmol/l) and presence (46 nmol/l) of forskolin were similar. In contrast, the effects of 0.01 and 0.1 μmol/l muscarine on spontaneous outflow of tritium were potentiated in the presence of forskolin and the concentration-response curve was shifted to the left (Figure 1, right panel).

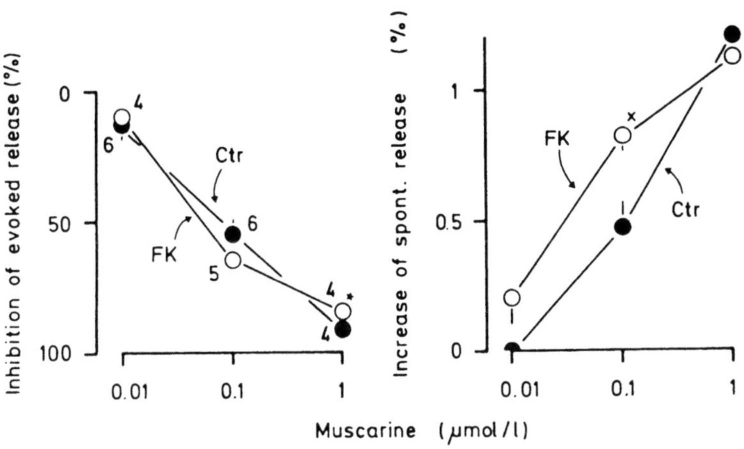

Figure 1. Concentration-response curves for the effects of muscarine on spontaneous and electrically-evoked release of tritium. Strips were stimulated twice (S1, S2; 0.1 Hz, 60 pulses). Forskolin (1 µmol/l) was added 20 min before S1, and muscarine 24 min before S2. Filled circles, control experiments; open circles, experiments in the presence of forskolin. The increase by muscarine of spontaneous release is given in per cent of the tissue content of tritium. Means ± SEM of the number of experiments indicated. Significance of difference from corresponding control value, ^{x} $p < 0.05$, * $p < 0.01$.

3. Muscarinic modulation of intestinal motility.

In control experiments the fluid volume expelled during successive periods of peristalsis remained constant up to 165 min after the set-up of the preparation. Very low concentrations of scopolamine, pirenzepine and dicyclomine caused significant increases of the fluid volume expelled during peristaltic activity. The maximal increases observed with scopolamine, pirenzepine and dicyclomine were reached only 45 - 60 min after the application of the drugs. Higher concentrations of the muscarine receptor antagonists caused the well-known inhibition of peristalsis. Table 3 shows the % changes of peristalsis caused by maximally effective concentrations of the antagonists. The pD_2 values were calculated from the biphasic concentration-response curves, and are summarized in Table 4.

Table 3. Increase and decrease by muscarine receptor antagonists of intestinal peristalsis.

Concentration (neg log mol/l)	% changes of fluid volume expelled Scopolamine	Pirenzepine	Dicyclomine
11	+29 ± 7 (4)*	+1 ± 7 (3)	+29 ± 5 (8)**
10	+21 ± 6 (6)	+43 ± 5 (13)**	+35 ± 4 (9)**
6	-60 ± 5 (5)**	-27 ± 5 (7)*	-50 ± 4 (4)**
5	-61 ± 3 (5)**	-60 ± 5 (5)**	-59 ± 8 (49**

Peristalsis was induced by increasing the intraluminal pressure by 500 Pa for 2 min, and the amount of fluid expelled was determined. Means ± SEM of the number of experiments given in parenthesis. Significance of difference from corresponding value obtained in control experiments, * $p < 0.05$, ** $p < 0.01$.

Table 4. pD_2 values for the effects of muscarine receptor antagonists on intestinal peristalsis.

	Scopolamine	Pirenzepine	Dicyclomine
Increase	11.5	10.5	12.0
Inhibition	8.2	5.9	6.5

Table 5. Effects of drugs on the pirenzepine-induced increase of intestinal peristalsis.

Drug (μmol/l)	% increase
-	43 ± 5 (13)
Tolazoline 1.0	32 ± 8 (6)
Naloxone 0.3	31 ± 2 (4)
Bicuculline 0.01	11 ± 5 (8)*
Bicuculline 0.1	1 ± 4 (6)**

Tolazoline, naloxone or bicuculline were added to the medium 60 min before the application of pirenzepine (0.1 nmol/l). Means ± SEM of the number of experiments in parenthesis. Significance of difference from the effect of pirenzepine alone, * $p < 0.01$, ** $p < 0.001$.

The twofold effects of pirenzepine were studied in the presence of various drugs. Table 5 shows that the increase in peristalsis remained unchanged in the presence of tolazoline or naloxone. However, the effect of pirenzepine was strongly reduced by bicuculline in concentrations which alone had no effect on peristalsis. The inhibition of peristaltic activity by 1 µmol/l pirenzepine was not affected in the presence of either tolazoline, naloxone or bicuculline (data not shown).

DISCUSSION

Dicyclomine was found to have a more than 50 times higher affinity to the neuronal M1 receptors than to the M2 receptors of the guinea-pig intestine. Our results are in line with radioligand binding studies which showed that dicyclomine was bound with higher affinity to M1 sites in the cerebral cortex than to M2 sites in the heart or bladder (Nilvebrant and Sparf, 1986; Giachetti et al., 1986). Thus, both the present functional study as well as previous binding experiments suggest, that not only pirenzepine, but also dicyclomine distinguishes between M1 and M2 receptors.

Blockade of intestinal muscarine receptors by subnanomolar concentrations of the antagonists caused an increase in peristaltic movements, whereas higher concentrations inhibited peristalsis. We suggest that these differential effects are mediated by blockade of M1 and M2 receptors. Pirenzepine was 10 times less potent than scopolamine in facilitating peristalsis (Table 4). This corresponds exactly to the difference in the antagonistic potencies of scopolamine and pirenzepine on myenteric M1 receptors (Table 1). On the other hand, equieffective concentrations of scopolamine and pirenzepine causing inhibition of peristalsis differed by 2.3 log units (Table 4), and a similar difference was found for the effects of both antagonists on neuronal or muscular M2 receptors (Table 1). Furthermore, dicyclomine was approximately equipotent with scopolamine in increasing peristalsis and in antagonizing the effects on the M1 receptors, but dicyclomine was about 30-100 times less potent than scopolamine on the M2 receptor and in inhibiting peristaltic activity. All this suggests that the enhancement of peristalsis by muscarine receptor antagonists is mediated by blockade of M1 receptors, whereas the inhibition is due to blockade of M2 receptors.

The present study implies that the acetylcholine which is released by an increase in intraluminal pressure (Schwörer et al., 1987), activates under physiological conditions M1 receptors which, in turn, leads to inhibition of peristalsis. The interaction experiments with tolazoline or naloxone demonstrate that neither endogenous opioids nor endogenous noradrenaline play a role in the inhibition of peristalsis caused by M1 receptor activation. However, the antagonism of the pirenzepine-induced increase of peristalsis by bicuculline suggests that endogenous GABA is released by activation of M1 receptors. GABA does not act directly on intestinal smooth muscle but affects motility via release of an unknown inhibitory transmitter (see review by Jessen et al., 1987).
The intraneuronal events that occur between activation of neuronal muscarine receptors and alterations in acetylcholine release are not known. Muscarine receptor agonists cause inhibition of adenylate cyclase activity in neural tissues (see McKinney and Richelson, 1984), and cAMP has therefore been implicated in the mechanism of autoinhibition of acetylcholine release. In support of this hypothesis, the spontaneous and the evoked release of acetylcholine was enhanced by 8-bromo-cAMP (Alberts and Stjärne, 1982), IBMX or 50 µmol/l forskolin (Reese and Cooper, 1984). Similarly, in the present experiments, forskolin caused a concentration-dependent increase in the electrically-evoked acetylcholine release. However, the inhibitory effects of muscarine were unaltered in the presence of forskolin, which suggests that adenylate cyclase is not involved in the M2 receptor mediated autoinhibition of acetylcholine release.
Forskolin also enhanced the spontaneous release of acetylcholine. The site of action is certainly not the cholinergic terminal itself but the cell body, since blockade of action potential propagation by tetrodotoxin prevented the effect of forskolin. Recent electrophysiological studies have shown that forskolin depolarizes myenteric neurones (AH/Type 2) and mimics all aspects of slow synaptic excitation in these neurones (Nemeth et al., 1986). Several transmitter substances also mimic the slow EPSP, e.g. acetylcholine acting through

myenteric M1 receptors (North et al., 1985). In the light of these observations it is interesting that the M1 receptor mediated increase in spontaneous acetylcholine release by low concentrations of muscarine was potentiated in the presence of forskolin (Figure 1). It is possible that the effect of muscarine is simply additive to that of forskolin. But the finding that the near-maximal enhancement of release by 1 µmol/l muscarine was not altered by forskolin might also indicate that both drugs act through a common mechanism.

Acknowledgments. We thank Miss Doris Wolf for expert technical assistance. This work was supported by the Deutsche Forschungsgemeinschaft. The paper contains parts of the M.D. Thesis of A.S., and the Dr. rer. nat. Thesis of K.-D.S.

REFERENCES

Alberts P, Stjärne L (1982) Secretion of ^{3}H-acetylcholine from guinea-pig ileum myenteric plexus is enhanced by 8-Br adenosine 3', 5'-cyclic monophosphate but not changed by 8-Br guanosine 3', 5'-cyclic monophosphate. Acta Physiol Scand 115: 269-272

Birdsall N J M, Hulme E C (1985) Multiple muscarinic receptors: further problems in receptor classification. In: Kalsner S (ed) Trends in autonomic pharmacology. Taylor and Francis, London, pp 17-34

Buckley N J (1985) Autoradiographic localization of muscarinic receptors in the gut. In: Lux G, Daniel EE (eds) Muscarinic receptor subtypes in the GI tract. Springer, Berlin, Heidelberg, New York, pp 1-13

Giachetti A, Giraldo E, Ladinsky H, Montagna E (1987) Binding and functional profiles of the selective M1 muscarinic receptor antagonists trihexyphenidyl and dicyclomine. Br J Pharmacol 89: 83-90

Holzer P, Lembeck F (1979) Effect of neuropeptides on the efficiency of the peristaltic reflex. Naunyn-Schmiedeberg's Arch Pharmacol 307: 257-264

Jessen K R, Mirsky R, Hills J M (1987) GABA as an autonomic neurotransmitter: studies on intrinsic GABAergic neurons in the myenteric plexus of the gut. TINS 10: 255-262

Kilbinger H (1984) Facilitation and inhibition by muscarinic agonists of acetylcholine release from guinea-pig myenteric neurones: mediation through different types of neuronal muscarinic receptors. TIPS 5 (Suppl): 49-52

Kilbinger H, Halim S, Lambrecht G, Weiler W, Wessler I (1984) Comparison of affinities of muscarinic antagonists to pre- and postjunctional receptors in the guinea-pig ileum. Eur J Pharmacol 103: 313-320

Kilbinger H, Nafziger M (1985) Two types of neuronal muscarine receptors modulating acetylcholine release from guinea-pig myenteric plexus. Naunyn-Schmiedeberg's Arch Pharmacol 328: 304-309

McKinnney M, Richelson E (1984) The coupling of the neuronal muscarinic receptor to responses. Ann Rev Pharmacol Toxicol 24: 121-146

Nemeth P R, Palmer J M, Wood J D, Zafirow D H (1986) Effects of forskolin on electrical behaviour of myenteric neurones in guinea-pig small intestine. J Physiol 376: 439- 450

Nilvebrant L, Sparf B (1986) Dicyclomine, benzhexol, and oxybutynine distinguish between subclasses of muscarinic binding sites. Eur J Pharmacol 123: 133-143

North R A, Slack B E, Surprenant A (1985) Muscarinic M1 and M2 receptors mediate depolarization and presynaptic inhibition in guinea-pig enteric nervous system. J Physiol 368: 435-452

Reese J H, Cooper J R (1984) Stimulation of acetylcholine release from guinea-pig ileal synaptosomes by cyclic nucleotides and forskolin. Biochem Pharmacol 33: 3007-3011

Schwörer H, Lauteren S, Holzer P, Kilbinger H (1987) Peristalsis releases acetylcholine (ACh) to the serosal side but not into the vasculature of the isolated guinea-pig ileum. Naunyn-Schmiedeberg's Arch Pharmacol 335: R 77

Starke K (1987) Presynaptic α-autoreceptors. Rev Physiol Biochem Pharmacol 107: 73-146

CHOLINERGIC-ADRENERGIC PRESYNAPTIC INTERACTIONS ON THE HEART IN THE MILLISECOND RANGE AS STUDIED WITH THE PULSE-TO-PULSE METHOD

E.Muscholl, Alice Habermeier-Muth, and Birgit Jung
Pharmakologisches Institut der Universität Mainz
Obere Zahlbacher Straße 67
D-6500 Mainz
Federal Republic of Germany

A method has been devised to study the interactions of cholinergic and adrenergic nerves in the rabbit atria, a tissue known to be densely innervated by both vagus and sympathetic nerve fibres. The atria were isolated and perfused via the ascending branches of the coronary arteries. The acetylcholine (ACh) stores were labelled with [14C]-choline and the noradrenaline (NA) stores with [3H]-NA. The radioactive compounds were separated on columns and determined by scintillation spectrometry.

Three periods of postganglionic sympathetic nerve stimulation (SNS1-SNS3) were applied (3Hz, 3 min) at intervals of 10 min, the first and third one serving as controls. During SNS2 the preganglionic right vagus fibres were stimulated also at 3 Hz for 3 min (VNS1) but each pulse preceded the corresponding SNS2 pulse by a fixed time interval (0.3-283 ms in 13 steps). When the sympathetic pulses were delayed for either 3-10 or 200-283 ms, with respect to the vagus pulses, the [3H]-NA overflow was significantly decreased compared to the overflow calculated from the control stimulations. At delay times of 0.3-1.7 or 30-167 ms the [3H]-NA overflow was unaltered. Peak inhibitions of release were observed at 3 and 233 ms delays.

NATO ASI Series, Vol. H19
Modulation of Synaptic Transmission and Plasticity
in Nervous Systems. Edited by G. Hertting and H.-C. Spatz

Atropine (1-300 nM) decreased and finally abolished the inhibition of [3H]-NA overflow caused by VNS1 at both the 3 and 233 ms delay times confirming that both inhibitory peak responses are mediated by a muscarine receptor. At the 100 ms delay time, atropine 300 nM had no effect suggesting that the "valley" between the two peaks of inhibition is not caused by a non-muscarinic facilitation of NA release.

The stimulation-evoked overflow of [14C]-choline plus [14C]-ACh at VNS1 which is a measure of the ACh released was similar at all delay times and approximately doubled in the presence of atropine, corroborating the existence of an inhibitory autoreceptor at the terminal vagus fibres. When allowance is made for the conduction times in the nervous pathways the two peaks of muscarinic inhibition occur when the terminal cholinergic fibres are excited either 20 or 250 ms before the adrenergic terminals.

The results obtained with [3H]-NA were confirmed when the overflow of endogenous NA was determined by HPLC and electrochemical detection. This refers to the atropine-sensitive inhibition of NA release when vagus pulses were applied either 3 or 233 ms before the sympathetic pulses, and to the failure of inhibition at the 100 ms delay time in the absence or presence of atropine.

The method employed resolves, within the ms range, presynaptic interactions between two kinds of neurons and has revealed a muscarinic response of an extremely short latency. Preliminary observations suggest a mediator role of a cylo-oxygenase product for the late inhibitory peak.

INTRODUCTION

On various heart preparations with the autonomic innervation left intact, vagus stimulation has been found to reduce the release of NA that is evoked by simultaneously applied sympathetic nerve stimulation (for review see Muscholl, 1980; Vanhoutte & Levy, 1980). When ACh acted at discrete sites such as axo-axonic synapses, rather than in a diffuse manner at distant adrenergic terminal fibres, the interval between the action potentials invading the vagal and the sympathetic nerve fibres should be crucial for the efficiency of the muscarinic inhibition of NA release. When ACh was released too briefly before initiation of the exocytotic process liberating NA, activation of the presynaptic muscarine receptor will occur too late for a significant modulation of this release to be effected. On the other hand, when the time interval between ACh and NA release became too long, the cholinergic transmitter might have diffused away, the muscarine receptor activation ceased and the following (unknown) mechanism inhibiting exocytosis been terminated before the process of NA release was initiated.

In previous studies in which the effects of VNS on evoked NA release had been quantified the vagus and sympathetic nerves were stimulated either at different frequencies (Löffelholz & Muscholl, 1970; Levy & Blattberg, 1976; Lavallée et al.,1978, Muscholl & Muth, 1982) or at the same frequency (of 3 Hz) but with separate electronic stimulators delivering the pulses independently (Muscholl & Muth, 1982). The significance of different intervals between the VNS pulse and the SNS pulse (pulse-to-pulse interval) has not yet been evaluated. In the present study the sympathetic stimuli were delayed for 0.3-283 ms with respect to the vagal stimuli, and the presynaptic effects quantified by determination of both the ACh and NA overflow rates. In order to obtain enough transmitter for the chemical analysis 540 stimuli were applied at 3 Hz in a single run, *i.e.* for 3 min. This frequency has been selected because a previous study had shown that the evoked NA overflow is in fact decreased by infusion of the muscarinic agonist, methacholine, or by VNS at 3 Hz (Muscholl & Muth, 1982). Furthermore, vago-sympathetic delay periods ranging from 0 to 333.3 ms can be employed at 3 Hz.

Preliminary accounts of some of the following observations have been presented (Muscholl & Habermeier-Muth, 1985; Jung & Muscholl, 1987).

METHODS

The dissection of the rabbit heart and the set-up of the isolated atria preparation perfused via the aorta has been described previously (Muscholl & Muth, 1982). Medium was Tyrode's solution of 36°C supplemented with 57 μM ascorbic acid and gassed with 5% carbon dioxide in oxygen. The apparatus is shown by Fig. 1.

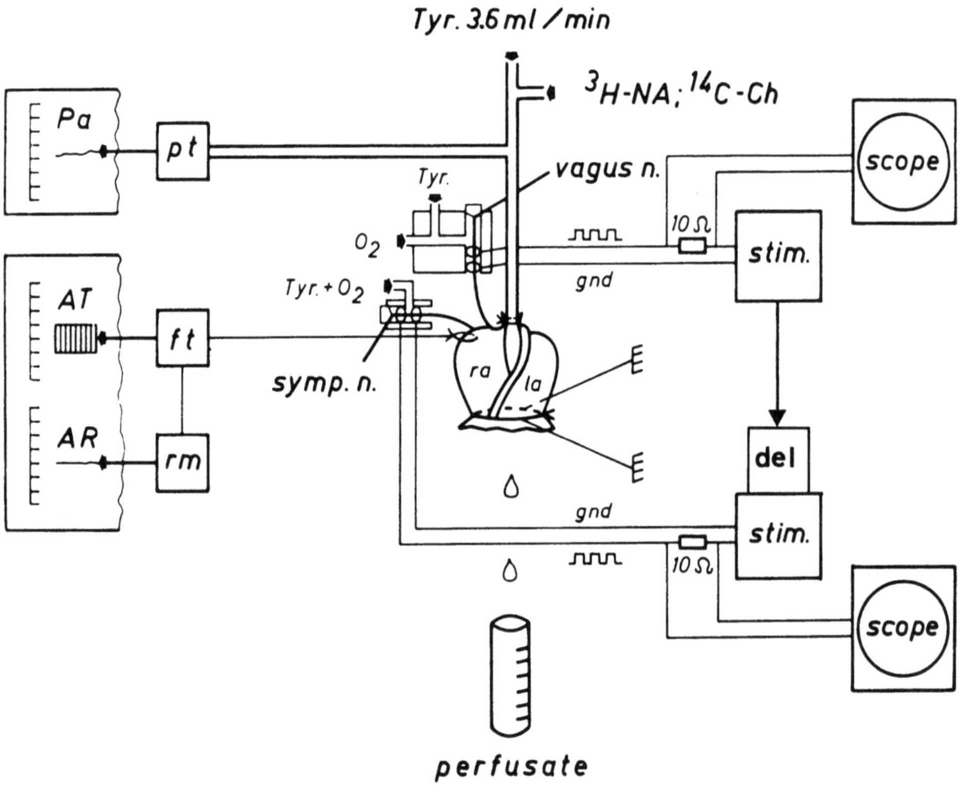

Fig. 1. Scheme of perfusion,stimulation and recording system of the isolated innervated perfused atria of the rabbit. *Tyr.*,Tyrode's solution, perfused at a constant flow of 3.6 ml/min; *ra* and *la*, right and left atrium; *pt*, pressure transducer; *Pa*, aortic pressure; *ft*, force transducer; *rm*, rate meter; *AT* and *AR*, atrial tension development and rate; *stim.*, stimulator delivering rectangular pulses of 1 ms (current strength displayed on oscilloscope); *del*, delay unit triggered by the vagus nerve (master) stimulator; *gnd*, grounded lead. The right vagus and sympathetic nerves were pulled into ring electrodes and were moistened by oxygen saturated Tyrode's solution.

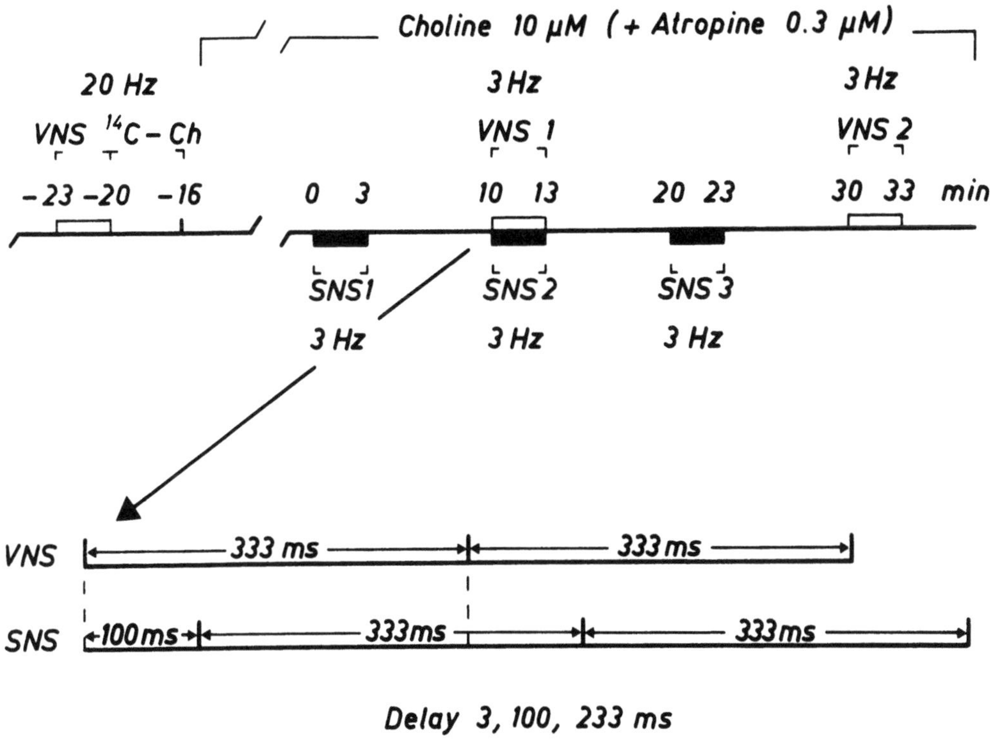

Fig. 2. Time schedule of [14C]-choline labelling procedure and electrical stimulation of the right sympathetic (SNS) and vagus nerves (VNS) of the rabbit atria.
Upper panel: Infusion of [14C]-choline for 4 min immediately after a conditioning VNS of 20 Hz for 3 min. Beginning at 0 time, 3 periods of SNS (pulses of 1 ms, 3 Hz, 8-10 mA for 3 min) at intervals of 10 min. VNS1 and VNS2 also with 1 ms pulses of 3 Hz for 3 min but at 5-6 mA.
Lower panel: VNS1 carried out simultaneously with SNS2 but sympathetic pulses delayed for a constant time interval in a single experiment. Present example is for 100 ms delay time. VNS2 carried out as a control, in the absence of SNS.

Mechanical performance of the spontaneously beating atria and perfusion pressure were recorded in order to survey the postsynaptic responses to SNS and VNS and to check the viability of the preparation.
Labelling of the transmitter stores with [3H]-NA and [14C]-choline was carried out as described previously (Muscholl & Muth, 1982). When endogenous NA was determined by HPLC with electrochemical detection (for details see Racké & Muscholl, 1986) only the ACh stores were labelled with [14C]-choline, spec. act. 52 mCi/mmol, final concentration 650 nM, infused for 4 min immediately after a conditioning VNS of 20 Hz for 3 min. After termination of the [14C]-choline infusion the atria were perfused with Tyrode's solution containing unlabelled choline

10 μM till the end of the experiment. The time schedule of these procedures and the following nerve stimulations (SNS1 to SNS3 at 10 min intervals, VNS1 simultaneously with SNS2, VNS2 10 min after SNS3) is shown by Fig. 2. When both nerves were stimulated simultaneously, the delay of the sympathetic pulses was kept constant in a single experiment but was varied from 0.3 till 283 ms (13 steps, see Table 1) in different preparations ([3H]-NA experiments). In the experiments using the HPLC method only 3, 100 and 233 ms delay times were tested (Fig.2).
The perfusates were collected in continuous 2 min samples from 2 min before SNS1 till 10 min after VNS2 was started. [3H]-NA was separated from its metabolites according to Graefe et al. (1973) using some modifications (Muscholl & Muth, 1982) which also permitted the isolation of [14C]-choline plus [14C]-ACh. The radioactivity of the effluents from the columns was measured by scintillation spectrometry in a Packard Tricarb Model 460C. For quench corrections, recovery figures, cross contaminations and spill-over of [3H] into the [14C] channel see Muscholl & Muth (1982).
The evoked release of [3H]-NA (and, in analogy, that of endogenous NA) was estimated in the following way. Resting output values were determined in the 2 min collection sample immediately before a SNS and 5-7 min after SNS3 (4 values per experiment). The resting outputs were assumed to decline linearly between actual measurements. The sum of [3H]-NA expressed as pmol per stimulation which exceeded the resting outputs calculated for the corresponding time period was considered as "stimulation-evoked overflow".
The evoked release of ACh can be estimated reasonably well (Lindmar et al.,1980; Muscholl & Muth, 1982) from the overflow of [14C]-choline plus [14C]-ACh collected during and immediately after VNS, minus the resting outputs determined as for [3H]-NA. The present method avoids cholinesterase inhibition or the addition of an inhibitor of choline uptake.
Calculations. The stimulation-evoked overflow of [3H]-NA (or endogenous NA) due to SNS2 and SNS3 was expressed as % of that caused by SNS1. As observed previously (Muscholl & Muth, 1982) and presently (see Table 1), in the control experiments (absence of VNS or drugs) the [3H]-NA overflow due to SNS2 corresponds closely to the geometric mean of the overflows measured at SNS1 and SNS3. Therefore, the "overflow expected from SNS2" was calculated from the overflows measured from SNS1 and SNS3 of the same experiment according to the term: (log % overflow at SNS1 + log % overflow at SNS3)/2. The effect of VNS carried out during SNS2 was calculated as the difference: log % overflow at SNS2 minus the expected value (see above). Thus, the effect of VNS is expressed as log value of the ratio: "overflow observed/overflow expected during SNS2".
Experiments in which the postsynaptic VNS1 effect was poor (< 1 pmol [14C]-compounds released), or the stimulated overflow of endogenous NA at SNS1 was < 4 ng, were discarded.
Means of n observations are given ± S.E. of mean. The statistical procedures were carried out according to Sachs (1984).
The sources of the radioactive compounds were NEN ([methyl-14C]-choline chloride; [3H]-NA, 3-22 Ci/mmol) and Amersham-Buchler ([3H]-NA, 13-14 Ci/mmol).

RESULTS

The effect of sympathetic nerve stimulation on [3H]-NA overflow

In the 16 control experiments the second and third nerve stimulations released 56.6 ± 2.9 and 33.1 ± 2.9 %, respectively, of the [3H]-NA that was released by SNS1. In order to account for different rates of decline in different preparations, the geometric mean of overflows evoked by SNS1 and SNS3 was regarded as the "expected value" for the overflow at SNS2 (see Methods). In the control group this value was very close to the actual overflow measured (Table 1). Therefore, in the following experiments the "expected value" served as an individual control value that was used for comparison with the overflow modified by VNS1.

Effect of vagus nerve stimulation on evoked [3H]-NA overflow

When the vagus impulses preceded those exciting the sympathetic nerves by a time interval of 3, 5.2 or 10 ms the [3H]-NA overflow was decreased but delay times of 0.3, 1.0, 1.7, 30, 100 and 167 ms were ineffective (Table 1). The peak inhibition occurred at 3 ms. However, there was a secondary peak at 233 ms. VNS at sympathetic delay times of 200-283 ms, on the whole, caused a significant inhibition of evoked [3H]-NA overflow (Table 1). In the following, the two peaks will be referred to as the "early" and the "late" peak of inhibition of NA release. The two peaks and the "valley" in between (representing a silent period in which VNS failed to elicit presynaptic effects on the adrenergic nerve terminals) might be due to corresponding fluctuations of ACh release. However, this is unlikely since in none of the individual "delay" groups the evoked overflow of [14C]-choline plus [14C]-ACh differed significantly from the common mean (analysis of variance, $p > 0.25$).
SNS accelerated rate of beat and increased tension development of the atria while VNS caused opposite responses. The postsynaptic responses to simultaneous VNS and SNS exhibited a complex interplay which could not be related to different delay times or, otherwise,be analysed in a meaningful manner.

Table 1. Inhibition by VNS of [3H]-NA overflow evoked by SNS. The sympathetic pulses were delayed for 0.3-283 ms with respect to the vagus pulses. For stimulation parameters see Fig. 2. Inhibition expressed as log (observed/expected overflow of NA). *No VNS*, controls without VNS showing that the log (observed/expected overflow of NA) did not differ significantly from zero and thus, is a good estimate of the "expected overflow".

Delay (ms)	n	log (obs./exp. overflow of NA) mean ± S.E.mean	ANOVA p values vs. *No VNS*
0.3	4	- 0.0315 ± 0.0116	n.s.
1.0	4	- 0.0381 ± 0.0109	n.s.
1.7	4	- 0.0584 ± 0.0283	n.s.
3.0	5	- 0.1222 ± 0.0111	
5.2	4	- 0.0662 ± 0.0267	< 0.005
10.0	5	- 0.0610 ± 0.0077	
30	5	- 0.0353 ± 0.0113	
100	5	- 0.0382 ± 0.0241	> 0.1
167	4	- 0.0375 ± 0.0237	
200	4	- 0.0720 ± 0.0151	
233	5	- 0.1081 ± 0.0078	
258	4	- 0.0706 ± 0.0188	< 0.005
283	4	- 0.0536 ± 0.0325	
333*)	4	- 0.0315 ± 0.0116	n.s.
No VNS	16	- 0.0046 ± 0.0077	--

*),identical with the 0.3 ms group. *ANOVA*, analysis of variance and linear contrasts according to Scheffé. Group comparisons with *No VNS* indicated by the brackets. *n.s.*, $p > 0.05$.

<u>Effects of atropine on the two inhibitory peaks</u>

The muscarine receptor antagonist was perfused from 15 min before SNS1 and maintained until the end of the experiment. The mechanical effects of SNS were unaltered as was the "expected value" of [3H]-NA overflow during SNS2 in control experiments without VNS. The cardioinhibitory action of VNS was blocked by atropine 300 nM; the log (observed/expected overflow of NA) was + 0.0046 ± 0.0101 (n=3) at the 3 ms sympathetic delay time, and - 0.0091 ± 0.0230 at the 233 ms delay (n=3), respectively. As neither value differed from zero or from the atropine controls (log value, +0.0028 ± 0.0265,n=3) the vagus-induced inhibitions of NA release, at both the early and the late peak, can be supposed to be mediated by a muscarine receptor. Concentrations of atropine lower than 300 nM (down to 1 nM) were increasingly less effective to inhibit the presynaptic action of VNS (results not shown). Finally, atropine 300 nM had no significant effect on the log (observed/expected overflow of NA) at a 100 ms delay (- 0.013 ± 0.014, n=4) thus excluding a vagally mediated, non-muscarinic facilitation of NA release which might have counteracted a single-peaked muscarinic inhibition and produced an erroneous two-peak response.

The fractional rates of overflow of [14C]-choline plus [14C]-ACh of the experiments of Table 1 (no drug) were 0.292 ± 0.005 %/min, n=57. Atropine (all concentrations and delay times) approximately doubled this value (0.543 ± 0.074 %/min, n=24, $p < 0.001$) confirming earlier evidence of a muscarinic inhibitory autoreceptor at the vagal fibres of the rabbit atria (Muscholl & Muth, 1982).

<u>Effect of vagus nerve stimulation on evoked overflow of endogenous NA and its blockade by atropine</u>

Tritium-labelled transmitter might be stored and released in a different way from endogenous NA (this refers to anatomical localizations or to compartments that may or may not be subject to presynaptic modulation). Therefore, it was investigated whether the early and the late peaks of muscarinic inhibition by VNS could be verified avoiding the procedure of labelling with [3H]-NA. Endogenous NA was determined by HPLC and electro-

chemical detection. The conditions of the earlier experiments were copied except for the [3H]-NA infusion. The protocol of the experiments is shown by Fig. 2.
Similar to the results of Table 1, VNS1 preceding SNS2 by either 3 or 233 ms significantly decreased evoked NA overflow but had no effect when the delay time was 100 ms (Fig. 3). When the results were calculated in an analogous way to Table 1, *i.e.* the log (oberved/expected overflow of endogenous NA) was

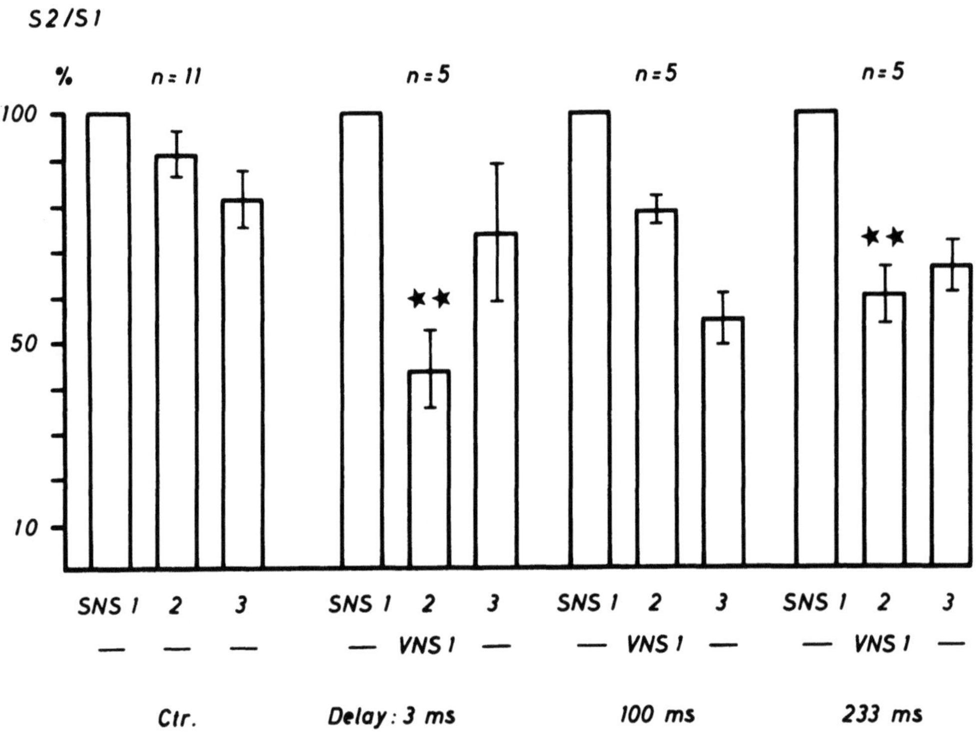

Fig. 3. Inhibition by vagus stimulation of the evoked release of endogenous NA depending on the pulse-to-pulse delay of sympathetic stimuli (3, 100 and 233 ms). Height of columns, evoked overflow of NA in response to SNS2 expressed as % of that of SNS1 (*S2/S1*), means of n observations ± S.E.M. For protocol see Fig. 2. **, Statistical difference from SNS2 of the controls (*Ctr.*, no VNS), $p < 0.01$, t test according to Dunnett.

determined, the significance levels of the 3 and 233 ms delay experiments were slightly higher than those of the S2/S1 ratios shown in Fig. 3. Likewise, the lack of statistical significance for the VNS1 effect in the 100 ms delay experiments remained true under the different conditions of calculation ($p > 0.3$).
In order to test the atropine-sensitivity of the vagally induced inhibition of evoked NA overflow the drug was perfused at 300 nM 15 min before SNS1 and maintained throughout the experiment. The S2/S1 ratios obtained for the 3, 100, and 233 ms sympathetic delay periods were 79.9 ± 9.0, 84.3 ± 7.1, and

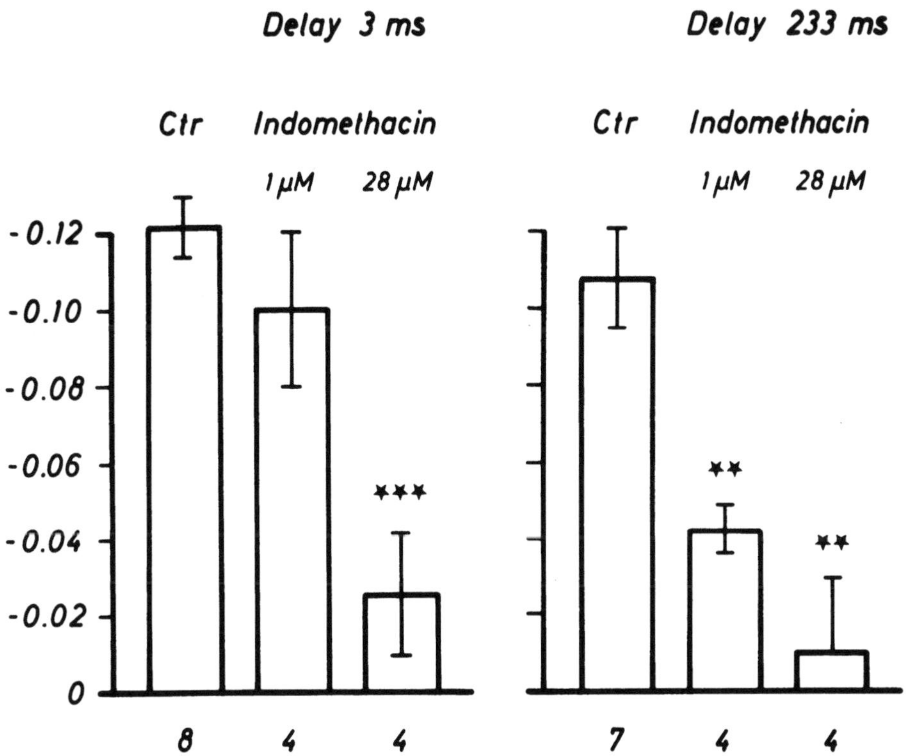

Fig. 4. Effect of indomethacin on the early and late peaks of inhibition of [3H]-NA release produced by VNS. Height of columns, log (observed/expected overflow of NA) calculated as for Table 1, ± S.E.M. n = figures at the base of columns. *Ctr*, controls, absence of indomethacin. Statistical difference from Ctr, **, $p < 0.01$; ***, $p < 0.001$, t test according to Dunnett.

84.8 ± 2.2 %, respectively (each n=3). There was no significant difference from the SNS2 control level obtained in the absence of vagus stimulation (90.8 ± 5.0%, n=11, see Fig. 3) or in a group of 3 experiments in which atropine 300 nM but no VNS was applied (86.5 ± 10.6 %).

As in the above experiments with [3H]-NA, the evoked overflow of [14C]-choline plus [14C]-ACh was similar in all groups (controls; 3, 100 and 233 ms delay) ruling out systematic differences among the individual groups of the total amount of ACh liberated which might be responsible for occurrence or lack of muscarinic inhibition of NA release. Atropine enhanced the overflow of [14C]-choline plus [14C]-ACh (9.90 ± 2.92 pmol, n=9, vs. 3.92 ± 0.63 pmol, n=15, $p < 0.02$), in agreement to the above experiments with [3H]-NA.

Effects of indomethacin on the two inhibitory peaks

For reasons to be pointed out in the DISCUSSION the late peak of inhibition of NA overflow induced by the VNS was assumed to be caused by a mediator substance liberated by activation of a muscarine receptor. Since one possible candidate for such an action may be a prostaglandin, it was tested whether indomethacin, an inhibitor of cyclo-oxygenase, had a differential effect on the late as opposed to the early peak.

The drug was perfused from 10 min before SNS1 till the end of an experiment. The [3H]-NA method was used to estimate the release of the adrenergic transmitter. As shown by Fig.4 a large concentration of indomethacin (28 µM) blocked the inhibitory action of VNS on [3H]-NA overflow at the 3 and 233 ms sympathetic delay times but 1 µM had a differential effect. The VNS response at 233 ms was decreased by about 60% ($p < 0.001$) whereas the response at 3 ms was not significantly affected ($p > 0.2$).

DISCUSSION

The aim of the present study was to examine the presynaptic modulation of NA release by vagus stimuli preceding the sympathetic impulses by a defined time interval. For this type of work a double innervated organ is required on which the two branches of the autonomic nervous system can be stimulated independently of each other. Several investigations carried out on the heart have shown that cholinergic-adrenergic interactions do exist (see INTRODUCTION). The presence of both pre- and postsynaptic muscarine receptors in the heart and the lack of receptor antagonists which are able to selectively inhibit presynaptic muscarine receptors exclude an indirect approach (which might be technically easier), namely using mechanical responses such as rate or tension development as indices of an altered transmitter release.

Therefore, a methodology was developed to estimate the rates of overflow of NA and ACh evoked by electrical stimulation on the perfused atria preparation. Perfusion of the tissue was preferred to incubation because rapid changes of transmitter release are more easily detected, autoinhibition, metabolism and re-uptake of NA are less likely to occur (Fuder et al., 1984) and the equilibration of drugs is quicker than in a superfused tissue. The following qualities of the preparation have been established (Muscholl & Muth, 1982): There is no NA release during VNS and no ACh release during SNS, the overflows of both transmitters are fully dependent on calcium ions and blocked by tetrodotoxin.

The main result of this study is the observation that vagus pulses significantly inhibit NA release when they are applied either 3 or 233 ms before the sympathetic stimuli. Since the vagus electrode excites preganglionic, and the sympathetic electrode postganglionic, nerve fibres, the corresponding intervals of the action potentials arriving at the terminal vagus and sympathetic fibres will be different from the intervals of stimuli *applied*. Taking into account the conduction velocities and length of nerve fibres, a time of 17 ms has been estimated by which the vagus pulse arrives earlier at the terminal than

the pulse elicited by SNS (calculation to be published elsewhere). Thus, the early peak of inhibition of NA release occurs when the vagus terminals are excited 20ms, and the late peak, when they are excited 250 ms before the adrenergic terminals. Nevertheless, the 20 ms time elapsing from the action potential initiating the release of ACh till a muscarinic peak response (modulation of exocytotic NA overflow) is extremely short, compared with a postsynaptic myocardial effect (Pott, 1979).
The causal role of ACh released by VNS for the modulation of NA release, and the mediation of the two inhibitory responses by a muscarine receptor is indicated by the concentration-dependent and full blockade of the effects by atropine; this excludes also a mediator role of an unknown transmitter co-released with ACh, an *electrotonic* interaction between adjacent cholinergic and adrenergic fibres, or an early nicotinic and late muscarinic response.
Apparently, ACh once released has two muscarinic actions differing in latency. There is the possibility of the early response being mediated by an axo-axonic synaptic transmission and the late one by a "lateral inhibitory" effect. This is, however, unlikely as physostigmine, a cholinesterase inhibitor which increases the "diffusion radius" of ACh, does <u>not</u> enhance the presynaptic effect of VNS (Muscholl & Muth, 1982), and the 250 ms delay time is too long to be explained by a diffusionial process. Instead, an alternative working hypothesis is proposed. The early peak may be caused by a channel-operated mechanism (potassium channel opening leading to hyperpolarization of the axonal membrane or calcium channel closure reducing the influx of the ion) that requires little time to be activated, and the late one by a mediator substance which is liberated by the muscarinic stimulus. On the rabbit heart prostaglandins of the E series have been shown to be released by VNS, and they are known to inhibit NA overflow (Junstad & Wennmalm, 1974). The preliminary observations of the discriminating effect of indomethacin 1 μM (Fig. 4) lend support to the above hypothesis but further evidence is required.

ACKNOWLEDGEMENTS. This work was supported by the Deutsche Forschungsgemeinschaft (Mu 118/15-3; 118/13).

REFERENCES

Fuder H, Bath F, Wiebelt H, Muscholl E (1984) Autoinhibition of noradrenaline release from the rat heart as a function of the biophase concentration.Effects of exogenous α-adrenoceptor agonists, cocaine, and perfusion rate. Naunyn-Schmiedebergs Arch Pharmacol 325:25-33

Graefe KH, Stefano FJE, Langer SZ (1973) Preferential metabolism of (-)-3H-norepinephrine through the deaminated glycol in the rat vas deferens. Biochem Pharmacol 22:1147-1160

Jung B, Muscholl E (1987) Influence of the time interval between vagus and sympathetetic nerve stimulations on the inhibition of endogenous noradrenaline release from perfused rabbit atria. Naunyn-Schmiedebergs Arch Pharmacol 335:Suppl R 83

Junstad M, Wennmalm A (1974) Release of prostaglandin from the rabbit isolated heart following vagal nerve stimulation or acetylcholine infusion. Br J Pharmacol 52:375-379

Lavallée M, De Champlain J, Nadeau RA, Yamaguchi N (1978) Muscarinic inhibition of endogenous myocardial catecholamine liberation in the dog. Canad J Physiol Pharmacol 56:642-649

Levy MN, Blattberg B (1976) The effect of vagal stimulation on the overflow of norepinephrine into the coronary sinus during cardiac sympathetic nerve stimulation in the dog. Circ Res 38:81-85

Lindmar R, Löffelholz K, Weide W, Witzke J (1980) Neuronal uptake of choline following release of acetylcholine in the perfused heart. J Pharmacol Exp Ther:215,710-715

Löffelholz K, Muscholl E (1970) Inhibition by parasympathetic nerve stimulation of the release of the adrenergic transmitter. Naunyn-Schmiedebergs Arch Pharmacol 267:181-184

Muscholl E (1980) Peripheral muscarinic control of norepinephrine release in the cardiovascular system. Am J Physiol 239:H713-H720

Muscholl E, Habermeier-Muth A (1985) Presynaptic inhibition by vagus nerve stimulation of noradrenaline release from the perfused rabbit atria preparation: muscarinic responses with short and long latencies. Naunyn-Schmiedebergs Arch Pharmacol 329:Suppl R 76

Muscholl E, Muth A (1982) The effect of physostigmine on the vagally induced muscarinic inhibition of noradrenaline release from the isolated perfused rabbit atria. Naunyn-Schmiedebergs Arch Pharmacol 320:160-169

Pott L (1979) On the time course of the acetylcholine-induced hyperpolarization in quiescent guinea-pig atria. Pflüg Arch 380:71-77

Racké K, Muscholl E (1986) Release of endogenous 3,4-dihydroxyphenylethylamine and its metabolites from the isolated neurointermediate lobe of the rat pituitary gland. Effects of electrical stimulation and of inhibition of monoamine oxidase and reuptake. J Neurochem 46:745-752

Sachs L (1984) Angewandte Statistik, 6th ed.Springer Berlin Heidelberg New York Tokyo

Vanhoutte PM, Levy MN (1980) Prejunctional cholinergic modulation of adrenergic neurotransmission in the cardiovascular system. Am J Physiol 238: H275-H281

NEW ASPECTS ON MODULATION OF SYMPATHETIC NEUROTRANSMISSION: BY CHANGE OF PROBABILITY OF SECRETION OF SINGLE MIXED QUANTA FROM TWO CLASSES OF NERVE VARICOSITIES

Lennart Stjärne[1], Etsuro Tanaka[1] and Peter Århem[2]
Departments of Physiology[1] and Neurophysiology[2]
Karolinska Institutet, S-10401 Stockholm
Sweden

INTRODUCTION

Exciting new findings in many areas force us to change radically our views on sympathetic neurotransmission. The purpose of this short review is to suggest some features of an emerging **new working model**, and to discuss in that context the basic secretory characteristics of individual nerve varicosities, and the transmission of impulses in some sympathetic neuro-effector junctions, and its pre- and postjunctional modulation.

NEW MODEL: SYMPATHETIC NERVES UTILIZE MULTIPLE TRANSMITTERS

Noradrenaline (NA) is not the exclusive mediator of all effects of sympathetic nerve stimulation (Burnstock 1976, 1986, Hökfelt et al 1986, Lundberg and Hökfelt 1986), as earlier thought (Fig. 1A). Individual sympathetic neurons in culture are able to express up to a dozen different putative messenger substances (Potter et al 1986), and in vivo also may store three classes of messengers (Fig. 1B): (**a**) A 'classical' transmitter, usually NA but occasionally acetylcholine, (**b**) a nucleotide, presumably ATP, and (**c**) one or several neuropeptides, for example, neuropeptide Y (NPY). These substances are stored in different combinations in two classes of transmitter organelles, the 'small' and the 'large' dense cored vesicles (SDVs and LDVs, respectively, Fig. 1B, Thuresson-Klein 1983). The varicosities seem to be able to 'choose' between secreting the contents of a SDV or a LDV, and hence to vary the composition of the 'transmitter cocktail' released, according to the nerve impulse pattern and the degree of activation of prejunctional autoreceptors (Lundberg and Hökfelt 1986).

The fact that many nerves utilize several transmitters seems to make the **'-ergic'** classification of neurons according to their first discovered transmitter (Dale 1934)

NATO ASI Series, Vol. H19
Modulation of Synaptic Transmission and Plasticity
in Nervous Systems. Edited by G. Hertting and H.-C. Spatz

obsolete. Many effects of sympathetic nerve activity are mediated by NA, but other effects by its cotransmitters (Fig. 1C). We think therefore it is misleading to continue to call a neuron '**noradrenergic**' just because it contains NA, or exerts some of its effects through NA (Stjärne and Åstrand 1985a, Stjärne 1986a,b, Stjärne and Lundberg 1986, Stjärne et al 1986).

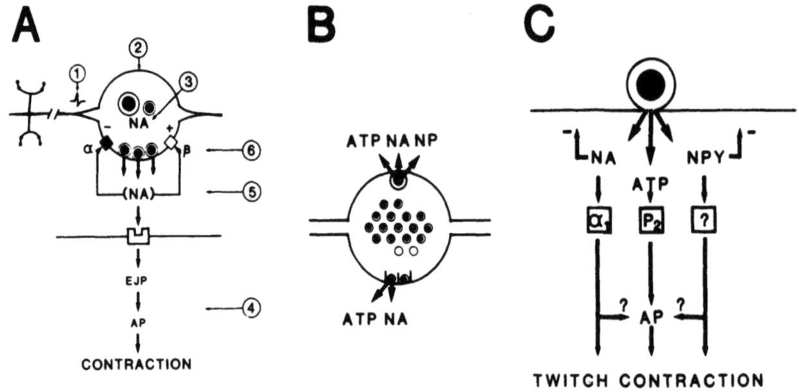

Fig. 1. (**A**) Six crucial assumptions in the **current model** ('**CM 1-6**') of sympathetic neurotransmission, and criticism: **CM1**: Every nerve impulse invades all varicosities and releases from each a graded number of NA quanta. **Criticism**: Direct evidence that **NA** is secreted in quanta is lacking. Although each varicosity may be invaded by every nerve impulse (Brock and Cunnane 1987), it responds extremely rarely and may secrete only single quanta (of **NA** and **ATP**; Cunnane and Stjärne 1984, Stjärne and Åstrand 1984). **CM2**: Sympathetic nerve varicosities form a single homogeneous population; all of them are junctionally related to effector cells. **Criticism**: No, most of them are widely separated from effector cells, and in some tissues they form **two** anatomically as well as functionally distinctly different populations (Bennett 1973). **CM3**: NA is the only sympathetic neurotransmitter. **Criticism**: No, NA is only one of several messengers in these nerves (see the text; Burnstock 1986). **CM4**: NA induces muscle contraction by causing excitatory junction potentials (EJPs) and muscle action potentials (APs). **Criticism**: NA does cause a slow depolarization in some tissues (Cheung 1982, Suzuki 1983), but never 'fast' EJPs, and in many tissues acts 'metabotropically', i.e., without depolarizing the cells at all (Bolton and Large 1986). **CM5**: NA accumulates **intrajunctionally**; the NA concentration in this biophase determines the magnitude of the pre- and postjunctional effects of sympathetic nerve stimulation (Bevan et al 1980). **Criticism**: That is probably not true even of NA. Since nerve impulses rarely activate the individual varicosity, neither NA nor any of its cotransmitter accumulate within individual neuro-effector gaps, by repeated release (Bevan et al 1984, Stjärne 1986a,b). **CM6**: The **intrajunctional** NA concentration is autoregulated by negative and positive feedback, mediated via prejunctional α_2- and β_2-adrenoceptors, respectively; the control via these autoreceptors operates directly on the varicosity from which the release occurred (Langer 1977, 1981, Starke 1977, 1981, Westfall 1977). **Criticism**: That is probably incorrect; autoregulation by prejunctional autoreceptors does not control ('stabilize') the NA concentration in individual junctional gaps but in an '**extrajunctional** biophase'. Further, autoinhibition does not operate on an impulse by impulse basis directly on the varicosity from which the transmitter was released (Cunnane and Stjärne 1984, Stjärne 1985).
(**B**) **In the new view sympathetic nerves store and release multiple transmitters**: The average sympathetic nerve varicosity may contain some 500

transmitter vesicles; there is a wide variation in this regard even in the same tissue (Hökfelt 1969). In small animals varicosities have a low proportion of LDVs (less than 5%); in large animals and in man LDVs may represent up to 30% of all vesicles (Thuresson-Klein 1983). NA is synthesized only in LDVs, which leak it into the cytosol from which SDVs take up the newly formed NA. The internal diameter of a LDV is about twice that of a SDV; hence the matrix volume (and storage capacity?) of a LDV may be about 8-fold larger than that of a SDV (Klein 1982). In some but not all sympathetic nerves LDVs in addition contain a neuropeptide (NP, Klein 1982, Lundberg and Hökfelt 1986). Sympathetic nerve varicosities lack well-defined preformed 'active zones' but may possess functional **'preferred release sites'** (Cunnane and Stjärne 1984). The varicosity can 'determine' the composition of the 'transmitter cocktail' it secretes: Impulses at low frequency tend to release the NA and ATP contents of SDVs, possibly from 'preferred release sites', but 'bursts' at high frequency, and block of prejunctional α_2-adrenoceptors, promote preferential release of NA, ATP and NP from LDVs (Lundberg and Hökfelt 1986), perhaps from **'random sites'** on the varicosity membrane (Zhu et al 1986).

(C) **In the new view sympathetic transmission involves interaction between several transmitters** (Burnstock 1986, Changeux 1986, Lundberg and Hökfelt 1986). The relative pre- and postjunctional importance of the different sympathetic cotransmitters varies in different tissues and species (Stjärne and Åstrand 1985a, Burnstock 1986, Stjärne and Lundberg 1986). (**a**) **NA** is mainly **'metabotropic'** (Eccles 1986), i.e. often exerts its pre- and postjunctional effects without depolarizing the target cells (Bolton and Large 1986). However, there is a variation in this regard; in some tissues NA causes a slow depolarization of effector cells and triggers muscle action potentials (Cheung 1982, Suzuki 1983). (**b**) **ATP** is the only truly **'ionotropic'** messenger in sympathetic nerves, i.e. the only one which causes rapid depolarization (EJPs) and action potentials (AP) in effector cells (Fedan et al 1981, Sneddon and Westfall 1984, Burnstock 1986). ATP acts mainly postjunctionally, via P_2-purine receptors and is not, or is only weakly autoinhibitory (Stjärne and Åstrand 1985a), but after degradation to adenosine may contribute to the prejunctional inhibition, via P_1-purine receptors (Burnstock 1986). The role of ATP as a mediator of the contractile response (i.e., of the **'electrical link'** in sympathetic neuromuscular transmission) is highly variable. In some tissues pharmacological 'elimination' of ATP-induced EJPs and muscle APs does not much reduce the contractile response to nerve stimulation; in other tissues the contractions become severely depressed (Stjärne and Åstrand 1985a). (**c**) The role played by **neuropeptides** is difficult to assess, since pharmacological blocking agents are lacking. However, **NPY** for example, seems to play a substantial role in sympathetic neurotransmission in some but not all tissues (Lundberg and Hökfelt 1986). In some tissues NPY causes a substantial depolarization; in others it acts without depolarizing target cells, i.e., 'metabotropically' (Neild 1987). As a sympathetic cotransmitter **endogenous** NPY would have long-lasting inhibitory or excitatory effects, judging from the action of the exogenous peptide. In some tissues the inhibitory prejunctional effects dominate over the excitatory postjunctional ones; here NPY would act as a **brake** on sympathetic neurotransmission. In other tissues the reverse is true; here NPY as a sympathetic cotransmitter would **amplify** the effects of sympathetic nerve stimulation, as well as of circulating NA (Lundberg and Hökfelt 1986, Stjärne and Lundberg 1986, Stjärne et al 1986, Neild 1987).

NEW MODEL: SYMPATHETIC NERVE VARICOSITIES OPERATE BOTH IN DIGITAL AND ANALOGUE MODE

In many neurobiological systems each individual release site (the bouton, or presynaptic grid, or active zone) has been reported to operate as an independent unit, activated intermittently and each time releasing a single quantum equal to the contents of one transmitter vesicle (for references, see Korn 1984), i.e. to function in a **digital** mode. As explained below, the same principle seems to apply to sympathetic nerves also.

(a) **Basically, sympathetic nerve varicosities function in digital mode**: This issue has been studied in detail in certain model tissues (e.g., the guinea-pig or mouse vas deferens). Here the smooth muscle cells are innervated by two, more or less distinctly separate, populations of varicosities (Fig. 2A). In a small subpopulation the varicosities (here termed **'junctional'**) lie very close to the smooth muscle cell membrane (gap width: less than 20 nm); the majority of varicosities (here termed **'nonjunctional'**) are widely separated from the effector cells (Burnstock and Costa 1975). Spontaneous release of single quanta of transmitter causes a phasic depolarization of smooth muscle cells, which can be recorded by an intracellular microelectrode as an excitatory junction potential (EJP, Fig. 2B). The transmitter causing this response is not NA, as earlier thought (Burnstock and Holman 1966), but ATP (Sneddon and Burnstock 1984, Sneddon and Westfall 1984) secreted from a 'junctional' varicosity as one component of the mixed transmitter quantum in a SDV (Fig. 1B; Stjärne and Åstrand 1984). Nerve impulses cause two classes of EJPs (see Fig. 2B): (1) The **'slow'** EJP: Each suprathreshold stimulus causes an EJP characterized by a slow time course (time to peak about 100 ms, total duration about 500 ms) and a relatively constant amplitude (after a period of facilitation during the first few stimuli). (2) The **'compound'** EJP: This response consists of a **'fast'** component superimposed on a 'slow' EJP; it is characterized by intermittency, fast time course (time to peak about 20 ms) and large amplitude variation (Bennett 1973, Stjärne and Åstrand 1984). The 'fast' EJP has a time course similar to that of spontaneous EJPs (Fig. 2B); both may therefore reflect release of single quanta from an **individual** 'junctional' varicosity. In contrast, the 'slow' EJP is thought to be caused by single (?) quanta from many 'nonjunctional' varicosities, i.e., a **'population response'** (Bennett 1973).

The 'compound' EJPs (or rather, the phasic peaks in the first time differential of their rising phase, termed the 'discrete events') have been used to study the secretory activity of individual ('junctional') sympathetic nerve varicosities in the guinea-pig and mouse vas deferens. Based on analysis by this method it has been proposed that the individual varicosity often fails to secrete transmitter in response to a nerve stimulus and that release occurs in quanta probably equal to the transmitter contents of a single vesicle (Blakeley and Cunnane 1979). During low frequency stimulation the average individual varicosity seems to be activated by only 0.2-3 % of the nerve impulses, each time releasing only a single quantum, the ATP contents of a SDV (Cunnane 1984, Cunnane and Stjärne 1984, Stjärne and Åstrand 1984; for a different view, see Blakeley et al 1984).

The electrophysiological method gives **direct** information concerning the secretion of the ATP component of sympathetic transmitter quanta from **'junctional'** varicosities, i.e. a small subpopulation of all sympathetic nerve varicosities. However, indirect evidence by study of the overflow of NA, which reflects the secretory behaviour of the entire varicosity population, are fully compatible with the possibility that all sympathetic nerve varicosities are intermittently activated (by about 1% of the nerve impulses) and then release single quanta composed of the NA (and ATP) content of an SDV (Folkow et al 1967, Folkow and Häggendal 1970, Stjärne 1985).

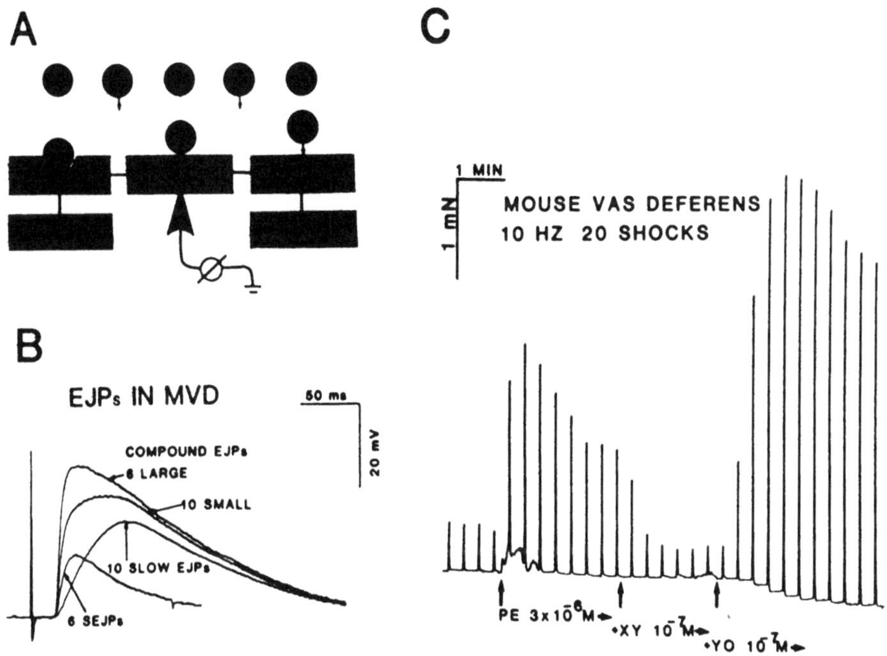

Fig. 2. (**A**) Cartoon (drawn approximately to scale) of a cross section through the mouse vas deferens, showing six electrically coupled smooth muscle cells and their innervation. Each cell is coupled to perhaps 15 neighbouring cells (Furness 1970) and innervated by 1-5 **'junctional'** and some 80 **'nonjunctional'** varicosities (Bennett 1973).Therefore an intracellular microelectrode may detect transmitter quanta from up to 75 'junctional' and 1200 'nonjunctional' varicosities (Stjärne 1987).

(**B**) **Averaged** records of 6 spontaneous EJPs (SEJPs), 10 evoked slow EJPs and 10 of the smaller and 6 of the larger evoked compound EJPs, in a single cell in the mouse vas deferens (**MVD**, recording by E. Stjärne). SEJPs are caused by spontaneous secretion of single quanta (Bennett 1973) of ATP (Stjärne and Åstrand 1984) from 'junctional' varicosities; note that they may cause a large depolarization (up to 20-30 mV or more, cf. Burnstock and Holman 1966, Bennett 1973). Nerve stimulation evokes two classes of stimulus-locked responses, the **slow** and the **compound** EJPs (Bennett 1973). All EJPs are abolished by desensitization of P_2-purine receptors by α-β-methylene ATP; hence they are all caused by the same transmitter, presumably ATP (Sneddon and Burnstock 1984, Sneddon and Westfall 1984, Stjärne and Åstrand 1984). **Slow EJPs** characteristically are non-intermittent and (after the first 6-10 shocks in a train) remarkably constant in amplitude (Stjärne and Åstrand 1984), presumably because they represent a 'population response', i.e., the release of ATP quanta from a relatively constant fraction (about 1%) of a large population of 'nonjunctional' varicosities. **Compound EJPs** occur only intermittently and vary markedly in amplitude (since the figure shows the **averages** of a selection of large and small compound EJPs, the amplitude variability of individual responses cannot be seen). The rising phase of compound EJPs is similar to that of the spontaneous EJPs; hence the compound EJP represents a **fast** EJP superimposed on a **slow** EJP. The fast component presumably represents the effect of single transmitter quanta from a single, or a few, 'junctional' varicosities (Bennett 1973).

(**C**) Effects of bath application of the α_1-adrenoceptor agonist phenylephrine ('**PE**'), alone or in combination with the α_2-adrenoceptor agonist xylazine ('**XY**') on the twitch contractions of the mouse vas deferens to electrical field stimulation (via platinum ring electrodes around the prostatic end, 10 Hz trains of 20 shocks, 90V/2 ms). Phenylephrine caused a small and transient increase in the resting tension and a large and long-lasting increase in the twitch contraction to nerve stimulation. These effects were not caused by an increase in transmitter secretion. They were blocked by the α_1-adrenoceptor antagonist prazosine (10^{-7}M, not shown). Further addition of xylazine inhibited the twitch contractions to 20 shock trains; this effect was abolished when the train length was increased to 50 shocks (not shown). The inhibitory effect was parallelled by depression of NA secretion and completely abolished by the α_2-adrenoceptor antagonist yohimbine ('**YO**').

As shown by extracellular recording of the **'nerve terminal spike'** (NTS) and of the **excitatory junction current** (EJC) in these tissues, the main reason for the extreme degree of secretory intermittency does not seem to be failure of the nerve impulse to invade the varicosity, but failure of its secretory mechanisms to respond to the nerve impulse (Brock and Cunnane 1987, see Fig. 3 and discussion below).

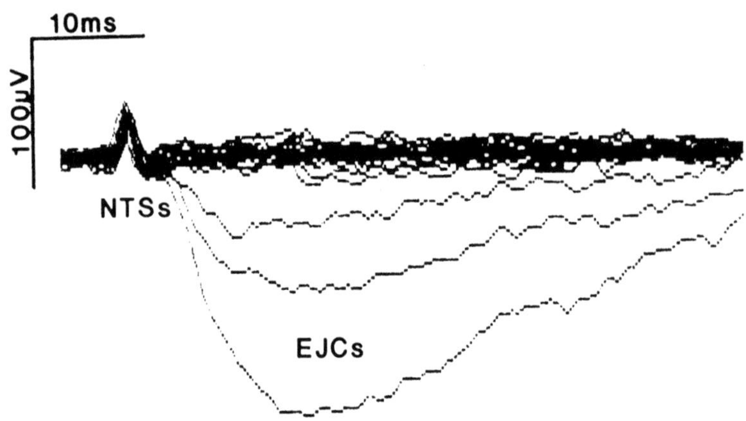

Fig. 3. Twenty superimposed traces showing the relationship between the **excitatory junction currents** ('**EJCs**'), presumably caused by single quanta of ATP, and the **nerve terminal spikes** ('**NTSs**') by which the release may have been triggered (cf. Brock and Cunnane 1987). Extracellular recording in the guinea-pig vas deferens (by E. Stjärne). Internal tip diameter of the recording microelectrode: 60 µm. Suprathreshold focal electrical stimulation of the postganglionic sympathetic nerves, with 20 shocks at 1 Hz. Shown are 3 evoked EJCs of different amplitude (representing single quanta from 3 different varicosities?) and 17 'failures'. Note that NTSs were closely similar whether or not they were followed by an EJC, suggesting that failure of invasion was not the cause of the frequent secretory failure in individual varicosities.

Note of caution: The proof is conclusive only if the recorded pattern reflects activity in a **single** fibre. However, the number of nerve fibres and varicosities in the membrane patch enclosed by the microelectrode is not known. Hence it cannot be totally excluded that the NTSs were caused by activity in more than one nerve fibre. Further refinement of the method is required to prove conclusively that the observed extremely low probability of transmitter secretion from individual varicosities is not due to failure of invasion by the nerve impulse.

(b) **Both digital and analogue local modulation of transmitter secretion from sympathetic nerve varicosities**: Since (at least at low impulse frequency) individual varicosities seem to represent 'monoquantal' units, alternating between failure and release of single SDV quanta, the local modulation of transmitter secretion from these nerves (Kirpekar 1975, Stjärne 1975, 1981, Langer 1977, 1981, Starke 1977, 1981, 1987, Westfall 1977, Vizi 1979) also may operate basically in a **'digital'** mode, by altering the **probability** that the nerve impulse will release a single SDV quantum from the individual varicosity (Cunnane and Stjärne 1984, Stjärne 1985), not the **number** of quanta released from the individual varicosity (Blakeley et al 1986, for a discussion, see Starke 1987).

However, as mentioned above modulation of transmitter secretion from sympathetic nerves also operates in an **'analogue'** mode: A change in the nerve impulse pattern and/or activation of prejunctional adrenoceptors alters the **composition of the 'transmitter cocktail'** released, presumably by influencing the selective secretion of the contents of a SDV or a LDV (Lundberg and Hökfelt 1986, Stjärne and Lundberg 1986).

NEW VIEW: TWO BIOPHASES AND TWO MODES OF TRANSMITTER ACTION

A. Two biophases: Current models of sympathetic neurotransmission (Fig 1A) mostly are illustrated by diagrams of events at **the** varicosity; i.e. often (unintentionally?) give the impression that all varicosities behave uniformly and that every nerve impulse causes each of them to secrete a graded number of transmitter quanta into a well-defined and narrow junctional gap (Langer 1977, 1981, Starke 1977, 1981, Westfall 1977, Bevan et al 1980, Blakeley et al 1984). As explained above, this seems to be highly misleading (see Fig. 2A). Each nerve impulse appears to activate only a small fraction (about 1%) of all varicosities, basically releasing single quanta (Cunnane and Stjärne 1984, Stjärne 1985). Further, only the few 'junctional' varicosities secrete the transmitter into narrow neuro-effector gaps, i.e., into an **intrajunctional biophase**; the transmitters from 'nonjunctional' varicosities, which make up the vast majority of sympathetic nerve varicosities in all tissues (Burnstock and Costa 1975), are secreted into open tissue space and hence exert their effects in a more or less homogeneous **extrajunctional biophase**.

B. Transmitters may act either intra- or extrajunctionally: The intra- and extrajunctional actions of **ATP** as a sympathetic cotransmitter may be distinguished by intracellular recording in single smooth muscle cells (in the guinea-pig or mouse vas deferens, Fig. 2B). The **'fast'** (spontaneous or stimulus-evoked) EJPs are thought to be caused by single transmitter (ATP) quanta from 'junctional' varicosities, acting **focally** on receptors in narrow neuromuscular gaps. In contrast, the **'slow'** EJPs are thought to be caused by transmitter (ATP) in quanta from many 'nonjunctional' varicosities, acting **diffusely** on receptors located all over the surface of smooth muscle cells (Bennett 1973, Stjärne and Åstrand 1984)

For the other sympathetic transmitters it has not been possible as yet to distinguish between the intra- and extrajunctional actions. However, for anatomical reasons all transmitters from the two classes of varicosities must be assumed to act in two different 'biophases', probably according to two opposite principles. **Intrajunctionally**, i.e., in the narrow gap between a 'junctional' varicosity and its target cell (width less than half the diameter of a SDV) the local concentration of transmitters initially presumably is equal to that inside the SDV; for NA it must be in the millimolar range (Stjärne 1975, Klein 1982). In spite of small fluctuations in the number of transmitter molecules released from a SDV, the transient initial peak concentration of transmitters in the gap may always be sufficient to saturate the available receptors, i.e. 'locally supramaximal', and hence cause a constant, **all-or-none** response. On this assumption 'identical' (intracellularly recorded) 'discrete events', or (extracellularly recorded) EJCs have been used, tentatively, as 'fingerprints' of the secretory activity in individual 'junctional' varicosities (Cunnane and Stjärne 1984, Brock and Cunnane 1987). In contrast, transmitters from 'nonjunctional' varicosities presumably exert their effects in a more or less homogeneous **extrajunctional** biophase, diffusing considerable distances in the extracellular space and mixing with transmitters from neighbouring varicosities. In this case the effective concentration of transmitters may be in the nano- (neuropeptides) to micromolar (NA) range, and the biological effects are **graded** in proportion to the degree of accumulation of the respective transmitter in this space (Schipper et al 1980).

C. Geometry determines if a messenger will act as 'trigger' or 'modulator': Currently neuronal messengers often are classified either as 'true transmitters' (e.g., NA, acetylcholine) or as 'modulators' (e.g., neuropeptides). However, such a classification is misleading. As shown by results of studies e.g., of the mouse vas deferens, one and the same substance may act either as a 'phasic transmitter', or as a pre- and/or postjunctional 'modulator', depending on the **geometry** of a neuro-effector junction.

In this tissue the rapid twitch contraction is caused by the combined effects of NA and ATP (Stjärne and Åstrand 1985a). Our hypothesis is that both, when released from 'junctional' varicosities, have a **triggering** function, i.e. induce rapid, phasic motor responses, and when released from 'nonjunctional' varicosities have a dual effect, namely (1) a **direct** effect on the target cells, causing a slow contraction, and (2) an **indirect** effect, by **modulation** of the pre- and postjunctional excitability of the neuro-effector junction. The argument will now be applied briefly to the three known 'putative transmitters' in these nerves, NA, ATP and neuropeptide Y (NPY).

(a) **ATP**: When released from the numerous 'nonjunctional' varicosities ATP is believed to cause a 'subthreshold' depolarization, the 'slow' EJP (Fig. 2B, Bennett 1973). This implies a transient increase in the **excitability** of the smooth muscle cells to all stimuli, i.e. a postjunctional **'modulatory'** action. In contrast, ATP in single quanta from individual 'junctional' varicosities causes a rapid and large local depolarization, the 'fast' component of the compound EJP. Against the background of increased excitability this may lead to generation of a muscle action potential, which is how ATP contributes to the twitch contraction.

(b) **NA**: The intrajunctional effects of the NA component of quanta from 'junctional' varicosities are difficult to analyze. However, the extrajunctional 'modulatory' pre- and postjunctional effects presumably exerted by endogenous NA in the low concentrations in which it occurs in the unspecialized extracellular space can be simulated by bath application of selective α_1– or α_2-agonists at concentrations too low to affect the smooth muscle cells directly (Fig. 2C, the figure legend describes the effects in more detail). Two opposite effects are seen: (1) A **potentiating action**: At concentrations too low to increase the resting tension NA may sensitize the smooth muscle cells to the transmitters released by subsequent nerve impulses (without *per se* increasing the amount released). This effect is blocked by prazosine and hence exerted via postjunctional α_1-adrenoceptors. (2) An **inhibitory action**: Via prejunctional α_2-adrenoceptors extrajunctional NA depresses further transmitter secretion, and hence neurotransmission. The net effect may vary with the tissue; in the mouse vas deferens the depressant prejunctional effect dominates, at low nerve impulse frequency. However, it is strongly counteracted by the facilitating effect of nerve impulses at high frequency; for example, the inhibitory effect of xylazine shown in Fig. 2C was abolished when the train length was increased to 40-50 shocks (not shown).

(c) **NPY**: Exogenous NPY also may have amplifying postjunctional and depressing prejunctional effects; the net effect varies with the tissue. In the mouse vas deferens the inhibitory prejunctional action is predominant and NPY depresses neurotransmission (Stjärne and Lundberg 1986, Stjärne et al 1986). However, e.g., in rat tail artery, NPY has practically no prejunctional inhibitory effects, but strongly contracts the smooth muscle by a direct effect, and also greatly amplifies the contractile response to nerve stimulation, as well as to exogenous NA (Neild 1987).

D. Conclusions: The **microanatomical physiology** of sympathetic neurotransmission is as yet poorly understood, and apparently varies from one tissue to the next. In a model tissue such as the mouse vas deferens neuromuscular transmission seems to require the following model: (1) All varicosities are rarely activated and (at least at low impulse frequency) each time release single mixed SDV quanta. (2) The majority of varicosities are widely separated from the target cells and transmitters from these varicosities act **extrajunctionally**, in two different ways: (**a**) At 'suprathreshold' concentrations they cause relatively sustained motor responses in the target cells (**direct** effect). (**b**) At much lower concentrations they 'set' the pre- and postjunctional excitability of the junction to other stimuli (**modulatory** effect). (3) Transmitters from the minority of varicosities in intimate contact with the target cells act **intrajunctionally**. When the excitability of the system is increased by the mechanism described above (2**b**) quanta from these varicosities may **trigger** rapid and transient motor responses (Stjärne 1987).

NEW MODEL: MECHANISMS UNDERLYING RESTRICTIVITY AND PLASTICITY OF TRANSMITTER SECRETION FROM SYMPATHETIC NERVE VARICOSITIES

As described above all nerve impulses may invade all varicosities (Brock and Cunnane 1987), but still the individual nerve impulse only rarely releases transmitter from a varicosity (Cunnane and Stjärne 1984). Sympathetic nerve terminals thus seem to have a **high safety factor for invasion**; and yet a **low probability for transmitter secretion**.

A. How is the invasion of varicosities ensured? The possibility has been considered that each nerve impulse may actively invade only part of the terminals of sympathetic nerve, and that individual varicosities are depolarized relatively rarely, and passively (Haefely 1972, Stjärne 1978). However, two lines of recent evidence favour the opposite alternative: (1) **Pharmacological evidence**: Veratridine, a Na^+ channel activating agent, which depolarizes e.g., rat brain synaptosomes by a tetrodotoxin-sensitive mechanism, causing Ca^{2+} influx and transmitter release (Blaustein et al 1981), also induces a tetrodotoxin-sensitive, in part Ca^{2+}-dependent, NA release from sympathetic nerves (Thoa et al 1973, Stjärne 1979). Hence sympathetic nerve varicosities, similarly to those in other 'boutons-en-passant' terminals, presumably possess voltage-gated Na^+ channels. (2) **Functional evidence**: The sympathetic 'nerve terminal spikes' recorded by extracellular electrodes, e.g., in the guinea-pig vas deferens (Fig. 3) are blocked by tetrodotoxin; hence even in the terminal region the nerve impulses depend mainly or exclusively on regenerative influx of Na^+ rather than of Ca^{2+} ions (Brock and Cunnane 1987). It seems likely therefore that impulses in sympathetic nerve terminals are actively conducted as regenerative 'Na^+ action potentials', possibly to the very end of each branch.

Interestingly, this does not seem to be the case in terminals of neurons in which the secretory area is restricted to the most distal region of each branch. Thus in 'cholinergic' motor terminals in striated muscle in the rat (Brigant and Mallart 1982), or frog (Mallart 1984) Na^+ channels have been reported to be lacking entirely, or at least to be absent except in the most proximal part of the terminals. Further, Na^+ channels do not seem to occur in 'synaptosomes' of the 'cholinergic' motor terminals innervating the electric organ of *Torpedo* (Dunant and Muller 1986). In these systems transmitter secretion presumably is triggered by impulses conducted electrotonically the short distance separating the release site from the site of action potential failure. Perhaps active impulses conduction by voltage gated Na^+ channels, into the release sites, occurs mainly or exclusively in **'boutons-en-passant'** terminals, in which such a mechanism is required to ensure that the nerve impulse will propagate beyond the most proximal site of release and reach the outermost end of each string of potentially secretory varicosities.

B. How is the secretory probability maintained at the 'desired' low level? Each nerve impulse should release enough transmitter to carry the signal to the effector cells, and 'plasticity' in the release mechanisms should allow some local prejunctional regulation of neurotransmission, but since excessive release would rapidly deplete the transmitter store, the release caused by individual nerve impulses should be severely restricted. The majority of the nerve impulses passing through a varicosity should do so without effectively challenging its release mechanisms. How is this balance between the short- and long-term interests of neurotransmission achieved?

(a) **Scope of the restrictive role of receptor-mediated autoinhibition**: Block of prejunctional α-adrenoceptors (e.g., by phentolamine) increases the nerve impulse-induced secretion from sympathetic nerve terminals by up to 5-fold (Fig. 4A), corresponding to an increase in the probability in the average varicosity from 0.01 to about 0.05. Hence inhibition by prejunctional autoreceptors contributes only to a marginal extent (by less than 5%) to the secretory restrictiveness of sympathetic nerve varicosities (Stjärne 1985).

(b) **Scope of restriction by activation of voltage-gated K^+ channels**: Substances known to block voltage-gated K^+ channels, e.g., tetraethylammonium (TEA) and/or 4-aminopyridine (4AP) increase nerve impulse-induced NA secretion much more strongly than α-blocking drugs (Fig. 4A,B). Both TEA and 4AP increase the Ca^{2+} influx into the terminals by increasing the duration of the action potential; possibly 4AP in addition may do so by a more direct effect (Thesleff 1980, Hille 1984, see also below under **D**).

In the guinea-pig vas deferens, for example, TEA (20 mM) + 4AP (1 mM) increased the secretory response to electrical nerve stimulation at 1Hz by 40-fold (Fig. 4B). On stimulation at (about) 0.1 Hz 'complete' block of K^+ efflux (by addition of TEA plus 4AP at the concentrations shown in Fig. 4) increased the output of ^{3}H-NA per shock to a level corresponding to one quantum per varicosity per shock, i.e., the theoretical maximum for a monoquantal system (Stjärne 1985, Stjärne and Åstrand 1985b). Conversely therefore activation of voltage-gated K^+ channels is likely to play a major role in maintaining the 'desired' low secretory probability in sympathetic nerve varicosities, under 'normal' conditions.

(c) Localization of the voltage-gated K^+ channels: Both for theoretical reasons (the regulator should be close to the effector system) and because other varicosities (rat brain synaptosomes) have been shown to possess (several classes of) voltage-gated K^+ channels (Bartschat and Blaustein 1985), it seems likely that the voltage-gated K^+ channels restricting sympathetic transmitter secretion are located in the varicosities themselves.

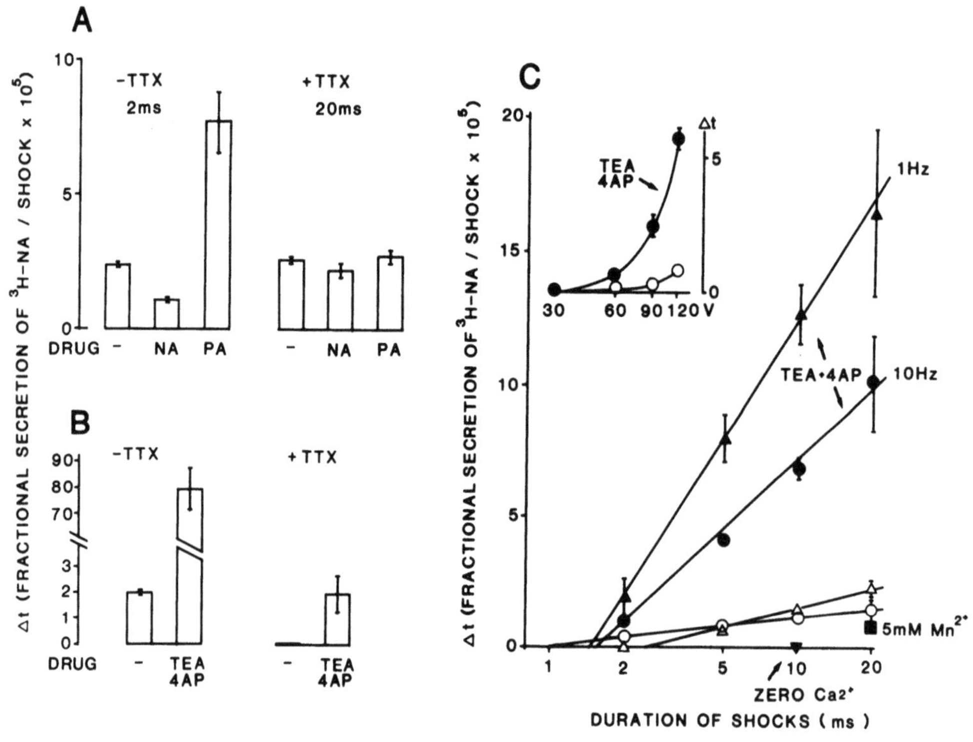

Fig. 4. Tetrodotoxin (TTX 0.5 μM) -sensitive and -resistant secretion of ^{3}H-NA from the sympathetic nerves of the guinea-pig vas deferens, evoked by electrical field stimulation (via ring electrodes at the top and bottom of the preparation) with trains of 300 shocks (120 V), at 1 or 10 Hz. The medium (Tris-buffered, O_2-gassed Tyrode solution, Ca^{2+} 1.8 mM, pH 7.2) contained desipramine (0.6 μM), normetanephrine (10 μM) and atropine (2.6 μM). (**A**) Effects of **NA** (1 μM) or phentolamine (**PA** 1 μM) on the secretory response to stimulation at 1 Hz in the absence of TTX (**'-TTX'**, pulse duration: 2 ms), or in the presence of TTX (**'+TTX'**, pulse duration: 20 ms). (**B**) Effects of tetraethylammonium (**TEA**, 20mM) + 4-aminopyridine (**4AP**, 1 mM), on the response to stimulation at 1 Hz in the absence or presence of TTX (pulse duration: 2 ms in both cases). (**C**) The secretory

response to stimulation at 1 or 10 Hz in the presence of TTX, as a function of the duration of pulses. The TTX-resistant secretion in the presence of TEA+4AP was Ca^{2+}-dependent; the secretory response to stimulation at 10 Hz was abolished in **'zero' Ca^{2+}** (no Ca^{2+} added, 0.1 mM EGTA present, pulse duration: 10 ms), and depressed by 92% by the Ca^{2+} channel blocker, **Mn^{2+}** (5 mM, pulse duration 20 ms). The results show that TEA+4AP did not change the 'duration threshold', but increased the secretory response equally throughout the pulse range tested (2-20 ms). **Inset:** Voltage-dependence of TTX-resistant secretion at 10 Hz, 5 ms). — All points represent at least 4 observations (mean ± SE).

One direct approach to analyze the presence of voltage-gated K^+ channels in varicosities would be to use voltage or patch clamp methods. However, that has not been technically feasible, due to the small size of sympathetic nerve varicosities. Moreover, such a demonstration would not in itself prove that the channels are involved in the regulation of transmitter secretion. Another approach would be to study isolated sympathetic varicosities ('synaptosomes') by biochemical methods. Again, that has not been technically feasible. A third approach would be to study the effect of e.g., K^+ channel blocking agents on the secretory response of varicosities to direct depolarization, i.e. that evoked by electrical stimulation when Na^+ channels have been blocked by tetrodotoxin (TTX).

This third approach has been employed by different groups, but the evidence reported is controversial. Both in the cat spleen (Kirpekar and Prat 1978) and mouse vas deferens (Illés et al 1983) it was possible to evoke a secretory response by increasing the intensity of electrical stimulation after blocking the nerve impulse induced release of NA by addition of TTX. In cat spleen the TTX-resistant secretory response was fully Ca^{2+}-dependent and strongly increased by TEA. In the presence (but not in the absence) of TEA stimulation with increasing intensity beyond a critical level caused the secretory response to grow 'explosively', suggesting that it was triggered by Ca^{2+} action potentials (Kirpekar and Prat 1978). In contrast, the TTX-resistant secretory response in the mouse vas deferens was not fully Ca^{2+}-dependent and not increased either by TEA or 4AP (Illés et al 1983).

The results of our own studies of the TTX-resistant secretion of ^{3}H-NA from the sympathetic nerves of the guinea-pig vas deferens shown in Fig. 4 in most respects support those reported by Kirpekar and Prat (1978). The secretory response to electrical field stimulation with shocks shorter than 1.5 ms was blocked by TTX, but stimuli of longer duration overcame the block and caused a secretory response which grew hyperbolically with the duration of stimuli (2-20 ms). TEA+4AP did not change the 'time threshold' but increased the TTX-resistant secretory response equally at all pulse durations. The growth of the TTX-resistant secretory response with the applied voltage was not 'explosive', either in the presence or in the absence of the K^+ channel blocking agents, as it was reported to be in cat spleen (Kirpekar and Prat 1978). Instead, the voltage-dependence of the secretory response in this preparation looked rather similar to that of the EPSPs evoked by passive depolarization of the presynaptic terminals in squid giant synapse (Llinás et al 1976).

In conclusion: Electrical field stimulation of these sympathetic nerves with shocks longer than 1.5 ms in duration depolarizes the varicosities sufficiently to evoke a 'normal' Ca^{2+}-dependent secretory response, even in the presence of TTX. The TTX-resistant secretory response is enhanced by TEA+4AP. The simplest explanation is that this action was exerted by block of voltage-gated K^+ channels in the varicosities.

One point of particular interest is the long duration of the depolarizing stimulus required to begin to (cause sufficient Ca^{2+} influx to begin to) trigger a secretory response in these sympathetic nerve varicosities. The **'lag time'** is remarkably long, when compared to that of the Ca^{2+} channels in other systems, e.g., in squid giant synapse where it is about 0.5 ms (Llinás et al 1976). We think that this may be one explanation of the low probability of transmitter secretion in the varicosities of the guinea-pig vas deferens.

C. Which are the roles of ionic channels in the physiological control of transmitter secretion? In this section we consider the roles of two **ionic** mechanisms, namely if the **physiological** control of the probability of transmitter secretion is mediated predominantly by regulation of Ca^{2+} or of K^+ channels (for a recent discussion of other aspects, see Starke 1987).

(a) Does autoinhibition primarily involve control of K^+ channels? Due to the small size of sympathetic nerve terminals the properties of their ionic channels are difficult to study directly. However, activation of α_2-adrenoceptors in the soma of sympathetic neurons promotes K^+ efflux and consequently hyperpolarizes the membrane and increases its conductance (Aghajanian 1981, North and Yoshimura 1984). If this effect occurs in the terminals also, it may be one of the mechanisms by which NA autoinhibits transmitter secretion. In that case agents blocking α-adrenoceptors (e.g., phentolamine) ought to have a 'TEA/4AP-like' effect.

The experiments in Fig. 4 show that this is not always the case. With shocks of 5 ms or less both phentolamine (not shown) and TEA+4AP (Fig. 4C) enhanced the TTX-resistant secretion, although much less strongly than the nerve impulse-induced secretion (see Fig. 4A,B). However, while TEA+4AP increased the TTX-resistant secretory responses equally at all pulse durations, the modulatory effects of (NA and) phentolamine were reduced at longer pulse durations and totally absent on stimulation with 20 ms pulses (Fig. 4A). Note that in this case the baseline, i.e., the secretory responses caused by stimulation in the absence or presence of TTX before addition of NA or phentolamine, were closely similar. Hence the differences in the modulatory effects of exogenous and endogenous NA on the TTX-sensitive and -resistant secretory responses cannot be explained in terms of differences in the concentrations of NA in the 'biophase' (Starke 1987).

Instead the finding that prolonged passive depolarization abolished the regulatory effects of α_2-adrenoceptor activation may provide a clue about the mechanism by which these autoreceptors inhibit the secretory mechanisms. The finding is reminiscent of the earlier observation in the same preparation, that an increase in the intensity of field stimulation (at 1 or 10 Hz, with 2 ms shocks, in the absence of TTX), overcame the inhibitory effect of α-adrenoceptor activation (Alberts et al 1981). Perhaps therefore activation of these receptors 'normally' limits the amplitude and/or duration of the depolarization of varicosities. The **'shutter'** effect may be exerted by activation of K^+ efflux, but presumably promotes K^+ efflux through other channels than those blocked by TEA and 4AP, judging from the differences between the effects of TEA/4AP and phentolamine (Fig. 4).

(b) Facilitation and autoinhibition regulate the 'utilization' of Ca^{2+}: The depression of transmitter secretion by activation of prejunctional adrenoceptors is often ascribed, rather vaguely, to restriction of the **'utilization'** of Ca^{2+} in the medium (Kirpekar 1975, Stjärne 1975, Langer 1977, Starke 1977, Westfall 1977). This important mechanism seems to be controlled by **two opposing factors**: It is restricted by activation of α_2-adrenoceptors and promoted by the facilitating effect of trains of nerve impulse at high frequency (Fig. 5).

It can be seen that the Ca^{2+} concentration in the medium required for a half-maximal secretory response to field stimulation (the **EC_{50}** value) was inversely related to the stimulation frequency, both in the absence and presence of phentolamine (see Alberts et al 1981, Stjärne 1981). In controls on stimulation at 1 or 8 Hz it was amazingly high, 41 or 18 mM, respectively. Phentolamine reduced the EC_{50} value to one third and hence improved the 'utilization' of Ca^{2+}, but still on stimulation at 1 Hz the EC_{50} was high (about 13 mM). However, in the presence of phentolamine the facilitating effect of stimulation at higher frequencies reduced it; at 8 Hz it was 1.1 mM, i.e., at about the 'physiological' Ca^{2+} level in extracellular fluid.

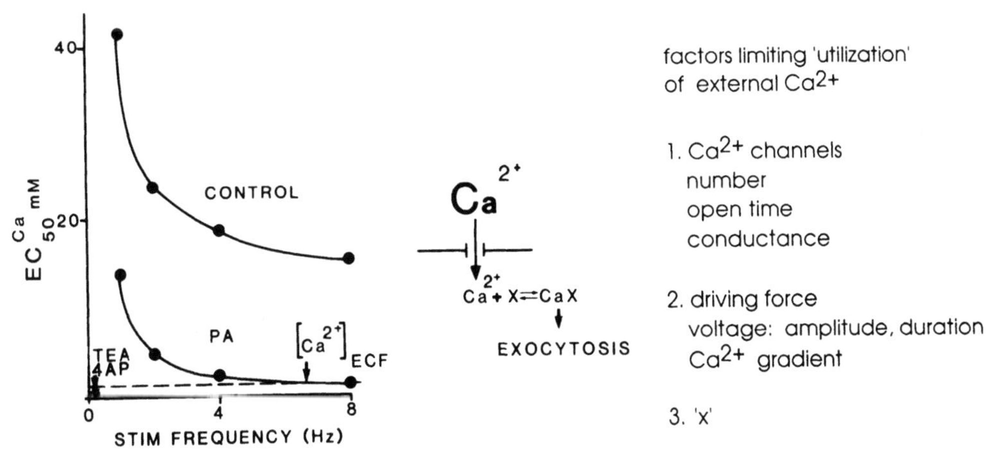

Fig. 5. The secretion of ^{3}H-NA (output per stimulus) evoked by electrical field stimulation of the sympathetic nerves of the guinea-pig vas deferens with trains of 300 shocks

increases hyperbolically with the Ca^{2+} concentration in the medium (in the range 1-8 mM), at each stimulation frequency in the range 1-8 Hz, both in the absence and presence of 1 μM phentolamine, added to block the α_2-adrenoceptor mediated autoinhibition. The **maximal** output per stimulus at each frequency, at 'infinitely' high Ca^{2+}, as well as the Ca^{2+} concentration required to increase the secretory response to its half-maximal value (**EC_{50}**), were calculated by double reciprocal plots (Alberts et al 1981, Stjärne 1981).

The **left panel** shows the frequency dependence of the **EC_{50}** values in this preparation, in the absence (**'CONTROL'**) as well as in the presence of phentolamine (**'PA'**), and also the EC_{50} value for the secretory response to stimulation at 0.0625 Hz in the presence of **TEA+4AP** (Stjärne 1985, Stjärne and Åstrand 1985b). The broken line marks the physiological Ca^{2+} concentration in the extracellular fluid (**ECF**).

The **middle** and **right** panel list the factors which may limit the 'utilization' of Ca^{2+} in the medium. The activity of K^+ channels regulates the Ca^{2+} influx by controlling the duration of depolarization of varicosities by nerve impulses and hence the open time of Ca^{2+} channels. The mechanisms by which activation of prejunctional α_2-adrenoceptors restricts the 'utilization' of external Ca^{2+} are not known, but may involve either (**a**) restriction of Ca^{2+} **influx**, directly via depression of the number or conductance of Ca^{2+} channels, or indirectly via promotion of K^+ efflux and hence shortening of the duration of depolarization of varicosities, or (**b**) restriction of the **'utilization'** of Ca^{2+} within the terminals, by depressing the hypothetical function **'X'** to which Ca^{2+} may have to bind to induce exocytosis (Stjärne 1975, Starke 1987)

Apparently therefore the 'desirably' **ineffective** utilization of Ca^{2+} to stray impulses at low frequency is the combined effect of (**1**) α_2-adrenoceptor-mediated autoinhibition by endogenous NA, and (**2**) lack of facilitation. Removal of the barrier effect of autoreceptor activation is not sufficient; the facilitating effect of increased nerve activity is also required to make external Ca^{2+} fully 'available' for the secretory mechanisms.

D. Ca^{2+} 'utilization' mainly depressed by K^+ channel activation? One consequence of activation of voltage-gated K^+ channels in the varicosities should be restriction of their ability to 'utilize' external Ca^{2+}. The results in Fig. 5 show that this was the case; the EC_{50} value for the secretory response in the presence of TEA and 4AP was extremely low (0.4 mM), even in the complete absence of facilitation (i.e., when stimulation was at less than 0.1 Hz).

E. Conclusions: (1) In sympathetic nerve terminals the presence of **voltage-gated Na^+ channels** in the varicosities may ensure the 'desired' high safety factor for active (?) impulse conduction throughout all parts of the terminals. (2) The presence of **voltage-gated K^+ channels** in the varicosities may ensure the 'desired' low probability for transmitter release (whether such channels occur in the thin sections between the varicosities is not known). Efflux of K^+ through these channels may repolarize the varicosity membrane so rapidly that Ca^{2+} channels barely begin to open (apparent lag time about 1.5 ms). If the action potential has a duration close to this value, most nerve impulses may pass through most varicosities without effectively challenging their release mechanisms. Activity in the K^+ channels represents a **'safety mechanism'**; it is only marginally influenced by physiological mechanisms but may be influenced extensively **pharmacologically**. (3) The **physiological** control of sympathetic transmitter secretion involves facilitation by nerve impulse trains and inhibition via α_2-autoreceptors. The latter may operate, at least in part, via promotion of K^+ efflux from the varicosities (North 1986; for a discussion of other mechanisms, see Starke 1987) but in that case involve different K^+ channels. (4) The physiological control ultimately concerns the **'utilization' of Ca^{2+}**, restricted by α_2-autoreceptors and promoted by facilitation, to an extent proportional to the frequency and length of the nerve impulse trains passing through the varicosity.

THE NEW WORKING HYPOTHESIS

As explained in this paper we think that the current model of sympathetic neurotransmission is challenged in so many aspects that modifications are insufficient; an entirely **new paradigm** is needed. Future work will show to what extent the working model we propose is viable.

ACKNOWLEDGEMENT

This work was supported by The Swedish Medical Research Council (project B87-14X-03027-18A). E. Tanaka is a vocational scholar from Kumamoto University (Japan), supported by The Rotary Foundation of Rotary International. Our thanks are due to Mrs Ingmarie Eriksson who carried out the overflow experiments and to Mrs Eivor Stjärne who performed the electrophysiological experiments and made the line drawings.

REFERENCES

Aghajanian GK (1981) Regulation of central noradrenergic cell firing: Role of α_2-adrenoceptors and opiate receptors. In: Stjärne L, Hedqvist P, Lagercrantz H and Wennmalm Å (eds). *Chemical Neurotransmission 75 years.* Academic Press, London: 273-284.

Alberts P, Bartfai T and Stjärne L (1981) Site(s) and ionic basis of α-autoinhibition and facilitation of [^{3}H]noradrenaline secretion in guinea-pig vas deferens. *J Physiol (Lond)* **312**: 297-334.

Bartschat DK and Blaustein MP (1985) Potassium channels in isolated presynaptic nerve

terminals from rat brain. *J Physiol (Lond)* **361**: 419-440.
Bennett MR (1973) Structure and electrical properties of the autonomic neuromuscular junction. *Phil Trans R Soc Lond B* **265**: 25-34.
Bevan JA, Bevan RD and Duckles SP (1980) Adrenergic regulation of vascular smooth muscle. In: Bohr DF, Somlyo AP and Sparks HV (eds). *Handbook of Physiology 2, The Cardiovascular System, Vol. 2, Vascular Smooth Muscle.* Amer Physiol Soc, Baltimore: 515-566.
Bevan JA, Tayo FM, Rowan RA and Bevan RD (1984) Presynaptic α-receptor control of adrenergic transmitter release in blood vessels. *Fed Proc* **43**: 1365-1370.
Blakeley AGH and Cunnane TC (1979) The packeted release of transmitter from the sympathetic nerves of the guinea-pig vas deferens: An electrophysiological study. *J Physiol (Lond)* **296**: 85-96.
Blakeley AGH, Mathie A and Petersen SA (1984) Facilitation at single release sites of a sympathetic neuroeffector junction in the mouse. *J Physiol (Lond)* **349**: 57-71.
Blakeley AGH, Mathie A and Petersen SA (1986) Interactions between the effects of yohimbine, clonidine and $[Ca]_0$ on the electrical response of the mouse vas deferens. *Br J Pharmacol* **88**: 807-814.
Blaustein MP, Nachshen DA and Drapeau P (1981) Excitation-secretion coupling: The role of calcium. In: Stjärne L, Hedqvist P, Lagercrantz H and Wennmalm Å (eds). *Chemical Neurotransmission 75 years.* Academic Press, London: 125-138.
Bolton TB and Large WA (1986) Are junction potentials essential? Dual mechanism of smooth muscle cell activation by transmitter released from autonomic nerves. *Quart J Exp Physiol* **71**: 1-28.
Brigant JL and Mallart A (1982) Presynaptic currents in mouse motor endings. *J Physiol (Lond)* **333**: 619-636.
Brock JA and Cunnane TC (1987) Relation between the nerve action potential and transmitter release from sympathetic postganglionic nerve terminals. *Nature* **326**: 605-607.
Burnstock G (1976) Do some nerve cells release more than one transmitter? *Neuroscience* **1**: 239-248.
Burnstock G (1986) The changing face of autonomic neurotransmission. *Acta Physiol Scand* **126**: 67-91.
Burnstock G and Costa M (1975) *Adrenergic Neurons.* Chapman and Hall, London.
Burnstock G and Holman ME (1966) Junction potentials at adrenergic synapses. *Pharmacol Rev* **18**: 481-493.
Changeux J-P (1986) Coexistence of neuronal messengers and molecular selection. *Prog Brain Res* **68**: 373-403.
Cheung DW (1982) Two components in the cellular response of rat tail arteries to nerve stimulation. *J Physiol (Lond)* **328**: 461-468.
Cunnane TC (1984) The mechanism of neurotransmitter release from sympathetic nerves. *Trends in Neurosci* **7**: 248-253.
Cunnane TC and Stjärne L (1984) Transmitter secretion from individual varicosities of guinea-pig and mouse vas deferens: Highly intermittent and monoquantal. *Neuroscience* **13**: 1-20.
Dale HH (1934) Nomenclature of fibres in the autonomic system and their effects. *J Physiol (Lond)* **80**: 10-11.
Dunant Y and Muller D (1986) Quantal release of acetylcholine evoked by focal depolarization at the *Torpedo* nerve-electroplaque junction. *J Physiol (Lond)* **379**: 461-478.
Eccles JC (1986) Chemical transmission and Dale's principle. *Prog Brain Res* **68**: 3-13.
Fedan JS, Hogaboom GK, O'Donnell JP, Colby J and Westfall DP (1981) Contribution by purines to the neurogenic response of the vas deferens of the guinea pig. *Eur J Pharmacol* **69**: 41-53.
Folkow B, Häggendal J and Lisander B (1967) Extent of release and elimination of noradrenaline at peripheral adrenergic nerve terminals. *Acta Physiol Scand Suppl* **307**: 1-38.
Folkow B and Häggendal J (1970) Some aspects of the quantal release of the adrenergic

transmitter. *Bayer-Symposium II.* Springer, Berlin: 91-97.
Furness JB (1970) The excitatory input to a single smooth muscle cell. *Pflügers Arch* **314**: 1-13.
Haefely W (1972) Electrophysiology of the adrenergic neuron. In: Blaschko H and Muscholl E (eds). *Catecholamines. Handbook of Experimental Pharmacology, Vol. 33.* Springer, Berlin: 661-725.
Hille B (1984) *Ionic Channels of Excitable Membranes.* Sinauer Associates Inc. Publishers, Sunderland Massachusetts.
Hökfelt T (1969) Distribution of noradrenaline storing particles in peripheral adrenergic neurons as revealed by electron microscopy. *Acta Physiol Scand* **76**: 427-440.
Hökfelt T, Holets VR, Staines W, Meister B, Melander T, Schalling M, Schultzberg M, Freedman J, Björklund H, Olson L, Lindh B, Elfvin L-G, Lundberg JM, Lindgren JÅ, Samuelsson B, Pernow B, Terenius L, Post C, Everitt B and Goldstein M (1986) Coexistence of neuronal messengers — an overview. *Prog Brain Res* **68**: 33-70.
Illes P, Meier C and Starke K (1984) Tetrodotoxin-resistant release of ^{3}H-noradrenaline from the mouse vas deferens by high intensity electrical stimulation. *Neuroscience* **11**: 715-721.
Kirpekar SM (1975) Factors influencing transmission at adrenergic synapses. *Prog Neurobiol* **4**: 163-210.
Kirpekar SM and Prat JC (1978) Effect of tetraethylammonium on noradrenaline release from cat spleen treated with tetrodotoxin. *Nature* **276**: 623-624.
Klein RL (1982) Chemical composition of the large noradrenergic vesicles. In: Klein RL, Lagercrantz H and Zimmermann H (eds). *Neurotransmitter Vesicles.* Academic Press, London: 133-150.
Korn H (1984) What central inhibitory pathways tell us about mechanisms of transmitter release. *Exp Brain Res Suppl* **9**: 201-224.
Langer SZ (1977) Presynaptic receptors and their role in the regulation of transmitter release. *Br J Pharmacol* **60**: 481-497.
Langer SZ (1981) Presynaptic regulation of the release of catecholamines. *Pharmac Rev* **32**: 337-362.
Llinás R, Izcak Z, Steinberg and Walton K (1976) Presynaptic calcium currents and their relation to synaptic transmission: Voltage clamp study in squid giant synapse and theoretical model for the calcium gate. *Proc Natl Acad Sci USA* **73**: 2918-2922.
Lundberg JM and Hökfelt T (1986) Multiple co-existence of peptides and classical transmitters in peripheral autonomic and sensory neurons — functional and pharmacological implications. *Prog Brain Res* **68**: 241-262.
Mallart A (1984) Presynaptic currents in frog motor endings. *Pflügers Arch* **400**: 8-13.
Neild TO (1987) Actions of neuropeptide Y on innervated and denervated rat tail arteries. *J Physiol (Lond)* **386**: 19-30.
North RA (1986) Receptors on individual neurones. *Neuroscience* **17**: 899-907.
North RA and Yoshimura M (1984) The actions of noradrenaline on neurones of the rat substantia gelatinosa *in vitro*. *J Physiol (Lond)* **349**: 43-55.
Potter DD, Matsumoto SG, Landis SC, Sah DWY and Furshpan EJ (1986) Transmitter status in cultured sympathetic principal neurons: plasticity, graded expression and diversity. *Prog Brain Res* **68**: 103-120.
Schipper J, Tilders FJH and Ploem JS (1980) A scanning microfluorimetric study on sympathetic nerve fibres: Intraneuronal differences in noradrenaline turnover. *Brain Res* **190**: 459-472.
Sneddon P and Burnstock G (1984) Inhibition of excitatory junction potentials in guinea-pig vas deferens by α, β-methylene-ATP: Further evidence for ATP and noradrenaline as cotransmitters. *Eur J Pharmacol* **100**: 85-90.
Sneddon P and Westfall DP (1984) Pharmacological evidence that adenosine triphosphate and noradrenaline are co-transmitters in the guinea-pig vas deferens. *J Physiol (Lond)* **347**: 561-580.
Starke K (1977) Regulation of noradrenaline release by presynaptic receptor systems. *Rev Physiol Biochem Pharmacol* **77**: 1-124.

Starke K (1981) Presynaptic receptors. *Ann Rev Phamacol Toxicol* **21**: 7-30.
Starke K (1987) Presynaptic α-autoreceptors. *Rev Physiol Biochem Pharmacol* **107**: 73-146.
Stjärne L (1975) Basic mechanisms and local feedback control of secretion of adrenergic and cholinergic neurotransmitters. In: Iversen LL, Iversen SD and Snyder SH (eds) *Handbook of Psychopharmacology. Vol. 6.* Plenum Press, New York: 179-233.
Stjärne L (1978) Facilitation and receptor-mediated regulation of noradrenaline secretion by control of recruitment of varicosities as well as by control of electro-secretory coupling. *Neuroscience* **3**: 1147-1155.
Stjärne L (1979) Presynaptic α-receptors do not depress the secretion of ^{3}H-noradrenaline induced by veratridine. *Acta Physiol Scand* **106**: 379-380.
Stjärne L (1981) On sites and mechanisms of presynaptic control of noradrenaline secretion. In: Stjärne L, Hedqvist P, Lagercrantz H and Wennmalm Å (eds). *Chemical Neurotransmission 75 years.* Academic Press, London: 257-272.
Stjärne L (1985) Scope and mechanisms of control of stimulus-secretion coupling in single varicosities of sympathetic nerves. *Clinical Science* **68**: (Suppl. 10), 77s-81s.
Stjärne L (1986a) New paradigm: Sympathetic neurotransmission by lateral interaction between secretory units? *News in Physiological Sciences* **1**: 103-106.
Stjärne L (1986b) New paradigm: sympathetic transmission by multiple messengers and lateral interaction between monoquantal release sites? *Trends in Neurosci* **9**: 547-548.
Stjärne L (1987) New paradigm: A digital model of feedback regulation of sympathetic neurotransmitter secretion. In: Vanhoutte PM (ed) Proc Symp "*Mechanisms of Vasodilatation*", Rochester, Minnesota. Raven Press. In press.
Stjärne L and Lundberg JM (1986) On the possible roles of noradrenaline, adenosine 5'-triphosphate and neuropeptide Y as sympathetic cotransmitters in the mouse vas deferens. *Prog Brain Res* **68**: 263-278.
Stjärne L and Åstrand P (1984) Discrete events measure single quanta of adenosine 5'-triphosphate secreted from sympathetic nerves of guinea-pig and mouse vas deferens. *Neuroscience* **13** : 21-28.
Stjärne L and Åstrand P (1985a) Relative pre- and postjunctional roles of noradrenaline and adenosine 5'-triphosphate as neurotransmitters of the sympathetic nerves of guinea-pig and mouse vas deferens. *Neuroscience* **14**: 929-946.
Stjärne L and Åstrand P (1985b) Site of action of presynaptic inhibition mediated via adrenoceptors. In: Szabadi E, Bradshaw CM and Nahorski SR (eds). *Pharmacology of Adrenoceptors.* Macmillan: 157-166.
Stjärne L, Lundberg JM and Åstrand P (1986) Neuropeptide Y — A cotransmitter with noradrenaline and adenosine 5'-triphosphate in the sympathetic nerves of the mouse vas deferens? A biochemical, physiological and electropharmacological study. *Neuroscience* **18**: 151-166.
Suzuki H (1983) An electrophysiological study of excitatory neuromuscular transmission in the guinea-pig main pulmonary artery. *J Physiol (Lond)* **336**: 47-59.
Thesleff S (1980) Aminopyridines and synaptic transmission. *Neuroscience* **5**: 1413-1419.
Thoa NB, Wooten GF, Axelrod J and Kopin IJ (1973) On the mechanism of release of norepinephrine from sympathetic nerves induced by depolarizing agents and sympatomimetic drugs. *Mol Pharmacol* **11**: 10-18.
Thuresson-Klein Å (1983) Exocytosis from large and small dense cored vesicles in noradrenergic nerve terminals. *Neuroscience* **10**: 245-252.
Vizi ES (1979) Presynaptic modulation of neurochemical transmission. *Prog Neurobiol* **12**: 181-290.
Westfall TC (1977) Local regulation of adrenergic neurotransmission. *Physiol Rev* **57**: 659-728.
Zhu PC, Thuresson-Klein Å and Klein RL (1986) Exocytosis from large dense cored vesicles outside the active synaptic zones of terminals within the trigeminal subnucleus caudalis: A possible mechanism for neuropeptide release. *Neuroscience* **19**: 43-54.

PARTICIPATION OF REGULATORY G-PROTEINS AND PROTEIN KINASE C IN THE MODULATION OF TRANSMITTER RELEASE IN HIPPOCAMPUS

Hertting, G., Wurster, S., Gebicke-Härter, P. and Allgaier, C.

Institute of Pharmacology, University of Freiburg, Hermann-Herder-Str. 5, 7800 Freiburg, Federal Republic of Germany.

1.) Regulatory G-proteins participate in the inhibition of noradrenaline (NA) and acetylcholine (ACh) release following activation of the respective autoreceptors (α_2, M_2). Islet-activating protein (IAP) and N-ethylmaleimide (NEM) increased the electrically evoked release of [^{3}H]NA and [^{3}H]ACh from hippocampal slices and reduced inhibition of release by the autoreceptor agonists clonidine and carbachol as well as its facilitation by the antagonists yohimbine and atropine. In synaptosomes prepared from rabbit hippocampus two polypeptides with molecular weights corresponding to those of $G_{i\alpha}$ and $G_{o\alpha}$ were [^{32}P]ADP-ribosylated by IAP in the presence of [^{32}P]NAD$^+$. Pretreatment of synaptosomes with NEM reduced the subsequent [^{32}P]ADP-ribosylation by IAP in a concentration-dependent manner.

2.) There is a functional interaction between the autoreceptor mechanisms and the A_1-adenosine receptor mechanisms. Inhibition of release of [^{3}H]NA and [^{3}H]ACh by the adenosine receptor agonist (-)N^6-(R-phenylisopropyl)adenosine ((-)PIA) was dependent on the concentration of the endogenous transmitter in the synaptic cleft. At lower concentrations of endogenous ACh or in the presence of yohimbine the action of (-)PIA was increased.

3.) Regulatory G-proteins seem to be involved also in the A_1-adenosine receptor-coupled mechanisms. NEM diminished inhibition of release of both [^{3}H]NA and [^{3}H]ACh by (-)PIA.

4.) Presynaptic protein kinase C (PKC) appears to play a

NATO ASI Series, Vol. H19
Modulation of Synaptic Transmission and Plasticity
in Nervous Systems. Edited by G. Hertting and H.-C. Spatz

crucial role in transmitter release. Activators of protein kinase C such as 12-O-tetradecanoylphorbol 13-acetate (TPA) or 4β-phorbol 12,13-dibutyrate (4β-PDB) increased the electrically evoked release of [³H]NA and [³H]ACh. The PKC inhibitor polymyxin B (PMB) decreased transmitter release and counteracted the phorbol ester-induced enhancement of release.

5.) PKC does not not participate in autoreceptor-mediated inhibition of transmitter release. Inhibition of [³H]NA release by clonidine and its facilitation by yohimbine was only indirectly changed by 4β-PDB by an increase of the concentration of endogenous NA in the synaptic cleft.

I. INTRODUCTION

Action potential-induced neurotransmitter release is initiated by an influx of Ca^{2+} through voltage-dependent Ca^{2+}-channels. Some possible molecular mechanisms which may occur in a nerve terminal upon Ca^{2+}-entry are drawn in Figure 1.

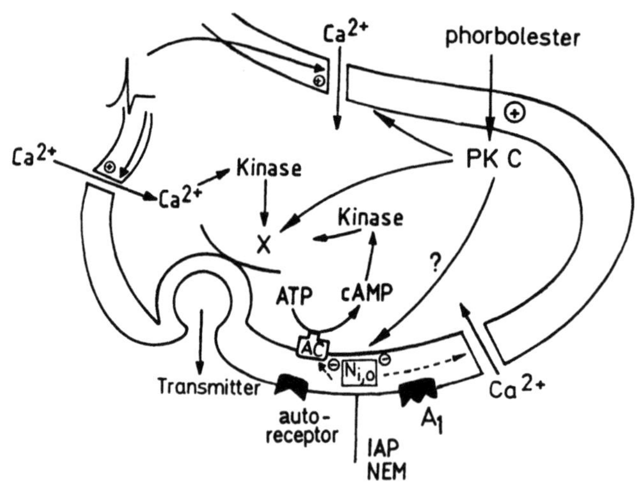

Fig.1: Possible mechanisms underlying action potential-induced transmitter release and its inhibition by presynaptic receptors. For detail see Introduction. In the case of the cholinergic nerve terminal of hippocampus guanylate cyclase is dicussed as autoreceptor-coupled effector system (Nordström & Bartfai, 1981; Marchi et al., 1981).

Protein kinases are activated by an increase of cytosolic Ca^{2+}. Subsequent phosphorylation of specific proteins leads to fusion of the transmitter storing granules with the plasma membrane and to exocytotic release of the transmitter. Modulation of release may occur either by varying the free intraneuronal Ca^{2+}-concentration (by actions on Ca^{2+}-influx or by redistribution of intracellular Ca^{2+}) and/or by regulating the activity of enzymes involved in the release mechanism.

One of these enzymes, the cAMP-dependent protein kinase seems to participate in depolarization-induced release of NA, since forskolin and fluoride (both activators of AC) as well as phosphodiesterase inhibitors and cAMP analogs have been shown to facilitate electrically evoked release of [^{3}H]NA (for review see Starke 1987). Therefore, activation of adenylate cyclase and subsequent elevation of intracellular cAMP seems to represent an important event in transmitter release. Conversely, it was reasonable to hypothesize that an inhibition of AC and resulting decrease of cAMP is some crucial step in α_2-autoreceptor-mediated inhibition of noradrenaline release. α_2-agonists have been reported to inhibit AC in many cell types (Jakobs et al., 1984) and circumstantial evidence has been presented that also by activation of presynaptic α_2-autoreceptors AC activity is inhibited (Schoffelmeer & Mulder, 1983; Mulder et al., 1984).

Receptor-coupled inhibition of adenylate cyclase is mediated by the regulatory guanine-nucleotide-binding protein Gi (Rodbell, 1980; Cooper, 1982). The transducer function of Gi can be cancelled either by ADP-ribosylation in the presence of islet-activating protein (IAP) (Murayama & Ui, 1983) or by alkylation with N-ethylmaleimide (NEM) (Jakobs et al., 1984). Both, IAP and NEM have been used to show the participation of a G protein in the signal transduction mechanism of presynaptic α_2-autoreceptors (Allgaier et al., 1985; 1986a; Fredholm et al., 1987). However, in addition to Gi, at least one (Go; Sternweis & Robishaw, 1984) or possibly two further IAP substrates (Katada et al., 1987) have been described in brain tissue. Therefore, these results do not provide conclusive evidence for the involvement of AC in presynaptic α_2-adrenoceptor-mediated inhibition of NA release, a question which remained to date open (Fredholm et al.,1987; Johnston et al., 1987). In this paper we give further data for a participation of regulatory G-proteins in the post receptor mechanisms of presynaptic receptors of noradrenergic and cholinergic neurons. In addition, the possibility that various receptors (auto- and heteroreceptors) of a certain nerve ending use a common pathway for signal transduction was investigated.

In previous studies transmitter release in hippocampus has been described to be increased by activation of protein kinase C (PKC) with the tumor-promoting phorbol esters 12-O-tetradecanoyl 13-acetate (TPA) and 4β-phorbol 12,13-dibutyrate (4β-PDB) (Allgaier & Hertting, 1986; Allgaier et al., 1986b; 1987b,c; Feuerstein et al., 1987; Versteeg & Florijn, 1987). These compounds are mimicking the action of the endogenous activator of PKC, diacylglycerol, which is transiently produced by enzymatic cleavage of inositolphospholipids (Nishizuka, 1984; Berridge, 1984). Hence, a possible participation of PKC in modulation of transmitter release via presynaptic receptors was investigated and the results are discussed herein.

II. METHODS

Superfusion experiments

Tissue slices (thickness: 0.35 mm), prepared from the middle part of rabbit hippocampus, were prelabelled with either [^{3}H]NA or [^{3}H]choline. Subsequently, the slices were superfused with physiological saline containing cocaine (30 μM), when [^{3}H]NA release was studied, and hemicholinium-3 (10 μM) in the case of [^{3}H]ACh release. During superfusion the slices were stimulated electrically (3 or 0.3 Hz, 24 mA, 2 ms, 5 V/cm) twice for 2 min each. The radioactivity in 5-min-fractions of the superfusate (5 ml each) and the radioactivity in the slices at the end of the experiment were determined by liquid scintillation counting. The outflow of tritiated compounds was computed as fractional rate of "tritium" outflow per 5 min, i.e. as radioactivity in a 5-min-superfusate-fraction divided by the radioactivity in the slice at the beginning of the fraction. The increase of the radioactivity in the superfusate following electrical stimulation represents the difference between total and basal radioacvtivity in the superfusate (S_1, S_2; expressed as a percentage of the radioactivity in the slice at the beginning of the respective stimulation period; for further detail see: Hertting et al., 1980; Allgaier et al., 1985; 1987a). The stimulus-evoked increase of radioactivity in the superfusate of slices prelabelled with [^{3}H]NA consisted to about 90% of [^{3}H]NA, when reuptake of NA was prevented by cocaine. By electrical stimulation of tissue preincubated with the [^{3}H]ACh precursor [^{3}H]choline exclusively [^{3}H]ACh is released, as shown by Richardson & Szerb (1974). This transmitter is almost immdiately hydrolyzed in the synaptic cleft by ACh esterase. Reuptake of the resulting [^{3}H]choline was prevented by hemicholinium-3.

All results are shown as arithmetic means ± SEM. The existence of differences between groups was tested with ANOVA. The significance of differences was determined by Student's t–test(two-tailed).

[^{32}P]ADP-ribosylation of synaptosomes pretreated with NEM.

Synaptosomes of rabbit hippocampus were prepared on a discontinuous Percoll gradient as described by Robinson & Lovenberg (1986). Synaptosomal fractions 3 and 4 of the Percoll gradient were resuspended in a Krebs-Ringer-buffer (KRB) and adjusted to a protein concentration of 2.25 mg/ml. In some experiments 100 µl of this suspension were added to 100 µl KRB containing NEM. Final concentration of NEM was between 1 and 1000 µM. NEM treatment was carried out at 37 °C and stopped after 30 min by addition of dithiothreitol. Synaptosomes were recovered by centrifugation. The pellet was resuspended in 50 mM Tris/HCl, 1 mM EDTA, pH 8.0. IAP-catalysed [^{32}P]ADP-ribosylation of synaptosomal proteins was performed according to the method of Kawai et al. (1986). Samples containing 150 µg protein were subjected to SDS-polyacrylamide gel electrophoresis on a 10% slab gel as described by Laemmli (1970). Radiolabelled proteins were visualized by autoradiography.

III. RESULTS AND DISCUSSION

Both NA and ACh release in rabbit hippocampus is inhibited by activation of the respective autoreceptors (α_2, M_2) (Strittmatter et al., 1982; Jackisch et al., 1984). Therefore, evoked transmitter release is increased by application of autoreceptor antagonists such as yohimbine in the case of α_2-adrenoceptors and atropine in the case of muscarine receptors. In addition to the autoreceptors both nerve terminals are endowed with A_1-adenosine receptors also mediating inhibition of stimulation-evoked transmitter release (Jackisch et al., 1984; 1985).

1.) Participation of G-proteins in modulation of transmitter release by presynaptic receptors

- Noradrenaline release

Pretreatment of hippocampal slices with IAP or NEM increased the electrically-evoked [^{3}H]NA release in a time- and concentration-dependent manner. Already after 6 h of incubation with IAP (8 µg/ml) [^{3}H]NA release was enhanced (Allgaier et al., 1985). In Figure 2 (left panel) the effects of IAP after an incubation period of 12 and 18 h are shown. In comparison to control slices IAP increased the evoked [^{3}H]NA release after 18 h by about 120%. There was virtually no effect on basal "tritium" outflow. NEM (30 µM) enhanced [^{3}H]NA release after 30 min of incubation by 80% (Fig. 2, right; Allgaier et al., 1986a). Basal outflow was also increased by NEM. In slices pretreated with IAP, the evoked [^{3}H]NA release was not further increased by NEM (Hertting et al., 1986).

In Figure 3 inhibition of [^{3}H]NA release by the α_2-adrenoceptor agonist clonidine (left) and facilitation of

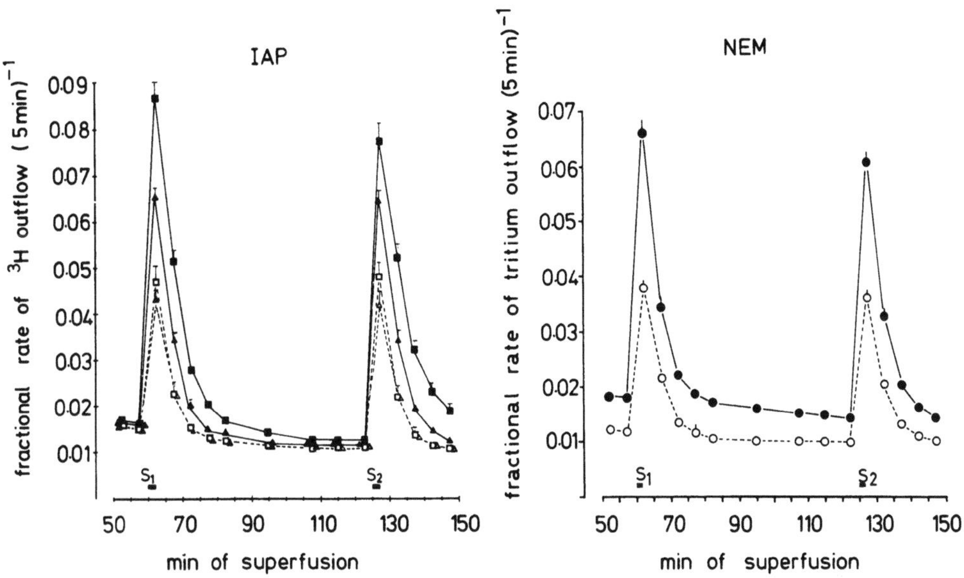

Fig. 2: Enhancement of evoked [^{3}H]NA release by IAP and NEM. Slices were either pretreated with IAP (8 µg/ml) for 12 (▼) or 18 h (■) or with NEM (30 µM) for 30 min (●) prior to labelling with [^{3}H]NA. Control slices (open symbols) were incubated under the same conditions but without IAP or NEM. Means ± SEM of the fractional rates are given. Number of experiments per group: 6-11.

release by the α_2-adrenoceptor antagonist yohimbine (right) are depicted. The effects of both drugs were strongly reduced or even abolished after pretreatment with IAP. Similar effects were seen following NEM treatment (Allgaier et al., 1986a). From these results it can be concluded that both IAP and NEM affect the α_2-adrenoceptor mechanism, probably at the level of regulatory G-proteins linked to the α_2-receptors. This conclusion is supported by results from experiments performed on synaptosomes of rabbit hippocampus. After treatment of the synaptosomes with IAP in the presence of [^{32}P]NAD two polypeptides with molecular weights corresponding to that of the α-subunits of Go (39,000) and Gi (41,000) were [^{32}P]ADP-ribosylated. After preincubation with NEM the subsequent IAP-induced [^{32}P]ADP-ribosylation was diminished in a concentration-dependent manner (Figure 4).

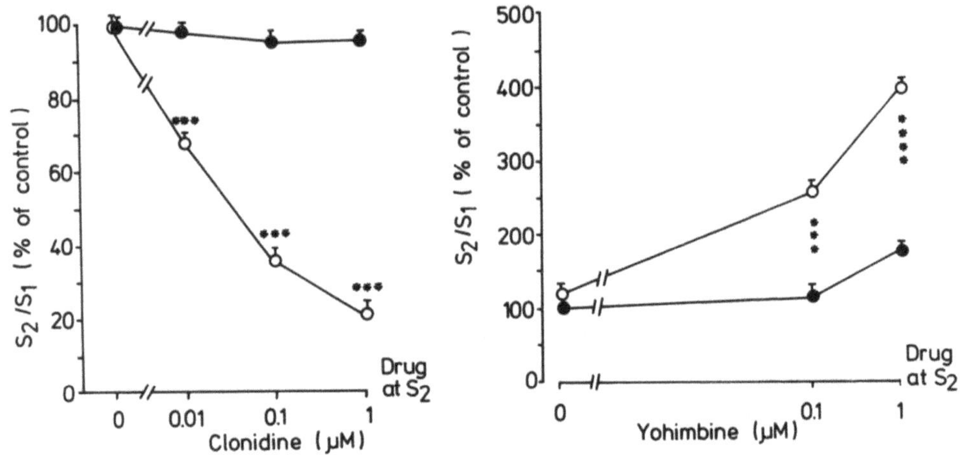

Fig. 3: Slices were pretreated with IAP (8 µg/ml; closed symbols) for 18 h and then labelled with [^{3}H]NA. Control slices (open symbols) were incubated under the same conditions but in the absence of IAP. During superfusion the slices were stimulated twice electrically at 3 Hz for 2 min each. Clonidine (left panel) or yohimbine (right) were added to the superfusion medium from 15 min before the second stimulation onwards. S_2/S_1 represents the ratio of [^{3}H]NA evoked by the two stimulations. Number of experiments/point:4-8. Significant differences from respective drug-free control: ***P < 0.001; ****P < 0.0001.

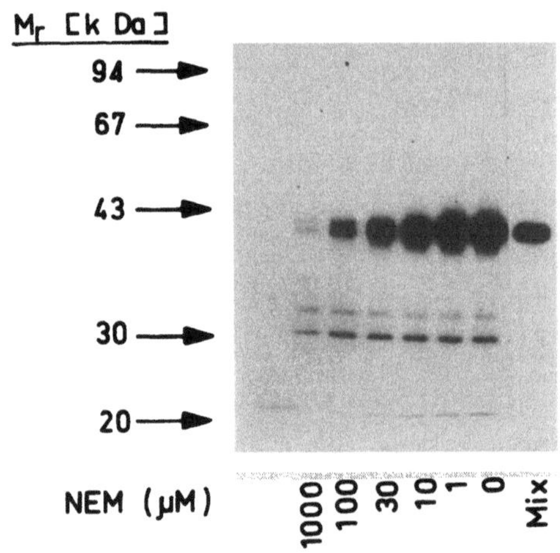

Fig. 4: [^{32}P]ADP-ribosylation of hippocampal synaptosomes pretreated with NEM. Experimental procedure as described in Methods. Following NEM treatment of synaptosomes [^{32}P]ADP-ribosylation of synaptosomal proteins was performed in the presence of IAP (5 µg/ml). The right lane represents [^{32}P]-labelled α_i and α_o subunits of a mix of purified G_i and Go.

As already described, evoked NA release is diminished by stimulation of A_1-adenosine receptors (Fredholm et al., 1983; Jackisch et al., 1985). This effect is counteracted by NEM similarly as already shown for α_2-adrenoceptors (Fredholm & Lindgreen, 1986; 1987; Allgaier et al., 1987a). In order to investigate the possibility that the α_2-adrenoceptors and the adenosine receptors utilize a common signal transduction mechanism, inhibition of NA release by the A_1-agonist (-)PIA was studied in the presence of the α_2-adrenoceptor antagonist yohimbine (Figure 5; Allgaier et al., 1987a). (-)PIA decreased the electrically evoked [^{3}H]NA release in a concentration-dependent manner. The action of (-)PIA was enhanced when the autoreceptor was blocked by yohimbine. NEM antagonized the effect of ()PIA under both conditions (Figure 5).

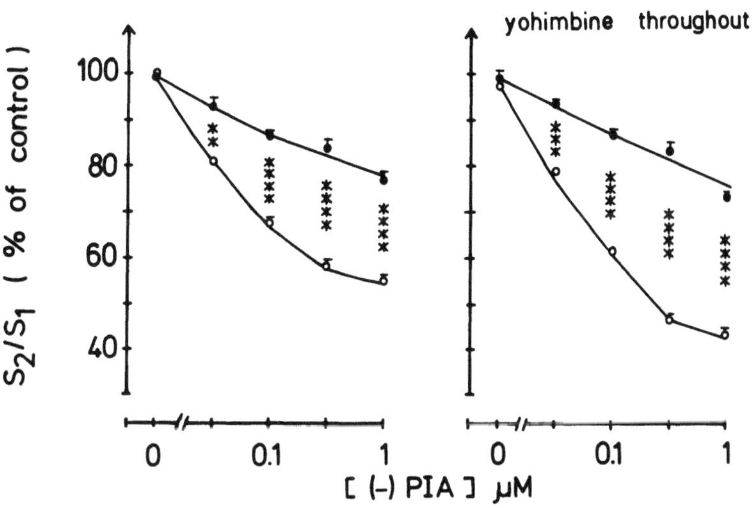

Fig. 5: Effect of NEM on the action of (-)PIA in the absence and presence of yohimbine. Slices were pretreated with NEM (30 µM; ●) for 30 min prior to labelling with [^{3}H]NA. Control slices (○) were incubated under the same conditions but without NEM. (-)PIA (0.03 - 1 µM) was added 15 min before the second stimulation. Yohimbine (0.1 µM) was present throughout superfusion. Significant differences of the action of (-)PIA between NEM-treated and control slices. **P < 0.01; ***P < 0.001; ****P < 0.0001.

Since the effect of the heteroreceptor agonist (-)PIA was dependent on the extent of autoinhibition and both auto- and heteroreceptor-mediated effects were restrained by NEM, a common pathway for both inhibitory signals, possibly at the level of G-proteins can be assumed (Figure 6). However, an interaction of both receptor mechanisms at another level, possibly at common ionic channels, would also explain the increased inhibitory effect of (-)PIA in the presence of yohimbine (Allgaier et al., 1987a).

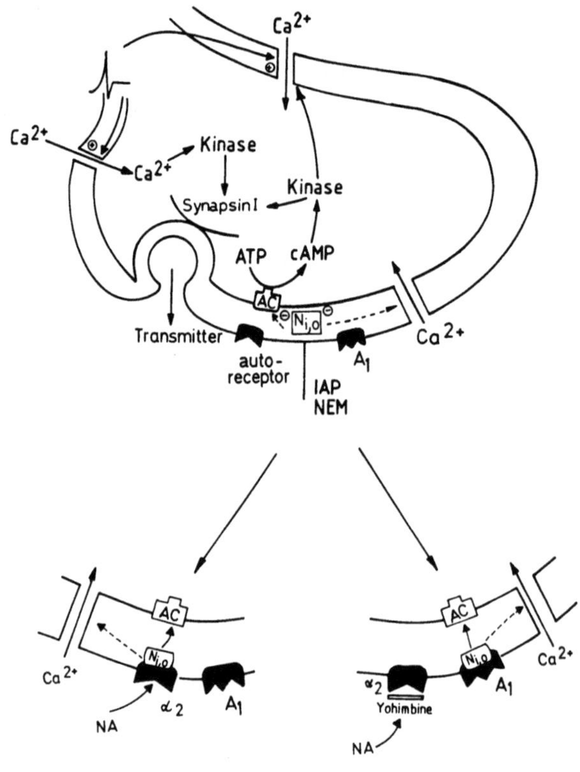

Fig. 6: Scheme of the possible interaction of the α_2-autoreceptor- and adenosine receptor-coupled mechanisms. Both receptor types may be coupled via G proteins (Gi, Go) to intracellular effector systems, like AC, or directly to Ca^{2+}-channels. Under the stimulation conditions used (3 Hz, 24 mA, 2 ms, 5 V/cm) the biophase concentration of endogenous NA exerted a strong autoinhibtion (see Figure 2, right panel). Therefore, the common signal transduction pathway was partially occupied by the autoreceptor signal and the response to the A_1-agonist was decreased. Following blockade of the autoreceptors by yohimbine the fraction of G proteins available for coupling with the A_1-receptor was increased, consequently the effect of (-)PIA was more pronouced.

- Acetylcholine release

In the following Tables (Table 1, 2) the effects of the muscarine receptor agonist carbachol, the muscarine receptor antagonist atropine and the adenosine receptor agonist (-)PIA on the electrically evoked [^{3}H]ACh release in NEM-treated and control slices are shown. In the presence of atropine the evoked [^{3}H]ACh release was increased by about 30% (Table 1).

Table 1: Effects of atropine and NEM on [^{3}H]ACh release under various conditions of autoinhibition

Drug (µM)	drug throughout (µM)	S_2/S_1	n
-	-	1.11±0.02	8
atropine (2)	-	1.50±0.02 ****[a]	6
NEM (30)	-	1.47±0.03 ****[a]	7
NEM/atropine	-	1.55±0.04 n.s.[b]	6
-	physostigmine (1)	1.18±0.03	6
atropine (2)	physostigmine (1)	5.22±0.31 ****[c]	5
NEM (30)	physostigmine (1)	2.85±0.11 ****[c]	6
NEM/atropine	physostigmine (1)	5.59±0.23 n.s.[d]	5

Hippocampal slices were superfused either in the presence of hemicholinium-3 (10 µM) or hemicholimium-3 and physostigmine (1 µM) and stimulated twice electrically during superfusion. NEM was added to the medium between the two stimulations for 30 min, atropine from 15 min before the second stimulation onwards. S_1±SEM in the presence of hemicholininium-3 was 2.98 ± 0.16% (n = 27); when physostigmine was present S_1±SEM was 0.99 ± 0.03 (n = 22). n = number of experiments/group. Significances given: [a], vs. drug-free control; [c] vs. physostigmine throughout; [b] NEM/atropine vs atropine [d] NEM/atropine vs. atropine in the presence of physostigmine. ****: P < 0.0001.

This increase was induced by a cancellation of a rather small autoinihibition. Treatment of the tissue slices with NEM between the first and the second stimulation period augmented the evoked [^{3}H]ACh release by a similar degree. Combined treatment with both NEM and atropine did not further increase the [^{3}H]ACh release.

Following inhibition of ACh esterase by physostigmine the concentration of endogenous ACh in the synaptic cleft was increased and hence a strong enhancement of release by atropine by about 340% was seen (Table 1). NEM only partially counteracted the autoinhibition and increased transmitter release by about 140%. Again the effects of atropine and NEM given in combination were not additive.

Carbachol (1 μM) inhibited [^{3}H]ACh release by about 40% (Table 2). NEM counteracted the action of carbachol indicating, as in the case of noradrenaline release, that also the muscarine autoreceptor operates via a G-protein.

In the presence of atropine (-)PIA inhibited the electrically evoked ACh release by almost 60% (Table 2). This effect of (-)PIA was abolished by NEM. In the presence of physostigmine when autoinhibiition was increased due to the increased biophase concentration, the inhibitory effect of (–)PIA was markedly decreased (data not shown). Blockade of the autoreceptors by atropine restored inhibition of release by (–)PIA.The interaction of auto- and heteroreceptor mechanisms modulating transmitter release has been described by several authors on different transmitter systems (Pedata et al., 1986; Limberger et al., 1986, Allgaier et al., 1987a).

Table 2: Effects of the muscarine receptor agonist carbachol and the adenosine receptor agonist (-)N^6-(R-phenylisopropyl)-adenosine ((-)PIA) on [^{3}H]ACh release.

Drug (μM)	drug (μM) throughout	S_2/S_1		net effect	n
-		1.13±0.02		-	4
carbachol (1.0)	-	0.68±0.02	****a	0.45±0.03	4
NEM (30.0)	-	1.48±0.02		-	4
NEM/carbachol	-	1.17±0.02	****b	0.31±0.03**e	4
-	atropine (1)	0.96±0.04		-	5
(-)PIA (0.3)	atropine (1)	0.41±0.05	****c	0.55±0.06	4
NEM (30.0)	atropine (1)	0.95±0.04		-	5
NEM/(-)PIA	atropine (1)	0.82±0.06	n.s.d	0.13±0.07**f	4

Tissue slices incubated with [^{3}H]choline were stimulated electrically twice during superfusion. In some experiments atropine was present throughout superfusion. NEM treatment was performed between the first and the second stimulation for 30 min. Carbachol or (-)PIA was added to the medium from 15 min before the second stimulation onwards. $S_1 \pm$ SEM was 3.06 ± 0.13 (n = 16); in the presence of atropine $S_1 \pm$ SEM amounted to 3.84 ± 0.38 (n = 18). Significances given: [a] vs. drug-free control; compared to the respective control [b] NEM/carbachol vs. NEM; [c] vs. atropine throughout; [d] NEM/(-)PIA vs. NEM in the presence of atropine; [e] net effect NEM/carbachol vs. carbachol; [f] net effect NEM/(-)PIA vs. ()PIA in the presence of atropine. *: $P < 0.05$; **: $P < 0.01$; ****: $P < 0.0001$.

The result that both the autoreceptor as well as the adenosine receptor mechanism were affected by NEM points to a competition of both receptors for a common pool of G-proteins. However, since some evidence has been provided that both presynaptic receptors exert inhibition of ACh release by restricting the utilization of external Ca^{2+} an interaction at this level is also possible.

2.) Participation of protein kinase C in evoked transmitter release

Activation of PKC by the biologically active phorbol esters 4β-PDB or TPA increased stimulus-evoked transmitter release, as has been shown e. g. for NA (Wakade et al., 1985; Allgaier & Hertting, 1986; Allgaier et al., 1986b; 1987c; Versteeg & Florijn, 1987; Nichols et al., 1987), ACh (Tanaka et al., 1986; Allgaier et al., 1987b; Nichols et al., 1987; Versteeg & Florijn, 1987) and 5-hydroxytryptamine (Feuerstein et al., 1987; Wang & Friedmann, 1987).

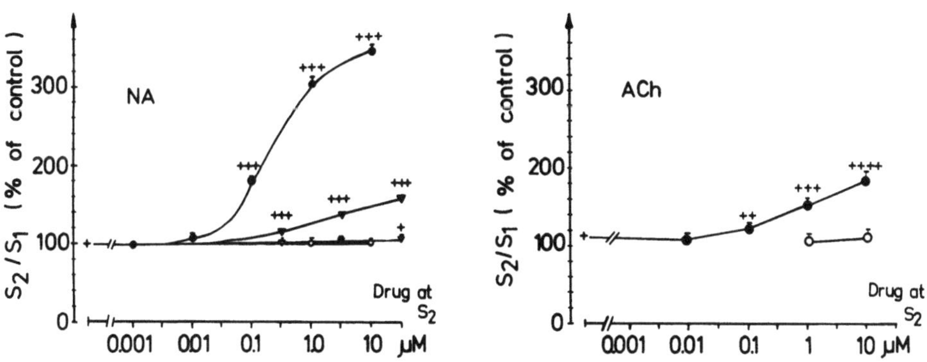

Fig. 7: Effects of phorbol esters on evoked [³H]NA and [³H]ACh release. During superfusion slices were stimulated twice electrically. 4β-PDB (●), TPA (▲), 4α-PDB (○), 4-O-methyl-TPA (△) were added 15 min before the second stimulation. Means of the ratios S_2/S_1 ± SEM are given. Number of experiments: 6-9. Significant differences from control: +$P < 0.05$; ++$P < 0.01$; +++$P < 0.001$; ++++$P < 0.0001$

Figure 7 shows the concentration-dependent augmentation of [^{3}H]NA and [^{3}H]ACh release in rabbit hippocampus by TPA and 4β-PDB. Inactive analogs of these phorbol esters (4-O-methyl-TPA or 4α-PDB) were ineffective on transmitter release. It can be seen that the more hydrophilic 4β-PDB increased [^{3}H]NA release in tissue slices to a much higher degree than TPA, in contrast to the phorbolester-induced activation of purified PKC, where both compounds were of about equal potency (Castagna et al., 1982). Enhancement of [^{3}H]ACh release by 4β-PDB was remarkably smaller than that of [^{3}H]NA release.

Compounds like polymyxin B (PMB) or 1-(5-isoquinolinesulfonyl)-2-methylpiperazine (H7) have been described as putative inhibitors of PKC (Kuo et al., 1983; Kawamoto & Hidaka, 1984). Both drugs counteracted not only the enhancement of release by the phorbol esters, but decreased also the evoked transmitter release per se (Allgaier & Hertting, 1986; Tanaka et al., 1986; Wakade et al., 1986; Versteeg & Ulenkate, 1987; Wang & Friedman, 1987). The latter result points to a possible participation of PKC in the chain of events of stimulus secretion-coupling. However, it should be taken into consideration that these inhibitors in contrast to the phorbol esters are by no means selective for PKC and may interfere with other protein kinases at the concentrations used.

As already described above, release of NA and ACh from hippocampal slices can be enhanced by a decrease of autoinhibition. This can be achieved either by blockade of the autoreceptors or by a functional inactivation of G-proteins involved in presynaptic receptor mechanisms. The facilitation of transmitter release by activation of PKC could be due to a functional impairment of autoinhibtion similar as described for IAP and NEM. It has been shown that phorbol esters phosphorylate G proteins by activating PKC (Jakobs et al., 1985). Therefore, the question was studied, if the enhancement of release resulted, at least in part, from a diminution of autoinhibition following activation of PKC.

The main findings from these experiments were

1.) Autoreceptor antagonists and phorbol esters increased stimulus-evoked release of both NA and ACh in an additive or

even over-additive manner (Allgaier et al., 1986b; 1987b,c).
2.) Inhibition of release by exogenous agonists was decreased due to the increase of the concentration of the endogenous transmitter in the synaptic cleft following activation of PKC by PDB (Allgaier et al., 1987c). A mere competition of both the endogenous and the exogenous agonist for the common receptor seems to be the reason for this effect. In the case of $[^3H]$NA release it has been seen that the inhibitory effect of clonidine, which was almost abolished in the presence of the phorbol ester at higher stimulation frequencies (3 Hz), was restored at lower stimulation frequencies (0.3 Hz) and hence lower biophase concentrations of NA.
3.) If the actual concentration of transmitter in the synaptic cleft is not enhanced during stimulation, since the transmitter is immediately metabolized before becoming operative at the autoreceptor, as in the case of ACh, the inhibitory net effects of exogenous agonists were not diminished in the presence of 4β-PDB. The same was true for release inhibition by heteroreceptor agonists (Allgaier et al., 1987b).

In conclusion, these results indicate an involvement of PKC in the mechanism of depolarization-induced transmitter release. However, a direct participation of this enzyme in modulation via presynaptic receptors seems to be unlikely.

References

Allgaier C, Feuerstein TJ, Jackisch R, Hertting G (1985) Islet-activating protein (pertussis toxin) diminishes α_2-adrenoceptor mediated effects on noradrenaline release. Naunyn-Schmiedeberg's Arch. Pharmacol. 331:235-239.

Allgaier C, Hertting G (1986) Polymyxin B, a selective inhibi bitor of protein kinase C, diminishes the release of noradrenaline and the enhancement of release caused by phorbol 12,13-dibutyrate. Naunyn Schmiedeberg`s Arch. Pharmacol. 334:218-221.

Allgaier C, Feuerstein TJ, Hertting G (1986a) N-ethylmaleimide (NEM) diminishes α_2-adrenoceptor mediated effects on noradrenaline release. Naunyn-Schmiedeberg's Arch. Pharmacol. 333:104-109.

Allgaier C, Von Kügelgen O, Hertting G (1986b) Enhancement of noradrenaline release by 12-O-tetradecanoyl-phorbol-13-acetate, an activator of protein kinase C. Eur. J. Pharmacol. 129:389-392.
Allgaier C, Hertting G, Von Kügelgen O (1987a) The adenosine receptor-mediated inhibition of noradrenaline release possibly involves a N-protein and is increased by α_2-autoreceptor blockade. Br. J. Pharmacol. 90:403-412.
Allgaier C, Daschmann B, Huang H, Hertting G (1987b) Protein kinase C and presynaptic modulation of acetylcholine release in rabbit hippocampus. Brit. J. Pharmacol. (in press).
Allgaier C, Hertting G, Huang HY, Jackisch R (1987c) Protein kinase C activation and α_2-autoreceptor-modulated release of noradrenaline. Br. J. Pharmacol. 92:161-172.
Berridge MJ (1984) Inositol trisphosphate and diacylglycerol second messengers. Biochem. J. 220:345-360.
Castagna M, Takai Y, Kaibuchi K, Sano K, Kikkawa U, Nishishizuka Y (1982) Direct activation of calcium-activated, phospholipid-dependent protein kinase by tumor-promoting phorbol esters. J. Biol. Chem. 257:7847-7851.
Cooper DMF (1982) Bimodal regulation of adenylate cyclase FEBS Lett. 138:157-163.
Feuerstein TJ, Allgaier C, Hertting G (1987) Possible involve- of protein kinase C (PKC) in the regulation of electrically evoked serotonin (5-HT) release from rabbit hippocampal slices. Eur. J. Pharmacol. 139:267-272.
Fredholm BB, Jonzon B, Lindgren E (1983) Inhibition of noradrenaline release from hippocampal slices by a stable adenosine analogue. Acta Physiol. Scand. 515:7-10.
Fredholm BB, Lindgren E (1986) Possible involvement of the Ni-protein in the prejunctional inhibitory effect of a stable adenosine analogue (R-PIA) on noradrenaline release in the rat hippocampus. Acta Physiol. Scand. 126:307-309.
Fredholm BB, Lindgren E (1987) Effects of N-ethylmaleimide and forskolin on noradrenaline release from rat hippocampal slices. Evidence that prejunctional adenosine and α-receptors are linked to N-proteins but not to adenylate cyclase. Acta Physiol. Scand. 130:95-105.
Hertting G, Zumstein A, Jackisch R, Hoffmann I, Starke K (1980) Modulation by endogenous dopamine of the release of acetylcholine in the caudate nucleus of the rabbit. Naunyn-Schmiedeberg's Arch. Pharmacol. 315:111-117.
Hertting G, Feuerstein TJ, Jackisch R, Allgaier C (1986) Possible involvement of the N_i protein of adenylate cyclase in the autoreceptor-mediated modulation of the evoked noradrenaline release. Sixth General Meeting of the European Society for Neurochemistry in Parague. Molecular Basis of Neural Function ed. by S. Tucek, S. Stipek, F. Stastny, J. Krivanek pp. 219.
Jackisch R, Werle E, Hertting G (1984) Identification of mechanisms involved in the modulation of release of noradrenaline in the hippocampus of the rabbit in vitro. Neurophamacol. 23:1363-1371.
Jackisch R, Fehr R, Hertting G (1985) Adenosine: an endogenous modulator of hippocampal noradrenaline release.

24:499-507.
Jakobs KH, Aktories K, Schultz G (1984) Mechanisms and components involved in adenylate cyclase inhibition by hormones. Adv. Cyclic Nucleotide Res. 17:135-143.
Jakobs KH, Bauer S, Watanabe Y (1985) Modulation of adenylate cyclase of human platelets by phorbol ester. Impairment of the hormone-sensitive inhibitory pathway. Eur. J. Biochem. 151:425-430.
Johnston H, Majewski H, Musgrave IF (1987) Involvement of cyclic nucleotides in prejunctional modulation of noradrenaline release in mouse atria. Br. J. Pharmacol. 91:773-781.
Katada T, Oinuma M, Kusakabe K, Ui M (1987) A new GTP-binding protein in brain tissues serving as the specific substrate of islet-activating protein, pertussis toxin. FEBS Lett. 213:353-358.
Kawamoto S, Hidaka H (1984) 1-(5-Isoquinolinesulfonyl)-2-me thylpiperazine (H-7) is a selective inhibitor of protein kinase C in rabbit platelets. Biochem. Biophys. Res. Commun. 125:258.
Kuo JF, Raynor L, Mazzei GJ, Schatzman RC, Turner RS, Kem WR (1983) Cobra polypeptide cytotoxin I and marine worm polypeptide cytotoxin A-IV are potent and selective inhibitors of phospholipid-sensitive Ca^{2+}-dependent protein kinase. FEBS Lett. 153:183-186.
Kawai Y, Whitsel C, Arinze IJ (1986) $NADP^+$ enhances cholera and pertussis toxin-catalyzed ADP-ribosylation of membrane proteins. J. Cyclic Nucleotide Res. 11:265-274.
Laemmli CK (1970) Cleavage of structural proteins during assembly of the head of bacteriophage T4. Nature 227:680-685.
Limberger N, Späth L, Hölting T, Starke K (1986) Mutual interaction between presynaptic α_2-adrenoceptors and opioid-receptors at the noradrenergic axons of rabbit brain cortex. Naunyn-Schmiedebergs's Arch. Pharmacol. 334:166-171.
Marchi M, Paudice P, Raiteri M (1981) Autoregulation of acetyl choline release in isolated hippocampal nerve endings. Eur. J. Pharmacol. 73:75-79.
Mulder AH, Frankhuyzen AL, Stoof JC , Wemer J, Schoffelmeer (1984) Catecholamine receptors, opiate receptors, and presynaptic modulation of neurotransmitter release in the brain. Catecholamines: Neuropharmacol. and Central Nervous System - Theoretical Aspects, Alan R Liss, Inc., New York, pp. 47-58
Murayama T, Ui M (1983) Loss of the inhibitory function of the guanine nucleotide regulatory component of adenylate cyclase due to its ADP-ribosylation by islet-activating protein, pertussis toxin, in adipocyte membranes. J. Biol. Chem. 258:3319-3326.
Nichols RA, Haycock JW, Wang JKT, Greengard P (1987) Phorbolester enhancement of neurotransmitter release from rat brain synaptosomes. J. Neurochem. 48:615-621.
Nishizuka Y (1984) The role of protein kinase C in cell surface signal transduction and tumour promotion. Nature 308:693-698.

Nordström Ö, Bartfai T (1981) 8-Br-cyclic GMP mimics activation of muscarinic autoreceptor and inhibits acetylcholine release from rat hippocampal slices. Brain Res. 213:467-471.

Pedata F, Giovannelli L, De Sarno P, Pepeu G (1986) Effect of adenosine, adenosine derivatives, and caffeine on acetylcholine release from brain synaptosomes: Interaction with muscarinic autoregulatory mechanisms. J. Neurochem. 46:1593-1598.

Richardson IW, Szerb JC (1974) The release of labelledacetylcholine and choline from cerebral cortical slices stimulated electrically. Br. J. Pharmacol. 52:499-507.

Robinson PJ, Lovenberg W (1986) Dopamine and serotonin in two populations of synaptosomes isolated by percoll gradient centrifugation. Neurochem. Int. 9:455-458.

Rodbell M (1980) The role of hormone receptors and GTP-regulaproteins in membrane transduction. Nat. 284:17-22.

Schoffelmeer ANM, Mulder AH (1983) ^{3}H-Noradrenaline release rat neocortical slices in the absence of extracellular Ca^{2+} and its presynaptic alpha$_2$-adrenergic modulation. A study on the possible role of cyclic AMP. Naunyn-Schmiederberg's Arch. Pharmacol. 323:188-192.

Starke K (1987) Presynaptic α-autoreceptors. Rev. Physiol. Biochem. Pharmacol. 107: 73-146.

Sternweis PC, Robishaw JD (1984 Isolation of two proteins with high affinity for guanine nucleotides from membranes of bovine brain. J. Biol. Chem. 259:13806-13813.

Strittmatter H, Jackisch R, Hertting G (1982) Role of dopamine receptors in the modulation of acetylcholine release in the rabbit hippocampus. Naunyn-Schmiedeberg's Arch Pharmacol. 321:195-200.

Tanaka C, Fujiwara H, Fujii Y (1986) Acetylcholine release from guinea pig caudate slices evoked by phorbol ester and calcium. FEBS Lett. 195:129-134.

Versteeg DHG, Florijn WJ (1987) Phorbol 12,13-dibutyrate enhances electrically stimulated neuromessenger release from rat dorsal hippocampal slices in vitro. Life Sci. 40:1237-1243.

Versteeg DHG, Ulenkate HJLM (1987) Basal and electrically stimulated release of [^{3}H]dopamine from rat amygdala slices in vitro: effects of 4β-phorbol 12,13-dibutyrate, 4α-phorbol 12,13-didecanoate and polymyxin B. Brain Res. 416:343-348.

Wakade AR, Malhotra RK, Wakade TD (1985) Phorbol ester, an activator of protein kinase C, enhances calcium-dependent release of sympathetic neurotransmitter. Naunyn-Schmiedeberg`s Arch Pharmacol. 331:122-124.

Wakade AR, Malhotra RK, Wakade TD (1986) Phorbol ester facilitates ^{45}Ca accumulation and catecholamine secretion by nicotine and excess K^{+} but not by muscarine in rat adrenal medulla. Nature 321-698-700.

Wang HY, Friedman E (1987) Protein kinase C: regulation of serotonin release from rat brain cortical slices. Eur. J. Pharmacol. 141:15-21.

ATTEMPTS TO CHARACTERISE DOPAMINE RECEPTOR-EFFECTOR MECHANISMS IN THE BRAIN

E. Kelly and S.R. Nahorski,
Department of Pharmacology and Therapeutics,
Medical Sciences Building,
University of Leicester,
University Road,
Leicester. LE1 7RH.
U.K.

In the central nervous system, communication between cells is considered to occur by means of chemical neurotransmitters. The neurotransmitter is released from a presynaptic neurone, traverses a synaptic cleft, and interacts with specific receptor sites on the membrane of the postsynaptic cell. This interaction modifies cell function either by directly opening or closing membrane ion channels, or by altering the intracellular concentration of second messenger molecules, such as cyclic AMP, inositol-1,4,5-trisphosphate (IP_3), or diacylglycerol. Dopamine is an example of a monoamine neurotransmitter that is thought to modify cell function by enhancing or suppressing the production of second messenger molecules. Here we describe our attempts to delineate the cerebral receptor-effector mechanisms for dopamine.

In 1972, dopamine was found to stimulate adenylate cyclase activity in homogenates of rat striatum (Kebabian et al 1972). Although this generated great interest in the possibility that this effector system might mediate all the physiological actions of dopamine in the brain, it eventually became clear that this was not the case, since some clinically effective dopamine antagonists, such as the substituted benzamides, were inactive in antagonising the dopamine stimulated enzyme activity (Spano et al 1978). This consequently led to the concept of multiple dopamine receptors, such that D-1 receptors are linked in a stimulatory fashion to adenylate cyclase with D-2 receptors unlinked to cyclase activity (Kebabian and Calne 1979). Around this time the D-2 receptor present in the intermediate lobe of the pituitary gland which mediates inhibition of α-MSH release was found to inhibit adenylate cyclase activity in this tissue (Munemura et al 1980). Before long cerebral D-2 receptors were also shown to suppress cyclic AMP formation (Stoof and Kebabian 1981) and accordingly the inhibition of striatal adenylate cyclase by D-2 receptors has been widely studied (Battaglia et al 1985; Cooper et al 1985; Lazareno et al 1985; Onali et al

NATO ASI Series, Vol. H19
Modulation of Synaptic Transmission and Plasticity
in Nervous Systems. Edited by G. Hertting and H.-C. Spatz

1985; Weiss _et al_ 1985; Kelly and Nahorski 1986a). As it stands at the present time, the classification of cerebral dopamine receptors includes D-1 receptors that stimulate adenylate cyclase, and D-2 receptors, at least a portion of which inhibit cyclase activity (for reviews see Stoof and Kebabian 1984; Kelly and Nahorski 1986b). On the basis of ligand binding studies, more extensive classifications of dopamine receptors have in the past been proposed (Seeman 1980). However, since these classifications are not based on functional effects they will not be considered further here, but we do acknowledge at least the possibility of dopamine receptor sub-types other than those discussed here.

The effects of dopamine receptor stimulation on adenylate cyclase activity can be measured in a number of ways. Apart from directly assessing enzyme activity in tissue homogenates, another method has been to measure the production of cyclic AMP in striatal slices, either by monitoring the efflux of cyclic AMP from tissue slices (Stoof and Kebabian 1981), or by measuring the cyclic AMP content of tissue slices (Lazareno _et al_ 1985). We have employed the latter method, and have been able to demonstrate that D-2 receptors inhibit D-1 stimulated cyclic AMP accumulation in striatal slices. This was done (Kelly and Nahorski 1986a) by incubating striatal slices with a selective D-1 agonist, SKF 38393 (Setler _et al_ 1978) in the presence or absence of a selective D-2 agonist, RU 24926 (Euvrard _et al_ 1979). Increasing concentrations of the D-2 agonist could inhibit the D-1 stimulation, and the D-2 inhibition could also be blocked by the selective D-2 antagonist, (-)-sulpiride (Kelly and Nahorski, 1986a). Further evidence for such an interaction was obtained by incubating slices with dopamine in the presence or absence of (-)-sulpiride. Under these conditions dopamine alone enhanced cyclic AMP accumulation, but in the presence of D-2 receptor blockade could now produce a much larger increase in cyclic AMP accumulation (Table 1), presumably due to the loss of inhibitory tone supplied by D-2 receptors on D-1 receptor function. Interestingly, in striatal slices we were only able to observe D-2 inhibition of D-1 stimulated cyclic AMP accumulation, since a D-2 agonist was unable to inhibit cyclic AMP accumulation due to isoprenaline, PGE_1, 2-Chloroadenosine or vasoactive intestinal polypeptide (Kelly and Nahorski 1987a). This suggests a selective inhibition of D-1 cyclase by D-2 receptors, although in primary culture of striatal neurones activation of D-2 receptors inhibits vasoactive intestinal polypeptide as well as D-1 stimulated cyclic AMP accumulation (Weiss _et al_ 1985). These differences

Table 1

Treatment	Cyclic AMP accumulation (pmoles/mg protein)
Basal	12.85 ± 1.64
Dopamine (50 μM)	22.75 ± 2.26
Dopamine+(-)-sulpiride (30 μM)	35.73 ± 1.50*
Basal	10.64 ± 1.20
Amphetamine (10 μM)	11.95 ± 0.76
Amphetamine + (-)-sulpiride (30 μM)	16.76 ± 2.20*

Effect of drugs on cyclic AMP accumulation in striatal slices, done in the presence of 1 mM isobutylmethylxanthine. Slices were incubated for 5 min in the presence or absence of (-)-sulpiride before addition of dopamine or amphetamine for a further 20 min (Kelly and Nahorski 1986a). The cyclic AMP content was measured by a protein binding assay. Values are means ± SEM from single experiments. Asterisks indicate values in presence of (-)-sulpiride significantly higher than without D-2 blockade. $P < 0.05$, Student's 't' test.

could relate to the presence of many different cell types in the slice preparation.

The hypothesis that in striatum, intracellular cyclic AMP is modulated by D-1 and D-2 receptors is further substantiated by experiments in which we have treated striatal slices with amphetamine to enhance the release of dopamine onto postsynaptic dopamine receptors (Kelly and Nahorski 1987c). Under these conditions, amphetamine produced a small increase in intracellular cyclic AMP accumulation which was greatly enhanced by addition of the D-2 selective antagonist (-)-sulpiride to the incubation medium (Table 1). Conversely, addition of SCH 23390, a selective D-1 antagonist (Hyttel 1983) abolished amphetamine's enhancement of cyclic AMP accumulation in the presence of (-)-sulpiride. Furthermore, pretreatment of animals with reserpine and α-methyl-p-tyrosine, or prior lesion of nigrostriatal neurones with 6-hydroxydopamine could block both the ability of amphetamine to release dopamine and to enhance cyclic AMP accumulation in the presence of (-)-sulpiride in striatal slices (Kelly and Nahorski 1987c). Together these studies indicate that in striatum, D-1 and D-2 receptors reciprocally regulate adenylate cyclase activity. At least a portion of D-1 and D-2 receptors appear to coexist on the same striatal neurone (Weiss _et al_ 1985), although the functional implications of such an arrangement are not known. Interestingly, a recent report indicates that D-1 and D-2 receptor stimulation exert opposite effects on GABA release in the dorsal striatum (Girault _et al_ 1986). Another possibility is that D-2 receptors serve to prevent over-stimulation of adenylate cyclase via D-1 receptors.

As with other neurotransmitter receptors linked to adenylate cyclase, dopamine receptors are thought to act via guanine nucleotide binding regulatory proteins termed G_s and G_i for stimulation and inhibition of adenylate cyclase respectively (for reviews see Gilman 1984; Levitzki 1987). The means by which dopamine receptors might interact with these GTP binding proteins has been described previously (Kelly and Nahorski 1986b). The investigation of the function of these GTP binding proteins has been greatly aided by the advent of toxins which selectively interact with and ADP-ribosylate G_s or G_i (for review see Dolphin 1987). Pertussis toxin ADP-ribosylates the α subunit of G_i and in effect blocks the receptor mediated inhibition of adenylate cyclase activity (Murayama and Ui 1983), and can thus be used to identify receptor responses mediated by a G_i protein. In a series of experiments we investigated whether intrastriatal injection of pertussis toxin could produce neurochemical and behavioural changes in the rat consistent with a loss of D-2 receptor function. One day folowing unilateral intrastriatal injection of 2 µg pertussis toxin, rats displayed ipsilateral postural asymmetries which intensified into ipsilateral circling upon peripheral administration of the dopamine agonist apomorphine (Kelly et al 1987b). This is consistent with a loss of D-2 receptor function in the toxin injected striatum. Furthermore, injections of pertussis toxin partially reversed the ability of a selective D-2 agonist, RU 24926, to inhibit cyclic AMP accumulation due to the D-1 agonist SKF 38393 in striatal slices (% D-2 inhibition of D-1 stimulation reduced from 53 ± 4 in control to 33 ± 4 in toxin treated striata, n = 8 separate experiments, $P < 0.05$ by paired 't' test). Similar findings have been reported recently by other groups (Fujita et al 1985; Olianas and Onali 1987). The toxin treatment also reduced high affinity agonist binding to ^{3}H-spiperone labelled D-2 receptors (Table 2) and, like (±)-sulpiride could reduce the

Table 2

	Control	Pertussis Toxin
Dopamine IC_{50} (M)	1.02 (± 0.33) x 10^{-8}	4.33 (± 0.90) x 10^{-8}*
Dopamine + GTP IC_{50} (M)	2.28 (± 0.40) x 10^{-7}	1.60 (± 0.25) x 10^{-7}

Dopamine displacement of ^{3}H-spiperone binding to striatal membranes in the presence or absence of pertussis toxin pretreatment: effect of 100 µM GTP. Method according to Kelly et al 1987b. Values are means ± SEM of 3 experiments. Asterisk indicates value in presence of pertussis toxin lower than control. $P < 0.05$, paired 't' test.

dopamine/DOPAC ratio in striatal tissue (Table 3). All these results together suggest the functional uncoupling of striatal D-2 receptors from a G_i protein, following ADP-ribosylation of the latter, and indicate that striatal D-2 receptors may function via cyclase inhibition to produce known behavioural and neurochemical effects. However, it is now known that

Table 3

	Dopamine/DOPAC
Saline	5.76 ± 0.65
(±)-Sulpiride (10 μg)	3.84 ± 0.28*
Bovine serum albumin (2 μg)	5.09 ± 0.45
Perussis toxin (2 μg)	3.79 ± 0.19*

Effect of intrastriatal (±)-sulpiride and pertussis toxin on dopamine metabolism, as measured by HPLC with electrochemical detection (Herdon et al 1985). Dopamine metabolism was estimated by measuring the ratio of tissue dopamine:DOPAC (3,4-Dihydroxyphenylacetic acid) content. (±)-Sulpiride was injected 1 hr previously, pertussis toxin 24 hr previously. Each value represents the mean ± SEM of 4-8 determinations. Asterisks indicate values where $P < 0.05$ compared to control side, paired 't' test.

another protein, G_o, is also a substrate for pertussis toxin (Sternweis and Robishaw 1984). Since the linkage of receptors to ion channels via a pertussis toxin sensitive G protein has recently been demonstrated (Holz et al 1986; Breitwieser and Szabo 1985), a possible effector linked to G_o and therefore linkage of D-2 sites to such an effector mechanism could be possible. With relation to the above experiments such an effector mechanism, like G_i and cyclase inhibition, would be susceptible to pertussis toxin treatment.

The possibility that some cerebral D-2 receptors may not link to cyclase inhibition has gained support from experiments we have carried out on tissue from the limbic system (nucleus accumbens and tuberculum olfactorium), which like the striatum has high dopamine content and substantial levels of both D-1 and D-2 receptors. Incubation of striatal or limbic slices with the D-1 agonist SKF 38393 enhanced cyclic AMP accumulation in both preparations (Table 4). However, whereas coincubation of the slices with the D-2 agonist LY 171555 inhibited the D-1 stimulation in striatum, it did not do so in limbic slices (Table 4). This indicates that in contrast to the striatum, the limbic forebrain D-2 receptors are not linked to D-1 stimulated adenylate cyclase activity. Also, the possibility that D-2 receptors in limbic forebrain might inhibit

Table 4

	Cyclic AMP accumulation (pmoles/mg protein)	
	Limbic	Striatum
Basal	13.2 ± 0.6	6.8 ± 0.2
SKF 38393 (1 μM)	20.2 ± 1.0	15.5 ± 0.8
SKF 38393 + LY 171555 (10 μM)	22.1 ± 1.6	10.5 ± 1.1*

Effect of selective D-1 and D-2 agonists on cyclic AMP accumulation in slices of limbic forebrain and striatum, in the presence of 1 mM isobutylmethylxanthine. Samples were preincubated in the presence or absence of LY 171555 for 5 min before addition of SKF 38393 for a further 20 min (Kelly and Nahorski 1987a). Values are means ± SEM in a single experiment. Asterisk indicates the value in the presence of the D-2 agonist significantly lower than in the presence of the D-1 agonist alone. $P < 0.05$, Student's 't' test.

cyclic AMP accumulation due to activation of other cyclase linked receptors appears unlikely, since LY 171555 did not inhibit cyclic AMP accumulation due to isoprenaline, vasoactive intestinal polypeptide or 2-Chloroadenosine (Kelly and Nahorski 1987a). It therefore remains a possibility that cerebral D-2 receptors may also be linked to effector mechanisms other than adenylate cyclase. Recently it was reported that D-2 receptor stimulation decreases the inositol trisphosphate level of rat striatal slices (Pizzi et al 1987). However in our experiments we have been unable to observe an effect of D-2 (or D-1) receptor stimulation on the accumulation of labelled inositol phosphates in striatal slices under basal conditions, or in the presence of the muscarinic agonist carbachol (Table 5). These results do not support the possibility that dopamine receptors modulate phosphoinositide metabolism in striatum.

Table 5

	Labelled inositol phosphates, cpm/150 μl slices		
	Basal	Carbachol (1 mM)	Carbachol + dopamine (100 μM)
IP_1	4369 ± 1142	14036 ± 2261	15238 ± 2525
IP_2	1092 ± 291	4842 ± 1056	4878 ± 955
IP_3	850 ± 206	2182 ± 434	2288 ± 432
IP_4	153 ± 25	1276 ± 200	1388 ± 409

Lack of effect of dopamine on carbachol stimulated accumulation of labelled inositol phosphates in slices of rat striatum. Slices were prelabelled with myo[2-^{3}H]inositol at 5 μCi/50 ul slices for 1 hr. Dopamine was added for 5 min prior to carbachol, which was added to the slices for 30 min. The inositol phosphates were extracted and separated according to the method of Batty et al 1985. Values are means ± SEM from 4 separate experiments.

As mentioned previously the existence of receptors linked to ion channels via a pertussis toxin sensitive G protein has recently been demonstrated (Holz _et al_ 1986; Breitwieser and Szabo 1985). Thus, along with $GABA_B$ and α_2 receptors (Holz _et al_ 1986) it is possible that D-2 receptors might also inhibit the opening of voltage sensitive Ca^{2+} channels. In experiments to address this possibility, we labelled synaptoneurosomal preparations of rat striatum (Hollingsworth _et al_ 1985) with the Ca^{2+} sensitive dye Fura 2 (2.5 μM; Grynkiewicz _et al_ 1985). Under these conditions the level of free intracellular Ca^{2+} ($Ca^{2+}{}_{int}$) lay between 90 and 180 nmolar. Addition of KCl (6-25 mM) or veratridine (1-100 μM) concentration dependently increased $Ca^{2+}{}_{int}$ in the synaptoneurosomal preparation. However prior (1 min) addition of the D-2 agonist LY 171555 (10 μM) failed to alter either basal $Ca^{2+}{}_{int}$ or the increase in $Ca^{2+}{}_{int}$ induced by 25 mM KCl or 30 μM veratridine (Table 6). These results suggest that striatal D-2 receptors do not modify the function of voltage-sensitive Ca^{2+} channels. However, it has recently been shown using slices of guinea-pig striatum loaded with Quin 2, D-2 receptor stimulation inhibits the KCl evoked increase in $Ca^{2+}{}_{int}$. So it is possible that D-2 receptors do modulate $Ca^{2+}{}_{int}$, however in our experiments the preparation stage for the synaptoneurosomes may have disrupted the linkage of D-2 receptors to Ca^{2+} channels.

Table 6

	Change in fluorescence (arbitrary units)
KCl (25 mM)	21 ± 1
KCl + LY 171555 (10 μM)	22 ± 1
Veratridine (30 μM)	37 ± 4
Veratridine + LY 171555 (10 μM)	35 ± 3

Lack of effect of D-2 activation on KCl and veratridine induced increase in $Ca^{2+}{}_{int}$ in striatal synaptoneurosomes loaded with 2.5 μM Fura 2. Excitation wavelength was 385 nm, emission wavelength 505 nm, such that an increase in $Ca^{2+}{}_{int}$ was recorded as a decrease in fluorescence (giving a better signal to noise ratio). The D-2 agonist or vehicle (water) was added 1 min prior to addition of the depolarising agent. Values are means ± SEM from a single experiment.

In conclusion, the firmest evidence to date indicates that D-1 and D-2 receptors exert reciprocal effects on the regulation of adenylate cyclase. These effects undoubtedly occur via GTP binding regulatory proteins termed G_s and G_i. The evidence that dopamine receptors regulate the hydrolysis of membrane phosphoinositides remains controversial and requires further careful study. It remains in our view a strong possibility that like some other neurotransmitter receptors, D-2 receptors may influence membrane ion channel function by a mechanism independent of adenylate cyclase but involving an as yet undefined G protein. Such a mechanism could be crucial for the modulation of transmitter release and synthesis.

Furthermore the purification of D-1 and D-2 sites and reconstitution with G-proteins will also be crucial to fully elucidate the underlying mechanisms of dopamine receptor function. Such a study has already provided evidence that in the anterior pituitary gland, D-2 receptors are associated with a G-protein that may be G_0 (Senogles et al 1987). Hopefully such studies will eventually define the exact nature of cerebral dopamine receptor-effector mechanisms, and thus a better understanding of dopaminergic neurotransmission.

Acknowledgement: We thank The Wellcome Trust for financial support, Jenny Bell for manuscript preparation, and Alan Willcocks, Jim Strupish and Ian Batty for assistance with the experimental work.

References

Battaglia G, Norman AB, Hess EJ, Creese I (1985) D-2 Dopamine receptor-mediated inhibition of forskolin-stimulated adenylate cyclase activity in rat striatum. Neurosci Lett 59: 177-182

Batty IR, Nahorski SR, Irvine RF (1985) Rapid formation of inositol 1,3,4,5-tetrakisphosphate following muscarinic receptor stimulation of rat cerebral cortical slices. Biochem J 232: 211-215

Breitwieser GE, Szabo G (1985) Uncoupling of cardiac muscarinic and β-adrenergic receptors from ion channels by a guanine nucleotide analogue. Nature 317: 538-540

Cooper DMF, Bier-Laning CM, Halford MK, Ahlijanian MK, Zahniser NR (1985) Dopamine, acting through D-2 receptors, inhibits rat striatal adenylate cyclase by a GTP-dependent process. Mol Pharmacol 29: 113-119

Dolphin AC (1987) Nucleotide binding proteins in signal transduction and disease. Trends in Neurosci 10: 53-57

Euvrard C., Premont J, Oberlander C, Boissier JR, Bockaert J (1979) Is dopamine-sensitive adenylate cyclase involved in regulating the activity of striatal cholinergic neurones? Naunyn-Schmiedeberg's Arch Pharmacol 309: 241-245

Fujita N, Nakahiro M, Fukuchi I, Saito K, Yoshida H (1985) Effects of pertussis toxin on D-2 dopamine receptor in rat striatum: evidence for coupling of N_i regulatory protein with D-2 receptor. Brain Res 333: 231-236

Gilman AG (1984) G-proteins and dual control of adenylate cyclase. Cell 36: 577-579

Girault JA, Spampinato U, Glowinski J, Besson MJ (1986) In vivo release of [^{3}H]γ-aminobutyric acid in the rat neostriatum - II. Opposing effects of D-1 and D-2 dopamine receptor stimulation in the dorsal caudate putamen. Neuroscience 19: 1109-1117

Grynkiewicz G, Poenie M, Tsien RY (1985) A new generation of Ca^{2+} indicators with greatly improved fluorescence properties. J Biol Chem 260: 3440-3450

Herdon H, Strupish J, Nahorski SR (1985) Differences between the release of radiolabelled and endogenous dopamine from superfused rat brain slices: effects of depolarizing stimuli, amphetamine and synthesis inhibition. Brain Res 348: 309-320

Hollingsworth EB, McNeal ET, Burton JL, Williams RJ, Daly JW, Creveling CR (1985) Biochemical characterisation of a filtered synaptoneurosome preparation from guinea pig cerebral cortex: cyclic adenosine 3':5'-monophosphate-generating systems, receptors and enzymes. J Neurosci 5: 2240-2253

Holz GG, Rane SG, Dunlop K (1986) GTP-Binding proteins mediate transmitter inhibition of voltage-dependent calcium channels. Nature 319: 670-672

Hyttel J (1983) SCH 23390 - the first selective dopamine D-1 antagonist. Eur J Pharmacol 91: 153-155

Kebabian JW, Petzold GL, Greengard P (1972) Dopamine-sensitive adenylate cyclase in caudate nucleus of rat brain and its similarity to the 'dopamine receptor'. Proc Nat Acad Sci 69: 2145-2149

Kebabian JW, Calne DB (1979) Multiple receptors for dopamine. Nature 277: 93-96

Kelly E, Nahorski SR (1986a) Specific inhibition of dopamine D-1 mediated cyclic AMP formation by dopamine D-2, muscarinic cholinergic, and opiate receptor stimulation in rat striatal slices. J Neurochem 47: 1512-1516

Kelly E, Nahorski SR (1986b) Dopamine receptor-effector mechanisms. Rev in the Neurosci 1: 35-54

Kelly E, Nahorski SR (1987a) Dopamine D-2 receptors inhibit D-1 stimulated cyclic AMP accumulation in striatum but not limbic forebrain. Naunyn-Schmiedeberg's Arch Pharmacol 335: 508-512

Kelly E, Willcocks AL, Nahorski SR (1987) Neurochemical and behavioural evidence that dopamine D-2 receptors in striatum couple to the N_i regulatory protein and inhibition of cyclic AMP accumulation. Naunyn-Schmiedeberg's Arch Pharmacol 335: 618-623

Kelly E, Nahorski SR (1987) Endogenous dopamine functionally activates D-1 and D-2 receptors in striatum. J Neurochem 49: 115-120

Lazareno S, Marriott DB, Nahorski SR (1985) Differential effects of selective and non-selective neuroleptics on intracellular and extra-cellular cyclic AMP accumulation in rat striatal slices. Brain Res 361: 91-98

Levitzki A (1987) Regulation of hormone-sensitive adenylate cyclase. Trends in Pharm Sci 8: 299-303

Munemura M, Cote TE, Tsuruta K, Eskay RL, Kebabian JW (1980) The dopamine receptor in the intermediate lobe of the rat pituitary gland: pharmacological characterization. Endocrinol 107: 1676-1683

Murayama T, Ui M (1983) Loss of the inhibitory function of the guanine nucleotide regulatory component of adenylate cyclase due to its ribo-sylation by islet activating protein, pertussis toxin, in adipocyte membranes. J Biol Chem 258: 3319-3326

Olianas MC, Onali P (1987) Pertussis toxin attenuates D-2 inhibition and enhances D-1 stimulation of adenylate cyclase by dopamine in rat striatum. J Neurochem 48: 1443-1447

Pizzi M, D'Agostini F, Da Prada M, Spano PF, Haefely WE (1987) Dopamine D-2 receptor stimulation decreases the inositol trisphosphate level of rat striatal slices. Eur J Pharmacol 136: 263-264

Seeman P (1980) Brain dopamine receptors. Pharmacol Rev 32: 229-313
Senogles SE, Benovic JL, Amlaiky N, Unson C, Milligan G, Vinitsky R, Spiegel AM, Caron MG (1987) The D-2 dopamine receptor of anterior pituitary is functionally associated with a pertussis toxin-sensitive guanine nucleotide binding protein. J Biol Chem 262: 4860-4867
Setler PE, Sarau HM, Zirkle CL, Saunders HL (1978) The central effects of a novel dopamine agonist. Eur J Pharmacol 50: 419-430
Spano PF, Govoni S, Trabucchi M (1978) Studies on the pharmacological properties of dopamine receptors in various areas of the central nervous system. Advan Biochem Psychopharmacol 19: 155-165
Sternweis PC, Robishaw JD (1984) Isolation of two major proteins with high affinity for guanine nucleotides. J Biol Chem 259: 13806-13813
Stoof JC, Kebabian JW (1981) Opposing roles for D-1 and D-2 dopamine receptors in efflux of cyclic AMP from rat neostriatum. Nature 294: 366-368
Stoof JC, Kebabian JW (1984) Two dopamine receptors: biochemistry, physiology and pharmacology. Life Sci 35: 2281-2296
Weiss S, Sebben M, Garcia-Sainz JA, Bockaert J (1985) D-2 Dopamine receptor-mediated inhibition of cyclic AMP formation in striatal neurones in primary culture. Mol Pharmaol 27: 595-599

POSTSYNAPTIC SIGNAL TRANSDUCTION IN NEUROBLASTOMA AND GANGLION CELLS: RECEPTOR-MEDIATED CONTROL OF K-CURRENTS

[1]D.A. Brown, [2]H. Higashida, [3]P.R. Adams and [4]N.V. Marrion
[1]MRC Neuropharmacology Research Group, Department of Pharmacology, University College London, Gower Street, London WC1E 6BT.

[2]Department of Pharmacology, Cancer Research Institute, Kanazaura University, Kanazaura 920, Japan.

[3]Department of Neurobiology and Behavior, State University of New York, Stony Brook, NY 11790, U.S.A.

[4]MRC Neuropharmacology Research Group, Department of Pharmacology, School of Pharmacy, 29/39 Brunswick Square, London WC1N 1AX.

Introduction

Communication within the nervous system is governed by the frequency and pattern of action potential discharges, and hence by variations in ionic current flow across the outer cell membrane. Thus, if a chemical neurotransmitter is to change the electrical activity of a recipient neurone, its biochemical action on the receptor has to be transduced into a change in ionic current flow.

Functionally, the transduction step from receptor activation to ionic current flow may be classified into two forms - DIRECT and INDIRECT. This functional division has a physical basis in the structure of the receptor.

Direct transduction - Here the "receptor" - that is, the binding site for the transmitter - is an integral component of a macromolecular complex which also incorporates a potential ion channel. As a result, binding of the transmitter induced a very rapid (< 100 μsec) conformational change in the protein complex allowing the transient opening of the ion channel. This form of transmission, with its short-latency, is ideally suited for rapid point-to-point communication, the transmitter effectively "closing the switch" in a hard-wired circuit. The number of chemical transmitter substances fulfilling this function appears rather limited: principal

NATO ASI Series, Vol. H19
Modulation of Synaptic Transmission and Plasticity
in Nervous Systems. Edited by G. Hertting and H.-C. Spatz

excitatory (i.e. cation-channel generating) transmitters in vertebrates are acetylcholine in the peripheral nervous system (acting on nicotinic receptors) and glutamate in the central nervous system, while glycine and GABA (γ-aminobutyric acid) fulfill inhibitory transmitter functions, the channel being anion-selective. A close homology between the structure of the receptors for acetylcholine, GABA and glycine has recently been shown (1,2) supporting this functional analogy.

Indirect transduction - In this case, the receptor protein does not contain a potential ion channel; instead, the receptor and ion channel comprise quite separate molecules. Thus, transduction of the signal received by the receptor (in the form of transmitter binding) into a change in ionic channel function involves a number of intermediary steps. Several characteristics distinguish this form of transmision from direct transmission.

(a) The intermediary steps in the transduction process confer a finite lag of some tens of milliseconds (or more) in the response to the transmitter. Thus, this form of transmission is ill-suited to rapid information processing but instead provides a form of fine-tuning, controlling neuronal excitability over periods of seconds or minutes.

(b) There is considerable flexibility in the response to a given transmitter, both in terms of which ion channels are affected, and in terms of the change in ion channel function - i.e. whether they are opened or closed or whether their gating properties are modified. Though certain constraints are emerging, depending on the structure of the individual receptors, coupling proteins (see below) and ion channels, at present these are insufficient to permit fully predictive rules.

(c) This form of transmission is operated by a remarkably wide variety of transmitters, such as acetylcholine (via muscarinic receptors), catecholamines and other monoamines, GABA (at $GABA_B$ receptors), and peptides, all acting on physically discrete receptors. However, available information suggests that these receptors may all conform to a generally homologous structure, distinct from the direct-transmitting ionophore-receptors, and constituting a second "super-family" of receptors (3).

(d) In most systems so far investigated, the immediate post-receptor transduction step comprises coupling with one of a family of GTP-binding proteins (G-proteins), resulting in dissociation of the G-protein into its constituent α and $\beta\gamma$ components and activation of its GTPase function (4).

Activation of the G-protein may then lead to ion channel modulation in one or more of the following ways:

(i) Activated G-protein subunits may couple directly to an ion channel, to open or close it.

(ii) The G-protein may activate or inhibit one of several intracellular enzyzmes, such as adenylate cyclase, cyclic nucleotide phosphodiesterase or phospholipase C, to modify the levels of soluble second messengers such as cyclic adenosine monophosphate, cyclic guanosine monophosphate, inositol phosphate or diacylglycerol.

(iii) The soluble messengers may interact directly with the channel, or, in turn, activate kinase enzymes, modifying the state of channel protein phosphorylation.

The inositol phosphate/diacylglycerol pathway -

This is a widely-used post-receptor transduction pathway in a variety of cell types (5). Here, activation of the receptor leads to the G-protein-coupled activation of phospholipase C, resulting in the hydrolysis of membrane phosphatidylinositol-4,5-bisphosphate (PIP_2) to two potential "second messengers" inositol-1,4,5-trisphosphate (IP_3) and diacylglycerol (DAG). The principal effect of IP_3 is to release Ca from the endoplasmic reticulum, which may act as a "tertiary messenger", in turn activating membrane Ca-dependent ion channels, triggering secretion or contraction, or activating Ca-dependent kinase or phosphatase enzymes. However, IP_3 may also activate membrane ion channels more directly (6), or be phosphorylated to another prospective messenger inositol-1,3,4,5-tetrakisphosphate (IP_4) (7), whose effects have yet to be fully defined. The other messenger, DAG, activates protein kinase C (C-kinase: ref. 8), which is potentially capable of phosphorylating, and modifying, a number of membrane ionic channels (9).

In the remainder of this article we consider the extent to which this "PI-system" is responsible for the modification of K-currents in neuroblastoma and ganglion cells by peptides and by cholinergic muscarinic agonists.

Membrane current changes mediated by receptor-triggered PIP_2 hydrolysis in neuroblastoma hybrid cells

NG108-15 neuroblastoma × glioma hybrid cells provide an interesting example of a "neuronal" cell-type capable of exhibiting a strong PI-response to certain peptides, most notably bradykinin (BK). Thus, application of BK in nanomolar concentrations rapidly accelerates PIP_2 hydrolysis (10) and produces a large, transient rise in intracellular IP_3 (11). This effect is accompanied by accelerated GTPase activity, implicating G-protein coupling (12). The G-protein involved is at least partly insensitive to pertussis toxin (PTX: 12). BK also raises intracellular Ca levels (13,14); this is probably due to the Ca-releasing action of the IP_3 formed from PIP_2 breakdown, since IP_3 itself releases Ca when injected into the cells (14) or from a microsomal preparation of these cells (15).

At the membrane level, BK produces a sequence of hyperpolarization followed by depolarization, accompanied by a decrease and increase respectively in spontaneous action potential discharge rate (10,16). Voltage clamp studies have revealed that these two potential changes result from the sequential modification of two quite separate K conductances (17). The initial hyperpolarization results from the activation of a conductance which is essentially voltage-independent, whereas the subsequent depolarization results primarily from the inhibition of a time- and voltage-dependent current closely resembling a current previously described in sympathetic neurones and termed the "M-current" (cf. ref. 18).

BK-induced outward current - The initial outward current induced by BK is probably a Ca-dependent K-current, activated by the transient rise in internal Ca resulting from the formation and Ca-releasing action of IP_3. The evidence for this is as follows. (a) A comparable current can be induced by the intracellular injection of either Ca^{2+} ions or IP_3. (b) The conductance underlying all 3 current responses (to BK, Ca and IP_3) appear similarly independent of voltage between -40 and -120 mV. (c) All 3 currents are equally blocked by the K-channel blocker d-tubocurarine (19) but are resistant to TEA at $<$ 5 mM. Current responses to BK and IP_3 slowly run down in Ca-free external solution: we interpret this to suggest depletion of a common internal Ca-store, rather than a dependence on accelerated Ca-influx as the immediate source of Ca. Even in normal Ca-solution, the initial response to BK shows pronounced tachyphyllaxis (i.e.

it fades rather rapidly with repeated applications); this does not apply to repeated IP_3 injections if given at intervals > 2 min, suggesting that tachyphyllaxis to BK results from some form of receptor desensitization.

Using cell-attached patch-electrodes, bursts of single-channel current have been detected accompanying the initial hyperpolarization produced by IP_3 or Ca-injections, or following application of BK to the cell membrane outside the patch (20). Channel opening was not secondary to the hyperpolarization since it was not induced by equivalent hyperpolarizations evoked by intracellular current injections. Most frequently-observed channels showed a linear single-channel current-voltage curve with a mean conductance (in 140 mM K^+) of about 40 pS. Opening probability did not show a strong or consistent dependence on membrane potential over the range -130 to +20 mV, in agreement with the lack of voltage-dependence of the conductance underlying the macroscopic current. Comparable channels were preserved in excised (inside-out) patches exposed to 0.1 - 1 μM Ca^{2+}. On the other hand such channels were rarely seen in patches exposed on the outside to d-tubocurarine. On the basis of these results, we feel fairly confident that current-flow through these 40 pS channels accounts for most of the initial outward current provoked by BK; although large (150-200 pS) channels resembling the Ca-dependent "BK-channels" seen in a variety of other cells (21) were detected in many patches, they were rarely triggered by IP_3 or Ca-injections sufficient to activate the macroscopic current.

Inward current - As indicated above (see ref. 17) this results from the inhibition of a K-current and hence is normally accompanied by a reduced membrane conductance (see also ref. 10). This cannot reflect a secondary inhibition of the Ca-dependent current activated during the initial hyperpolarization since (a) the conductance decrease showed strong voltage-dependence and (b) the inward current persisted when the initial outward current was blocked with d-tubocurarine (see Fig. 3 in ref. 17). Instead, it appears to be due to inhibition of a quite separate, Ca-independent K-current, the M-current.

One possibility addressed was whether this resulted from a secondary action of the IP_3 generated by BK. Tests with intracellularly-injected IP_3 suggest this to be unlikely: although such injections frequently produced a secondary inward current, this was accompanied by an increased conductance (unlike the response to BK), during which the time-dependent current relaxations induced by hyperpolarizing voltage steps (characteristic of M-current deactivation) persisted. A similar inward current (with increased conductance) also occurred following injection of the 1,3,4,5-tetrakisphosphate and 1,3,4-trisphosphate derivatives of IP_3 (22). (It has recently been reported (23) that the secondary depolarization produced by BK in neuroblastoma NIE-115 cells is accompanied by an increased membrane conductance, and more closely resembles the response to IP_4 injections. This we (24) have not been able to confirm under voltage-clamp conditions).

An alternative possibility is that the inward current might results from the effect of DAG, the other second messenger generated by PIP_2 hydrolysis. This possibility is supported by the observations that phorbol dibutyrate (PDBu), an alternative activator of C-kinase (8), also produced an inward current similar in form to that induced by BK, at the same time occluding the inward current response to BK (17). In contrast, PDBu enhanced the initial outward current produced by BK. Subsequent experiments revealed that PDBu also enhanced the outward current response to IP_3 or Ca injections (ref. 24 and unpublished), and potentiated 40 pS channel opening produced by IP_3 injections (Higashida & Brown, unpublished). This suggests an interesting (and unexpected) potential positive-feedback effect of DAG on the response to IP_3-production, probably located at the level of the K-channels opened by the Ca released by IP_3 rather than on (say) the metabolism or action of IP_3 itself.

Thus, these tests are clearly compatible with the view that the electrophysiological action of BK on NG108-15 cells result from receptor-mediated hydrolysis of PIP_2 and the consequential actions of the two soluble messengers IP_3 and DAG. However, the evidence cannot be regarded as watertight, in the sense that it depends on replication tests rather than on selective blockade of the individual components of the response.

Cholinergic agonist action on neuroblastoma hybrid cells and on sympathetic neurones

In contrast to BK, muscarinic agonists such as methacholine were observed not to induce an outward current in NG108-15 cells but solely to generate an inward current (25). This resembled that produced by BK in that it was accompanied by a decreased conductance, primarily affecting a voltage-dependent component of K-conductance. Thus the action of muscarinic agonists on these cells appears essentially identical to (though weaker than) their action on sympathetic neurones, i.e. they inhibit the voltage-dependent K-current I_M (17).

How, then, is this muscarinic response transduced? So far, we have two clues available. Firstly, the receptors appear to be coupled to a GTP-binding protein since, in whole-cell clamped (internally-dialyzed) ganglion cells, application of the "irreversible" G-protein activators $GTP_\gamma S$ or Gpp(NH)p to the cell interior also inhibited I_M and sensitized the residual I_M to inhibition by muscarine (26); conversely, GDPβS reduced the sensitivity of I_M to muscarine (26). Secondly, as in NG108-15 cells, the response of ganglion cells to muscarine can be replicated (in part, at least) by the C-kinase activator PDBu (27). Thus, application of 0.1-1 μM PDBu inhibited I_M by up to about 40% in bullfrog sympathetic ganglion cells. Further, the residual current becomes much less sensitive to inhibition by muscarine or by the peptide receptor-activator LHRH, while remaining fully susceptible to the (presumed) direct channel-blocking action of Ba. (Interestingly, the Ca-activated K-current in these cells, which is partly-inhibitable by muscarinic agonists (28) was not clearly depressed by PDBu. This accords with effects on neuroblastoma cells (29) but contrasts with hippocampal cells where PDBu inhibits the Ca-activated current while sparing I_M (30). Analogous currents in different cell types clearly vary in their susceptibility to changes induced by C-kinase-mediated phosphorylation, as is apparent from other studies (9).)

How might these results be interpreted? Muscarinic agonists have been reported to accelerate PIP_2-breakdown in sympathetic neurones (31-33), so one interpretation might be that I_M-inhibition results from a G-protein-linked hydrolysis of PIP_2 resulting in DAG-production, and that the action of phorbol esters is truly imitative of this natural effect of DAG. However, there are two stumbling blocks to the interpretation. (i) There is no firm evidence that muscarinic agonists accelerate PIP_2 breakdown in NG108-15 cells (34), where their action is equally well imitated by PDBu. (ii) If DAG were formed as the "second messenger" from PIP_2, should there not be an equimolar production of IP_3, and should not this then induce a Ca-activated K-current like that produced in NG-108-15 cells by BK?

There are several ways of side-stepping these obstacles. For example, we might suppose that the rate of PIP_2 breakdown in NG108-15 cells is too slow to produce an appropriate pulse of IP_3 yet sufficient to yield an adequate concentration of DAG to activate C-kinase. Alternatively, in ganglion cells, the Ca-releasing effect of IP_3 may be less pronounced. (Indeed, we have some evidence for this, in that iontophoretic injections of IP_3 cells, or inclusion of IP_3 in the intracellular WHC-pipette up to 100 μM, produced much less outward current than that anticipated from the experiments on NG108-15 cells.) Or perhaps DAG is formed from some other source than PIP_2 by muscarinic agonists (35). On the other hand, we must also face the possibility that the ability of phorbol esters to replicate a receptor-mediated event may not be indicative of a normal role for DAG - a possibility which applies with equal force to the numerous other studies (9) of this type.

Conclusions

The response of NG108-15 cells provides a very clear and interesting example of a neuronal membrane response to a receptor clearly coupled to PIP_2 breakdown, which has allowed us to explore some of the facets of this type of action in fair detail However, although similar events undoubtedly occur in some neurosecretory cells when challenged with appropriate agonists (36), at present the precise role played by this pathway in governing the membrane current effects of transmitters in normal vertebrate neurones is still somewhat uncertain. Thus, to date, there seems to be no example of a transmitter-induced K-current in a normal (non-secretory) vertebrate neurone which is clearly mediated by IP_3, and the evidence for a role for DAG must remain at best circumstantial in the absence of specific blockers. It may well be that both agents exert their most significant effects at secretory terminals (37), on the release of transmitter, rather than via modifications of membrane currents and consequent changes in excitability.

This work was supported by the U.K. Medical Research Council, the U.S. Public Health Service and the Fogarty International Center of the National Institutes of Health.

References

1. GRENNINGLOH, G., RIENITZ, A., SCHMITT, B., METHFESSEL, C., ZENSEN, M., BAYREUTHER, K., GRUNDELFINGER, E.O. & BETZ, H. (1987) The strychnine-binding subunit of the glycine receptor shows homology with nicotinic acetylcholine receptors. *Nature* **328**, 215-200.
2. SCHOFIELD, P.R., DARLISON, M.G., FUJITA, N., BURT, D.R., STEPHENSON, F.A., RODRIGUEZ, H., RHEE, L.M., RAMACHANDRAN, J., REALE, V., GLENCORSE, T.A., SEEBURG, P.H. & BARNARD, E.A. (1987) Sequence and functional expression of the $GABA_A$ receptor shows a ligand-gated receptor super-family. *Nature* **328**, 221-227.
3. HALL, Z.W. (1987) Three of a kind: the β-adrenergic receptor, the muscarinic acetylcholine receptor, and rhodopsin. *Trends in Neurosci.* **10**, 99-101.
4. DOLPHIN, A. (1987) Nucleotide binding proteins in signal transduction and disease. *Trends in Neurosci.* **10**, 53-57.
5. BERRIDGE, M.J. & IRVINE, R.F. (1984) Inositol trisphosphate, a novel second messenger in cellular signal transduction. *Nature* **312**, 315-321.
6. KUNO, M. & GARDNER, P. (1987) Ion channels activated by inositol 1,4,5-trisphosphate in plasma membrane of human T-lymphocytes. *Nature* **326**, 301-304.
7. BATTY, I.R., NAHORSKI, S.R. & IRVINE, R.F. (1985) Rapid formation of inositol 1,3,4,5-tetrakisphosphate following muscarinic receptor stimulation of rat cerebral cortical slices. *Biochem. J.* **232**, 211-215.
8. NISHIZUKA, Y. (1986) Studies and perspectives of protein kinase C. *Science* **283**, 305-312.
9. KACZMAREK, L. (1987) The role of protein kinase C in the regulation of ion channels and neurotransmitter release. *Trends in Neurosci.*, **10**, 30-33.
10. YANO, K., HIGASHIDA, H., INOUE, R. & NOZAWA, Y. (1984) Bradykinin-induced rapid breakdown of phosphatidyl inositol 4,5-bisphosphate in neuroblastoma × glioma hybrid NG108-15 cells. *J. biol. Chem.* **259**, 10201-10207.
11. YANO, K., HIGASHIDA, H., HATTORI, H. & NOZAWA, Y. (1985) Bradykinin-induced transient accumulation of inositol trisphosphate in neuron-like cell line NG108-15 cells. *FEBS Lett.* **181**, 403-406.
12. GRANDT, R., GREINER, C., ZUBIN, P. & JAKOBS, K.H. (1986) Bradykinin stimulates GTP hydrolysis in NG108-15 membranes by a high-affinity, pertussis toxin-insensitive GTPase. *FEBS Lett.* **196**, 279-283.
13. REISER, G. & HAMPRECHT, B. (1985) Bradykinin causes a transient rise of intracellular Ca^{2+}-activity in cultured neuronal cells. *Pflüg. Arch.* **405**, 260-264.
14. OSUGI, T., UCHIDA, S., IMAIZUMI, T. & YOSHIDA, H. (1986) Bradykinin-induced intracellular Ca^{2+} elevation in neuroblastoma × glioma hybrid NG108-15 cells: relationship to the action of inositol phospholipid metabolites. *Brain Res.* **379**, 84-89.
15. JEAN, T. & KLEE, C.B. (1986) Calcium modulation of inositol 1,4,5-trisphosphate-induced calcium release from neuroblastoma × glioma hybrid (NG108-15) microsomes. *J. biol. Chem.*, **261**, 16414-16420.

16. REISER, G. & HAMPRECHT, B. (1982) Bradykinin induces hyperpolarizations in rat glioma cells and in neuroblastoma x glioma hybrid cells. *Brain Res.* **239**, 191-199.
17. HIGASHIDA, H. & BROWN, D.A. (1986) Two polyphosphatidylinositide metabolites control two K^+ currents in a neuronal cell. *Nature* **323**, 333-335.
18. BROWN, D.A. & ADAMS, P.R. (1980) Muscarinic suppression of a novel voltage-sensitive K^+ current in a vertebrate neurone. *Nature* **283**, 673-676.
19. COOK, N.S. & HAYLETT, D.G. (1985) Effects of apamin, quinine and neuromuscular blockers on calcium-activated potassium channels in guinea-pig hepatocytes. *J. Physiol.* **358**, 373-394.
20. BROWN, D.A., FURUYA, K. & HIGASHIDA, H. (1987) K-channels in NG108-15 rodent neuroblastoma x glioma cells activated by intracellular injections of inositol-1,4,5-trisphosphate (IP_3). *J. Physiol.* (in press).
21. MARTY, A. (1983) Ca^{2+}-dependent K^+-channels with large unitary conductance. *Trends in Neurosci.* **6**, 262-265.
22. HIGASHIDA, H. & BROWN, D.A. (1986) Membrane current responses to intracellular injections of inositol 1,3,4,5-tetrakisphosphate and inositol 1,3,4-trisphosphate in NG108-15 cells. *FEBS Lett.* **208**, 283-286.
23. TERTOOLEN, L.G.J., TILLY, B.C., IRVINE, R.F. & MOOLENAAR, W.H. (1987) Electrophysiological responses to bradykinin and microinjected inositol polyphosphates in neuroblastoma cells. *FEBS Lett.* **214**, 365-369.
24. HIGASHIDA, H. & BROWN, D.A. (1987) Bradykinin inhibits potassium (M) current in NIE-115 neuroblastoma cells. *FEBS Lett.* (in press).
25. BROWN, D.A. & HIGASHIDA, H. (1987) A note on voltage- and Ca-activated K-currents in mouse neuroblastoma x rat glioma hybrid cells. *J. Physiol.* (in press).
26. MARRION, N.V. (1987) Probable role of a GTP-binding protein in mediating M-current inhibition by muscarine in rat sympathetic neurones. *J. Physiol.* (submitted).
27. BROWN, D.A. & ADAMS, P.R. (1987) Effects of phorbol dibutyrate on M-currents and M-current inhibition in bullfrog sympathetic neurons. *Cell Molec. Neurobiol.* (in press).
28. PENNEFATHER, P., LANCASTER, B., ADAMS, P.R. & NICOLL, R.A. (1985) Two distinct Ca-dependent K-currents in bullfrog sympathetic ganglion cells. *Proc. Natl. Acad. Sci.* **82**, 3040-3044.
29. BROWN, D.A. & HIGASHIDA, H. (1988) Inositol-1,4,5-trisphosphate and diacylglycerol mimic bradykinin-effects on mouse neuroblastoma x rat glioma hybrid cells. *J. Physiol.* (in press).
30. MALENKA, R.C., MADISON, D.V., ANDRADE, R. & NICOLL, R.A. (1986) Phorbol esters mimic some cholinergic actions in hippocampal pyramidal neurons. *J. Neurosci.* **6**, 475-480.
31. BONE, E.A., FRETTEN, P., PALMER, S., KIRK, C.J. & MICHELL, R.H. (1984) Rapid accumulation of inositol phosphates in isolated rat superior cervical ganglia exposed to V_1-vasopressin and muscarinic cholinergic stimuli. *Biochem. J.* **221**, 803-811.
32. PATTERSON, B.A. & VOLLE, R.L. (1984) Muscarinic receptors and [^{3}H] inositol incorporation in a rat sympethatic ganglion. *J. aut. nerv. Syst.* **10**, 69-72.
33. HORWITZ, J., TSYMBALOV, S. & PERLMAN, R.L. (1984) Muscarine increases tyrosine 3-monoxygenase activity and phospholipid metabolism in the superior cervical ganglion of the rat. *J. Pharmac. exper. Ther.* **229**, 577-582.

34. HARDEN, T.R. *et al.* (1986) Characteristics of two biochemical responses to stimulation of muscarinic cholinergic receptors. *Trends in Pharmacol. Sci.* (Suppl). 14-18.
35. MAJERUS, P.W. *et al.* (1986) The metabolism of phosphoinositide-derived messenger molecules. *Science*, **234**, 1519-1526.
36. DUFY, B., JAKEN, S. & BARKER, J.L. (1987) Ca^{2+}-dependent PKC-activation mimicks delayed effects of TRH on clonal pituitary cell excitability. *Endocrinology* (in press).
37. HIGASHIDA, H. (1988) Acetylcholine release induced by bradykinin, inositol 1,4,5-trisphosphate and phorbol dibutyrate in neuroblastoma × glioma hybrid cells. *J. Physiol.* (in press).

NEUROTRANSMITTERS, ION CHANNELS AND SECOND MESSENGERS IN THE HIPPOCAMPUS

R. C. Malenka, D. V. Madison, P. Dutar, R. Andrade, and R.A. Nicoll
Departments of Pharmacology and Physiology
University of California
San Francisco, California 94143-0450
U.S.A.

The mechanisms by which neurotransmitter receptors are coupled to ion channels has been the focus of a great deal of research since the discovery that neurotransmitters can alter neuronal excitability by acting on voltage-dependent, as well as voltage-independent, ion channels. Using the *in vitro* hippocampal slice preparation we have examined the role of second messenger systems and GTP-binding proteins (G-proteins) in mediating the actions of a variety of neurotransmitters found in the hippocampus. Specifically we have examined the actions of norepinephrine (NE), acetylcholine (ACh), serotonin (5-HT), and baclofen (a $GABA_B$ receptor agonist).

NOREPINEPHRINE BLOCKS ACCOMMODATION OF PYRAMIDAL CELL SPIKE DISCHARGE VIA INCREASES IN CYCLIC AMP

Initial studies of the action of NE in the CNS suggested that it may not conform to the classical view of neurotransmitter action, since NE-induced changes in membrane potential rarely exceeded a few millivolts. In contrast, extracellular single unit recordings suggested that NE could enhance the response of neurons to glutamate (Rogawski and Aghajanian, 1980; Woodward et. al., 1979), an action which could not easily be explained simply by the transmitter-induced opening of ligand-gated ion channels. We therefore examined the actions of NE, using intracellular recording from pyramidal cells in *in vitro* rat

NATO ASI Series, Vol. H19
Modulation of Synaptic Transmission and Plasticity
in Nervous Systems. Edited by G. Hertting and H.-C. Spatz

hippocampal slices, to find a mechanism which could explain the interactions between NE and glutamate.

In response to iontophoretically applied glutamate or long (500-700 msec) depolarizing current pulses pyramidal cells typically exhibit a high-frequency discharge which slows and then stops within a few 100 msec of the onset of the depolarization (Fig. 1A and 1C). This accommodation of spike discharge is primarily due to the activation of a calcium-activated potassium conductance [$G_{K(Ca)}$] (Madison and

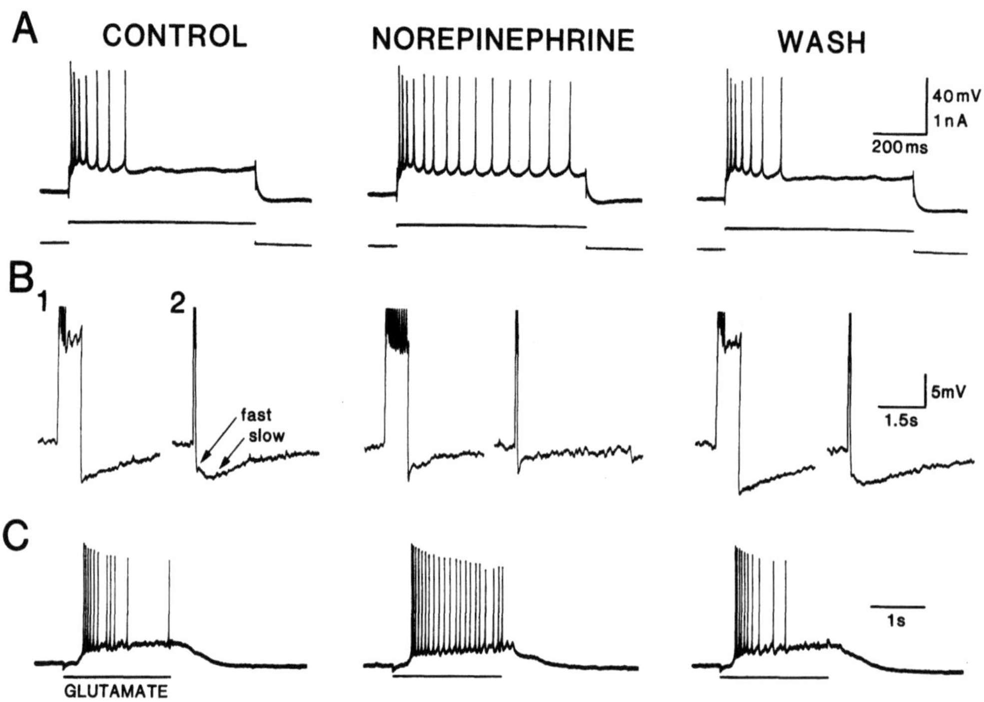

Figure 1. NE blocks calcium-activated potassium responses and prolongs spike discharge in hippocampal pyramidal cells. The responses in A-C are all from the same cell. A. The response of a cell to a depolarizing current pulse demonstrates that NE (10 μM) reversibly reduces accommodation of spike discharge. NE had no effect on membrane potential. B. Simultaneously recorded chart records of the responses shown in A (B1) and responses to a short (100 msec) depolarizing current pulse (B2). The middle record shows that NE blocks the slow AHP. C. Records of the cell's responses to iontophoretically applied glutamate (30 nA; 1 M). (From Madison and Nicoll, 1982)

Nicoll, 1984). The intracellularly recorded correlate of $G_{K(Ca)}$ is the slow afterhyperpolarization (AHP) which is elicited by giving a short (60-100 msec) depolarizing current pulse (Fig. 1B). NE (1-10 μM) markedly increases the number of spikes elicited by glutamate iontophoresis or depolarizing current pulses (Fig. 1A and 1C). A NE-induced reduction in $G_{K(Ca)}$ is the mechanism underlying this action since NE always concomitantly decreases or eliminates the slow AHP (Fig. 1B) (Madison and Nicoll, 1982; 1986a). A similar action of NE on the slow AHP has been reported by Haas and Konnerth (1983).

The action of NE on accommodation and the slow AHP is mediated by a beta-adrenergic receptor since the beta-antagonist, propranolol (10 μM), blocks NE's actions whereas the alpha-antagonist, phentolamine (20 μM), does not. Furthermore, the alpha-agonist, phenylephrine, was ineffective at 10 μM while the beta-agonist, isoproterenol, was more potent than NE at decreasing accommodation and the slow AHP (Madison and Nicoll, 1986a).

In most systems, the effects of beta-adrenergic receptor stimulation are mediated by activation of adenylate cyclase and the consequent increase in cyclic adenosine monophosphate (cAMP). Application of the membrane permeable analog of cAMP, 8-bromoadenosine-3',5'-cyclic monophosphate (1 mM), blocked both the slow AHP and the accommodation of spike discharge elicited by depolarizing current pulses (Fig. 2). Similar results were obtained when forskolin, an activator of adenylate cyclase, was applied to the slice, or when the recording microelectrode was filled with cAMP which was allowed to diffuse into the cell (Madison and Nicoll, 1986b). Furthermore, application of the phosphodiesterase inhibitors, IBMX or Ro20-1724, potentiated the actions of NE on the slow AHP. Together these results indicate that NE's actions on the slow AHP are mediated by beta-receptor stimulation of adenylate cyclase and the subsequent increase in intracellular cAMP levels (Madison and Nicoll, 1986b).

NE and cAMP could exert their actions on the slow AHP by reducing the influx of calcium through voltage-dependent calcium channels. However, neither NE nor 8-bromoadenosine cAMP reduce calcium spikes and NE does not reduce inward barium current carried by calcium channels (Madison and Nicoll, 1986a,b). Therefore NE and cAMP do not appear to reduce calcium influx and must exert their actions on $G_{K(Ca)}$

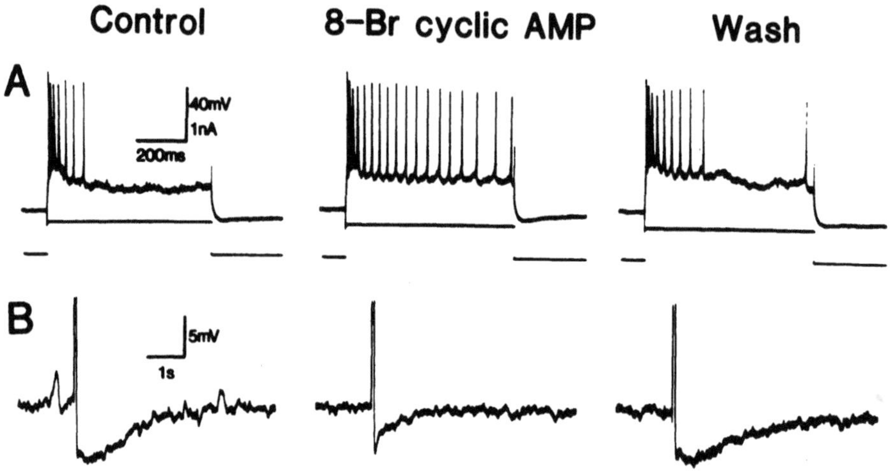

Figure 2. 8-Br cyclic AMP reduces accommodation. All records are from the same cell. A. The response of the cell to a long depolarizing current pulse shows that 8-Br cAMP (1 mM) reversibly decreases accommodation. B. The slow AHP elicited by a short (70 msec) depolarizing current pulse is blocked concomitantly. (From Madison and Nicoll, 1986b).

at some point subsequent to calcium entry, perhaps by activating the cAMP-dependent protein kinase which could then phosphorylate proteins composing the channel itself or closely associated with the channel.

ACETYLCHOLINE'S ACTIONS MAY BE MEDIATED BY INCREASES IN PHOSPHATIDYLINOSITOL (PI) TURNOVER

The cholinergic agonist carbachol, an hydrolysis resistant analogue of acetylcholine, has three distinct muscarinic actions on K^+ channels in hippocampal pyramidal cells. At low concentrations (1 μM) it blocks the calcium dependent slow AHP and accommodation (Fig. 3) (Benardo and Prince, 1982; Cole and Nicoll, 1984; Madison et al., 1987) and a

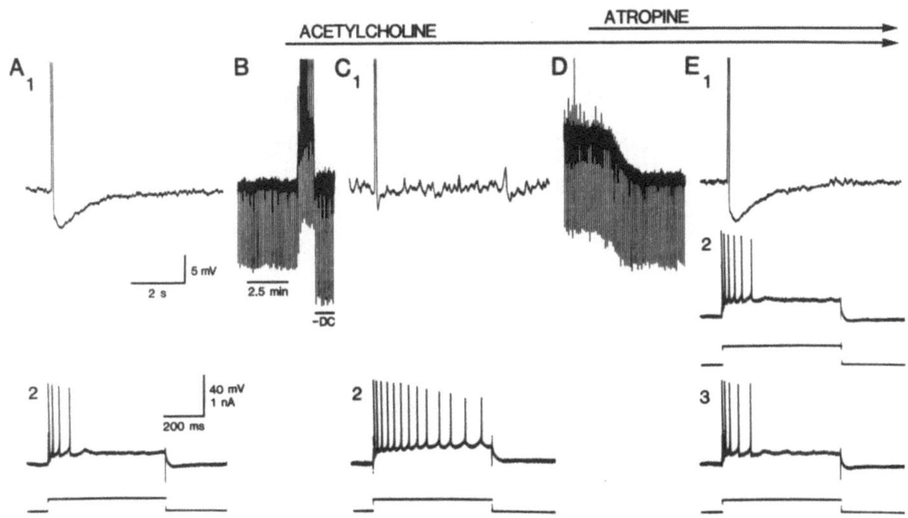

Figure 3. Actions of ACh on hippocampal pyramidal cells. All responses are from the same cell. A. 1. The slow AHP in response to a 60 msec depolarizing current pulse is shown along with the accommodation of spike discharge (2) elicited by a long current pulse. Application of ACh (200 μM; B) depolarized the membrane and increased the cell's input resistance. C. Blockade of the AHP (1) and accommodation (2) by ACh. Application of atropine (0.5 μM) reversed the effect of ACh on membrane potential (D), the AHP (E1) and accommodation (E2, E3). The current pulse in trace E3 was identical to those in A2 and C2 while the one in trace E2 was increased to match the depolarization evoked in the presence of ACh (trace C2). (From Cole and Nicoll, 1983).

background leak K^+ conductance (Benardo and Prince, 1982; Madison et al., 1987) which results in a depolarization of the cell; at approximately ten-fold higher concentrations it blocks the M-current (Halliwell and Adams, 1982; Madison et al., 1987). Synaptically released ACh induces a slow EPSP and a blockade of the slow AHP (Madison et al., 1987).

In a variety of systems, including mammalian CNS, muscarinic receptor activation leads to an increase in phosphatidylinositol (PI) turnover, a second messenger system by which many receptors seem to mediate their intracellular actions (Nishizuka, 1984). PI turnover leads to the production of inositol trisphosphate (IP_3) and diacylglycerol, an activator of the phospholipid and calcium-dependent protein kinase known as protein kinase C (PKC). To test whether the ACh-induced blockade of the slow AHP and accommodation might be mediated by PI turnover and the consequent activation of PKC we applied phorbol esters

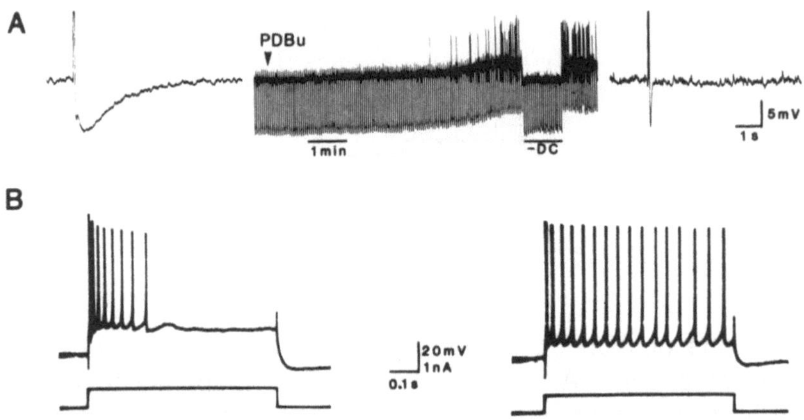

Figure 4. Phorbol-12,13-dibutyrate (PDBu) blocks accommodation and the slow AHP. A. The slow AHP elicited by a short (60 msec) depolarizing current pulse is blocked by 10 μM PDBu which also caused a small depolarization with little change in membrane input resistance. Constant current hyperpolarizing current pulses were given every 3 sec while monitoring the change in membrane potential caused by PDBu. B. The cell's response to a long (650 msec) current pulse. Records were taken within 1 min of the records of the AHP in A. (From Malenka et al., 1986a).

(PEs), compounds which directly activate PKC. Phorbol-12,13-dibutyrate, 12-deoxyphorbol,13-isobutyrate, or phorbol-12,13-diacetate all blocked the slow AHP and the accommodation of spike discharge elicited by depolarizing current pulses (Fig. 4) (Baraban et.al., 1985; Malenka, et al., 1986a). Application of phorbol analogs which do not activate PKC had minimal effects suggesting that the blockade of the slow AHP and accommodation by PEs is due to the activation of PKC. These results are consistent with the hypothesis that the actions of ACh on accommodation and the slow AHP might be due to the muscarinic stimulation of PI turnover. However PEs have several actions which do not mimic the actions of ACh including the enhancement of synaptic transmission (Malenka, et al., 1986b; 1987) and the blockade of a voltage-dependent chloride current (Madison et al., 1986). Furthermore PEs do not block the M-current (Malenka et al., 1986a).

To further examine the possible role of PI turnover in mediating ACh's actions in the hippocampus, a series of cholinergic agonists, that have been classified according to their ability to stimulate PI turnover (Fisher et al., 1983), were applied to hippocampal slices. Oxotremorine-M (oxo-M) and carbachol, which are full agonists for stimulating PI turnover, completely blocked the M-current at 20 μM and blocked the slow AHP at 1 μM. Weak agonists for stimulating PI turnover, pilocarpine, oxotremorine, and arecoline, had little effect on the M-current but were able to block the AHP (Dutar and Nicoll, 1987).

Since the M-current is not blocked by activation of PKC and only those agonists which strongly stimulate PI turnover blocked the M-current, we examined the possibility that IP_3 might be involved in the blockade of M-current by ACh. Consistent with this hypothesis was the finding that intracellular injection of IP_3 caused a reproducible depression of the M-current (Dutar and Nicoll, 1987).

The question remains as to whether ACh's postsynaptic actions in the hippocampus are mediated by the muscarinic stimulation of PI turnover. One possible interpretation of our results is that relatively low levels of diacylglycerol, generated by muscarinic receptor activation in the hippocampus, are sufficient to activate enough PKC to block the slow AHP while relatively larger amounts of IP_3, also generated by PI turnover, block the M-current. Alternatively it is possible that the blockade of the slow AHP by ACh is not mediated by PI turnover but by some other intracellular messenger and that only the M-current blockade by ACh is mediated by PI turnover. IP_3 in most systems is believed to exert its actions by increasing intracellular calcium concentrations (Berridge and Irvine, 1984) and elevated intracellular calcium has been shown to reduce the M-current in bullfrog sympathetic ganglion cells (Tokimasa, 1985). Although we cannot exclude a role for calcium in IP_3's action, chelating intracellular calcium with EGTA or BAPTA does not prevent the muscarinic blockade of the M-current.

SEROTONIN AND $GABA_B$ RECEPTORS ARE COUPLED TO THE SAME POTASSIUM CHANNELS VIA A G-PROTEIN

Historically, neurotransmitters were thought to mediate their actions by opening ligand-gated ion channels which were otherwise entirely silent. The conformation of the ionophore was presumably directly altered by the receptor ligand interaction. In the hippocampus a number of neurotransmitters (eg. serotonin and GABA acting on $GABA_B$ receptors) have been found to cause membrane hyperpolarizations by increasing the membrane conductance to potassium (Segal, 1980; Newberry and Nicoll, 1985; Andrade and Nicoll, 1987). At first glance these actions would conform to the classical action of neurotransmitters. However, it is conceivable that these receptors share the same potassium channels. If this were true it would raise the possibility that some indirect mechanism couples receptor to channel, allowing, as was the case for NE and ACh, two or more different neurotransmitter receptors to affect the same class of ion channel. We have attempted to examine this issue by comparing the actions of serotonin (5-HT), acting on 5-HT1a receptors, and baclofen, a $GABA_B$ receptor agonist.

Application of either 5-HT or baclofen causes a large membrane hyperpolarization which is due to a selective increase in potassium. Spiperone, completely blocks this action of 5-HT at a time when the same cell responds normally to baclofen (Fig. 5A). This clearly indicates that the two compounds exert their actions by acting on distinct receptors. If different mechanisms account for these actions it would be expected that the currents generated by the two agonists would summate. Figure 5B shows that 5-HT's and baclofen's actions do not summate. Instead, application of maximal concentrations of the two compounds together generate a current no different from that generated by 5-HT alone. This indicates that the two receptors are linked to the same class of potassium channels (Andrade et. al., 1986).

How might the two receptors be coupled to the same channel? Both receptors have been reported to be capable of inhibiting adenylate cyclase. However neither bath application of 8-bromo-cAMP nor intracellular injection of cAMP had any affect on the 5-HT- or baclofen-

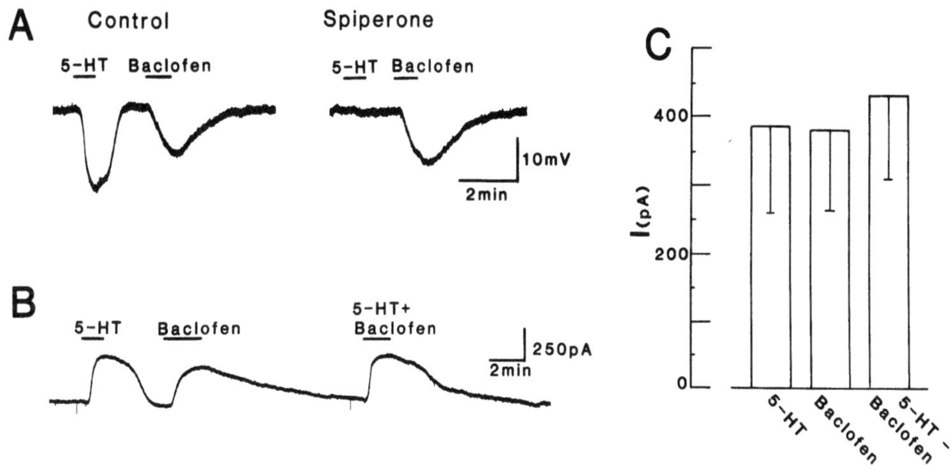

Figure 5. 5-HT and baclofen hyperpolarize hippocampal pyramidal cells by separate receptors but through a common conductance mechanism. A. Spiperone (5 μM) blocks the hyperpolarization elicited by 5-HT (10 μM) but not the hyperpolarization elicited by baclofen (10 μM). B. Continuous voltage clamp of the outward currents caused by application of 5-HT (30 μM) or baclofen (30 μM) first separately then together. The response evoked by both agonists is no larger than that evoked by 5-HT alone. C. Summary of 5 experiments similar to that shown in B. The bracket represents the standard deviation for the responses. (From Andrade et. al., 1986).

induced hyperpolarizations, indicating that changes in cAMP levels do not mediate their effects. 5-HT has also been linked to PI turnover. This might result in receptor-mediated increases in IP_3 and an increase in intracellular calcium which could open one of the two calcium-dependent potassium conductances found in hippocampal pyramidal cells (Lancaster and Nicoll, 1987; Storm, 1987). This potential mechanism has been ruled out by demonstrating that 5-HT and baclofen exert their normal hyperpolarizing actions when intracellular calcium has been chelated with EGTA or BAPTA (thus blocking the slow AHP which is mediated by one type of calcium-activated potassium channel), or following application of TEA which blocks the other type of calcium-activated potassium channel (Andrade et. al., 1986). PKC could not mediate the hyperpolarizing actions of 5-HT or baclofen since its activation by PEs causes a small depolarization (Fig. 4), not a hyperpolarization (Malenka et al., 1986a).

Although we have ruled out a role for cAMP, calcium or PI turnover in mediating the hyperpolarizing actions of 5-HT or baclofen, the binding of both compounds is sensitive to the presence of GTP suggesting that a G-protein may be involved in their actions. We performed a variety of experiments to test this hypothesis. Pertussis toxin ADP-ribosylates and inactivates some G-proteins. Intraventricular injection of pertussis toxin 3 days prior to the acute experiment dramatically reduced or eliminated the hyperpolarizing responses to 5-HT or baclofen (Fig. 6). Resting membrane potential, input resistance, and spike amplitude were not affected by pertussis toxin. Pertussis toxin also did not have any effect on the NE-induced decrease in the slow AHP (Fig. 6B2), an action due to increasing cAMP via stimulation of a pertussis toxin-insensitive G-protein (Andrade et. al., 1986).

The effects of intracellularly injecting hydrolysis-resistant analogs of GTP also support the hypothesis that a G-protein is involved in the actions of 5-HT and baclofen. When cells were recorded with microelectrodes containing guanosine 5'-O-(2-thiophosphate) (GDPβS), a GTP analog known to bind to G-proteins and inhibit their actions, the responses to 5-HT and baclofen were reduced. Conversely, cells recorded with microelectrodes containing guanosine 5'-O-13-thiotri-phosphate (GTPγS), a GTP analog which binds to and activates G-proteins, had more negative resting potentials than control cells and lower input resistances. This suggests that GTPγS was activating the potassium channels normally opened by 5-HT or baclofen. Consistent with this interpretation, application of 5-HT or baclofen to these cells had no further effect on membrane potential or input resistance.

The final evidence which supports the hypothesis that 5-HT and baclofen are coupled to the same potassium channels via a G-protein comes from the finding that PEs block the hyperpolarizations induced by 5-HT or baclofen. Application of phorbol analogs which do not activate PKC had no effect, suggesting that it is the activation of PKC which is responsible for the inability of the agonists to open the potassium channels. This action of PKC is likely due to its ability to phosphorylate and block the actions of some G-proteins (Jakobs et al., 1985; Katada et al., 1985).

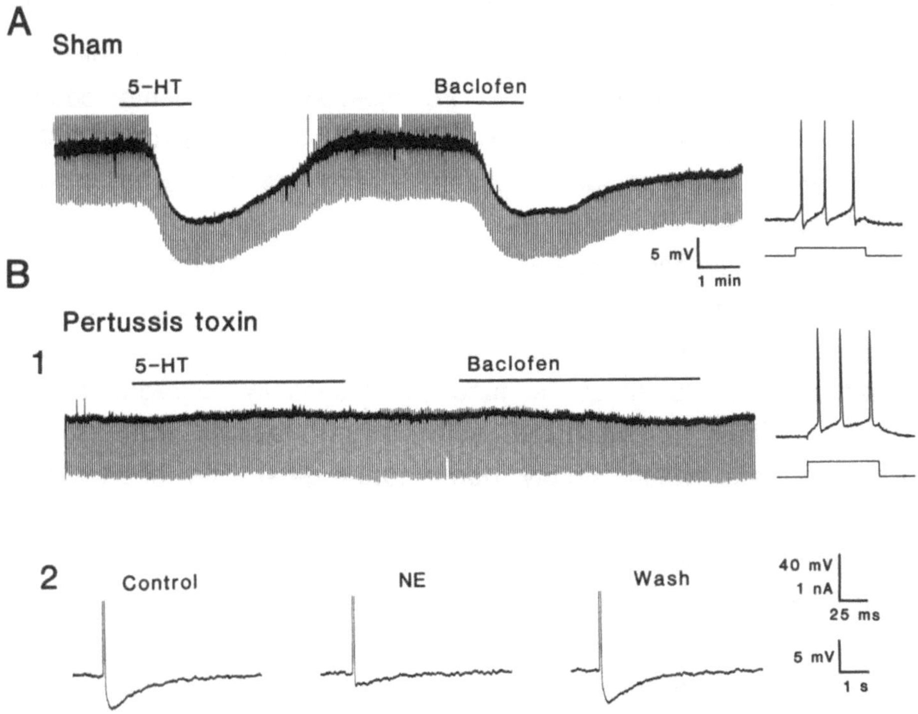

Figure 6. Pertussis toxin blocks the actions of 5-HT and baclofen. A. Responses to bath applied 5-HT (30 μM) and baclofen (30 μM) are shown from a cell recorded in a slice taken from an animal which had been injected with bovine serum albumin (1.5 μg) (Sham). Downward deflections are responses to hyperpolarizing constant current pulses. Action potentials from this cell are shown to the right. B1. 5-HT and baclofen responses from a cell recorded in a slice taken from an animal which had been injected with pertussis toxin (1.5 μg). Action potentials from this cell are to the right. B2. This same cell responds in a typical fashion to NE (5 μM) (Andrade et al., 1986)

CONCLUSIONS

We have presented studies that indicate that neurotransmitter receptors can be coupled to the opening or closing of ion channels using a variety of mechanisms. Different neurotransmitter receptors can be coupled to the same ion channel using distinct diffusable second messenger systems, as may be the case for the actions of ACh and NE on

$G_{K(Ca)}$ or using the same coupling protein, as is the case for 5-HT and baclofen. Other neurotransmitter receptors are also likely to be coupled to the ion channels discussed here. For example, any receptor coupled to the stimulation of adenylate cyclase would also affect $G_{K(Ca)}$, and other receptors which open potassium channels (e.g. adenosine) are also likely to use G-proteins.

The use of second messenger systems and G-proteins to mediate the actions of neurotransmitters has several advantages. Diffusable intracellular messengers would allow the activation of a receptor on one part of a cell to affect the function of spatially segregated parts of the cell. Their actions are also longer lasting than those of neurotransmitters which simply open ligand-gated ion channels. By utilizing second messengers or G-proteins activation of neurotransmitter receptors could also have effects on cell function that are not immediately apparent using electrophysiological techniques, for example, changes in the activity level of intracellular enzymes. The complexity of neurotransmitter actions that has become apparent over the last several years may be a major contributing factor to the enormous plasticity and information processing capabilities inherent in mammalian brain.

REFERENCES

Andrade R, Malenka RC, Nicoll RA (1986) A G protein couples serotonin and $GABA_B$ receptors to the same potassium channels in hippocampus. Science 234:1261-1265

Andrade R, Nicoll RA (1987) Pharmacologically distinct actions of serotonin on single pyramidal neurones of the rat hippocampus recorded *in vitro*. J Physiol, in press

Baraban JM, Snyder SH, Alger BE (1985) Protein kinase C regulates ionic conductance in hippocampal pyramidal neurons: electrophysiological effects of phorbol esters. Proc Natl Acad Sci 82:2538-2542

Benardo LS, Prince DA (1982) Ionic mechanisms of cholinergic excitation in mammalian hippocampal pyramidal cells. Brain Res 249:333-344

Berridge MJ, Irvine RF (1984) Inositol trisphosphate, a novel second messenger in cellular signal transduction. Nature 312:315-321

Cole AE, Nicoll RA (1983) Acetylcholine mediates a slow synaptic potential in hippocampal pyramidal cells. Science 221:1299-1301

Cole AE, Nicoll RA (1984) Characterization of a slow cholinergic postsynaptic potential recorded *in vitro* from rat hippocampal pyramidal cells. J Physiol 352:173-188

Dutar P, Nicoll RA (1987) A role for phosphatidylinositol (PI) turnover in the muscarinic blockade of the M-current in hippocampal pyramidal cells. Neurosci Abst 13, in press.

Fisher SK, Klinger PD and Agranoff BW (1983) Muscarinic agonist binding and phospholipid turnover in brain. J Biol Chem 258:7358-7363

Haas HL, Konnerth A (1983) Histamine and noradrenaline decrease calcium-activated potassium conductance in hippocampal pyramidal cells. Nature 302:432-434

Halliwell JV and Adams PR (1982) Voltage-clamp analysis of muscarinic excitation in hippocampal neurons. Brain Res 250:71-92

Jakobs KH, Bauer S, Watanabe Y (1985) Modulation of adenylate cyclase of human platelets by phorbol esters. Eur J Biochem 151:425-430

Katada T, Gilman AG, Watanabe Y, Bauer S, Jakobs KH (1985) Protein kinase C phosphorylates the inhibitory guanine nucleotide-binding regulatory component and apparently suppresses its function in hormonal inhibition of adenylate cyclase. Eur J Biochem 151:431-434

Lancaster B, Nicoll RA (1987) Properties of two calcium-activated hyperpolarizations in rat hippocampal neurones. J Physiol 389:187-204

Madison DV, Nicoll RA (1982) Noradrenaline blocks accommodation of pyramidal cell discharge in the hippocampus. Nature 299:636-638

Madison DV, Nicoll RA (1984) Control of the repetitive discharge of rat CA1 pyramidal neurones *in vitro*. J Physiol 354:319-331

Madison DV, Nicoll RA (1986a) Actions of noradrenaline recorded intracellularly in rat hippocampal CA1 pyramidal neurones, *in vitro*. J Physiol 372:221-244

Madison DV, Nicoll RA (1986b) Cyclic adenosine 3',5'-monophosphate mediates beta-receptor actions of noradrenaline in rat hippocampal pyramidal cells. J Physiol 372:245-259

Madison DV, Lancaster B and Nicoll RA (1987) Voltage clamp analysis of cholinergic action in the hippocampus. J. Neurosci. 7:733-741

Madison DV, Malenka RC and Nicoll RA (1986) Phorbol esters block a voltage-sensitive chloride current in hippocampal pyramidal cells. Nature 321:695-697

Malenka RC, Madison DV, Andrade R, Nicoll RA (1986a) Phorbol esters mimic some cholinergic actions in hippocampal pyramidal neurons. J Neurosci 6:475-480

Malenka RC, Madison DV, Nicoll RA (1986b) Potentiation of synaptic transmission in the hippocampus by phorbol esters. Nature 321:175-177

Malenka RC, Ayoub GS, Nicoll RA (1987) Phorbol esters enhance transmitter release in rat hippocampal slices. Brain Res 403:198-203

Newberry NR, Nicoll RA (1985) Comparison of the action of baclofen with gamma-aminobutyric acid on rat hippocampal pyramidal cells *in vitro*. J Physiol 360:161-185

Nishizuka Y (1984) The role of protein kinase C in cell surface signal transduction and tumor promotion. Nature 308:693-698

Rogawski MA, Aghajanian GK (1980) Modulation of lateral geniculate neurone excitability by noradrenaline microiontophoresis or locus coeruleus stimulation. Nature 287:731-734

Segal M (1980) The action of serotonin in the rat hippocampal slice preparation. J Physiol 303:423-439

Storm JF (1987) Action potential repolarization and a fast afterhyperpolarization in rat hippocampal pyramidal cells. J Physiol 385:733-759

Tokimasa T (1985) Intracellular Ca^{2+}-ions inactivate K^{+}-current in bullfrog sympathetic neurons. Brain Res 337:386-391

Woodward DJ, Moises HC, Waterhouse BD, Hoffer BJ, Freedman R (1979) Modulatory actions of norepinephrine in the central nervous system. Fed Proc 38:2109-2116

A FAMILY OF NEUROTRANSMITTER RECEPTORS COUPLE TO A POTASSIUM CONDUCTANCE

R. Alan North
Vollum Institute
Oregon Health Sciences University
3181 SW Sam Jackson Park Road
Portland, Oregon 97201, U.S.A.

It has been well known for several years that an important inhibitory mechanism in mammalian nervous tissue is an increase in the conductance of the membrane to chloride ions (Coombs *et al.* 1955). The two major substances involved in this mechanism are GABA and glycine and it is now known that their action is exerted through membrane receptors which incorporate the chloride ion channel (Grenningloh *et al.* 1987; Schofield *et al.* 1987). The role of potassium conductance increase in neuronal inhibition has been less clearly demonstrated except for some notable work in the invertebrate nervous system (Kehoe, 1972) and, particularly, in the mechanism of vagal inhibition of the heart (Trautwein & Dudel, 1958). Whereas the effects of agents that increase chloride conductance may be either excitatory or inhibitory, depending on the absolute magnitude of the chloride equilibrium potential, an increase in the cell conductance to potassium ions is universally inhibitory if it is sufficiently strong. This chapter describes some examples of synaptic inhibition mediated by potassium conductance increase in mammalian neurones, and points out that this mechanism of neuronal inhibition seems likely to be common to the action of agonists at an increasing number of neurotransmitter receptors.

The receptors

A number of neurotransmitter receptors have been shown to increase membrane potassium conductance in neurones (North et al. 1987). With the exception of the μ and δ subtypes of opioid receptor, only one receptor

NATO ASI Series, Vol. H19
Modulation of Synaptic Transmission and Plasticity
in Nervous Systems. Edited by G. Hertting and H.-C. Spatz

subtype for each naturally occurring transmitter has so far been associated with the potassium conductance increase. For example, muscarinic agonists at the cardiac M_2 type of receptor result in an increase in potassium conductance whereas agonists at the glandular M_2 receptor or at the M_1 receptor have different effects on neurones; dopamine D_2, but not D_1, receptors are coupled potassium channels, as are 5-HT_1 receptors, though not 5-HT_2 or 5-HT_3. Table 1 lists some of the receptor types that had been shown to affect potassium conductance in neurones.

The degree to which these receptors have been characterized exhibits a considerable range. The cardiac M_2 receptor is the only one for which the primary amino acid sequence is currently known. The μ, δ, α_2 and D_2 receptors have all been characterized by Schild analysis in conditions in which potassium conductance increase was measured. The lack of effective selective antagonists for A_1, somatostatin, 5-HT_1 and $GABA_B$ receptors has limited further definition of these subtypes, but selective agonists in general support the designations. In abdominal ganglia of *Aplysia*, acetylcholine, dopamine and histamine receptors are also coupled to potassium channels (Sasaki & Sato 1987).

The potassium conductance

The properties of the potassium conductance increased by agonists at the receptors mentioned above has been studied in considerable detail in cardiac tissue; acetylcholine (acting on muscarinic receptors) or adenosine act on the heart to increase the probability of opening of potassium channels that have a unit conductance of about 50 pS, and a mean open time of 1.6 ms (Sakmann *et al.* 1984; Kurachi *et al.* 1986; Yatani *et al* 1987a). Somatostatin and acetylcholine open channels with very similar properties in tumor cells derived from the anterior pituitary (Yatani et al. 1987b). These individual channels exhibit marked inward rectification; that is to say their conductance to potassium ions is very much higher for inward currents than for outward currents. Indeed, in some studies it was not possible to measure the opening of individual channels when the current flow was in the outward direction, presumably because the unit events were below the resolution of the current measurement.

Table 1. Some receptors that couple to potassium conductances. Evidence for involvement of G-protein comes either from block by pertussis toxin, or activation by GTP-γ-S.

Agonist (receptor)	Tissue	G-protein?	Synaptic potential	Reference
Acetylcholine (M_2)	Guinea-pig cardiac muscle	--	Yes	del Castillo & Katz, 1955
		--	--	Soejima & Noma, 1984
		Yes	--	Pfaffinger et al., 1985
		Yes	--	Breitweiser & Szabo, 1985
Acetylcholine (M_2)	Rat nucleus parabrachialis	--	--	Egan & North, 1986
Acetylcholine (?)	Rat thalamus reticularis	--	--	Prince & McCormick, 1986
Acetylcholine (?)	Frog sympathetic C cells	--	Yes	Dodd & Horn, 1983
Acetylcholine (M_2)	Mudpuppy cardiac ganglion	--	Yes	Hartzell et al., 1977
Noradrenaline (α_2)	Guinea-pig submucous plexus	Yes	Yes	North & Surprenant, 1985
Noradrenaline (α_2)	Rat locus coeruleus	--	Yes	Egan et al., 1983
		--	--	Williams et al., 1985
Noradrenaline (α_2)	Rat substantia gelatinosa	--	--	North & Yoshimura, 1984
Noradrenaline (α_2)	Guinea-pig myenteric plexus	--	--	Morita & North, 1981
Dopamine (D_2)	Rat substantia nigra	--	--	Lacey et al., 1987
		Yes	--	Innis & Aghajanian, 1987
Dopamine (D_2)	Cultured human pituitary adenoma	--	--	Israel et al., 1985
Dopamine (?)	Rat intermediate pituitary	--	Yes	MacVicar & Pittman, 1986
5-hydroxytryptamine ($5HT_1$)	Rat dorsal raphe	--	--	Sprouse & Aghajanian, 1987
5-hydroxytryptamine ($5HT_1$)	Rat dorsal raphe	--	Yes	Williams, unpublished
5-hydroxytryptamine (?)	Rat dorsal raphe	--	Yes	Yoshimura & Higashi, 1985
5-hydroxytryptamine (?)	Rat hippocampus	Yes	--	Andrade et al., 1987
5-hydroxytryptamine (?)	Rat hippocampus	--	--	Colino & Halliwell, 1987
5-hydroxytryptamine ($5HT_1$)	Guinea pig myenteric plexus	--	--	Galligan, unpublished
γ- aminobutyric acid ($GABA_B$)	Rat hippocampus	Yes	--	Andrade et al., 1987
		--	Yes	Newberry & Nicoll, 1984
γ- aminobutyric acid ($GABA_B$)	Rat dorsal raphe	--	--	Williams, unpublished
Adenosine (?A_1)	Mouse striatum	--	--	Trussell & Jackson 1985
Adenosine (?A_1)	Guinea-pig cardiac cells	--	--	Kurachi et al., 1980
Somatostatin	Guinea pig submucous plexus	Yes	--	Mihara et al., 1987a
		--	(?)	Mihara et al., 1987b
Somatostatin	Cultured GH_3 pituitary cells	--	--	Yamashita, 1986
Somatostatin	Cultured GH_3 pituitary cells	Yes	--	Yatani et al., 1987b
Somatostatin	Rat locus coerulus cells	Yes	--	Inoue et al., 1987
Enkephalin (μ)	Rat locus coeruleus	Yes	--	Williams & North, 1984
		Yes	--	North et al., 1987
Enkephalin (?μ)	Rat substantia gelatinosa	--	--	Yoshimura & North, 1983
Enkephalin (?μ)	Guinea-pig myenteric plexus	--	--	Morita & North, 1982
Enkephalin (μ)	Mouse dorsal root ganglion	--	--	Werz & Macdonald, 1983
Enkephalin (μ)	Rat nucleus parabrachialis	--	--	Christie, unpublished
Enkephalin (δ)	Guinea pig submucous plexus	Yes	--	North et al., 1987

Measurements of membrane currents in intact cells resulting from application of the agonists has also generally shown that the conductance shows inward rectification. This was first explicitly demonstrated by Kehoe (1972) in her studies of the action of acetylcholine and dopamine in *Aplysia* and has now been demonstrated for a number of other receptor-operated potassium channels. These include those affected by μ and δ opioid agonists, by somatostatin, and by α_2 agonists in neurones of the nucleus locus coeruleus of the rat and submucous plexus of the guinea pig (North *et al.* 1987; Mihara *et al.* 1987). In these cases the conductance of the cell membrane that is increased by the agonist allows potassium ions to pass into the cell approximately three times more readily than they pass out of the cell. The degree of inward rectification in the potassium conductance is not so marked as that which is observed in the classical inward rectifiers of skeletal muscle or tunicate egg cells (Adrian, 1964; Katz, 1949; Hagiwara & Takahashi, 1964). One consequence of this is that the outward currents close to the resting potential of the cell are still quite substantial and, in any event, sufficiently large to cause prominent hyperpolarizations of the membrane. However, the potassium conductance affected by these agonists does show several similarities to inward rectifier potassium conductances in skeletal muscle and tunicate egg cells. These similarities are a high sensitivity to extracellular barium (10-30 μM is sufficient to remove the inward rectification), gating and permeation by rubidium, and a voltage-dependent sensitivity to extracellular caesium (which blocks inward current at hyperpolarized potentials, but does not affect outward current).

Adenosine and 5-HT have also been said to increase an inwardly rectifying potassium conductance, but it is important to differentiate between rectification in the conductance of the entire cell and rectification in the conductance of that potassium channel which is opened by the agonist. It may be better to reserve the term inward rectification for a defined potassium conductance, rather than for a whole cell, since in the whole cell this behaviour often results from opening of a variety of ion channels as a function of membrane potential which are unrelated to the particular potassium conductance activated by the agonists. The magnitude of the membrane currents increased by adenosine, 5-HT and the opioids, all reached a maximum at about -50 mV and then become smaller with membrane depolarization but it is not yet known whether this reflects the properties of the individual channels affected by the agonists, or whether it simply results from the very large increase in other outward currents which occur at these less negative membrane potentials (Trussell & Jackson, 1985; Andrade *et al.* 1986; North & Williams, 1985).

The coupling mechanism

It is now thought likely that the receptors listed in Table 1 couple to their potassium channels by a process that involves a guanosine 5'-(O-triphosphate) (GTP) binding protein, but probably not a diffusible cytoplasmic second messenger. Overall support for this assertion comes from a variety of approaches, but the main evidence for the lack of a diffusible second messenger comes from patch-clamp studies. These are essentially of two kinds. In the cell-attached configuration the agonist is effective to open individual potassium channels only when it is included inside the recording pipette and not when it is applied to the remainder of the cell (Soejima & Noma, 1984; Yatani *et al.* 1987*a*; 1987*b*). Secondly, when a small patch of membrane is excised from the cell in the outside/out configuration, the agonist is still effective in opening the individual channels within the patch (Soejima & Noma 1984; Yatani *et al.* 1987*a*,*b*; Kurachi *et al.* 1986). This kind of evidence is so far most advanced for the actions of acetylcholine on potassium channels in the heart in which case it can be shown that the same channels are activated even in the complete absence of muscarinic agonist, by the addition of the suitably purified and activated α subunit of the GTP binding protein (increasingly known as α_K, though possibly identical to α_i). In this kind of experiment, the endogenous coupling G-protein is first inactivated, by taking advantage of its sensitivity to pertussis toxin, and then purified subunits of the G-protein can be added directly to the inner surface of an excised membrane patch. Such experiments show that only the alpha subunit of G_i is effective to open the channels (Codina *et al.* 1987).

At the whole cell level the evidence for this coupling mechanism is necessarily less complete, although in a number of instances the effectiveness of agonists to increase the potassium conductance of the cell has been shown to be blocked by pretreatment with pertussis toxin (Andrade *et al.* 1986; Innis & Agahajanian, 1987; Aghajanian & Wang 1986). These experiments are not entirely satisfactory since in general it is necessary to treat the animal with pertussis one to two days prior to removal of tissue for recording and in some cases the subsequent recording was made by extracellular rather than intracellular methods. Nonetheless, these experiments provide evidence consistent with the hypothesis that the inhibition of cell firing (which is known to result from potassium conductance increase) mediated by α_2 ligands, 5-HT_1 ligands and $GABA_B$ agonists does require a pertussis sensitive GTP binding protein. The hyperpolarization of locus coeruleus neurones by somatostatin in neuronal cell cultures can also be abolished by pretreatment of the cultures with pertussis toxin (Inoue *et al.* 1987).

The third line of evidence, which bears on the coupling mechanism, comes from experiments in which nonhydrolyzable derivatives of GTP have been injected into the cell. Since hydrolysis of GTP is necessary for the termination of the action of agonists which act through G-proteins, it would be expected that substitution for GTP by GTP-γ-S would have relatively little effect on the properties of the cell in the absence of agonists and would make permanent the otherwise reversible effects of the agonist when it is applied. Such results have been observed in the case of somatostatin, δ-opioid and α_2 ligands in the neurones of the guinea pig submucous plexus (Mihara *et al.* 1987; North *et al.* 1987) and for the somatostatin in the rat locus coeruleus in culture (Inoue *et al.* 1987). In most brain neurones in the slice preparation in which this experiment has been attempted, the GTP-γ-S in the recording electrode generally caused a progressive hyperpolarization of the cell, even in the absence of any applied agonist. This seems likely to be due to the presence of endogenous agonists in the environment of the cell, perhaps released from terminals and other elements within the intact brain slice. Nonetheless, even in these experiments, subsequent addition of agonists to cells penetrated with electrode that contained GTP-γ-S was largely ineffective in producing any membrane hyperpolarization (Andrade *et al.* 1986; North *et al.* 1987). The results are therefore widely interpreted to indicate that a GTP binding protein is necessary for the effect.

A further approach to the characterization of the G-protein which is necessary to couple between these receptors and the potassium channel, has been to study the action of phorbol esters, and muscarinic agonists which are thought to act through stimulation of C kinase. Experiments of this kind in the hippocampus of the rat have indicated that the coupling between receptor (in this case $5HT_1$, adenosine, or $GABA_B$) and potassium channel can be blocked by treatment with phorbol esters (Andrade *et al.* 1986; Worley *et al.* 1987). This has been interpreted to suggest that some element in the coupling process is a substrate for C kinase phosphorylation; G_i in platelets has been shown to be phosphorylated by C kinase. It remains to be seen how widespread is this effect; in the rat locus coeruleus and the guinea pig submucous plexus, pretreatment with phorbol esters had no effect on the potassium conductance increase evoked by agonists at α_2, somatostatin (submucous plexus) or α_2, μ and $GABA_B$ receptors (locus coeruleus) (North *et al.* 1987).

Synaptic transmission

Despite the very considerable amount of information dealing with the linkage between receptor and potassium channel, in only a few instances has a synaptic potential been shown unequivocally to be mediated by a transmitter acting on one of these receptors. The best evidence so far comes from cardiac tissue (Hill-Smith & Purves, 1978; Hartzell *et al.* 1977) and the invertebrate nervous system (Kehoe, 1972). In these cases the synaptic potential mediated by this mechanism has a latency to onset of about 100 ms and an overall time course of 1-2 s.

In the nerve cells of the heart of the mudpuppy, a synaptic potential is mediated by acetylcholine that results from an increase in the membrane potassium conductance (Hartzell, *et al.* 1977). Although the receptor was not characterized in these experiments, it seems likely to belong to the M_2 class. This synaptic potential rises with a latency of 60-100 ms and has an overall duration of approximately 1 s. The time course of this synaptic response is remarkably similar both in latency, time to peak and overall duration to that of the synaptic potential mediated by noradrenaline acting on α_2 receptors that has been recorded from the neurones of the guinea pig submucous plexus (Hirst & McKirdy 1975; North & Surprenant, 1985). In these cases the receptors (M_2 and α_2) can be clearly distinguished, but the similarity in time course of the current flow which follows receptor activation speaks strongly to a common intramembrane mechanism for the mediation of the response. A synaptic potential in hippocampal pyramidal cells also results from an increase in the membrane potassium conductance and since this can be mimicked by agonists at $GABA_B$ receptors, this may reflect another synaptic potential in the same category (Newberry & Nicoll, 1984). The lack of antagonists at $GABA_B$ receptors has so far precluded positive identification of the synaptic transmitter. A synaptic hyperpolarization of similar time course is also mediated by 5-HT within the dorsal raphe nucleus (Yoshimura & Higashi, 1985).

Conclusions

Most mammalian neurotransmitters act on a number of distinct receptor subtypes. In many cases, one of these receptor subtypes is coupled to a unique class of potassium conductance which appears to be quite distinct from other potassium conductances in the membrane. Activation of these receptors causes membrane hyperpolarization. Direct synaptic inhibition by this mechanism has so far been demonstrated for acetylcholine acting at muscarinic receptors, for noradrenaline acting at α_2 receptors and in a less complete way, for 5HT acting at $5HT_1$ receptors and GABA acting at $GABA_B$ receptors. It seems likely that these receptor proteins belong to a family that couple to potassium channels and therefore mediate a highly effective form of neuronal inhibition. The intervention of a GTP binding protein between receptor and channel clearly provides the possibility for further integration within the neurone. The G-protein itself may have effects on the cell other than potassium conductance increase; for example, it may couple to quite different channels. The presence of the G-protein provides a site at which the coupling between receptor and channel can be modulated by other intracellular metabolites resulting from the concomitant action of other transmitters.

This work was supported by grants from the United States Department of Health and Human Services, MH40416, DA03160, DK32979, DA03161.

References

Adrian RH (1964) The rubidium and potassium permeability of frog muscle membrane. J Physiol Lond 175:134-159.

Aghajanian GK, Lakoski JM (1984) Hyperpolarization of serotonergic neurons by serotonin and LSD; studies in brain slices showing increased K conductance. Brain Res 305:181-190.

Aghajanian GK, Wang YY (1986) Pertussis toxin blocks the outward currents evoked by opiate and α_2 agonists in locus coeruleus neurons. Brain Res 371:390-394.

Andrade R, Malenka RC, Nicoll, RA (1986) A G-protein couples serotonin and $GABA_B$ receptors to the same channels in hippocampus. Science 234:1261-1265.

Breitwieser GE, Szabo G (1985) Uncoupling of cardiac muscarinic and β-adrenergic receptors from ion channels by a guanine nucleotide analogue. Nature 317:538-540.

Codina J, Yatani A, Grenet D, Brown AM, Birnbaumer L (1987) The α subunit of the GTP binding protein G_k opens atrial potassium channels. Science 236:442-445.

Colino A, Halliwell JV (1987) Differential modulation of three separate K-conductances in hippocampal CA1 neurons by serotonin. Nature 328:73-77.

Coombs JS, Eccles JC, Fatt P (1955) The inhibitory suppression of reflex discharges from motoneurones. J Physiol Lond 130:396-413.

Del Castillo J, Katz B (1955) Production of membrane potential changes in the frog's heart by inhibitory nerve impulses. Nature 175:1035.

Dodd J, Horn JP (1983) Muscarinic synaptic inhibition of sympathetic C neurones in the bullfrog. J Physiol Lond 334:271-291.

Egan TM, Henderson G, North RA, Williams JT (1983) Noradrenaline mediated synaptic inhibition in locus coeruleus neurones. J Physiol Lond 345:477-488.

Egan TM, North RA (1986) Acetylcholine hyperpolarizes central neurones by acting on a muscarinic receptor subtype. Nature 319:405-407.

Grenningloh G, Rienitz A, Schmitt B, Methfessel C, Zenson M, Beyreuther K, Gundelfinger ED, Betz H (1987) The strychnine-binding subunit of the glycine receptor shows homology with nicotinic acetylcholine receptors. Nature 328:215-220.

Hagiwara S, Takahashi K (1974) The anomalous rectification and cation selectivity of the membrane of a starfish egg cell. J Membr Biol 18:61-80.

Hartzell HC, Kuffler SW, Stickgold R, Yoshikami D (1977) Synaptic excitation and inhibition resulting from direct action of acetylcholine on two types of chemoreceptors on individual amphibian parasympathetic neurones. J Physiol Lond 271:817-846.

Hill-Smith I, Purves RD (1978) Synaptic delay in the heart: an ionophoretic study. J Physiol Lond 279:31-54.

Hirst GDS, McKirdy HC (1975) Synaptic potentials recorded from some neurones of the submucous plexus of the guinea-pig small intestine. J Physiol Lond 249:369-385.

Innis RB, Aghajanian GK (1987) Pertussis toxin blocks autoreceptor-mediated inhibition of dopaminergic neurons in rat substantia nigra. Brain Research 411:139-143.

Inoue M, Nakajima S, Nakajima Y (1987) Pertussis toxin abolishes somatostatin-induced enhancement of inward rectification in brain neurones. J Physiol Lond (in press).

Israel JM, Jaquet P, Vincent JD (1985) The electrical properties of isolated human prolactin-secreting adenoma cells and their modification by dopamine. Endocrinology 117:1448-1455.

Katz, B (1949) Les constantes electrique de la membrane du muscle. Arch Sci Physiol 2:285-299.

Kehoe J (1972) Ionic mechanisms of a two component cholinergic inhibition in Aplysia neurones. J Physiol Lond 225:85-114.

Kurachi Y, Nakajima T, Sugimot T (1986) On the mechanism of activation of muscarinic K^+ channels by adenosine in isolated atrial cells: involvement of GTP-binding proteins. Pflugers Archives 407:264-274.

Lacey MG, Mercuri N, North RA Dopamine acts on D_2 receptors to increase potassium conductance in neurones of the rat substantia nigra zona compacta. J Physiol Lond (in press)

MacVicar BA, Pittman QJ (1986) Novel synaptic responses mediated by dopamine and γ-aminobutyric acid in neuroendocrine cells of the intermediate pituitary. Neurosci Lett 64:35-40.

McCormick DA, Prince DA (1986) Acetylcholine induces burst firing in thalamic reticular neurones by activating a potassium conductance. Nature 319:402-405.

Mihara S, Nishi S, North RA, Surprenant A (1987) A nonadrenergic, noncholinergic slow inhibitory postsynaptic potential in neurones of the guinea-pig submucous plexus. J Physiol Lond 390:335-355.

Mihara S, North RA, Surprenant A (1987) Somatostatin increases an inwardly rectifying potassium conductance in guinea-pig submucous plexus neurons. J Physiol Lond 390:357-365.

Morita K, North RA (1981) Clonidine activates potassium conductance in myenteric neurones. Br J Pharmac 74:419-428.

Morita K, North RA (1982) Opiate activation of potassium conductance: inhibition by calcium ions. Brain Res 242:145-150.

Newberry NR, Nicoll RA (1983) A bicuculline-resistant inhibitory post-synaptic potential in rat hippocampal pyramidal cells in vitro. J Physiol Lond 348:239-254.

North RA, Surprenant A (1985) Inhibitory synaptic potentials resulting from α_2-adrenoceptor activation in guinea-pig submucous plexus neurones. J Physiol Lond 358:17-32.

North RA, Williams JT, Surprenant A, Christie MJ (1987) μ and δ receptors belong to a family of receptors that are coupled to potassium channels. Proc Natl Acad Sci USA 84:5487-5491.

North RA, Williams JT (1985) On the potassium conductance increased by opioids in rat locus coeruleus neurones. J Physiol Lond 364:265-280.

North RA, Yoshimura M (1984) The actions of noradrenaline on neurones of the rat substantia gelatinosa in vitro. J Physiol Lond 349:43-55.

Peralta EG, Winslow JW, Peterson GL, Smith DH, Ashkenazi A, Ramachandran J, Schimerlik MI, Capon DJ (1987) Primary structure and biochemical properties of an M_2 muscarinic receptor. Science 236:600-605.

Pfaffinger PJ, Martin JM, Hunter DD, Nathanson NM, Hille B (1985) GTP-binding proteins couple cardiac muscarinic receptors to a K channel. Nature 317:536-538.

Sakmann B, Noma A, Trautwein W (1984) Acetylcholine activation of single muscarinic K^+ channels in isolated pacemaker cells of mammalian heart. Nature 303:250-253.

Sasaki K, Sato M (1987) A single GTP-binding protein regulates K^+-channels coupled with dopamine, histamine and acetylcholine receptors. Nature 325:259-262.

Schofield PR, Darlison MG, Fujita N, Burt DR, et al (1987) Sequence and functional expression of $GABA_A$ receptor shows a ligand-gated receptor super-family. Nature 328:221-227.
Soejima M, Noma A (1984) Mode of regulation of the ACh-sensitive K channel by the muscarinic receptor in rabbit atrial cells. Pflugers Archives 400:424-431.
Sprouse JS, Aghajanian GK (1987) Electrophysiological responses of serotoninergic dorsal raphe neurons to $5\text{-}HT_{1A}$ and $5\text{-}HT_{1B}$ agonists. Synapse 1:3-9.
Trautwein W, Dudel J (1958) Zum Mechanismus der Membranwirkung des Acetylcholin an der Herzmuskelfaser. Pflugers Arch ges Physiol 266:324-334.
Trussell LO, Jackson MB (1985) Adenosine-activated potassium conductance in cultured striatal neurons. Proc Natl Acad Sci USA 82:4857-4861.
Werz MA, Macdonald RL (1983) Opioid peptides with differential affinity for mu and delta receptors decrease sensory neuron calcium-dependent action potentials. J Pharmac exp Ther 227:394-402.
Williams JT, Henderson G, North RA (1985) Characterization of the α_2-adrenoceptors which increase potassium conductance in rat locus coeruleus neurones. Neuroscience 14:95-102.
Worley PF, Baraban JM, McCarren M, Snyder SH, Alger BE (1987) Cholinergic phosphatidylinositol modulation of inhibitory, G protein-linked, neurotransmitter actions: electrophysiological studies in rat hippocampus. Proc Natl Acad Sci USA 84:3467-3471.
Yamashita N, Shibuya N, Ogata E (1986) Hyperpolarization of the membrane potential caused by somatostatin in dissociated human pituitary adenoma cells that secrete growth hormone. Proc Natl Acad Sci USA 83:6198-6202.
Yatani A, Codina J, Brown AM, Birnbaumer L (1987a) Direct activation of mammalian atrial muscarinic potassium channels by GTP regulatory protein G_K. Science 235:207-211.
Yatani A, Codina J, Sekura RD, Birnbaumer L, Brown AM (1987b) Reconstitution of somatostatin and muscarinic receptor mediated stimulation of K channels by isolated G_K protein in clonal rat anterior pituitary cell membranes. Molec Endocrin (in press).
Yoshimura M, Hisgashi H (1985) 5-Hydroxytryptamine mediates inhibitory postsynaptic potentials in rat dorsal raphe neurons. Neurosci Lett 53:69-74.
Yoshimura M, North RA (1983) Substantia gelatinosa neurones in vitro hyperpolarized by enkephalin. Nature 305:529-530.

Table 1. Some receptors that couple to potassium conductances. Evidence for involvement of G-protein comes either from block by pertussis toxin, or activation by GTP-γ-S.

Agonist (receptor)	Tissue	G-protein?	Synaptic potential	Reference
Acetylcholine (M_2)	Guinea-pig cardiac muscle	—	Yes	del Castillo & Katz, 1955
		—	—	Soejima & Noma, 1984
		Yes	—	Pfaffinger et al., 1985
		Yes	—	Breitweiser & Szabo, 1985
Acetylcholine (M_2)	Rat nucleus parabrachialis	—	—	Egan & North, 1986
Acetylcholine (?)	Rat thalamus reticularis	—	—	Prince & McCormick, 1986
Acetylcholine (?)	Frog sympathetic C cells	—	Yes	Dodd & Horn, 1983
Acetylcholine (M_2)	Mudpuppy cardiac ganglion	—	Yes	Hartzell et al., 1977
Noradrenaline (α_2)	Guinea-pig submucous plexus	Yes	Yes	North & Surprenant, 1985
Noradrenaline (α_2)	Rat locus coeruleus	—	Yes	Egan et al., 1983
		—	—	Williams et al., 1985
Noradrenaline (α_2)	Rat substantia gelatinosa	—	—	North & Yoshimura, 1984
Noradrenaline (α_2)	Guinea-pig myenteric plexus	—	—	Morita & North, 1981
Dopamine (D_2)	Rat substantia nigra	—	—	Lacey et al., 1987
		Yes	—	Innis & Aghajanian, 1987
Dopamine (D_2)	Cultured human pituitary adenoma	—	—	Israel et al., 1985
Dopamine (?)	Rat intermediate pituitary	—	Yes	MacVicar & Pittman, 1986
5-hydroxytryptamine ($5HT_1$)	Rat dorsal raphe	—	—	Sprouse & Aghajanian, 1987
5-hydroxytryptamine ($5HT_1$)	Rat dorsal raphe	—	Yes	Williams, unpublished
5-hydroxytryptamine (?)	Rat dorsal raphe	—	Yes	Yoshimura & Higashi, 1985
5-hydroxytryptamine (?)	Rat hippocampus	Yes	—	Andrade et al., 1987
5-hydroxytryptamine (?)	Rat hippocampus	—	—	Colino & Halliwell, 1987
5-hydroxytryptamine ($5HT_1$)	Guinea pig myenteric plexus	—	—	Galligan, unpublished
γ- aminobutyric acid ($GABA_B$)	Rat hippocampus	Yes	—	Andrade et al., 1987
		—	Yes	Newberry & Nicoll, 1984
γ- aminobutyric acid ($GABA_B$)	Rat dorsal raphe	—	—	Williams, unpublished
Adenosine ($?A_1$)	Mouse striatum	—	—	Trussell & Jackson 1985
Adenosine ($?A_1$)	Guinea-pig cardiac cells	—	—	Kurachi et al., 1980
Somatostatin	Guinea pig submucous plexus	Yes	—	Mihara et al., 1987a
		—	(?)	Mihara et al., 1987b
Somatostatin	Cultured GH_3 pituitary cells	—	—	Yamashita, 1986
Somatostatin	Cultured GH_3 pituitary cells	Yes	—	Yatani et al., 1987b
Somatostatin	Rat locus coerulus cells	Yes	—	Inoue et al., 1987
Enkephalin (μ)	Rat locus coeruleus	Yes	—	Williams & North, 1984
		Yes	—	North et al., 1987
Enkephalin (? μ)	Rat substantia gelatinosa	—	—	Yoshimura & North, 1983
Enkephalin (? μ)	Guinea-pig myenteric plexus	—	—	Morita & North, 1982
Enkephalin (μ)	Mouse dorsal root ganglion	—	—	Werz & Macdonald, 1983
Enkephalin (μ)	Rat nucleus parabrachialis	—	—	Christie, unpublished
Enkephalin (δ)	Guinea pig submucous plexus	Yes	—	North et al., 1987

INDUCTION AND MAINTENANCE OF LONG-TERM POTENTIATION IN THE HIPPOCAMPUS

T.V.P. Bliss, M.L. Errington and M.A. Lynch
Division of Neurophysiology and Neuropharmacology
National Institute for Medical Research
Mill Hill
London NW7 1AA
UK

INTRODUCTION

There is increasing evidence to suggest that at least some types of learning are encoded in the mammalian brain by a form of synaptic plasticity that may be broadly described as Hebbian, in that underlying changes in synaptic strength occur as the result of coincident pre- and post-synaptic activity (Hebb, 1949; Singer, 1987; Ito, 1987). A widely-studied example of this sort of plasticity is the enduring enhancement of synaptic transmission known as long-term potentiation (LTP), which is produced with beguiling ease by the application of brief trains of high-frequency stimulation to any of the main excitatory pathways of the hippocampal formation (Bliss and Lomo, 1973; Teyler and DiScenna, 1987; Bliss and Lynch, 1988). In the last few years a concensus has developed regarding the cellular events leading to the initiation of LTP; in this view, LTP develops, in susceptible synapses, whenever there is a conjunction of synaptic activity and strong postsynaptic depolarization, conditions which allow the opening of the voltage-dependent ion channel associated with the NMDA subtype of glutamate receptor. A similar concensus on how the potentiated state is maintained has yet to emerge, with evidence for persistent changes in transmitter release and in synaptic morphology, possibly reflecting changes in protein phosphorylation, redistribution of calcium stores, and changes in protein synthesis. In this chapter we present a summary of the evidence which has led to our present understanding of the mechanisms responsible for the induction and maintenance of LTP.

NATO ASI Series, Vol. H19
Modulation of Synaptic Transmission and Plasticity
in Nervous Systems. Edited by G. Hertting and H.-C. Spatz

INDUCTION OF LTP

The traditional method of inducing LTP is by tetanic stimulation, and an example of an experiment in which tetani of increasing intensity were given to the perforant path of an anaesthetized rat is shown in Fig. 1. The graph illustrates a critical feature of LTP produced in this way - the existence of an intensity or 'cooperativity' threshold (McNaughton et al., 1978).

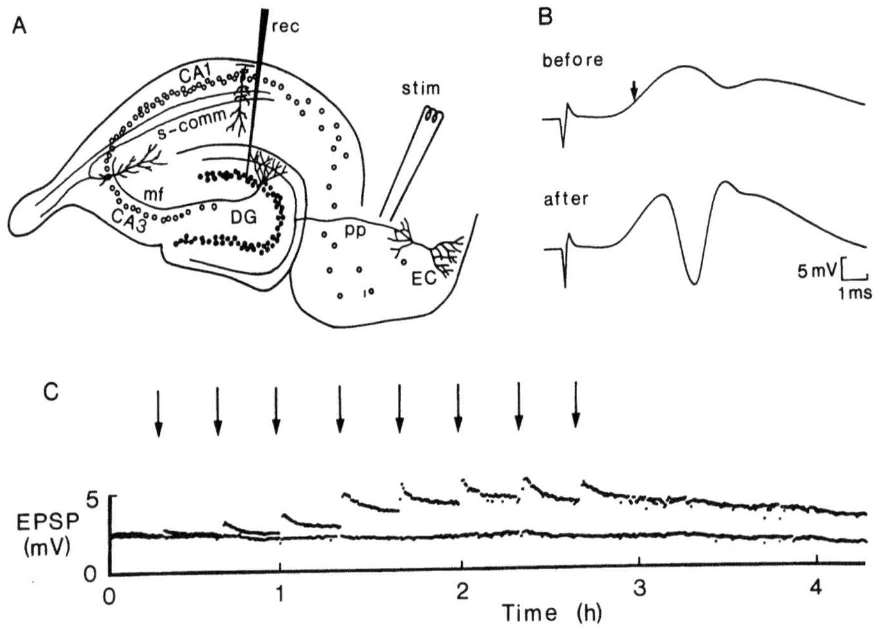

Fig. 1. Long-term potentiation in the dentate gyrus of the anaesthetized rat. A. Diagram of hippocampal formation showing stimulating electrode placed to excite perforant path fibres, and recording electrode in granule cell body layer of the dentate gyrus. The other principal excitatory pathways are also depicted (pp perforant path; DG dentate gyrus; mf mossy fibres; CA1, CA3 pyramidal cell fields; s-comm Schaffer-commissural projection). B. Field potentials recorded from the granule cell layer in response to single test shocks delivered to the perforant path before tetanic stimulation and after the last of the sequence of tetani shown in C. Note the potentiation both of the early synaptic component of the response (arrow), and of the population spike signalling the synchronous discharge of granule cells. C. Experiment in which stimulating and recording electrodes were placed bilaterally. The graph plots the amplitude of the synaptic component (population EPSP) of the evoked response. On one side only, high-frequency trains of increasing strength were delivered (through the same stimulating electrode as used for the test shocks) at the times indicated by arrows. LTP develops after the second or third train. (Adapted from Bliss, Goddard and Riives (1983): J Physiol 334: 475-491).

Below the cooperativity threshold, high-frequency trains result only in short-term potentiation, lasting for a few minutes. The reason for this threshold became clear when later work established that the induction of LTP was under the control of the NMDA subtype of glutamate receptor whose associated channel is normally blocked by physiological concentrations of Mg^{2+} ions. The blockade by Mg^{2+} is voltage-dependent and is relieved when the membrane is strongly depolarized (Nowak et al., 1984). Weak trains are therefore ineffective because they fail to produce sufficient depolarization. The first demonstration of the role of the NMDA receptor in the initiation of LTP came from the work of Collingridge et al., 1983), who showed that the specific NMDA antagonist D-aminophosphonovalerate (APV), which was without effect on responses evoked by single shocks, nevertheless blocked induction of LTP in the CA1 region of the hippocampal slice. The drug had no effect on previously established LTP. Similar effects were subsequently found in the dentate gyrus in vivo (Morris et al., 1986; Errington et al., 1987; see Fig. 5), although, interestingly, in the mossy fibre projection to CA3, the terminal region of which has a relatively low concentration of NMDA receptors, LTP is not blocked by APV (Harris and Cotman, 1986).

The voltage-dependency of the NMDA-linked channel provides the molecular basis for an induction rule for LTP (Wigstrom and Gustafsson, 1985; see Bliss and Lynch, 1988): a synapse will sustain an increment of potentiation if, and only if, it is active at a time when the region of dendrite on which it is located is strongly depolarized. Note that this formulation provides an immediate explanation not only for cooperativity but also for the important properties of associativity (by which a weak input, not by itself able to sustain LTP, is enabled to do so if combined with tetanic stimulation via a strong, convergent input (McNaughton et al., 1978; Levy and Steward, 1979; Barrionuevo and Brown, 1983; Wigstrom and Gustafsson, 1983)), and input specificity (potentiation is confined to those inputs which were active during tetanic stimulation (Andersen et al., 1977; Lynch et al., 1977)). Recently, three laboratories have published direct experimental verification of this rule for the Schaffer-commissural pathway: LTP can be produced intracellularly in CA1 pyramdial cells by the repeated application of single shocks in conjunction with strong depolarizing pulses delivered to the target cell; single shocks or depolarizing pulses alone have no effect (Wigstrom et al., 1986; see Fig. 2; Kelso et al., 1986; Sastry et al., 1986). 'Postsynaptic' LTP (i.e., potentiation of the probability of a test shock firing an impaled cell), can also be produced by single shocks delivered during dendritic iontophoresis of glutamate (Hvalby et al, 1987); again, neither procedure alone is effective.

Several studies testify to the importance of Ca^{2+} in the genesis of LTP. Intracellular buffering of cytosolic Ca^{2+} with the chelator EGTA blocks the induction of LTP in the injected cell (Lynch et al., 1983), while brief exposure to high extracellular concentrations of Ca^{2+} induces a form of

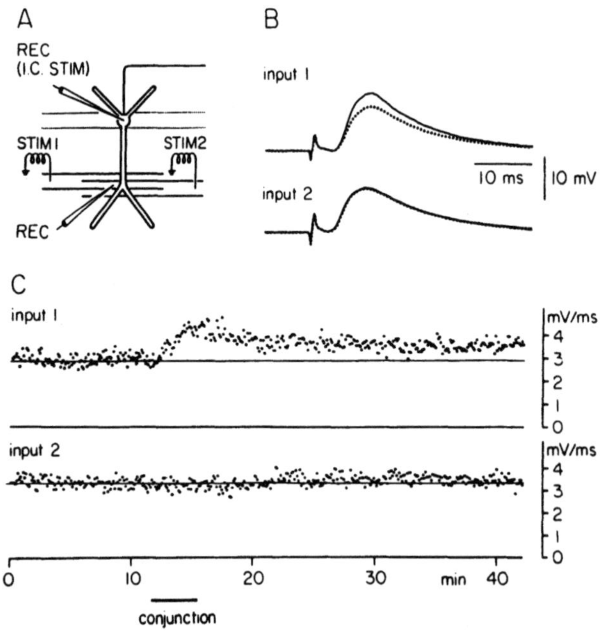

Fig. 2. Intracellular LTP can be induced by single afferent volleys timed to coincide with the injection of depolarizing current in the postsynaptic cell. A. Diagram of a pyramidal cell in area CA1 impaled with an intracellular electrode. Stimulating electrodes are positioned to activate two non-overlapping, converging inputs. B. EPSPs evoked before (solid traces) and after (dotted traces) the conjunction period indicated by the bar in C, during which 100 msec depolarizing pulses were delivered through the intracellular electrode in conjunction with stimuli to input 1. C. Plot of the slopes of the EPSP for the two pathways in response to test stimuli delivered alternately to inputs 1 and 2. LTP was induced only in the input in which there was coincidence of afferent activity and electrically-induced postsynaptic depolarization. (From Wigstrom et al., 1986).

LTP whose origins appear to be similar to that of tetanus induced-LTP (Turner et al., 1982; Bliss et al., 1986b; Reymann et al, 1986). NMDA-linked channels are permeable to Ca^{2+} (MacDermott et al., 1986), and it seems likely that entry of Ca^{2+} through these channels is the triggering event for the induction of LTP. However, direct evidence for this hypothesis is not yet available. Fig. 3 provides in schematic form a summary of the involvement of the two classes of glutamate receptor in synaptic transmission in the hippocampus; responses evoked by single shocks are mediated by kainate/quisqualate receptors, which are not affected by NMDA antagonists; during strong high-frequency stimulation, the postsynaptic membrane becomes sufficiently depolarized to relieve the Mg^{2+} block of the NMDA receptor-linked channel, allowing Ca^{2+} to enter the cell and trigger the biochemical cascade which leads within a few tens of seconds, at most, to enhanced transmission in those (and only those) synapses which were active during the period of intense dendritic depolarization.

MAINTENANCE OF LTP

Once initiated, how is the increase in synaptic efficacy maintained? In principle, three possibilities can be envisaged: a presynaptic mechanism leading to an increase in transmitter release, a postsynaptic mechanism, resulting in an increased coupling between transmitter and postsynaptic response (e.g., an increase in receptor number, or changes in spine morphology), or some combination of the two (e.g., overall growth of potentiated synapses). We consider the evidence for these candidates in turn, with emphasis on our own work on the correlation between LTP and transmitter release.

The first indication that LTP is associated with increased transmitter release came from an *in vitro* experiment by Skrede and Malthe-Sorenssen (1981) in which the resting and stimulated release of preloaded, radiolabelled [^{3}H]-D-aspartate from hippocampal slices was found to be increased following relatively prolonged tetanic stimulation. In an extensive series of experiments in our laboratory, we have made use of two techniques to study the relation between LTP and transmitter release: combined recording of evoked potentials and perfusion of the molecular layer of the dentate gyrus in the anaesthetized rat via a push-pull cannula, with

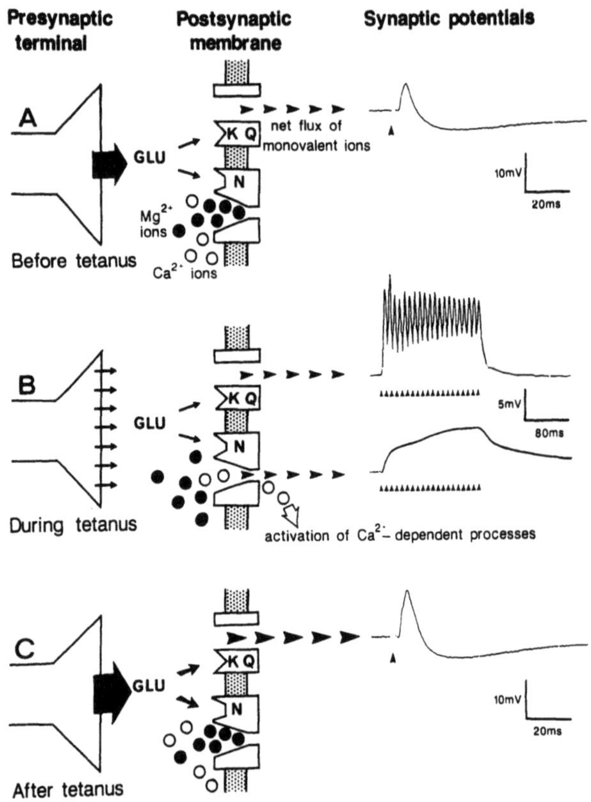

Fig. 3. Scheme of events underlying the induction of LTP.
A. Glutamate released during low-frequency stimulation acts on kainate/quisqualate (KQ) and NMDA (N) receptors. The latter are blocked by Mg^{2+}ions, so the EPSP is mediated principally by monovalent cations permeating KQ channels. B. During tetanic stimulation, sufficient depolarization occurs to lift the Mg^{2+} blockade of N channels. The synaptic potential now contains an APV-sensitive NMDA component, seen as a slow EPSP (lower trace on right, produced by subtraction of responses in the presence and absence of APV). Ca^{2+} enters the cell through the open NMDA channel, triggering a Ca^{2+}-dependent process which leads (C) to a sustained potentiation of synaptic transmission, in part via an increase in transmitter release. (From Collingridge and Bliss (1987): Trends in Neurosci 10: 288-293).

subsequent assay of the perfusate for endogenous or radiolabelled amino acid transmitter candidates; and <u>ex vivo</u> experiments in which LTP is first induced in the anaesthetized animal, the brain then removed and the hippocampus chopped into prisms for analysis of calcium-dependent, K^+-stimulated release of preloaded, radiolabelled transmitter candidates. In both types of experiment we have found that LTP is associated with an increase in transmitter release, measured for 1-4 hours after induction in the case of <u>in vivo</u> experiments (Dolphin, et al., 1982; Bliss et al., 1986;

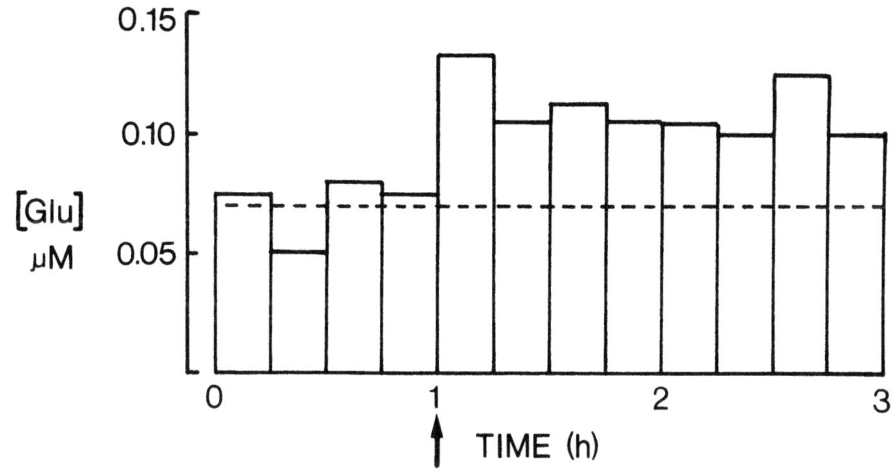

Fig. 4. LTP is associated with an increase in glutamate release. In this experiment, 15 min fractions were collected during push-pull perfusion of the dentate gyrus in an anaesthetized rat, and analyzed for endogenous glutamate. LTP was induced by a single high-frequency train delivered at the time indicated by the arrow. See Bliss et al. (1986) for further details of methodology.

Errington et al., 1987), or 45 min after induction in the case of ex vivo experiments (Feasey et al., 1986; Bliss and Lynch 1988) - a period recently extended to 3 days in an ex vivo experiment employing animals with chronically-implanted electrodes (Bliss et al., 1987). An example of the increase in concentration of endogenous glutamate in the perfusate following the induction of LTP in an anaesthetized rat is shown in Fig. 4. The increased release of glutamate is specifically related to the induction of LTP, and not to some other consequence of high-frequency stimulation, since no increase occurs in the dentate gyrus if induction is prevented by coincident tetanic stimulation of the inhibitory commissural input (Bliss et al., 1986), or by perfusion with APV (Errington et al., 1987; see Fig 5) or with the lipoxyenase inhibitor nordihydroguaiaretic acid (NDGA; Bliss et al., 1988). We have seen similar, but smaller and less persistent LTP-associated increases in the concentration of aspartate in hippocampal perfusates, but not in the concentrations of glycine or glutamine. Concentrations of GABA have generally been too low to be measured confidently.

Assuming that the enhanced concentration of glutamate in hippocampal perfusates following tetanic stimulation reflects an increase in

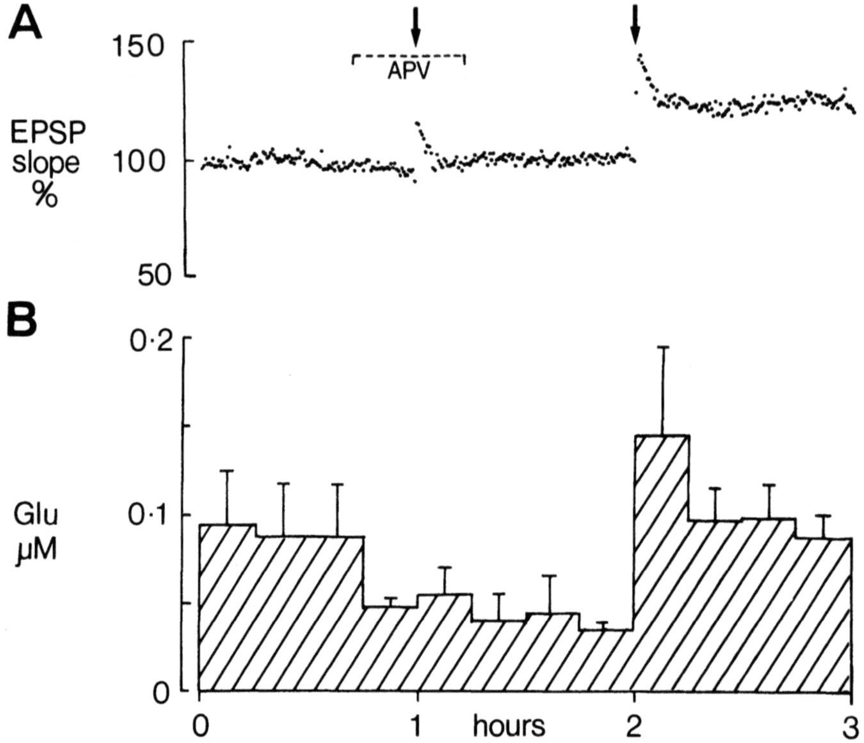

Fig. 5. APV prevents LTP and the associated increase in glutamate release. The synaptic region of the dentate gyrus was perfused in anaesthetized rats. When APV was included in the perfusion medium the synaptic response to single shocks was not affected, but the induction of LTP was prevented. Following washout, a repetition of the same high-frequency train (arrows) led to LTP and a sustained increase in glutamate release. (From Errington et al., 1987).

activity-dependent release from potentiated terminals (see Bliss et al., 1986 for a discussion of this point), what factors might contribute to the increase? One possibility, given the extreme sensitivity of the release process to Ca^{2+}, is a small but sustained increase in the concentration of intraterminal Ca^{2+}. Consistent with this possibility, an increase in the concentration of cytosolic Ca^{2+}, measured with the Ca^{2+}-binding dye Indo-1 (Lynch et al., 1988), and an increase in the influx of $^{45}Ca^{2+}$ (Agoston and Kuhnt, 1986), have been reported in synaptosomes prepared from potentiated hippocampal tissue. A possible source for the elevated level of intraterminal Ca^{2+} is suggested by the accelerated turnover of the phosphatidylinositol cycle in potentiated tissues, including synaptosomes, with a consequent increase in inositol phosphates (Lynch et al., 1988), one of which, inositol triphosphate, can release Ca^{2+} from non-mitochondrial stores in hippocampal synaptosomes (Lynch, M., unpublished observations). It is likely that Ca^{2+} exerts its influence on synaptic function, as on other cellular processes, through the activation of protein kinases, and it is therefore significant that calcium-dependent protein kinase C injected directly into CA1 pyramidal cells causes a prolonged potentiation of transmission (Hu et al., 1987), as does extracellular application of phorbol-12,13 dibutyrate, an exogenous activator of PKC (Malenka et al., 1986). Antagonists to PKC can block the induction and/or maintenance of LTP (Reymann et al., 1987; Madison et al., 1988). Furthermore, the phosphorylation of a presynaptically located substrate protein for PKC, variously known as F1, B50 and GAP43, is increased following the induction of LTP, and the degree of phosphorylation is related to the degree of LTP (Lovinger et al., 1985; Schrama et al., 1986). PKC thus appears to be implicated both pre- and post-synaptically in the maintenance of LTP. Another protein kinase, with a predominanty postsynaptic distribution, which has been discussed in the context of LTP is calcium/calmodulin dependent protein kinase; this has the interesting property that when autophosphorylated its activity becomes largely independent of calcium - it could therefore act as a molecular switch, activated by the transient inward surge of calcium generated during the inductive phase of LTP and thereafter in its autophosphorylated form carrying the 'memory' of that event (Kennedy, 1987).

An early theory of the mechanism of LTP attributed the sustained enhancement of synaptic efficacy to an irreversible increase in the number of glutamate receptors (Baudry and Lynch 1984). The interpretation of the binding assay used to determine receptor number has been questioned, however, and there is as yet no compelling biochemical or physiological evidence that increases in receptor number contribute to LTP (Bliss and Lynch, 1988).

Several groups have reported a change in the shape and/or number of synapses at various times after the induction of LTP (reviewed in Bliss and Lynch, 1988). Lynch and Baudry (1984) have proposed that during a tetanus the influx of calcium through voltage-dependent Ca^{2+} channels in the postsynaptic membrane activates calcium-dependent proteases leading to a degradation of spectrin-like cytoskeletal proteins and a consequent irreversible effect on synaptic morphology. However, whether changes in spine shape can account quantitatively for the observed changes in synaptic coupling remains in doubt (Brown et al., 1988).

It would be surprising if LTP, which can last for at least several weeks, did not involve changes at the level of gene expression or protein synthesis (Goelet et al., 1986). A beginning has been made at addressing this question by looking for LTP-related changes in the release of proteins into hippocampal perfusates (Duffy et al., 1981; Bliss et. al., 1987) and at the expression of specific proteins - for example, the nuclear protein encoded by the proto-oncogene, c-fos, which turns out not be related to LTP (Hunt et al, 1987). Protein synthesis inhibitors have been reported to affect both the induction and maintenance of LTP (see Bliss and Lynch, 1988), but the evidence from different laboratories is often conflicting, and no final conclusion can be drawn at present.

In this chapter we have briefly surveyed the evidence in support of the view that, on the one hand, the initiating event in LTP is an influx of Ca^{+} through NMDA receptor-linked channels in the postsynaptic membrane, while, on the other hand, the maintenance of LTP is due, at least in part, to a sustained increase in transmitter release. If this is an accurate description, the question arises as to how information is transferred from the postsynaptic side to the presynaptic side of the synapse to trigger the biochemical events culminating in enhanced transmitter release. One possibility is that the message is carried by a retrograde plasticity factor, as proposed several years ago by Griffiths (1966). Possible candidates for such a factor include released proteins (Duffy et al., Bliss et al., 1987) and membrane lipids or their metabolites (Piomelli et al., 1987; Bliss et al., 1988).

REFERENCES

Andersen P, Sundberg SH, Sveen O, Wigstrom H (1977): Specific long-lasting potentiation of synaptic transmission in hippocampal slices. Nature 266: 736-737.

Barrionuevo G, Brown TH (1983): Associative long-term potentiation in the hippocampal slices. Proc Natl Acad Sci USA 80: 7347-7351.

Bliss TVP, Dolphin AC, Errington ML, Fazeli MS (1987): Increase in concentration of specific extracellular proteins during long-term potentiation in the dentate gyrus of the rat. J Physiol 388: 49P.

Bliss TVP, Douglas RM, Errington ML, Lynch MA (1986): Correlation between long-term potentiation and release of endogenous amino acids from dentate gyrus of anaesthetized rats. J Physiol 377: 391-408.
Bliss TVP, Errington ML, Laroche S, Lynch MA (1987): Increase in K+-stimulated, Ca2+-dependent release of [3H]glutamate from rat dentate gyrus three days after induction of long-term potentiation. Neurosci Lett 83: 107-112.
Bliss TVP, Errington ML, Lynch MA (1986): Calcium-induced long-term potentiation in the dentate gyrus is accompanied by a sustained increase in glutamate release. In Hicks TP, Lodge D (eds): Excitatory Amino Acid Transmitters. New York: Alan R Liss, Inc., pp 337-340.
Bliss TVP, Errington ML, Lynch MA, Williams JH (1988): The lipoxygenase inhibition nordihydroguaiaretic acid (NDGA) blocks the induction of both tetanus-induced and calcium-induced long-term potentiation in the hippocampus of the rat. Pflug Arch Suppl. In press.
Bliss TVP, Lomo T (1973): Long-lasting potentiation of synaptic transmission in the dentate area of the anaesthetized rabbit following stimulation of the perforant path. J Physiol 232: 331-356.
Bliss TVP, Lynch MA (1988): Long-term potentiation of synaptic transmission in the hippocampus: properties and mechanisms. In Landfield PW, Deadwyler SA (eds): Long-term potentiation: From Biophysics to Behavior. New York: Alan R Liss Inc, pp 3-72.
Brown TH, Chang VC, Ganong AH, Keenan CL, Kelso SR (1988). Biophysical properties of dendrites and spines that may control the induction and expression of long-term synaptic potentiation. In Landfield PW, Deadwyler SA (eds): Long-term potentiation: From Biophysics to Behavior. New York: Alan R Liss Inc, p. 197.
Collingridge GL, Kehl SJ, McLennan, H (1983): Excitatory amino acids in synaptic transmission in the Schaffer collateral-commissural pathway of the rat hippocampus. J Physiol 334: 33-46.
Dolphin AC, Errington ML, Bliss TVP (1982): Long-term potentiation of the perforant path <u>in vivo</u> is associated with increased glutamate release. Nature 297: 496-498.
Duffy C, Teyler TJ, Shashoua VE (1981): Long-term potentiation in the hippocampal slice: evidence for stimulated secretion of newly synthesized proteins. Science 212: 1148-1151.
Errington ML, Lynch MA, Bliss TVP (1987): Long-term potentiation in the dentate gyrus: induction and increased glutamate release are blocked by D(-)aminophosphonovalerate. Neuroscience 20: 279-284.
Feasey KJ, Lynch MA, Bliss TVP (1986): Long-term potentiation is associated with an increase in calcium-dependent, potassium-stimulated release of [14C]glutamate from hippocampal slices: an ex vivo study in the rat. Brain Res 364: 39-44.
Goelet P, Castelluci VF, Schacher S, Kandel ER (1986): The long and the short of long-term memory - a molecular framework. Nature 322: 419-422.
Griffiths J.S. (1966): A theory of the nature of memory. Nature 211: 1160-1163.
Harris EW, Cotman CW (1986): Long-term potentiation of guinea pig mossy fiber responses is not blocked by N-methyl-D-aspartate antagonists. Neurosci Lett 70: 132-137.
Hebb DO (1949): The Organization of Behavior. New York: Wiley.
Hu G-Y, Hvalby O, Walaas SI, Albert KA, Skjeflo P, Andersen P, Greengard P. (1987): Protein kinase C injection into hippocampal pyramidal cells elicits features of long-term potentiation. Nature 328: 426-429.
Hunt SF, Pini A, Errington ML, Bliss TVP, Evan G (1987): Induction of c-fos protein in neurons of the central nervous system. Abstr Neurosci Soc 13: 1597.

Hvalby O, Lacaille J-C, Hu G-Y, Andersen P (1987): Postsynaptic long-term potentiation follows coupling of dendritic glutamate application and synaptic activation. Experientia 43: 599-601.
Ito M (1987): Characterization of synaptic plasticity in the cerebellar and cerebral neocortex. In Changeux J-P, Konishi M (eds): The Neural and Molecular Bases of Learning. Dahlem Konferenzen. Chichester: Wiley, pp 263-278.
Kelso SR, Ganong AH, Brown TH (1986): Hebbian synapses in hippocampus. Proc Natl Acad Sci USA 83: 5326-5330.
Kennedy MB (1987): Neuronal biochemical regulatory mechanisms. In Changeux J-P, Konishi M (eds): The Neural and Molecular Bases of Learning. Dahlem Konferenzen. Chichester: Wiley, pp 137-150.
Levy WB, Steward O (1979): Synapses as associative memory elements in the hippocampal formation. Brain Res 175: 233-245.
Lovinger DM, Akers RF, Nelson RB, Barnes CA, McNaughton BL, Routtenberg A (1985): A selective increase in phosphorylation of protein F1, a protein kinase C substrate, directly related to three day growth of synaptic long-term enhancement. Brain Res 343: 137-143.
Lynch G, Baudry M (1984): The biochemistry of memory: a new and specific hypothesis. Science 224: 1057-1063.
Lynch G, Dunwiddie T, Gribkoff V (1977): Heterosynaptic depression: a postsynaptic correlate of long-term potentiation. Nature 266: 737-739.
Lynch G, Larson J, Kelso S, Barrionuevo G, Schottler, F (1983): Intracellular injections of EGTA block induction of hippocampal long-term potentiation. Nature 305: 719-721.
Lynch MA, Clements MP, Errington ML, Bliss TVP (1988): On the mechanism of increased transmitter release in LTP: measurement of calcium concentration and phosphatidylinositol turnover in synaptosomes. In HL Haas and G. Buzsaki (eds): Synaptic Plasticity in the Hippocampus. Heidelberg: Springer Verlag, pp 110-113.
Lynch MA, Clements MP, Errington ML, Bliss TVP (1988): Increased hydrolysis of phosphatidylinositol-4-5-bisphosphate in long-term potentiation. Neurosci Lett: In press.
MacDermott A, Mayer ML, Westbrook GL, Smith SJ, Barker JL (1986): NMDA-receptor activation increases cytoplasmic calcium concentration in cultured spinal cord neurones. Nature 321: 519-522.
McNaughton BL, Douglas RM, Goddard GV (1978): Synaptic enhancement in fascia dentata: cooperativity among coactive afferents. Brain Res 157: 277-293.
Madison DV, Malinow R, Tsien RW (1988): Inhibitors of protein kinase C block long-term potentiation in rat hippocampus. J. Physiol (Proc): In press.
Malenka RC, Madison DV, Nicoll RA (1986): Potentiation of synaptic transmission in the hippocampus by phorbol esters. Nature 321: 175-177.
Morris RGM, Anderson E, Lynch GS, Baudry M (1986): Selective impairment of learning and blockade of long-term potentiation by an N-methyl-D-aspartate receptor antagonist, AP5. Nature 319: 774-776.
Nowak L, Bregestovski P, Ascher P, Herbet A, Prochiantz A. (1984): Magnesium gates glutamate-activated channels in mouse central neurones. Nature 307: 462-465.
Piomelli D, Volterra A, Dale N, Siegelbaum SA, Kandel ER, Schwartz JH and Belardetti F (1987): Lipoxygenase metabolites of arachidonic acid as second messengers for presynaptic inhibition of Aplysia sensory cells. Nature 328: 38-43.
Reymann KG, Matthies H (1987): Hippocampal long-term potentiation is blocked by polymixin B - an inhibitor of protein kinase C. Neurosci Suppl 22: S410.

Reymann KG, Matthies HK, Frey U, Vorobyev VS, Matthies H (1986): Calcium-induced long-term potentiation in the hippocampal slice: characterization of the time course and conditions. Brain Res Bull 17: 291-296.
Sastry BR, Goh JW, Auyeung A (1986): Associative induction of posttetanic and long-term potentiation in CA1 neurons of rat hippocampus. Science 232: 988-990.
Singer W (1987): Activity-dependent self-organization of synaptic connections as a substrate of learning. In Changeux J-P, Konishi M (eds): The Neural and Molecular Bases of Learning. Dahlem Konferenzen. Chichester: Wiley, pp 301-335
Schrama LH, De Graan PNE, Wadman WJ, Gispen WH (1986): Long-term potentiation and 4-aminopyridine-induced changes in protein- and lipid-phosphorylation in the hippocampal slice. In Gispen WH, Routtenberg A (eds): Function of Neural Phosphoproteins. Prog Brain Res 69: Amsterdam: Elsevier.
Skrede K, Malthe-Sorenssen D (1981): Increased resting and evoked release of transmitter following repetitive electrical tetanization in hippocampus: a biochemical correlate to long-lasting synaptic potentiation. Brain Res 208: 436-441.
Turner RW, Baimbridge KG, Miller JJ (1982): Calcium-induced long-term potentiation in the hippocampus. Neuroscience 7: 1411-1416.
Teyler TJ, DiScenna P (1987): Long-term potentiation. In Cowan WM, Shooter EM, Stevens CF, Thompson, RF (eds): Annual Review of Neuroscience 10: 131-161.
Wigstrom H, Gustafsson B (1983): Heterosynaptic modulation of long-lasting potentiation in the hippocampal slice. Acta Physiol Scand 119: 455-458.
Wigstrom H, Gustafsson B (1985): On long-lasting potentiation in the hippocampus: a proposed mechanism for its dependence on coincident pre- and postsynaptic activity. Acta Physiol Scand 123: 519-522.
Wigstrom H, Gustafsson B, Huang Y-Y, Abraham WC (1986): Hippocampal long-lasting potentiation is induced by pairing single afferent volleys with intracellularly injected depolarizing current pulses. Acta Physiol Scand 126: 317-319.

INITIATION OF LONG-TERM POTENTIATION: DEPENDENCY ON PARAMETERS OF STIMULATION AND EXTERNAL CALCIUM.

R. Anwyl,
Department of Physiology,
Trinity College,
Dublin 2,
Ireland.

Following a brief period of high frequency stimulation of presynaptic axons, there is a period of enhanced synaptic transmission in the hippocampus of the vertebrate brain. This enhancement includes a short term potentiation which lasts for a few minutes, and long-term potentiation (LTP) which lasts for several hours or even days. In the present study, the effects of different parameters of stimulation and external calcium have been investigated on the initiation of LTP and potentiation in the stratum radiatum of CA1 of the hippocampal slice in vitro.

Co-operativity

In the fascia dentata of the hippocampus, weak tetanic stimulation was found to only generate short-term potentiation. LTP only appeared if the stimulus strength was raised so as to generate an excitatory post-synaptic potential (epsp) of 5.6 mV[1]. The appearance of LTP at a threshold stimulation intensity led to the theory of cooperativity. This stated that collaboration of a large number of afferent fibres or participating neurons was necessary to produce LTP. In the present study, the generation of LTP has been investigated at different stimulus strengths. Tetanic stimulation was carried out using 5-20 trains of 5-9 stimuli at 200 Hz, with a 2 sec interval between each train. The stimulus intensity was adjusted to produce an initial epsp of the first train of the tetanus varying from 300 μV, the smallest which could be measured accurately, up to 900 μV,

NATO ASI Series, Vol. H19
Modulation of Synaptic Transmission and Plasticity
in Nervous Systems. Edited by G. Hertting and H.-C. Spatz

which was threshold for spike generation.

LTP was reliably generated at stimulation intensities producing epsp amplitudes of 300-900 μV. Moreover, the magnitude of LTP increased as the stimulus intensity was raised. Thus LTP, expressed as the percentage increase in a test epsp of 300 μV, averaged 7.7± 1.4%, 19±3.2% and 65±8.3% with the stimulus intensities set to produce initial epsp amplitudes of 300 μV, 400 μV and 900 μV respectively. These results demonstrate firstly that there is no distinct threshold amplitude of the epsp necesary to produce LTP, but that LTP is graded, increasing in ampitude as the stimulus intensity is raised. The generation of LTP of very small just suprathershold amplitude epsp's demonstrates that the amount of cooperativity of presynaptic afferents or post-synaptic neurons minimal and the possibility is raised that stimulation of single afferent fibres may produce LTP.

Effect Ca on LTP.

Extracellular calcium has previously been shown to be very important in the production of LTP in the hippocampus. Thus tetanic stimulation in the absence of Ca did not produce long-term potentiation[2], and decreasing external Ca concentration, was found to decrease the probability of LTP being initiated[3]. In the present studies, we have carried out a quantitative study on the dependency of LTP initiation on the extracellular Ca concentration. We have also determined whether there is any change in the Ca cooperativity participating in transmitter release following the production of LTP.
The magnitude of LTP evoked by a short series of trains of stimuli was studied in external Ca concentrations between 0.7-2.5 mM. As altering external Ca concentration changed the amplitude of the epsp through an effect on transmitter release, the stimulus intensity was always adjusted so as to evoke a test low frequency epsp of 800 μV. Thus as the

external Ca concentration was raised, a smaller number of presynaptic axons were stimulated.
The magnitude of LTP was highly dependent on the Ca concentration of the perfusate, with LTP increasing as the Ca concentration of the perfusate was raised. The threshold for LTP was 0.8 mM, measuring 5±2% (mean±SEM, n=5). A logarithmic plot of the amplitude of LTP against the external Ca concentration showed a very steep relationship (slope 7) between Ca concentrations of 0.8-1.0 mM. Thus LTP increased to 38±%, n=5, in 1.0 mM Ca. The increase in LTP was much less steep at higher concentrations, reaching a maximum of 65±4%, n=6, in 2 mM Ca. LTP declined to 38±6%, n=5 in 2.5 mM Ca.

Although previous studies have established the importance of Ca in the initiation of LTP[8], there have been no quantitative studies on the dependency of the magnitude of LTP on extracellular Ca. The present studies show clearly that the amplitude of LTP increases as the external Ca concentration is raised. The threshold Ca concentration for production of LTP was 0.8 mM, very similar to the threshold for production of an epsp (0.7-0.8 mM). This suggests that the Ca sensitivities of the biochemical processes responsible for transmitter release and for LTP are very similar. LTP showed a very steep increase between Ca concentrations of 0.8-1.0 mM, with a slope greater than 7.0 in the double logarithmic plot of LTP versus external Ca. This may be caused by a cooperative action of Ca ions at the Ca binding site responsible for LTP initiation. The much shallower increase in LTP at higher Ca concentrations may be due to saturation of Ca ions at this site. Alternatively, perhaps Ca acts at two different sites to establish LTP. We do not believe that the very steep dependency of LTP on Ca at low Ca concentrations is caused by the present methodology i.e. the increasing of stimulus as Ca was lowered in order to keep the epsp a constant amplitude. Increasing the intensity of stimulation causes an enhancement of LTP if Ca is kept constant[19] (Mulkeen & Anwyl, unpublished results), and therefore as Ca was decreased in the present experiments the measurements of LTP were most likely to be greater than if the stimulation current could have been kept constant. Thus the present experiments underestimate the true steepness of the curve relating LTP to extracellular Ca. The enhancement The enhancement of LTP by increasing extracellular Ca is most likely to be caused by an enhancement of Ca influx into neurons. The most generally accepted theory at present is that the Ca influx is via post-synaptic NMDA receptors [4].

<u>Effect of LTP on Ca cooperativity of transmitter release.</u>

In these experiments, the epsp was always evoked at a low frequency (0.02Hz). The epsp amplitude was adjusted to 300uV in 2.0 mM Ca and then the amplitude of the epsp measured following an increase or decrease in Ca concentration. Changes in the epsp were expressed as the percentage alteration relative to the amplitude in 2 mM Ca. The amplitude of the epsp was found to be highly dependent on the Ca concentration in the perfusate. Thus the epsp amplitude always increased as the Ca concentration was raised, and decreased as the Ca concentration was lowered, with the threshold for the generation of the epsp being 0.7-0.8 mM. A double logarithmic plot of the relationship between the epsp amplitude and the external Ca concentration gave a maximum slope of 2.1 between 0.8-1.0 mM Ca. This indicates that the epsp amplitude is related to the second power of the external Ca concentration at low Ca levels. At higher Ca concentration, the slope declined, probably due to saturation at the release site. Thus the slope was 1.3 between 1.0-1.5 mM Ca, and 0.92 between 1.5-2.5 mM Ca.

The relationship between the epsp amplitude and external Ca concentration was also determined following LTP. The LTP was iniated by a series of high frequency trains in 2 mM Ca, with the amplitude of LTP produced averaging 65 4%, n = 5.

Following initiation of LTP in 2 mM Ca, the slope of the regression line was 2.1 between Ca concentrations of 0.8-1.0 mM, 1.2 between 1.0-1.5 mM Ca, and 1.1 between 1.5-2.0 mM Ca. These slopes are not significantly different from those before induction of LTP. No change in the presynaptic compound action potential occurred over the range of external Ca used in these studies. The slope of the regression line relating the log epsp amplitude and the log external Ca had a maximum value of 2.1 between Ca concentration of 0.8-1.0 mM. A similar low power relationship between the epsp and external Ca has previously been found in the hippocampal and spinal cord slices[5,6]. However, at the neuromuscular junction and squid giant synapse, higher power relationship values of 2.5-4.0 have been found[7,8,9]. A power relationship greater than unity implies a cooperative action of Ca ions participating in transmitter release,such as more than one Ca ion being necessary to activate a release site. A lower value of the power relationship in the central nervous system implies greater efficiency of the release mechanism, and it has previously been suggestedthat a low power function may be due to a relatively small Ca influx compared with the intracellular Ca concentration of the presynaptic terminals [10]. The slope of the regression line relating the epsp amplitude to external Ca was not changed following the production of LTP, thus demonstrating that the generation of LTP does not cause a change in the Ca cooperativity of transmitter release. This is in contrast to the presynaptic short-lasting increase in the epsp caused by facilitation and potentiation in which a considerable reduction occurred in the power relationship between the epsp amplitude and the external Ca concentration[9]. One possible mechanism for such an increased sensitivity of release to Ca, namely a decrease in the cooperative action of Ca ions participating in transmitter release, can be excluded by the results of the present study.

The generation of short-term potentiation and LTP by stimulation with either a single train or a series of short trains was also compared. Increasing the number of stimuli in a single tetanus given at 200 Hz caused large increases in the initial amplitude of potentiation, P(o), and the decay time constant, T P. A maximum P(o) of 60-100% and T P of 60-80 sec was attained following 50-100 stimuli. No LTP was produced. These results are identical to those at the neuromuscular junction (Magleby & Zengel, 1975). However a tetanus in the form of short high frequency trains at 0.5 sec intervals (each train 5 stimuli at 200 Hz) resulted in potentiation with a smaller initial amplitude and a longer time course. Thus 25 trains initiated maximal P(o) of 26.6% (21.6-56.4) median (range) n=7, and T P of 189 sec (130-350) n=7. Long-term potentiation measuring 8-10% was initiated in 70%, and 30% of experiments following 13 & 25 trains of stimuli respectively. Moreover, potentiation was usually absent when LTP was evoked. These results suggest the presence of two types of short-term potentiation, one with a large initial amplitude and short time decay time, the other with a small initial amplitude and relatively long time decay time.

Perfusion of N-methyl aspartate (NMA) increased potentiation but not LTP. Control stimulation in slices using slightly larger amplitude epsp's (400 μV) which always evoked LTP, produced P(o) of 40.6%, T P of 157 sec and LTP of 21.3%, n=11. In slices perfused with 10 μM NMA, P(o) was significantly increased to 64.5% (P 0.005), T P to 255 sec (P 0.005) and LTP was unchanged at 23.5% (n=11). These results suggest that NMA increases short-term potentiation due to an influx of Ca ions, as it is known that NMA reception have a high permeability to Ca.

1. Naughton, B.L. Douglas, R.M. & Goddard, G.V. (1978) Brain Res 157: 277-293.

2. Wigstrom, H., Swann, J.W. and Anderson, P. (1979) Acta Physiol. Scand 105: 126-128.

3. Dunwiddie, T. V. and Lynch, G. Brain Res: (1979) 169 103-110.

4. Collingridge, G.L. & Bliss, T.V.P. (1987) Trends in Neuroscience 10: 288-293.

5. Dingledine, R and Somjen, G. (1981) Brain Res 207: 218-222.

6. Kuno, M. and Takahashi, T. J. (1984) Physiol 376: (1984) 543-553.

7. Dodge, F.A. and Rahamimoff, R. J. (1967) Physiol. 193: 419-432.

8. Jenkinson, D.H. (1957) J. Physiol 138: 434-444.

9. Stanley, E.F. (1986) J. Neuroscience 6: 782-789.

10. Errington, M.L., Lynch, M.A & Bliss TVP. (1986) Neuroscience 20: 279-286.

THE ROLE OF PROTEIN PHOSPHORYLATION IN LONG-TERM POTENTIATION

L. V. Dekker
P. N. E. De Graan
F. M. J. Heemskerk
L. H. Schrama
A. B. Oestreicher
P. Schotman
and W. H. Gispen

Division of Molecular Neurobiology
Rudolf Magnus Institute for Pharmacology
Laboratory for Physiological Chemistry
and Institute of Molecular Biology
and Medical Biotechnology

University of Utrecht
Padualaan 8
3584 CH Utrecht
Nederlande

INTRODUCTION

Long-term potentiation (LTP) is a phenomenon in which long-lasting changes in synaptic efficacy are induced by brief stimuli. It is often used as a model to investigate the mechanisms of information storage in the brain. The molecular events underlying LTP appear to occur in three phases: the initiation, development and maintenance of LTP. It is becoming clear that activation of the N-methyl-D-aspartate (NMDA) receptor-channel complex is important for the initiation and the development of certain types of LTP (for review see Collingridge and Bliss, 1987). However, there is no consensus on the molecular events responsible for the maintenance of LTP.

NATO ASI Series, Vol. H19
Modulation of Synaptic Transmission and Plasticity
in Nervous Systems. Edited by G. Hertting and H.-C. Spatz

Most likely three mechanisms contribute to the maintenance of LTP: (i) a sustained increase in the release of excitatory amino acids (L-glutamate and L-aspartate) from the presynaptic terminal (Skrede and Malthe-Sörensen, 1981; Dolphin et al., 1982; Errington et al., 1987); (ii) an increase in the number of functional neurotransmitter receptors at the postsynaptic membrane (Lynch and Baudry, 1984); (iii) changes in synaptic morphology (Chang and Greenough, 1984).

Several lines of evidence point to a crucial role for Ca^{2+} during LTP. Extracellular Ca^{2+} is required for the induction of LTP, since lowering the extracellular Ca^{2+} concentration prevents LTP (Dunwiddie and Lynch, 1979). Moreover, a temporary increase in the extracellular Ca^{2+} concentration induces LTP-like effects, which can be blocked by specific NMDA antagonists (Turner et al., 1982; Bliss et al., 1987). The site of Ca^{2+} action presumably is intracellular, since injection of EGTA prevents LTP (Lynch et al., 1983). A role for Ca^{2+} in the maintenance of LTP is substantiated by the observation that there is a long-lasting increase in $^{45}Ca^{2+}$ uptake (Agoston and Kuhnt, 1986) and intracellular ionic Ca^{2+} concentration in synaptosomes prepared from potentiated hippocampal tissue (Lynch et al., 1987). Although these and other studies clearly establish the importance of Ca^{2+} for LTP, it is not known what the intracellular site(s) of action of Ca^{2+} is (are).

Many effects of the intracellular messenger Ca^{2+} in the brain are mediated through Ca^{2+}-dependent protein kinases, such as the Ca^{2+}/calmodulin-dependent protein kinase type II (CaMK II) and the Ca^{2+}/phospholipid-dependent protein kinase C (PKC; Nairn et al., 1985). The role of these two kinases in neurotransmitter release has recently attracted much attention (Kaczmarek, 1987). In the present paper we focus on the role of protein phosphorylation in neurotransmitter release during LTP-like phenomena. First we will summarize evidence for the role of a 52 kDa coated vesicle phosphoprotein in LTP. In the second part we introduce the neuron-specific PKC substrate protein B-50 (GAP-43, F1) and discuss the role of PKC and B-50 in LTP.

ROLE OF A 52 kDa COATED VESICLE PHOSPHOPROTEIN IN LTP

A brief tetanus (15 pulses/sec, 15 sec), applied to the perforant path-granule cell subsystem of hippocampal slices, induces significant changes in the phosphorylation of a 52 kDa protein and slight but non-significant changes in B-50 phosphorylation (Bär et al., 1980; Lopes da Silva et al., 1982). The degree of phosphorylation of the two protein bands is unaffected when the frequency of stimulation is reduced and the duration increased, a condition not leading to LTP (Bär et al., 1980). Tetanization of the perforant path in the absence of extracellular Ca^{2+}, thereby preventing LTP and neurotransmission (Dunwiddie and Lynch, 1979), fails to generate changes in phosphorylation of the 52 kDa protein, indicating a relationship between the observed changes in protein phosphorylation and synaptic transmission.

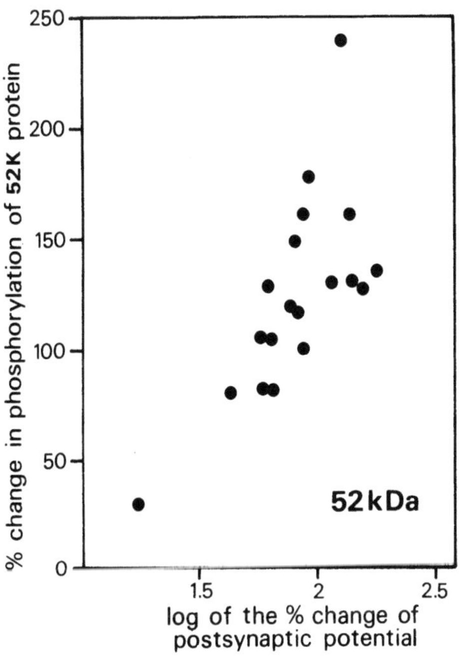

Fig. 1. Correlation between the degree of 52 kDa protein phosphorylation in hippocampal SPM employing a post-hoc phosphorylation assay 10 min after tetanization of the perforant path (15 pulses/sec, 15 sec) and the amplitude of the post-synaptic potential recorded at the granular cell layer of the fascia dentata.

Most of our initial studies were devoted to the 52 kDa protein since this protein displays the most consistent changes in post hoc protein phosphorylation following tetanization. We attempted to establish a quantitative correlation between the changes in amplitude of the post-synaptic potential (PSP) measured extracellularly, and the extent of phosphorylation of the 52 kDa protein (Tielen et al., 1983). A semi-logarithmic plot of the percentual change in 52 kDa phosphorylation versus the change in amplitude of the PSP per individual slice, fits a straight line with a correlation coefficient of 0.71 ($p < 0.005$; Fig. 1). These data indicate that there may be a quantitative correlation between LTP and 52 kDa phosphorylation.

Brain coated vesicles contain a major 52 kDa phosphoprotein (Pauloin et al., 1982) sharing phosphorylation properties with the 52 kDa protein in synaptic plasma membranes (SPM): its phosphorylation is insensitive to cyclic-AMP, cyclic-GMP and Ca^{2+} or to the Ca^{2+}-binding protein calmodulin (Bär et al., 1982; Pauloin et al., 1984). The data summarized in Table I show that the 52 kDa protein in rat brain SPM and in coated vesicles have similar biochemical characteristics (for further details see Schrama et al., 1986[a]) and that a polyclonal antibody directed against pp50 isolated from calf brain coated vesicles cross-reacts with the 52 kDa phosphoprotein in rat brain SPM (Schrama, Pauloin and Jollès, unpublished results). Therefore, we conclude that the 52 kDa protein and pp50 are identical.

Table I. Comparison between the 52 kDa phosphoprotein in SPM and pp50 in coated vesicles.

		52 kDa	pp50
Molecular weight on 11% SDS-PAGE		52 kDa	52 kDa
Sensitivity of phosphorylation to:	cAMP	0*	0
	calcium	0	0
	calmodulin	0	0
Phosphate acceptor amino acids		Ser, Thr	Ser, Thr
Isoelectric point range		9.0-6.5	9.0-6.5
Peptides formed after mapping with S. <u>aureus</u> protease V8		43,33,20 kDa	43,33,20 kDa
Immunostaining with pp50 antibody+		+	+

*= phosphorylation not sensitive to modulator
+= immunoreaction with the polyclonal pp50 antibody was determined on blots of SPM using an alkaline phosphatase-based staining procedure.

Coated vesicles and coated pits have been implicated in a number of intracellular processes, including presynaptic membrane recycling after transmitter release (Heuser and Rees, 1973; Kadota and Kadota, 1982; Miller and Heuser, 1984). Thus, enhanced transmitter release after tetanization (Bliss and Dolphin, 1982) may lead to an increase in coated pit and coated vesicle activity to retrieve excess presynaptic membrane material (Fig. 2).

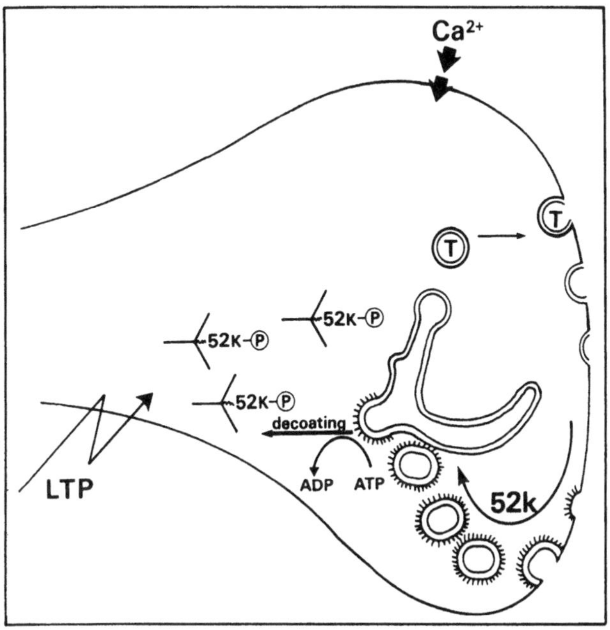

Fig. 2. Suggested role for the phosphorylation of the 52 kDa protein in coated vesicle-mediated membrane recycling and LTP.

The coated vesicle is thought to fuse with large intracellular cisternae, whereby the coat is removed from the coated vesicle (Heuser and Reese, 1973; Miller and Heuser, 1984). Preliminary data from Pauloin and Jollès (1986) suggest that the phosphorylation of pp50 can destabilize the interaction between certain coat proteins and the vesicle (Fig. 2). This idea is substantiated by the fact that the activation of pp50 kinase by the light chains of the coat protein clathrin, promotes pp50 phosphorylation (Pauloin and Jollès, 1984) and induces the dissociation of clathrin from the vesicles (Schmid et al., 1984). Therefore we propose that the changes in 52 kDa protein phosphorylation after LTP might reflect an increase in membrane retrieval concomitant with an increase in neurotransmitter release as a result of LTP.

NEURON-SPECIFIC PKC SUBSTRATE B-50

A second, well-characterized synaptosomal protein, of which the phosphorylation has been shown to be affected by tetanization is B-50 (for historical summary of research on B-50 see Table II).

Table II. Summary of research on B-50.

1976	B-50 phosphorylation sensitive to ACTH
1980	Purification of B-50: MW 48 kDa, IEP 4.5
1980	Purification of B-50 kinase
1980	Inverse relation B-50 phosphorylation PIP_2 labeling
1982	B-50 kinase = Protein kinase C
1982	B-50 is nervous tissue specific
1985	Microheterogeneity of B-50
1985	B-50 is mainly presynapticly localized (EM)
1985	Feedback role of B-50 in PPI respons
1985	B-50 in nerve growth cones
1986	B-50 levels increase during nerve regeneration
1986	B-50 = F_1 = GAP43 = pp46
1987	cDNA sequence GAP43, B-50 and F1

Under appropriate conditions it undergoes marked phosphorylation when SPM are incubated with [γ-^{32}P]-ATP (Zwiers et al., 1976). B-50 is a 48 kDa, acidic protein (IEP 4.5) that appears to be intimately associated with the plasma membrane, since it can only be solubilized in the presence of detergent (Zwiers et al., 1980). In view of its hydrophilic nature it is not likely to be an intrinsic membrane protein (Nielander et al., 1987). The purified protein displays microheterogeneity upon isoelectric focussing in a narrow pH gradient (pH 3.5-5.0): it resolves into four distinct protein spots, which are partially interconvertible by exhaustive phosphorylation or dephosphorylation (Zwiers et al., 1985). This indicates that B-50 contains more than one phosphorylation site, a notion which is supported by the fact that the primary structure of B-50 indicates three potential phosphorylation sites (Nielander et al., 1987).

Neonatal rat brain has been shown to be extremely rich in endogenous B-50 phosphorylation (Oestreicher et al., 1982). B-50 is associated with subcellular fractions enriched in growth cones (De Graan et al., 1985) and present in high amounts in outgrowing neurites in neonatal brain (Oestreicher and Gispen, 1986), in crushed peripheral neurons (Verhaagen et al., 1986) and in dorsal root ganglia in primary culture. Biochemical, immunological and recent genetical studies show that B-50 is identical to GAP43 (Karns et al., 1987; Nielander et al., 1987), pp46 (Meiri et al., 1986) and

F1 (Gispen et al., 1986; Rosenthal et al., 1987).

The endogenous kinase phosphorylating B-50 in SPM has been shown to be indistinguishable from the Ca^{2+}/phospholipid dependent PKC (Aloyo et al., 1982,1983). Of several kinases tested only PKC is able to phosphorylate purified B-50. In SPM B-50 phosphorylation can be stimulated by active phorbol diesters and by the membrane permeable diacylglycerol derivative dioctanoyl glycerol (Eichberg et al., 1986; De Graan et al., 1988), and inhibited by the PKC inhibitor polymixin B (De Graan et al., in preparation). Moreover, addition of purified PKC to SPM results in a marked increase in B-50 phosphorylation. Phosphorylation of heat-inactivated SPM with purified PKC results in an even stronger increase in B-50 phosphorylation (De Graan et al., 1988). From these data it is clear that the endogenous kinase phosphorylating B-50 in SPM is PKC.

In order to be able to measure the degree of B-50 phosphorylation in vivo, we developed a quantitative immunoprecipitation assay for B-50 (De Graan et al., in preparation). Intact hippocampal slices were loaded with [^{32}P]-orthophosphate and subsequently B-50 was immunoprecipitated from the solubilized slice homogenates with a polyclonal anti-B-50 antiserum. Immunoprecipitation of B-50 from a phosphorylated crude synaptosomal preparation (Fig. 3, lane 1) with crude antiserum results in a single phosphoband in the immunoprecipitate (lane 2), which disappears when an excess of cold B-50 was added during the immunoprecipitation (lane 3). Similar results are obtained when affinity-purified IgGs are used in the absence (lane 4) or presence (lane 5) of an excess of cold B-50.

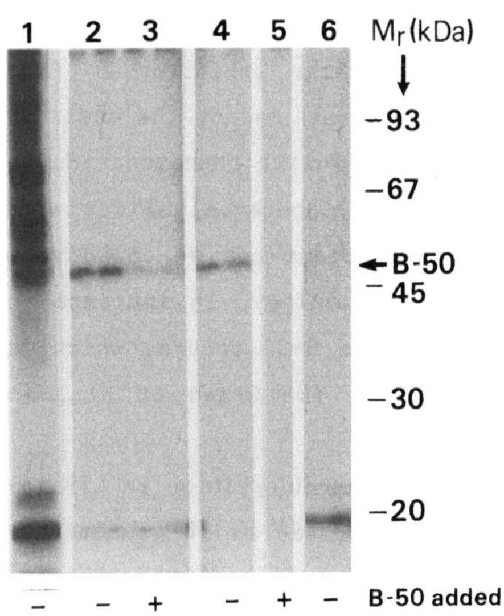

Fig. 3. SDS-PAGE of immunoprecipitates of [^{32}P]-labeled crude synaptosomal membranes. 1: total protein subjected to immunoprecipitation. 2-3: anti B-50 antiserum. 4-5: anti-B-50 IgGs. 6: affinity column effluent.

With this immunoprecipitation assay we have shown that active phorbol diesters stimulate the degree of B-50 phosphorylation in hippocampal slices (De Graan et al., in preparation). Since it is known that phorbol diesters stimulate neurotransmitter release and can mimic certain forms of LTP, we used the hippocampal slice system to study a possible involvement of B-50 in LTP.

ROLE OF PKC AND B-50 (F1/GAP43) IN LTP

Initial experiments designed to investigate the effect of tetanization of hippocampal slices on the phosphorylation of synaptic phosphoproteins revealed a non-significant increase (19%, $2p < 0.1$) in the degree of B-50 phosphorylation (Bär et al., 1980). After tetanization <u>in vivo</u>, Lovinger et al. (1985) reported an increase in F1 phosphorylation in the CA1 region measured with a post hoc <u>in vitro</u> assay. In their system the degree of F1 phosphorylation seems to be correlated with the degree of potentiation measured <u>in vivo</u> and the increase in F1 phosphorylation is maintained for at least 3 days. Concomitant with the increase in F1 phosphorylation they measured a tetanization-induced translocation of PKC activity from the cytosol to the plasma membrane (Akers et al., 1986). We have shown that tetanization of hippocampal slices under more standardized conditions results in an increase in B-50 phosphorylation which is correlated with the degree of potentiation (Schrama et al., 1986[b]).

Recently, it has been shown in the hippocampus that phorbol diesters induce LTP-like phenomena (Malenka et al., 1986; Hu et al., 1987) and prolong tetanus-induced LTP (Routtenberg et al., 1986). The involvement of PKC in the molecular mechanism of LTP is further substantiated by the observation that polymixin B, a relatively selective inhibitor of PKC, prevents the maintenance of LTP (Reymann et al., 1988) and by the observation that intracellular injection of PKC induces LTP-like phenomena (Hu et al., 1987). We confirmed that short treatment of hippocampal slices with phorbol diesters induced persistent LTP-like phenomena and we could show that concomitant with these changes in electrophysiology, an increase in the degree of phosphorylation of the PKC substrate B-50 occurs, which is not reversed upon removal of the phorbol diester (De Graan et al., in preparation).

It is not clear what the role of PKC-mediated phosphorylation in LTP is. One possibility is that PKC is involved in the mechanism of neurotransmitter release (for reviews see Nishizuka, 1986; Kaczmarek, 1987). Phorbol diesters have been shown to enhance neurotransmitter release in a variety of systems (Allgaier et al., 1986; Zurgil et al., 1986; Nichols et al., 1987; Versteeg and Ulenkate, 1987) including the hippocampus (Versteeg and Florijn, 1986; Malenka et al., 1987). We have recently shown that the increase in neurotransmitter release in the hippocampus is correlated with the degree of phosphorylation of B-50 (Dekker et al., in preparation). K^+-induced neurotransmitter release is parallelled by an increase in B-50 phosphorylation (Fig. 4a). This increase can be inhibited by the PKC inhibitor polymixin B (Fig. 4a), whereas the B-50 phosphorylation under normal K^+ was not affected. Under similar conditions polymixin B attenuates electrically stimulated neurotransmitter release in rat amygdala slices, whereas basal release is unaffected (Versteeg and Ulenkate, 1987). Moreover, in hippocampal slices stimulation of B-50 phosphorylation and neurotransmitter release by phorbol diesters is much more pronounced under K^+-stimulated than under basal conditions (Fig. 4b; Versteeg and Florijn, 1986).

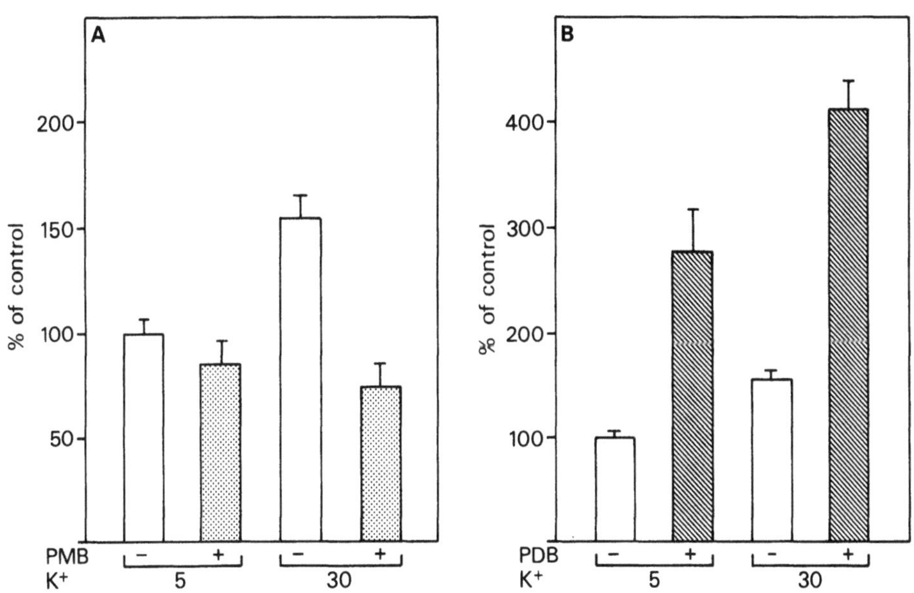

Fig. 4. Effect of polymyxin B (panel A) and PDB (panel B) on B-50 phosphorylation in K^+-depolarized and non-depolarized rat hippocampal slices. For experimental details see Dekker et al. (in preparation).

It is not clear how PKC is activated during LTP or under high potassium conditions. In general, PKC is thought to be activated by diacylglycerol which is generated together with inositol trisphosphate by the receptor-mediated hydrolysis of phosphatidylinositol 4,5-bisphosphate (PIP_2) (for recent review see Nishisuka, 1986). In a number of systems we have shown that there is an inverse relationship between the degree of B-50 phosphorylation and the labeling of PIP_2 (for a recent review see De Graan et al., 1986). Based on this evidence we have proposed that B-50 phosphorylation plays an important modulatory role in the synthesis of PIP_2 in the plasma membrane (Gispen et al., 1985). In line with this hypothesis we have shown in a semi-purified system that phosphorylated B-50 inhibits PIP_2 labeling (Van Dongen et al., 1985). Thus, B-50 phosphorylation seems to exert a negative feedback control on receptor-mediated PIP_2 hydrolysis.

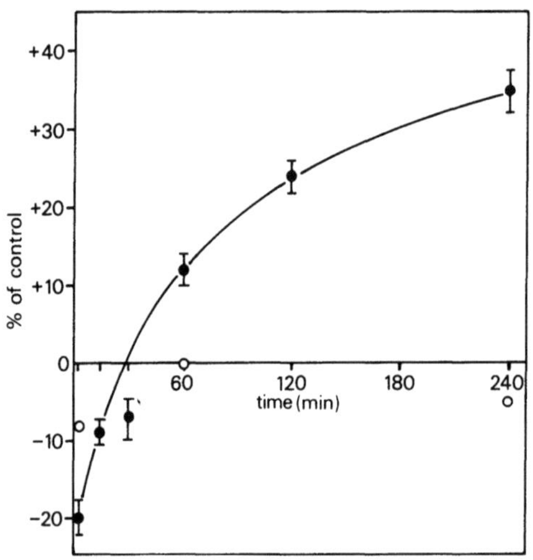

Fig. 5. Changes in PIP_2 labeling in rat hippocampal slices after tetanic stimulation of perforant path fibers. X-axis gives interval between application of tetanus and start of 10 min incubation with labeled phosphate. Data are expressed as % of control. Bars indicate the SEM. For experimental details see Bär et al. (1984).

LTP-induced stimulation of hippocampal slices induces a time-dependent change in PIP_2 labeling: at short intervals after tetanization a decrease of about 20% is observed; after longer time intervals there is a gradual increase in PIP_2 labeling (Fig. 5, filled symbols). A low stimulation frequency, which does not produce LTP, does not affect PIP_2 labeling (Fig. 5, open symbols; Bär et al., 1984). So, we have evidence that both protein and lipid phosphorylation change during LTP. However, at this moment it is not clear whether there is a causal relationship between these two phenomena.

CONCLUDING REMARKS

In the present paper we reviewed the significance of two presumably presynaptic phosphoproteins in neurotransmitter release during LTP-like phenomena in rat hippocampal slices. LTP-related changes in synaptic transmission are accompanied by changes in the degree of phosphorylation of the 52 kDa coated vesicle protein (pp50) and the B-50 (GAP43, F1) protein present in vesicle and presynaptic membranes. In keeping with the literature, we assume that the phosphostate of the 52 kDa protein somehow is involved in presynaptic membrane retrieval following exocytosis. The B-50 protein may be involved in presynaptic modulation of transmembrane signal transduction, in the release process of certain transmitters or in presynaptic plasticity as such. At present it is unclear what the nature of the relationship between the observed neurochemical changes is with respect to LTP. They may reflect only a temporal correlate of enhanced neurotransmitter release. Further work is in progress to delineate the function of this protein in more detail.

ACKNOWLEDGEMENT

This research was supported by CLEO-TNO grants A47 and A66, and by MEDIGON grant 900-548-072 of the Dutch Organization for the Advancement of Pure Research (ZWO). The authors wish to thank Drs. A. Pauloin and J. Jolles for the gift of the polyclonal antibody directed against pp50 from calf brain coated vesicles, Drs. B. Melchers, F.H. Lopes da Silva and D.H.G. Versteeg for helpful advise.

REFERENCES

Allgaier C, Von Kugelgen O and Herrting G (1986) Enhancement of noradrenaline release by 12-O-tetradecanoyl phorbol-13-acetate, an activator of protein kinase C. Eur. J. Pharmacol. 129:389-392.

Agoston DV and Kuhnt U (1986) Increased calcium uptake of presynaptic terminals during long-term potentiation in hippocampal slices. Exp. Brain Res. 62:663-668

Akers RF, Lovinger DM, Colley PA, Linden DJ and Routtenberg A (1986) Translocation of protein kinase C activity may mediate hippocampal long-term potentiation. Science 231:587-589

Aloyo VJ, Zwiers H and Gispen WH (1982) B-50 protein kinase and kinase C in rat brain. Prog. Brain Res. 56:303-315

Aloyo VJ, Zwiers H and Gispen WH (1983) Phosphorylation of B-50 protein by calcium-activated, phospholipid-dependent protein kinase and B-50 protein kinase. J. Neurochem. 41:649-653

Bär PR, Schotman P, Gispen WH, Lopes da Silva FH and Tielen AM (1980) Changes in synaptic membrane phosphorylation after tetanic stimulation in the dentate area of the rat hippocampal slice. Brain Res. 98:478-484

Bär PR, Tielen AM, Lopes da Silva FH, Zwiers H and Gispen WH (1982) Membrane phosphoproteins of rat hippocampus: sensitivity to tetanic stimulation and enkephalin. Brain Res. 245:69-79

Bär PR, Wiegant F, Lopes da Silva FH and Gispen WH (1984) Tetanic stimulation affects the metabolism of phosphoinositides in hippocampal slices. Brain Res. 321:381-385

Bliss TVP and Dolphin AC (1982) What is the mechanism of long-term poten-

tiation in the hippocampus? TINS 5: 289-290
Bliss TVP, Errington ML and Lynch MA (1987) Calcium-induced long-term potentiation in the dentate gyrus is accompanied by a sustained increase in glutamate release. In: Excitatory amino acid transmission (eds: TP Hicks, D Lodge and H McLennan) Alan R Liss Inc., New York. pp. 337-340.
Chang F-LF and Greenough WT (1984) Transient and enduring morphological correlates of synaptic activity and efficacy change in the rat hippocampal slice. Brain Res. 309:35-46
Collingridge GL and Bliss TVP (1987) NMDA receptorstheir role in long-term potentiaiton. TINS 10: 288-293
De Graan PNE, Van Hooff COM, Tilly BC, Oestreicher AB, Schotman P and Gispen WH (1985) Phosphoprotein B-50 in nerve growth cones from fetal rat brain. Neurosci. Lett. 61:235-241
De Graan PNE, Oestreicher AB, Schrama LH, and Gispen WH (1986) Phosphoprotein B-50: localization and function. Prog. Brain Res. 69:37-50
De Graan PNE, Dekker LV, De Wit M, Schrama LH and Gispen WH (1988) Modulation of B-50 phosphorylation and polyphosphoinositide metabolism in synaptic plasma membranes by protein kinase C, phorbol diesters and ACTH. J. Recep. Res. in press
Dolphin AC, Errington ML and Bliss TVP (1982) Long-term potentiation of the perforant path in vivo is associated with increased glutamate release. Nature (Lond.) 297:496-498
Dunwiddie TV and Lynch G (1979) The relationship between extracellular calcium concentrations and the induction of long-term potentiation. Brain Res. 169: 103-110
Eichberg J, De Graan PNE, Schrama LH and Gispen WH (1986) Dioctanoylglycerol and phorbol diesters enhance phosphorylation of phosphoprotein B-50 in native synaptic plasma membranes.Biochem. Biophys. Res. Comm. 136:1007-1012
Errington ML, Lynch MA and Bliss TVP (1987) Long-term potentiation in the dentate gyrus: induction and increased glutamate release are blocked by D(-)aminophosphonovalerate. Neuroscience 20:279-284
Gispen WH, Van Dongen CJ, De Graan PNE, Oestreicher AB and Zwiers H (1985) The role of phosphoprotein B-50 in phosphoinositide metabolism in brain synaptic plasma membranes. In: Inositol and phosphoinositides (Eds. JE Blaesdale, G Hauser and J Eichberg) Humana Press, Clifton, N.J. pp. 399-413
Gispen WH, De Graan PNE, Chan SY and Routtenberg A (1986) Comparison between the neural acidic protein B-50 and F1.Prog. Brain Res. 69:382-384
Heuser J and Reese T (1973) Evidence for recycling of synaptic vesicle membrane during transmitter release at the frog neuromuscular junction. J. Cell Biol. 57:315-344
Hu G-Y, Hvalby O, Walaas SI, Albert KA, Skjeflo P, Andersen P and Greengard P (1987) Protein kinase C injection into hippocampal pyrimidal cells elicits features of long term potentiation.Nature (Lond.) 328:426-429
Kaczmarak LK (1987) The role of protein kinase C in regulation of ion channels and neurotransmitter release. TINS 10:30-34
Kadota T and Kadota K (1982) Membrane retrieval by macropino cytosis in presynaptic terminals during transmitter release in cat synaptic ganglia in situ. J. Elec. Microsc. 31:73-80
Karns LR, Ng S-C, Freeman JA and Fishman MC (1987) Cloning of complementary DNA for GAP-43, a neuronal growth-related protein. Science 236:597-600
Lopes da Silva FH, Bär PR, Tielen AM and Gispen WH (1982) Plasticity in synaptic transmission of membranebound protein phosphorylation. Prog. Brain Res. 55:369-377
Lovinger DM, Akers RF, Nelson RB, Barnes CA, McNaughton BL and Routtenberg

A (1985) A selective increase in phosphorylation of protein F1, a protein kinase C substrate, directly related to three day growth of long term synaptic enhancement. Brain Res. 343:137-143

Lynch G, Larson J, Kelso S, Barrionuevo G and Schottler F (1983) Intracellular injections of EGTA block induction of hippocampal long-term potentiation. Nature (Lond.) 305:719-721

Lynch G and Baudry M (1984) The biochemistry of memory: a new and specific hypothesis. Science 224:1057-1063

Lynch MA, Clements MP, Errington ML and Bliss TVP (1987) Increase in intrasynaptosomal calcium in hippocampal long-term potentiation. Neurosci. Lett. Suppl. 29: S98

Malenka RC, Madison DV and Nicholl RA (1986) Potentiation of synaptic transmission in the hippocampus by phorbol ester. Nature (Lond.) 321:175-177

Malenka RC, Ayoub GS and Nicoll RA (1987) Phorbol esters enhance transmitter release in rat hippocampal slices. Brain Res. 403:198-203

Meiri KF, Pfenninger KH and Willard MB (1986) Growth-associated protein, GAP-43, a polypeptide that is induced when neurons extend axons, is a component of growth cones and corresponds to pp46, a major polypeptide of a subcellular fraction enriched in growth cones. Proc. Natl. Acad. Sci. USA 83:3537-3541

Miller TM and Heuser JE (1984) Endocytosis of synaptic vesicle membrane at the frog neuromuscular junction. J. Cell Biol. 98:685-689

Nairn AC, Hemmings Jr HC and Greengard P (1985) Protein kinases in the brain. Ann. Rev. Biochem. 54:931-976

Nichols RA, Haycock JW, Wang JKT and Greengard P (1987) Phorbol ester enhancement of neurotransmitter release from rat brain synaptosomes. J. Neurochem. 48:615-621

Nielander HB, Schrama LH, Van Rozen AJ, Kasparaitis M, Oestreicher AB, De Graan PNE, Gispen WH and Schotman P (1987) Primary structure of the neuron-specific protein B-50 is identical to growth-associated protein GAP-43. Neurochem. Res. Comm. in press

Nishisuka Y (1986) Studies and perspectives of protein kinase C. Science 233:305-312

Oestreicher AB, Zwiers H, Gispen WH and Roberts S (1982) Characterization of infant rat brain cerebral cortical membrane proteins phosphorylated in vivo:identification of the ACTH-sensitive phosphoprotein B-50. J. Neurochem. 39:683-692

Oestreicher AB and Gispen WH (1986) Comparison of the immunocytochemical distribution of phosphoprotein B-50 in the cerebellum and hippocampus of immature and adult rat brain. Brain Res. 375:267-279

Pauloin A and Jollés P (1984) Internal control of the coated vesicle pp50-specific kinase complex. Nature (London) 311:265-267

Pauloin A, Bernier I and Jollés P (1982) Presence of cyclic nucleotide Ca^{2+}-independent protein kinase in bovine brain coated vesicles. Nature (Lond.) 298:574-576

Pauloin A, Loeb J and Jollés P (1984) Protein kinase(s) in bovine brain coated vesicles. Biochim. Biophys. Acta 799:238-245

Reymann KG, Frey U, Jork R and Matthies H (1988) Polymixin B, an inhibitor of protein kinase C, prevents the maintenance of synaptic long-term potentiation in hippocampal CA1 neurons. Brain Res. in press

Rosenthal A, Chan SY, Henzel W, Haskell C, Kuang W-J, Chen E, Wilcox JN, Ullrich A, Goeddel DV and Routtenberg A (1987) Primary structure and mRNA localization of protein F1, a growth-related protein kinase C substrate associated with synaptic plasticity. EMBO J. in press

Routtenberg A, Colley P, Linden D, Lovinger D, Murakami K and Sheu F-S (1986) Phorbol ester promotes growth of synaptic plasticity. Brain Res.

378:374-378
Schmid SL, Brael WA, Schlossman DM and Rothman JE (1984) A role for clathrin light chains in the recognition of clathrin cages by "uncoating ATPase". Nature (Lond.) 311:228-231
Schrama LH, De Graan PNE, Zwiers H and Gispen WH (1986a). Comparison of a 52-kDa phosphoprotein from synaptic plasma membranes related to long-term potentiation and the major coated vesicle phosphoprotein. J. Neurochem. 47:1843-1848.
Schrama LH, De Graan PNE, Wadman WJ, Lopes da Silva FH and Gispen WH (1986b) Long-term potentiation and 4-aminopyridine-induced changes in protein and lipid phosphorylation in the hippocampal slice. Prog. Brain Res. 69:245-257
Skrede K and Malthe-Sörensen D (1981) Increasing resting and evoked release of transmitter following repetitive electrical tetanization in hippocampus:a biochemical correlate to long-lasting synaptic potentiation. Brain Res. 208:436-441
Tielen AM, De Graan PNE, Mollevanger WJ, Lopes da Silva FH and Gispen WH (1983) Quantitative relationship between posttetanic biochemical and electrophysiological changes in rat hippocampal slices. Brain Res. 277:189-192
Turner RW, Baimbridge KG and Miller JJ (1982) Calcium-induced long-term potentiation in the hippocampus. Neuroscience 7:1411-1416
Van Dongen CJ, Zwiers H, De Graan PNE and Gispen WH (1985) Modulation of the activity of purified phosphatidylinositol 4-phosphate kinase by phosphorylated and dephosphorylated B-50 protein. Biochem. Biophys. Res. Comm. 128:1219-1227
Verhaagen J, Van Hooff COM, Edwards PM, De Graan PNE, Oestreicher AB, Schotman P, Jennekens FGI and Gispen WH (1986) The kinase C substrate protein B-50 and axonal regeneration. Brain Res. Bull. 17:737-741
Versteeg DHG and Florijn WJ (1986) Phorbol 12,13-dibutyrate enhances electrically stimulated neuromessenger release from rat dorsal hippocampal slices in vitro. Life Sci. 40:1237-1243
Versteeg DHG and Ulenkate HJLM (1987) Basal and electrically stimulated release of [^{3}H]-noradrenaline and [^{3}H]-dopamine from rat amygdala slices in vitro:effects of 4β-phorbol 12,13-dibutyrate, 4α-phorbol 12,13-didecanoate and polymixin B. Brain Res. 416:343-348
Zurgil N, Yarom M and Zisapel N (1986) Concerted enhancement of calcium influx, neurotransmitter release and protein phosphorylation by a phorbol ester in cultured brain neurons. Neuroscience 19:1255-1264
Zwiers H, Veldhuis HD, Schotman P and Gispen WH (1976) ACTH, cyclic nucleotides and brain protein phosphorylation in vitro. Neurochem. Res. 1:669-677
Zwiers H, Schotman P and Gispen WH (1980) Purification and some characteristics of an ACTH-sensitive protein kinase and its substrate protein in rat brain membranes. J. Neurochem. 34:1689-1699
Zwiers H, Verhaagen J, Van Dongen CJ, De Graan PNE and Gispen WH (1985) Resolution of rat brain synaptic phosphoprotein B-50 into multiple forms by two-dimensional electrophoresis:evidence for multisite phosphorylation. J. Neurochem. 44:1083-1090

A POSSIBLE ROLE FOR EPENDYMIN IN HIPPOCAMPAL PLASTICITY

W. Seifert, H. Terlau, R. Schmidt*

Department of Neurobiology
Max-Planck-Institut für biophysikalische Chemie
Am Fassberg
3400 Göttingen
FRG

By using specific antibodies directed against β and γ-ependymin we have demonstrated the presence of these proteins or closely related ependymin-like proteins in hippocampal neurons from embryonic rat (E 18) in cell culture by immunocytochemistry. The same antibodies were used to test the possible effects of ependymin-antiserum on LTP in the hippocampal slice preparation. We were unable to see an effect at early times of LTP induction and maintenance (up to 1 hour).
The occurrence of this memory-related protein in the hippocampus - a brain structure involved in the consolidation of long-term memory - is remarkable. Our failure to demonstrate inhibition of LTP by a specific antiserum at early times, does not exclude the possibility that such effects might become significant at later times. Indeed the time course of ependymin-synthesis and secretion in goldfish would suggest this possibility.

* R. Schmidt - Dept. of Zoology, J.W. Goethe-University, Frankfurt, FRG

NATO ASI Series, Vol. H19
Modulation of Synaptic Transmission and Plasticity
in Nervous Systems. Edited by G. Hertting and H.-C. Spatz

INTRODUCTION

It has been known for many years that protein biosynthesis is necessary for the consolidation of long-term memory (1,2,3). Since the discovery of long-term potentiation (LTP) in the hippocampus (4), this limbic structure has become a favourite brain area for studies on synaptic plasticity and memory, although the evidence that LTP is indeed a basic process for the formation of long-term memory is still lacking. Nevertheless, considering all the evidence from experimental and neuropsychological work, the hippocampus is a good candidate area in the mammalian brain to look for a "molecular imprint" of the memory trace (5,6). It has been demonstrated also for LTP that this process is blocked by inhibition of protein biosynthesis suggesting that de novo protein synthesis is required for establishing long-term changes in synaptic transmission (7,8).

Work in V. Shashoua's laboratory has demonstrated that 3 specific brain proteins - α, β and γ ependymin with molecular weights of 37, 32 and 26 kDa - are synthesized with an increased rate when goldfishes had to learn a new pattern of swimming behavior (9). The β- and γ-proteins are acidic glycoproteins found in the cytoplasm and in the extracellular fluid. They were purified and used to prepare specific antisera. These antibodies inhibited the formation of long-term memory when injected into goldfish brain ventricles (10,11). Subsequently a sensitive radioimmunoassay for the quantitative analysis of goldfish ependymins (12) was used to demonstrate that ependymin-like immunoreactivity is not only expressed in other fish but also in rat brain. Interestingly the highest amount of ependymin-like proteins was measured in cultured neuronal cells derived from embryonic rat hippoccampus (13).

The demonstration of ependymin-like proteins in the rat hippocampus stimulated our interest to study possible effects of ependymin-antiserum on long-term potentiation in the hippocampal slice preparation.

RESULTS AND DISCUSSION

Several years ago we had established in our laboratory a serum-free cell culture system from the developing rat hippocampus (14,15,16).

These "neuronal cultures" are derived from hippocampus of 18 day old rat embryos at which time the pyramidal cells have reached their postmitotic phase while the granular neurons of the dentate gyrus have not yet proliferated. Thus this culture system consists predominantly of pyramidal neurons and a certain percentage of basket neurons since both cell types develop in the prenatal phase. Non-neuronal cells - mostly astrocytes - are present at a low percentage of less than 10%. This culture system has been characterized in our laboratory using various methods of immunocytochemistry, autoradiography and electronmicroscopy (14).

This in-vitro system of the hippocampus was used to check for the presence of ependymin-like proteins by immunocytochemistry with ependymin antibodies. Fig. 1 shows the result of such an experiment. Indirect immunofluorescence with ependymin antiserum and FITC-conjugated goat anti-rabbit IgG stained neurons with pyramidal appearance brightly. Astrocytes exhibit only weak background staining, they were identified by immunohistochemical staining with anti-GFAP (not shown here). The immunoreactivity in neurons was distributed throughout the cytoplasm, including all processes, while the cell nuclei remained unstained.

A more detailed description of these studies has been given elsewhere (13). It is interesting to note that other neuronal cell types - such as cerebellar neurons - did not reveal cross-reactivity with goldfish brain ependymins. This is in agreement with quantitative measurements with a highly sensitive radioimmunoassay for ependymin where the highest values of cross-reactivity were found for the embryonic hippocampus, both in dissected brain and in neuronal cell cultures (13).

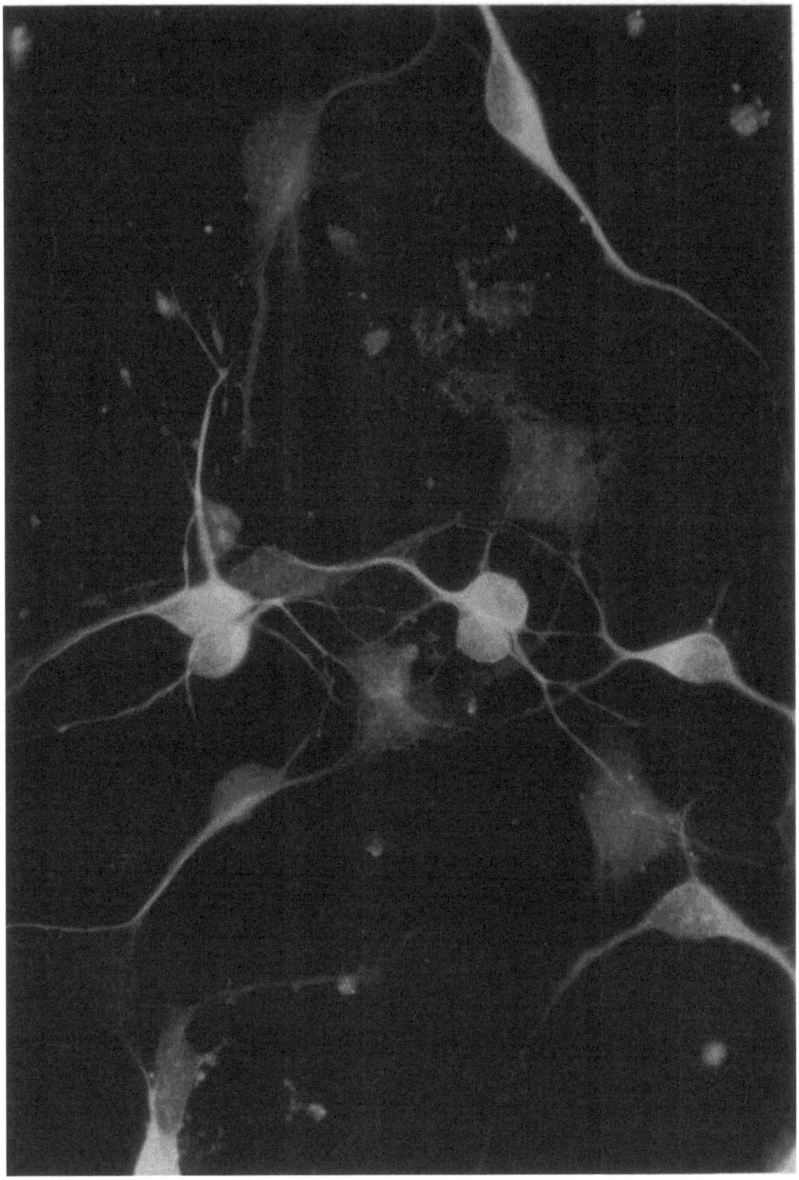

Fig. 1) Immunohistofluorescence staining of hippocampus pyramidal neurons with anti-ependymin antiserum. Neurons were derived from embryonic rat hippocampus (prenatal day 18) and kept in culture for 3 days. Intracellular staining: cells were fixed and opened with ethanolic acetic acid and incubated with anti-ependymin. FITC-conjugated goat anti-rabbit IgG was used as fluorochrome. All neurons with pyramidal appearance stained brightly for ependymin-like immunoreactivity. Astrocytes exhibited only weak background staining.

In view of these results and the requirement for de novo protein biosynthesis both in memory consolidation and in LTP, it seemed interesting to investigate the possible involvement of ependymin in synaptic plasticity of the rat hippocampus. The mechanism of long-term potentiation (LTP) has not yet been elucidated in every detail, but it seems clear that for the induction of LTP following tetanic stimulation the NMDA-receptors for the excitatory neurotransmitter glutamate on the post-synaptic site are required.
On the other hand there is good evidence that an increased release of glutamate from the pre-synaptic site is necessary for the maintenance of LTP. Finally increased calcium influx has been observed both post- and pre-synaptically. The necessity for some sort of feedback-communication from the post- to the pre-synaptic site of tetanically excited LTP-synapses has been postulated in form of a soluble factor (T. Bliss, these Proceedings). Therefore we have started experiments in our laboratory with ependymin antiserum to investigate if ependymin could be such a soluble molecule. The fact that it is found in high concentration in pyramidal neurons of the hippocampus, that its synthesis is increased during memory consolidation, that it is secreted as an extracellular protein and that it is probably a calcium-binding protein (R. Schmidt, personal communication) certainly makes ependymin a good candidate for a protein molecule mediating synaptic plasticity in the hippocampus.

For our experiments we have used extracellular recording in the CA1 - pyramidal cell layer of the hippocampal slice following the induction of LTP by tetanic stimulation of the Schaffer collateral-commissural pathway, as described in Fig. 2. Ependymin antiserum was applied to the slices and supernatant of heat-inactivated ependymin antiserum was used as a control. The size of the amplitude of the population spike was measured at various times following tetanic stimulation in presence of active or inactive antiserum. As shown in Fig. 2 for the time points of 1 min., 15 min. and 30 min. there was no significant difference between experimental values and controls.

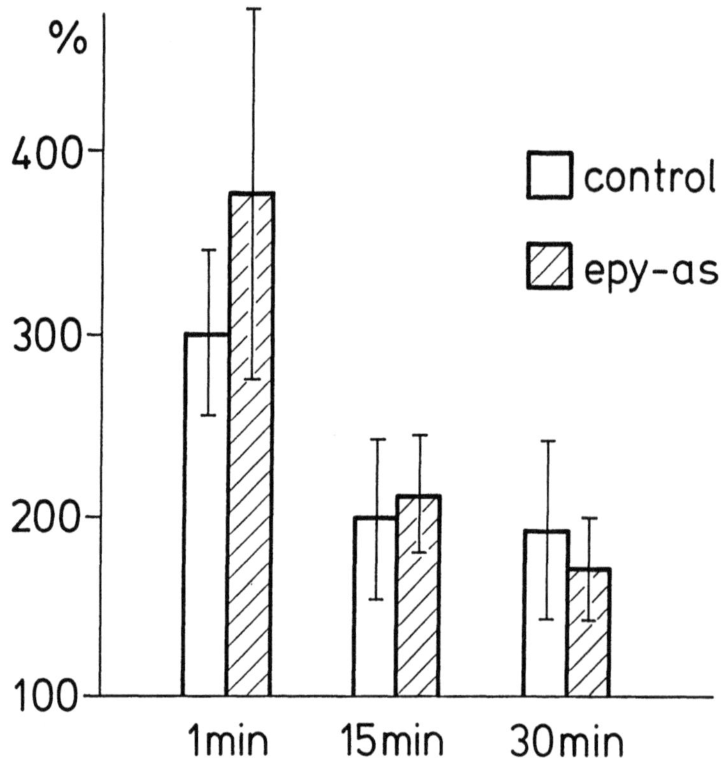

Fig. 2) Lack of influence of ependymin-antiserum on LTP

Preparation of slices and electrophysiological recording were as desribed (Wierazsko and Seifert, Brain Res. 371 (1986) 305-313). The extracellular recording electrode was placed in the CA_1-cell body layer, while the stimulating electrode was placed in the stratum radiatum of the CA_3-region. Either the antiserum (epy-as) or the supernatant of the heated (10 min. at 95°C) and centrifuged (10 min. at 18000 g) antiserum (control) was applied by pressure ejection (Picospritzer, General Valve, Fairfield, NJ.) to the surface of the slice in the dendritic layer at the recording electrode. Pipette ejection pressures and times were selected to supply 300 to 400 um diameter droplets in air. Three or four ejections were performed spaced 1 min. apart. Tetanic stimulation (100 Hz for 2 sec.) was applied 15 min. after ejection of the antiserum or the control. Paired slices were used in 6 different experiments. The 100% on the ordinate corresponds to the average value of the amplitude of the population spike during 10 min. before tetanic stimulation. For low frequency stimulation 7 stimuli per minute were applied. The values of the figure (bars signify +- S.E.M.) are given for 1, 15 and 30 min. after tetanic stimulation.

The initial increase at 1 min. for both values is due to post-tetanic potentiation, while the values at 15 min. and 30 min. can be considered as a measure of LTP. Thus in contrast to our expectations we could not demonstrate in these experiments an inhibition of LTP induction or maintenance for up to one hour after tetanic stimulation.

There are possible technical reasons for this lack of an influence by ependymin antiserum: the antiserum which we used might have lost its activity or might not have reached the local synaptic area of the LTP-stimulated synapses in the dendritic field of CA1 pyramidal neurons. Both these reasons did most likely not apply as far as we can judge.

The other possibility is that ependymin - as suggested by these experiments - has no influence on the initial development of LTP, but may still become crucial at a later stage. There are in fact indications that LTP in itself might not be a homogeneous process but rather - as suggested by the use of various inhibitors of proteinkinase C, RNA- and protein - synthesis - a heterogeneous process with a different underlying biochemistry at different time stages.

Indeed the time course of biosynthesis and extracellular secretion of ependymins in the goldfish brain would suggest that it might take several hours to find ependymin in the extracellular space (12).

Therefore we believe that the experiments reported here do not exclude an influence of ependymin on LTP at later times. It might well be necessary and worthwile to look for a possible role of ependymin in synaptic plasticity of the hippocampus at a different time window.

REFERENCES

1.Agranoff BW (1981) Learning and memory: biochemical approaches. In GJ Siegel, RW Albers, BW Agranoff and R Katzmann (eds). Basic Neurochemistry, 3rd edn., Little, Brown and Co. Boston

2.Dunn AJ (1980) Neurochemistry of learning and memory: an evaluation of recent data. Annu Rev Psychol 31:343-390

3.Shashoua VE (1982) Molecular and cell biological aspects of learning: toward a theory of memory. Adv Cell Neurobiol 3:97-141

4.Bliss TVP, Lomo T (1973) Long-lasting potentiation of synaptic transmission in the dentate area of the anaesthetized rabbit following stimulation of the perforant path. J Physiol London, 232:331-356

5.Seifert W (ed) (1983) Neurobiology of the Hippocampus. Academic Press, London

6.Changeux JP, Konishi M (eds) (1987) The neuronal and molecular bases of learning. J. Wily & Sons, Chichester, New York

7.Stanton PK, Sarvey JM (1984) Blockade of long-term potentiation in rat hippocampal CA_1 region by inhibitors of protein synthesis. J of Neuroscience 4:3080-3088

8.Deadwyler SA, Dunwiddie T, Lynch G (1987) A critical level of protein synthesis is required for long-term potentiation. Synapse 1:90-95.

9.Shashoua VE (1976) Brain metabolism and the acquisition of new behaviours. I. Evidence for specific changes in the pattern of protein synthesis. Brain Res 111:347-364

10.Schmidt R (1986) Biochemical participation of glycoproteins in memory consolidation after two different training pradigms in goldfish. In H Matthies (ed) Learning and memory. Adv. in the Biosciences 59:213-222, Perganon Press, Oxford

11.Shashoua VE, Moore ME (1978) Effect of antisera to ß and goldfish brain proteins on the retention of a newly acquired behavior. Brain Res 148:441-449

12.Schmidt R (1987) Changes in subcellular distribution of ependymins in goldfish brain induced by learning. J of Neurochemistry 48:1870-1878

13.Schmidt R, Löffler F, Müller HW, Seifert W (1986) Immunological cross-reactivity of cultured rat hippocampal neurons with gold fish brain proteins synthesized during memory consolida tion. Brain Res 386:245-257

14. Seifert W, Ranscht B, Fink HJ, Förster F, Beckh S, Müller HW (1983) Development of hippocampal neurons in cell culture: A molecular approach. In: W Seifert (ed) Neurobiology of the Hippocampus. Academic Press, London, pp 109-135

15. Seifert W, Müller HW (1984) Neuron-glia interaction in mammalian brain. In: DW Barnes, DA Sirbasku, GH Sato (eds) Cell Culture Methods for Molecular and Cell Biology 4:67-77, Alan Liss Inc, New York

16. Müller HW, Beckh S, Seifert W (1984) Neurotrophic factor for central neurons. Proc Natl Acad Sci USA 81:1248-1252

MODULATION OF QUANTAL SYNAPTIC RELEASE BY SEROTONIN AND FORSKOLIN IN CRAYFISH MOTOR NERVE TERMINALS

J. Dudel
Physiologisches Institut
Technische Universität München
Biedersteiner Str. 29
8000 München 40
Federal Republic of Germany

In crustacea, serotonin is released into the haemolymph and acts as a neurohormone and modulator at central and peripheral neurons (see Kravitz et al. 1983, Harris-Warrick and Kravitz 1984). Serotonin applied to neuromuscular junctions increases excitatory postsynaptic potentials (Grundfest and Reuben 1961, Dudel 1965, Wheal and Kerkut 1976, Enyeart 1981, Glusman and Kravitz 1982, Fischer and Florey 1983, Dixon and Atwood 1985). The increased synaptic potentials are due to enhanced release of transmitter quanta from motor nerve terminals (Dudel 1965, Glusman and Kravitz 1982, Dixon and Atwood 1985). These presynaptic effects of serotonin were suggested to reflect improvement of the action potential in the terminal nerve branch (Dudel 1965), or increases of the Ca concentration in the terminal (Glusman and Kravitz 1982, Dixon and Atwood 1985). These investigations have elicited release by action potentials and employed bath application of serotonin. Effects on the action potential and its conduction could not be distinguished from changes in the coupling between depolarization and release. In the present study excitation of the nerve terminals could be blocked depolarizing them with graded pulses, and serotonin was applied only to the terminal under investigation.

The preparation was the extensor of the dactyl of the first walking legs of crayfish (mostly Astacus astacus). The muscle

---

Supported by the Deutsche Forschungsgemeinschaft, SFB 220

NATO ASI Series, Vol. H19
Modulation of Synaptic Transmission and Plasticity
in Nervous Systems. Edited by G. Hertting and H.-C. Spatz

was isolated and fixed in a chamber where it was superfused with bathing solution. Quantal synaptic currents (Fig.1) were recorded by means of a macro-patch-clamp electrode which had an opening of about 20 µm which was placed on a suitable terminal. Depolarizing current pulses could be applied to the terminal through the recording electrode. These pulses elicited quantal releases (Dudel 1983) which were counted automatically. The recording electrode was perfused with variable solutions, exchanging the volume at the tip more than 300 times per second (Dudel 1988a). In most experiments (Figs.2 to 5) the superfusate of the bath contained no calcium and 16 mM magnesium ($0Ca_b$, 16 mM Mg_b). This prevented possible release from regions of the terminal outside the electrode and favoured a low resting Ca concentration within the terminals (Dudel 1988a).

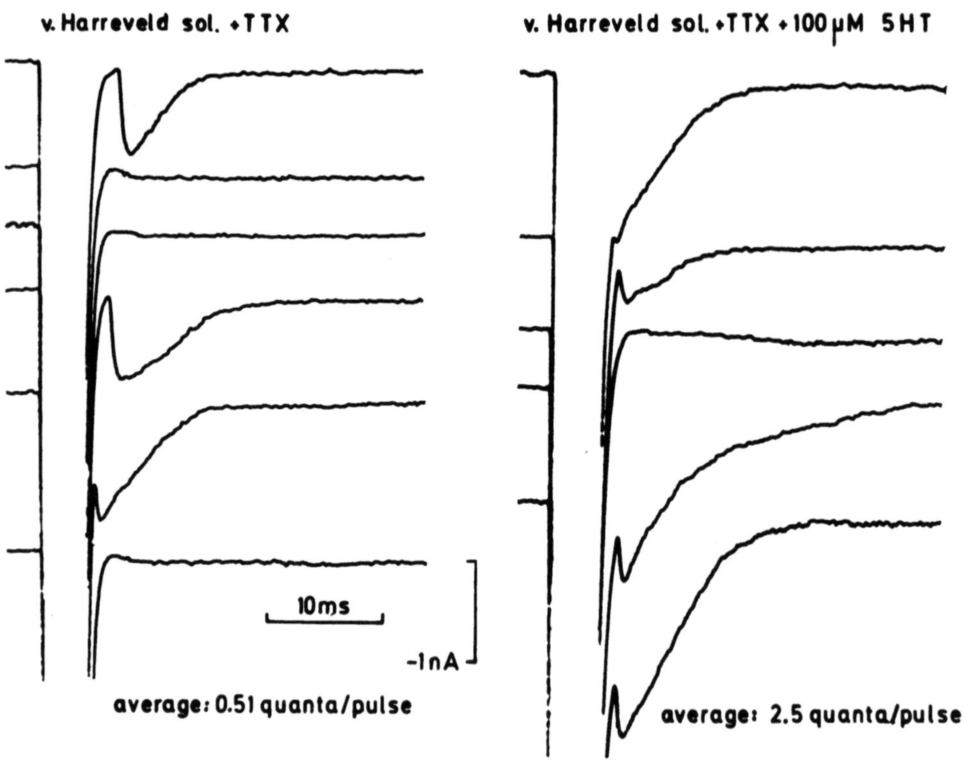

Fig.1: Samples of quantal currents recorded at one terminal. The electrode was perfused with van Harreveld solution containing 0.2 µM tetrodotoxin for the recordings in the left hand panel, to which 100 µM serotonin (5 HT) were added for the recordings in the right hand panel.

In the recordings of Fig.1, bath and electrode were perfused with physiological (v. Harreveld) solution. The perfusate of the electrode contained in addition 0.2 μM tetrodotoxin (TTX) to prevent possible excitation of the terminal. In the controls in the left hand panel, the depolarizing pulses (-0.4 μA, 2 ms) released 0.5 current-quanta/pulse, on average. In the recordings of the right hand panel, 100 μM serotonin (5 HT) was added to the perfusate of the electrode, and the same pulses elicited on average 2.5 quanta/pulse, i.e. 5 times more than in the control. Such increases in the release rate of quanta on application of serotonin have also been seen in the studies cited above.

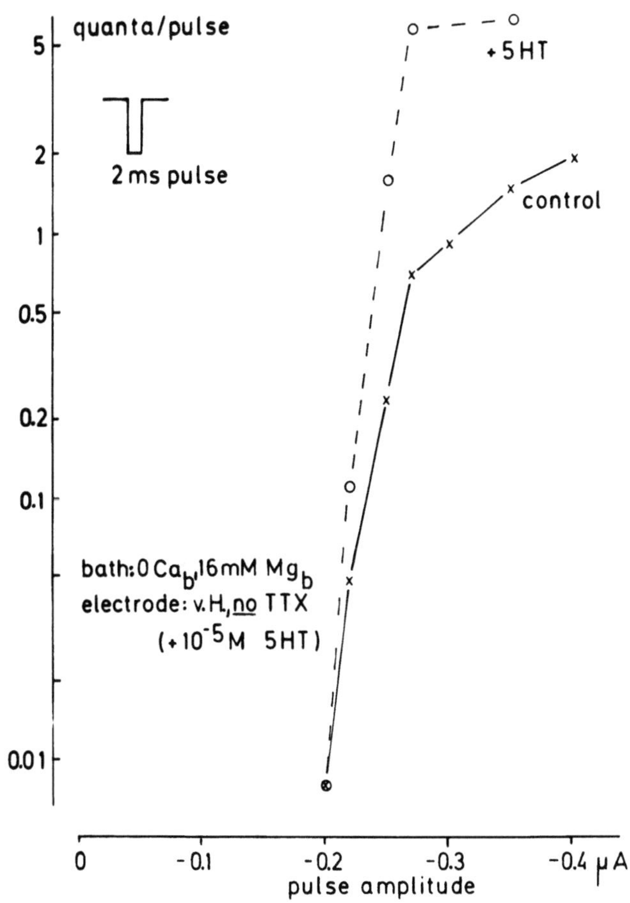

Fig.2: Rate of release (ordinate) in dependence on the amplitude of 2 ms depolarizing pulses (abscissa). In the control (x) the electrode was perfused with van Harreveld solution without TTX, to which later 10 μM serotonin was added (o).

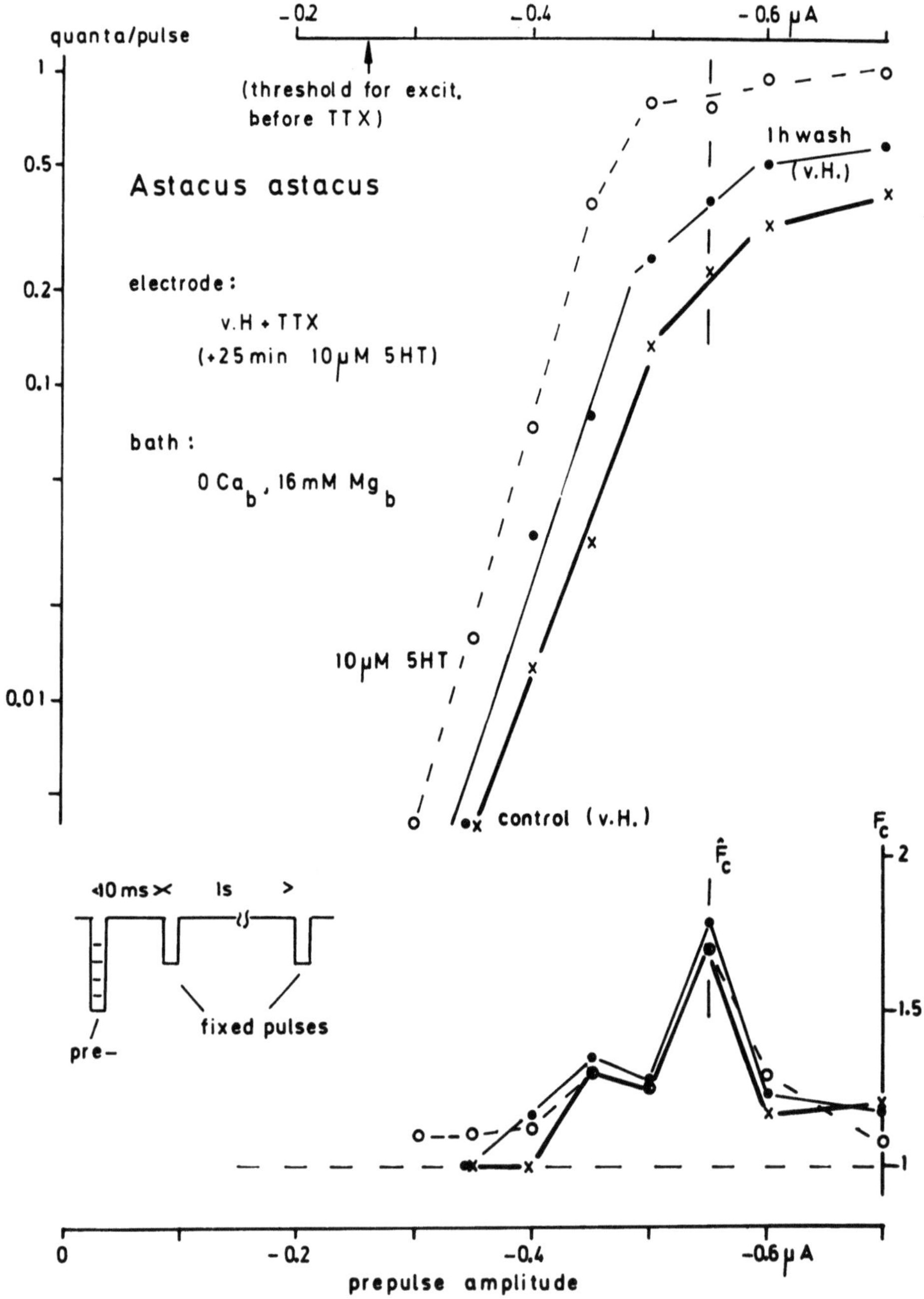

Fig.3: Dependence of release (left ordinate) and of fixed pulse facilitation, F_c (right ordinate, lower curves), on prepulse amplitudes. The pulse program is shown in the inset. In the control (x) the electrode was perfused with v. Harreveld solution +TTX, then 10 μM serotonin was added (o), and finally pure v. Harreveld solution +TTX was perfused again for 1 h (•).

Serotonin seems to have two effects: one on excitability of the terminal axon, another on the release process. In order to study excitability, in the experiment of Fig.2 the electrode was perfused with v. Harreveld solution without TTX. In the controls, on increasing the amplitude of the depolarization pulse first releases were seen at -0.2 µA, the rate of releases increasing steeply up to -0.27 µA and less steeply for larger pulses. When 100 µM 5 HT was added to the perfusate of the electrode, the release rate increased even more, by a factor of 600, between -0.20 and -0.27 µA pulse amplitude, reaching a saturation level at 6 quanta/pulse. It appears that in presence of 5 HT the terminal was excited by pulses $<$ -0.22 µA, and that one of the effects of serotonin is an increased excitability.

In further experiments excitations were prevented by TTX in the perfusate of the electrode. Under these conditions the depolarizing pulses almost exclusively shift the extracellular potential, and the potential shifts are proportional to the current and complete within less than 0.1 ms (Katz and Miledi 1967). In Fig.3, in control solution the rate of release (left ordinate) increased steeply in depolarizations between -0.35 and -0.5 µA, and less steeply for larger depolarizations. 10 µM serotonin added to the perfusate of the electrode raised the rate of release through the whole range of depolarizations by factors of 3 to 10. When the electrode perfusate was again control solution (wash), after one hour the rate of release had not fully returned to the control level.

Serotonin is known to activate intracellular adenylate cyclase which increases cyclic nucleotides like cAMP, as shown for a lobster preparation by Goy et al. (1984). Another activator of adenylate cyclase is forskolin (Seamon and Daly 1983, Middleton et al. 1986) which also in lobster imitates the respective biochemical effects of serotonin (Goy et al. 1984). In the present study, forskolin raised quantal release similar to serotonin (compare e.g. Figs.3 and 5). The actions of serotonin and of forskolin do not depend critically on Ca inflow, as shown already by Glusman and Kravitz (1982) for serotonin. In

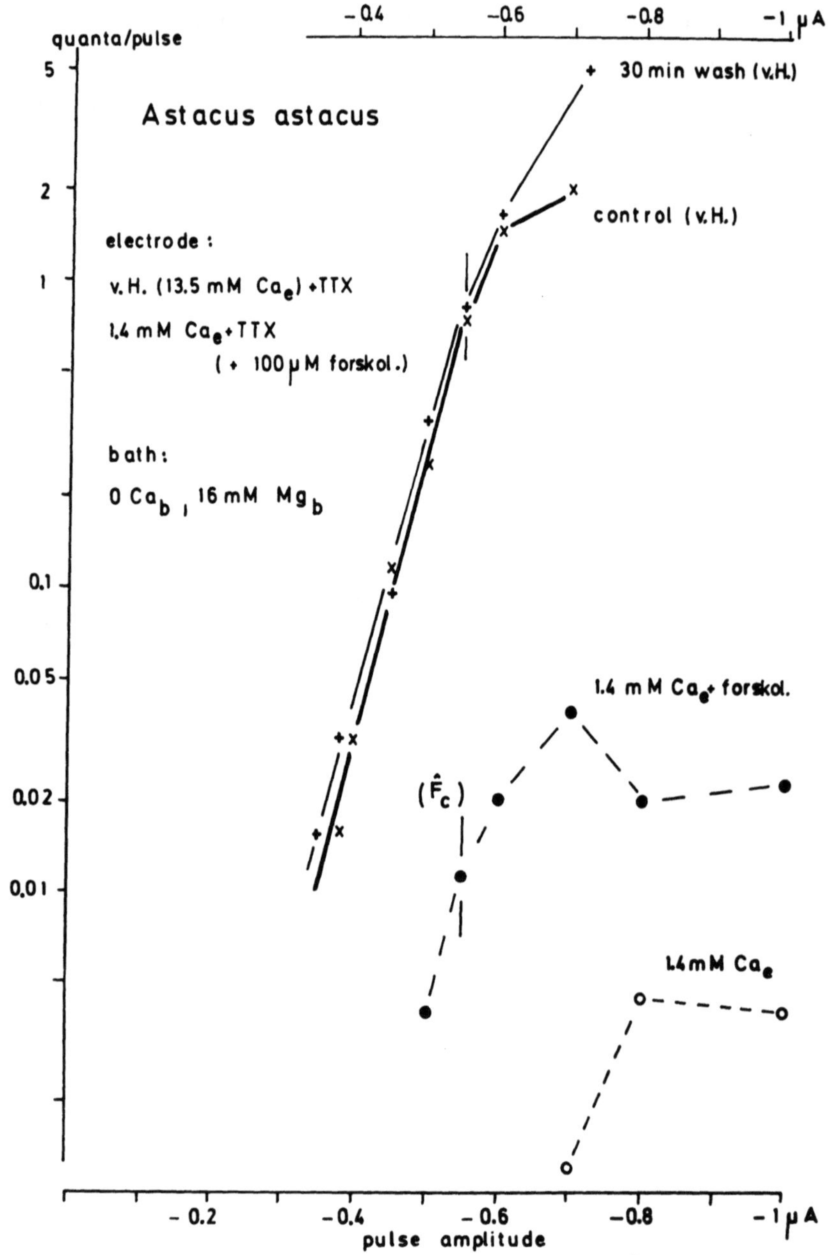

Fig.4: Dependence of release on pulse amplitude. The experiment started with the control (x), then Ca in the electrode perfusate was reduced from 13.5 mM (v. H.) to 1.4 mM Ca_e (o), then 100 μM forskolin was added to the perfusate (•), and finally the electrode was perfused again with v. Harreveld solution without forskolin for 30 min (+). In the 3 curves in which it could be measured $\hat{F}_c$ (Fig.3) is indicated by vertical lines.

the experiment of Fig.4, reduction of the Ca concentration in the electrode from the control level (v.H., 13.5 mM Ca_e) to 1.4 mM Ca_e depressed the rate of release by a factor of at least 1500. Addition of forskolin to the perfusate of the electrode, however, produced the typical increase of the rate of release by factors of 4 to 30. The radical experiment in 0 Ca_e, which could exclude Ca inflow as essential for the action of serotonin and forskolin, is not feasible, since the quantum size is diminished postsynaptically in low Ca_e. The openings of glutamate activated membrane channels (which produce the synaptic currents) are blocked at low Ca_e, becoming extremely rare at $Ca_e < 1$ mM (Hatt et al. 1987). When Ca_e was instead raised to 150 mM, replacing Na_e completely (not illustrated), the rate of release increased by a factor of about 5 compared to control in the whole range of depolarizations. Also under these conditions serotonin or forskolin increased the rate of release, although apparently less than under control conditions.

For constant depolarization pulses, the increase of release by serotonin or forskolin could be either due to a raise in the intracellular Ca concentration, Ca_i, or a change in another, non-Ca factor controlling release. We therefore have to estimate changes in Ca_i after application of serotonin. Facilitation of release by repeated depolarizations is generally accepted to be the consequence of 'residual Ca_i' remaining from Ca inflow during preceding depolarizations (Katz and Miledi 1968). Facilitation thus can be used as an indicator of Ca_i. In one type of twin-pulse facilitation a prepulse is varied, and fixed pulses of relatively small amplitude follow the prepulse with short (10 ms) and long (1 s) delay (see inset in Fig.3). The rate of release after the first fixed pulse is facilitated by residual Ca_i from the prepulse, while release after the second fixed pulse is not facilitated. Fixed pulse facilitation, F_c, then is the rate of release after the first fixed pulse divided by the rate of release after the second fixed pulse. F_c indicates the amount of Ca inflow during the prepulse. If F_c is plotted in dependence on the amplitude of the prepulse (Fig.3, right ordinate), F_c has a peak, $\hat{F}_c$, at a depolarization level below saturation of release. This peak has been observed generally (Dudel et al. 1983, H. Parnas et al. 1986, Zucker and Landò 1986, Dudel 1986, 1988a) and is interpreted to represent the maximum of Ca inflow at depolarizations to about 0 mV. Remarkably, F_c as a function of the amplitude of depolarization was not affected by serotonin, although the rate of release, after the prepulse increased up to tenfold (Fig.3). For Ca inflows far below the saturation level for the control of release F_c also depends strongly on the resting Ca_i, Ca_{ir}, F_c decreasing when Ca_{ir} increases (Dudel 1986, 1988a). No decrease of F_c was observed in the experiment of Fig.3 for low depolari-

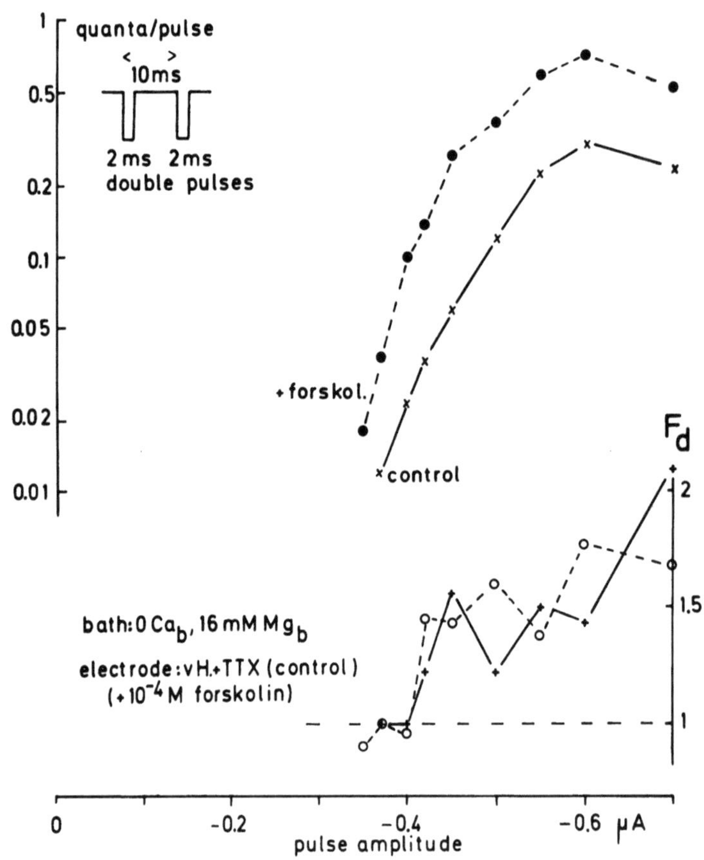

Fig.5: Dependence of release (left ordinate) and of double-pulse facilitation, F_d (right ordinate, lower curves), on amplitude of the depolarization pulses. Pulse program in inset. The electrode was perfused with v. Harreveld solution +TTX in the control (crosses), and then 100 μM forskolin was added to the perfusion (circles).

zations (and Ca inflows). The results of the F_c measurements in Fig.3 and in other such experiments thus are incompatible with a rise in Ca inflow or in Ca_{ir} as the basis for increased release due to serotonin or forskolin.

Similar conclusions can be drawn from convential double-pulse facilitation experiments. As shown in the inset of Fig.5, two identical depolarization pulses are applied to the terminal at short interval (10 ms). The facilitated rate of release after the second pulse divided by the rate of release after the first (control) pulse gives the double-pulse facilitation, F_d. Plotted as a function of Ca inflow during the pulses, F_d reaches a maximum when the slope of the release/Ca-inflow relation has a maximum also, and declines when Ca inflow approaches the saturation level of release. In addition, F_d depends steeply on the resting Ca_i, Ca_{ir}, declining when Ca_{ir} increases (I. Parnas et al. 1982, Dudel 1986, 1988b). In all experiments in which F_d was determined, F_d was unaffected by the presence of serotonin or forskolin, although the release was multiplied in amplitude. An example is given in Fig.5 for the action of forskolin. Also Fischer and Florey (1983) found no effect of serotonin on facilitation of release by a pair of action potentials. Analogous to the results on F_c, the measurements of F_d give no indication of a rise in Ca inflow or of Ca_{ir} during increased release due to serotonin or forskolin.

In conclusion, the results of this study support two mechanisms of the modulation of synaptic activity by serotonin at the neuromuscular junction of crayfish. Firstly, serotonin seems to increase the excitability of terminals which is blocked by TTX. Serotonin thus may improve the voltage dependent sodium conductance. Such an effect may explain also the small depolarizations of the terminal observed by Dixon and Atwood (1985), the increased speed of conduction of action potentials in preterminal axons and changes in extracellularly recorded action potentials near the terminals reported by Dudel (1965), and the repetitive axonal discharges seen by Finger and Martin (1986).

The second mechanism of action of serotonin improves the rate of release elicited by a given depolarization of the terminal. This improvement of release is not due to an increase of intracellular Ca concentration. Apparently serotonin promotes another coupling factor between depolarization of the terminal and release of transmitter quanta. Since serotonin and forskolin have identical effect, it seems probable that the promotion of the coupling factor is mediated by activation of adenylate cyclase and consequent phosphorylation of a controlling protein. Such phosphorylations have been observed by Goy et al. (1984). It may be speculated that the non-Ca coupling factor which is promoted by serotonin and forskolin is identical with the non-Ca potential dependent factor which controls release in addition to Ca_i, which was suggested on different lines of evidence (Dudel et al. 1983, Dudel 1984, 1986, 1988a, H. Parnas et al. 1986).

Also in molluscs serotonin leads to improvement of excitatory transmission. The mechanisms of action seem to be different from that concluded here for crayfish. Klein et al. (1986) found a decrease of a K^+ conductance mediated by cAMP, and Paupardin-Tritsch et al. (1986) in addition an increase in Ca inflow.

REFERENCES

Dixon D, Atwood HL (1985) Crayfish motor nerve terminal's response to serotonin examined by intracellular microelectrode. J Neurobiol 16:409-424

Dudel J (1965) Facilitatory effects of 5-hydroxy-tryptamine on the crayfish neuromuscular junction. Naunyn-Schmiedebergs Arch exp Path Pharm 249:515-528

Dudel J (1983) Graded all-or-nothing release of transmitter quanta by local depolarizations of nerve terminals on crayfish muscle? Pflügers Arch 398:155-164

Dudel J (1984) Control of quantal transmitter release at frog's motor nerve terminals. II. Modulation by de- or hyperpolarizing pulses. Pflügers Arch 402:235-243

Dudel J (1986) Dependence of double-pulse facilitation on amplitude and duration of the depolarization pulses at frog's motor nerve terminals. Pflügers Arch 406:449-457

Dudel J (1988a) Calcium dependence of quantal release triggered by graded depolarization pulses to nerve terminals on crayfish muscle. Submitted for publication

Dudel J (1988b) Calcium and depolarization dependence of facilitation of synaptic release at nerve terminals on crayfish muscle. Submitted for publication

Dudel J, Parnas I, Parnas H (1983) Neurotransmitter release and its facilitation in crayfish muscle. VI. Release determined by both, intracellular calcium concentration and depolarization of the nerve terminal. Pflügers Arch 399:1-10

Enyeart J (1981) Cyclic AMP, 5 HT, and the modulation of transmitter release at the crayfish neuromuscular junction. J Neurobiol 12:505-513

Finger W, Martin C (1986) Repetitive axonal discharges elicited by serotonin and intracellular adenosine 3',5'-cyclic monophosphate in the crayfish neuromuscular junction. Neurosci Lett 72:295-299

Fischer L, Florey E (1983) Modulation of synaptic transmission and excitation-contraction coupling in the opener muscle of the crayfish, Astacus leptodactylus, by 5-hydroxytryptamine and octopamine. J exp Biol 102:187-198

Glusman S, Kravitz E (1982) The action of serotonin on excitatory nerve terminals in lobster nerve-muscle preparations. J Physiol (Lond) 325:223-241

Goy MF, Schwarz TL, Kravitz EA (1984) Serotonin-induced protein phosphorylation in a lobster neuromuscular preparation. J Neurosci 4:611-626

Grundfest H, Reuben JP (1961) Neuromuscular synaptic activity in lobster. 'Nervous inhibition', Proccedings of an International Symposium, Ed. E. Florey, Oxford, Pergamon Press p 92-104

Harris-Warrick R, Kravitz EA (1984) Cellular mechanisms for modulation of posture by octopamine and serotonin in the lobster. J Neurosci 4:1976-1993

Hatt H, Franke C, Dudel J (1987) Calcium dependent gating of the L-glutamate activated, excitatory synaptic channel in crayfish muscle. Pflügers Arch (in press)

Katz B, Miledi R (1967) The release of acetylcholine from nerve endings by graded electric pulses. Proc Roy Soc B 167:23-38

Katz B, Miledi R (1968) The role of calcium in neuromuscular facilitation. J Physiol (Lond) 195:481-492

Klein M, Hochner B, Kandel ER (1986) Facilitatory transmitters and cAMP can modulate accommodation as well as transmitter release in Aplysia sensory neurons: Evidence for parallel processing in a single cell. Proc Natl Acad Sci 83:7994-7998

Kravitz EA, Beltz B, Glusman S, Goy M, Harris-Warrick R, Johnston M, Livingstone M, Schwarz T (1984) The well-modulated lobster: the roles of serotonin, octopamine, and proctolin in the lobster nervous system. Pesticide Biochem Physiol 22:133-147

Middleton P, Jaramillo F, Schuetze SM (1986) Forskolin increases the rate of acetylcholine receptor desensitization at rat soleus endplates. Proc Natl Acad Sci USA 83:4967-4971

Parnas H, Dudel J, Parnas I (1986) Neurotransmitter release and its facilitation in crayfish. VII. Another voltage dependent process beside Ca entry controls the time course of phasic release. Pflügers Arch 406:121-130

Parnas I, Parnas H, Dudel J (1982) Neurotransmitter release and its facilitation in crayfish muscle. V. Basis for synapse differentiation of the fast and slow type in one axon. Pflügers Arch 395:261-270

Paupardin-Tritsch D, Hammond C, Gerschenfeld HM (1986) Serotonin and cyclic GMP both induce an increase of the calcium current in the same identified molluscan neurons. J Neurosci 6:2715-2723

Seamon KB, Daly JW (1983) Forskolin, cyclic AMP and cellular physiology. TIPS 4:120-123

Wheal HV, Kerkut GA (1976) The pre- and post-synaptic actions of 5-HT in crustacea. Comp Biochem Physiol 54C:67-70

Zucker RS, Landò L (1986) Mechanism of transmitter release: voltage hypothesis and calcium hypothesis. Science 231:574-579

NEUROTRANSMITTER-INDUCED REGULATION OF VOLTAGE-DEPENDENT CALCIUM CURRENT IN IDENTIFIED SNAIL NEURONS

D. Paupardin-Tritsch, C. Hammond, R. Harris-Warrick* and H.M Gerschenfeld
Laboratoire de Neurobiologie
Ecole Normale Supérieure
46 Rue d'Um, 75230 Paris 05
France

Introduction

Recent studies using the methods of molecular biology and the patch clamp techniques have deeply modified our wiew of the mechanism of action of neurotransmitters in ionic channels. Accordingly, the relationship between neurotransmitter receptors and ionic channels appear now to be of three main classes:

1. The neurotransmitter receptor and the ionic channel gated by it are parts of an unique protein molecule. The receptor site activation by the neurotransmitter induces a conformational change in this protein molecule that results in the opening of the ionic channel. Exemples of this model are the nicotinic cholinergic receptor of vertebrate neuromuscular junctions and fish electroplaques (see Changeux et al., 1984; Noda et al., 1984) and both the GABA (Schofield et al., 1987) and glycine (Grenningoh et al., 1987) receptors of vertebrate central nervous system synapses (see also, Barnard et al., 1987).

2. The neurotransmitter receptor and the ionic channel gated by it are different and separate protein molecules. The functional coupling between them is mediated by an heterotrimer GTP-binding protein (G protein). The muscarinic receptor of the vertebrate atrial muscle fibres intervening in the vagal inhibition of the heart beating is an exemple of this model. The activation of this muscarinic receptor protein by acetylcholine

*Present adress: Section of Neurobiology and Behavior, Cornell University, Ithaca N.Y. 14850, USA

NATO ASI Series, Vol. H19
Modulation of Synaptic Transmission and Plasticity
in Nervous Systems. Edited by G. Hertting and H.-C. Spatz

induces the opening of a K^+ channel (Trautwein and Dudel, 1958), via a G protein (Pfaffinger et al., 1985; Yatani et al., 1987). The 40 KDa α subunit of this G protein would be responsible for the K^+ channel activation (Codina et al. 1987)

3. The neurotransmitter receptor activation regulates or modulates the function of ionic channels which are not gated by the transmitter itself but by changes in the transmembrane potential. An exemple of this mechanism is the neurotransmitter-induced modulation of Ca^{2+} channels (see review, Hoffman et al., 1987).

Voltage-dependent Ca+ channels play an important role in the contraction of cardiac muscle fibres and in the regulation of neuronal excitability and transmitter release at synapses. Neurotransmitters have been reported to either increase or decrease the voltage dependent Ca^{2+} current and the mechanisms of these transmitter induced regulations appear to be extremely diverse.

The exemple by excellence of a neurotransmitter-induced increase of Ca^{2+} current is the β-adrenergic modulation of the Ca^{2+} current in cardiac muscle fibres. β-adrenergic agonists stimulate adenylate cyclase, via a Gs protein, the resulting cAMP increase (Reuter et al., 1983, Brum et al. 1983) activates the cAMP-dependent protein kinase (Osterrieder et al., 1983; Brum et al., 1983) which phosphorylates the Ca^{2+} channels and/or some associated protein/s, thus increasing the number of Ca^{2+} channels (Reuter et al., 1983; Brum et al., 1984) and changing the kinetics of these channels (Bean et al., 1984; Bean et al., 1984). A similar cAMP-dependent increase in Ca^{2+} current is induced by noradrenaline in the soma of hippocampal neurons (Gray and Johnston, 1987).

Neurotransmitter induced reductions of the voltage-dependent Ca^{2+} current have been observed in neurons. Thus noradrenaline decreases the duration of the Ca^{2+}-dependent spike recorded in the soma of both rat sympathetic ganglion neurons (Horn and McAfee, 1982; Galvan and Adams, 1982) and chick embryo dorsal root sensory neurons in culture (Dunlap & Fischbach, 1980). The ionic mechanism of this noradrenaline effect on both types of neurons is the same and consists in a decrease of the Ca^{2+}-current, without apparent alteration of the Ca^{2+} current kinetics (Dunlap & Fischbach, 1980 ; Galvan & Adams,

1982). GABA, serotonin, dopamine, enkephalin and somatostatin were also found to decrease the duration of the action potential of the chick embryo dorsal root neurons in culture (Dunlap & Fischbach, 1980; Canfield & Dunlap, 1984).

The mechanisms of these neurotransmitter-induced decreases of Ca^{2+} current has not been yet completely elucidated. On one hand, in both the noradrenaline and GABA-induced decreases in Ca^{2+} current of chick embryo dorsal root ganglion neurons in culture, it has been shown that a G protein is involved and that this G protein is ADP-rybosilated by the toxin of Bordetella pertussis (PTX), since PTX blocks both these neurotransmitter-induced reductions of Ca^{2+} current (Holz et al., 1986). It has also been recently shown that PTX blocks the decrease in Ca^{2+} current induced by opiate peptides in NG 108-15 cells (neuroblastoma x glioma hybrid cell line). This PTX effect was reversed by the intracellular application of two different G proteins, G_i and G_o, G_o being ten times more powerful in restoring the response to the opiate peptides than G_i (Hescheler et al., 1987). These experiments, in which a patch-clamp whole cell recording was used, suggest that G_o may intervene by coupling directly the opiate receptors to the voltage-dependent, Ca^{2+} channels. Nevertheless, in similar recordings from chick embryo dorsal root ganglion neurons, Rane and Dunlap (1986) have also observed that the extracellular application of a diacylglycerol analog, oleylacetylglycerol (OAG), evokes a decrease of Ca^{2+} current in chick embryo dorsal root ganglion cells in culture, suggesting that the activation of protein kinase C may be also involved in a neurotransmitter-induced decrease of the Ca^{2+} current in these neurons.

In this paper, we will shortly review some of our own results concerning the neurotransmitter-induced modulation of the voltage dependent Ca^{2+} current in identified neurons of the snail Helix aspersa . The soma of these neurons discharge Ca^{2+}-dependent action potentials resulting from the activation of voltage-dependent Ca^{2+} channels. To analyze these somatic action potential we revealed the Ca^{2+}-dependent plateau component was revealed by bathing the neurons in a saline containing 30 mM of tetraethylammonium (TEA). To study the voltage-dependent Ca^{2+} current in voltage-clamped neurons, we bathed them in a saline containing 1 µM tetrodotoxin and 30 mM TEA, and in which Ca^{2+} was replaced by Ba^{2+}. Ba^{2+} besides permeating the Ca^{2+} channels blocks some of the K^+ currents. To confirm that the neurotransmitters actually affected the Ca^{2+} current we studied the current-voltage relationship in single neurons and ruled out a participation of K^+ currents loading the neurons with Cs^+ using the ionophore nystatin (Tillotson, 1979). The Ca^{2+} channels modulated by the neurotransmitters in snail neurons are of the L-type, according to the parameters and modalities of their activation. The reader will find a detailed account of the experimental procedures used and a full discussion of these results in the papers quoted at the beginning of each section.

The serotonin (5-HT)-induced increase of voltage dependent calcium current in identified snail neurons. Role of cGMP and cGMP-dependent protein kinase (cGMP-PK) (see Paupardin-Tritsch et al, 1986 a,b).

In a group of identified neurons located in the ventral face of the parietal ganglion of Helix bathed in a saline containing 30 mM TEA, 5-HT (1 µM) was found to evoke an increase of the duration of the somatic Ca^{2+}- dependent action potential. This effect showed a latency of seconds and was reversed by washing-out the neurotransmitter. When 5-HT was continuously applied for 15 minutes or more, no desensitization was observed. 5-HT (1 µM) did not affect any TEA-insensitive, Ca^{2+}-independent outward current component in the identified neurons. In contrast, the application of 5-HT (1 µM) induced a 25-30 % increase in Ca^{2+} current amplitude in these neurons (Fig. 1,a_1).

We explored next the involvement of a second messenger in this 5-HT-induced response. Having in mind the effect of β-adrenergic agonists on cardiac muscle fibers discussed above, we examined first whether cAMP was involved in the 5-HT response. This was not the case since neither the intracellular injection of cAMP in the ventral neurons nor the extracellular application of forskolin, affected the Ca^{2+} current or its modulation by 5-HT. We could also rule out the participation of intracellular Ca^{2+} in this 5-HT response, since loading the cells with intracellularly injected EGTA, did not affect the 5-HT induced decrease of Ca^{2+} current.

The most interesting results were obtained with cGMP. The intracellular injection of this nucleotide in the identified ventral neurons was found to mimic the 5-HT-induced increase in duration of the Ca^{2+}-dependent action potential and to induce an increase in the Ca^{2+} current (Fig. 1,a_2). That 5-HT and cGMP affected the same population of Ca^{2+} channels was shown by the results of "occlusion" experiments, thus when a prolonged 5-HT application evoked a maximal increase in Ca^{2+} current, the intracellular injection of cGMP became ineffective, and vice-versa, when the cGMP injection evoked a maximal increase in Ca^{2+} current, no further effect of a 5-HT application was observed.

The increase in endogenous of cGMP by the compound zaprinast (M&B 22948 100 μM), which is a specific inhibitor of the cGMP-dependent phosphodiesterase (Windquist et al, 1984) also evoked an increase in the amplitude of the Ca^{2+} current in these identified ventral neurons. By reasons still unclear, however, we could not observe a potentiation of the effects of 5-HT on the Ca^{2+} current with zaprinast.

These results indicate that the 5-HT-induced increase in Ca^{2+} current in the identified ventral neurons of Helix was mediated by cGMP. Two important questions remained to be answered :

a) Does the activation of the 5-HT receptor result in the stimulation of a membrane-bound guanylate cyclase (Windquist et al., 1984, see) or does 5-HT stimulate the cGMP synthesis through a mechanism involving the activation of phospholipase AII and/or the formation of arachidonic acid (see Goldberg and Haddox, 1977).

b) Does cGMP increase the Ca^{2+} current by a direct action on the Ca^{2+} channels from the inner surface of the neuronal membrane, as it activates cationic channels in vertebrate phoreceptors (Fesenko et al., 1986), or does it regulate the Ca^{2+} current by activating a cGMP-dependent protein kinase (cGMP-PK) which would phosphorylate the Ca^{2+} channels and/or some protein/s associated to them ?

We have tried to answer these questions experimentally. Concerning the formation of cGMP, the activation of phospholipase AII does not appear to be a necessary step for the induction of the cGMP-dependent, Ca^{2+} current increase by 5-HT. The application of either 4-bromophenacyl bromide, an inhibitor of phospholipase AII in other molluscan neurons (Piomelli et al., 1987) or arachidonic acid did not affect the 5-HT-induced increase of Ca^{2+} current (Paupardin-Tritsch and Gerschenfeld, unpublished).

The second question was answered by a series of experiments performed in collaboration with P. Greengard and A.C. Nairn (Rockefeller University, New York). In preliminary trials, we found that the intracellular injection of the preparation of purified cGMP-PK in the identified ventral neurons had no apparent effect on the Ca^{2+} current. The explanation was that in these experiments we injected the whole enzyme and not an isolated catalytic subunit as in the studies on the cAMP-dependent protein kinase (see Nestler and Greengard, 1983; Levitan, 1985). Therefore, we needed to activate the cGMP-PK either by increasing the endogenous cGMP once the cGMP-PK was injected or by the addition of a low concentration of cGMP (1 μM) prior to its intracellular injection. Both procedures were successful. One one hand, the effect of low concentrations of 5-HT and zaprinast, which induced an increase in the endogenous cGMP, was markedly potentiated when their application was preceded by an intracellular injection of cGMP-PK. On the other hand, the intracellular injection of cGMP-PK previously activated by 1 μM cGMP evoked an increase in Ca^{2+} current in the identified ventral neurons (Fig. 1,b_1). When the activated cGMP-PK was released into the cytoplasm of the ventral neurons from a "leaky" intracellular micropipette, the amplitude of the Ca^{2+} current was increased up to 200-300% (Fig. 1,b_2). In contrast, if the activated cGMP-PK was inactivated by heating prior to its intracellular injection, no effect on the Ca^{2+} current was observed.

These results confirm that the 5-HT-induced increase in Ca^{2+} current in the identified ventral neurons is mediated by cGMP which activated the cGMP-PK which, in turn, phosphorylates the Ca^{2+} channels and/or some protein/s associated with them.

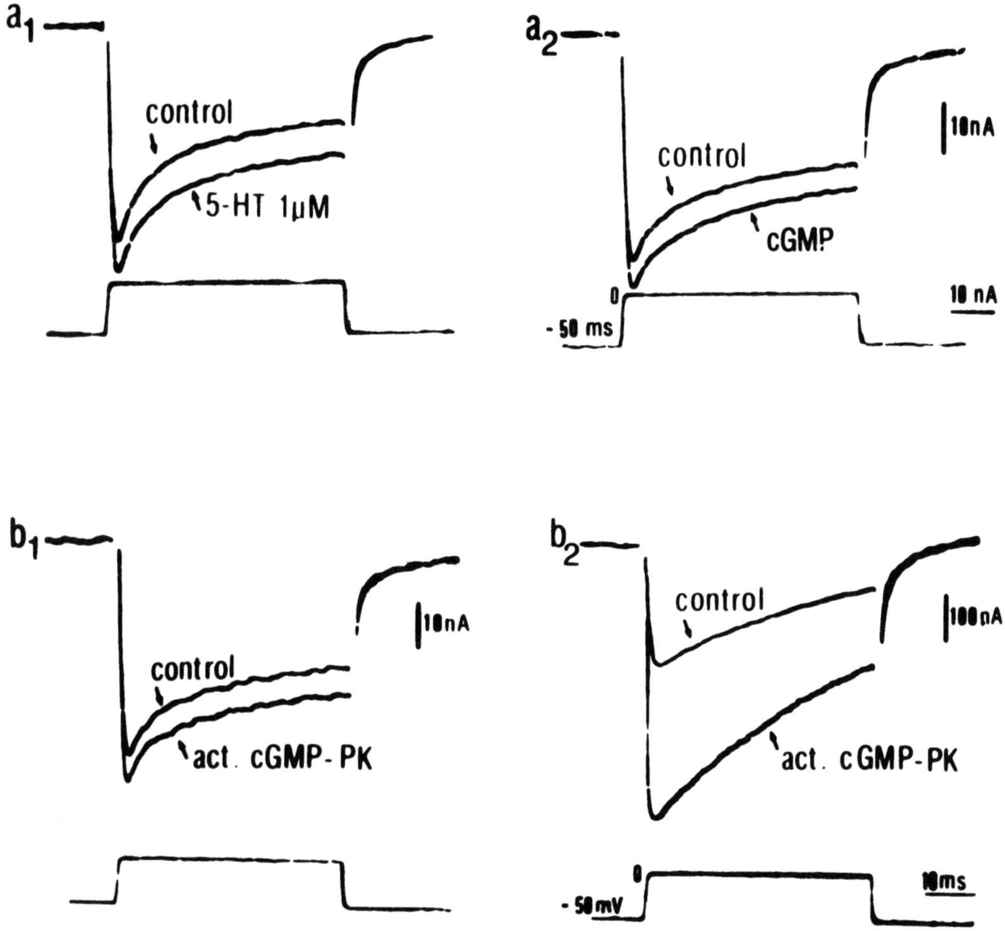

Fig. 1. 5-HT induced increase in voltage-dependent calcium current in identified snail ventral neurons is mimicked by intracellular injections of either cGMP or cGMP-dependent protein kinase. The voltage-clamped cells, bathed in a TTX-TEA containing saline in which calcium was replaced by barium were held at -50 mV and depolarized by pulses of 60 ms duration to 0 mV.

a1. A comparison of the calcium currents recorded in a ventral neuron before (control) and after application of 5-HT (1 μM).

a2. In another ventral neuron, the intracellular pressure-injection of cGMP from a micropipette filled with a 0.1 M solution of a cGMP monosodium salt increases the calcium current.

b1, In a different ventral neuron, the intracellular injection of cGMP-PK activated by addition of 1 μM cGMP (act. cGMP-PK), elicits an increase of calcium current.

b2, The intracellular release of activated cGMP-PK into another ventral neuron from a "leaky" micropipette induces a 150% increase of the peak amplitude of the calcium current.

The octapeptide 26-33 of cholecystokinin (CCK 8) decreases irreversibly the voltage dependent calcium current of identified snail neurons . Role of protein kinase C (Hammond et al., 1987)

A CCK-like peptide has been detected immunocytochemically in identified neurons of the molluscs *Helix* (Osborne et al., 1982) and *Aplysia* (Ono, 1986). In mammals, CCK 8 has been reported to decrease reversibly the Ca^{2+} dependent action potential in neurons of the dorsal horn of the spinal cord (Willets et al., 1985).

In snail neurons D2 and F77, located in the dorsal face of the parietal ganglia, CCK 8 (1-5 nM) was also found to decrease the duration of their somatic Ca^{2+}-dependent action potential. This effect had a peculiar time course: it showed a long latency (some minutes) and could not be reversed by a prolonged wash-out of the peptide (Fig 2).

CCK 8 (1-5 nM) was also found to induce a marked decrease of the Ca^{2+} current which took 7-9 minutes to reach a plateau and could not be reversed by prolonged wash out (Fig 2, curve and inset)). CCK 8 did not apparently alter any TEA- insensitive, Ca^{2+}-independent, K^+ current in these neurons.

The long latency of the CCK 8-induced decrease in Ca^{2+} current suggested the intervention of a second messenger mechanism in its generation. The intracellular injection of either cAMP or cGMP in the D2 and F77 neurons, did not alter the Ca^{2+} current of these cells or the CCK 8-induced decrease of this current. In contrast, a prolonged intracellular injection of EGTA prior to the application of CCK 8, either depressed or suppressed the peptide response. However, when the intracellular injection of EGTA was performed after CCK 8 application, no effect on the CCK 8-induced response was observed.

These experiments suggested that cytosolic Ca^{2+} played some role in the mechanism of the CCK 8 response. In consequence, we explored whether the phospholipid intracellular messengers resulting from the activation of phospholipase C, inositol trisphosphate ($InsP_3$) and diacylglycerol, (see Berridge, 1986 ; Nishizuka, 1986) were involved in the action of CCK 8. We first injected $InsP_3$ intracellularly in the

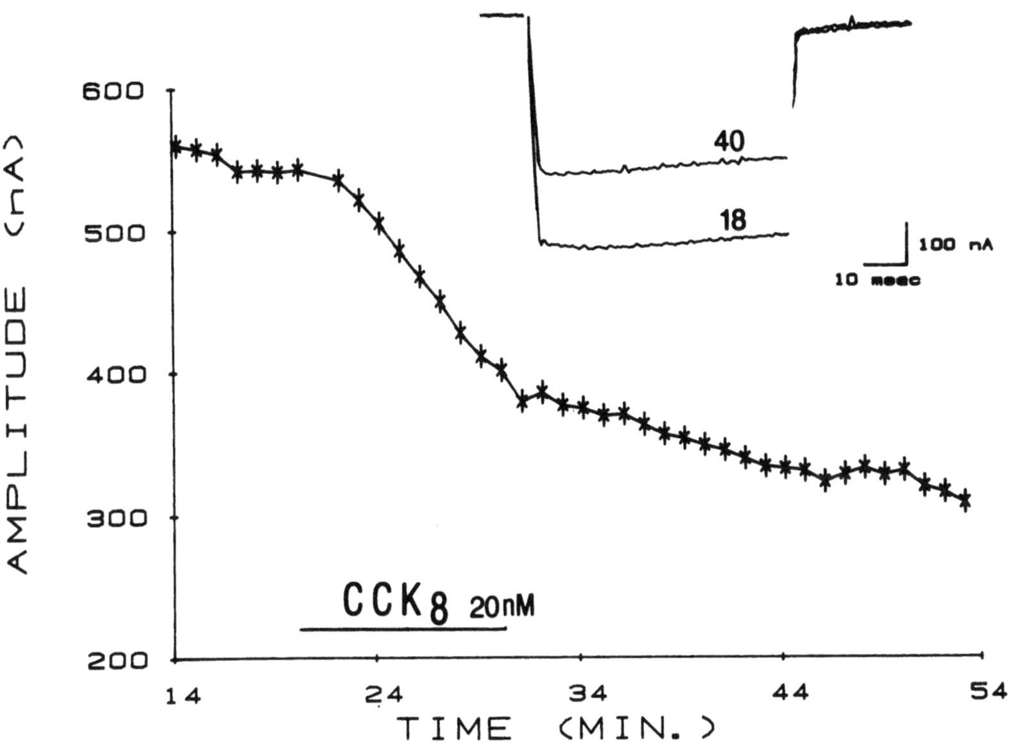

Fig. 2. CCK 8-induced decrease in voltage-dependent calcium current in a F77 snail neuron. The voltage clamp experiments followed the same protocol as in Fig. 1. The calcium current was measured each minute by averaging three values of its amplitude recorded 40 msec after the onset of the depolarizing pulse. An extracellular application of CCK 8 (20 nM) induces a slowly developping decrease in calcium which reaches a plateau in 9-10 minutes and it is not reversed by a 25 minutes wash-out of the peptide. Inset: a comparison between the calcium current recorded before (Minute 18) and after the application and washout of CCK 8 (Minute 40).

D2 and F77 neurons but the results of these injections gave rather inconsistent results. Moreover, they did not affect the CCK 8 response. The cytosolic Ca^{2+} concentration necessary for the response to CCK 8 appears then to be independent of an excess of $InsP_3$.

We examined next the possible involvement of diacylglycerol (DAG), the other product of phospholipase C activation which, in the presence of Ca^{2+}, activates and translocates protein Kinase C. The extracellular application of OAG, a DAG analogue, at low concentrations (250 nM-1 μM) did evoke a decrease in the Ca^{2+} current in the D2 and F77 neurons. This effect of OAG was also depressed by the previous intracellular injection of EGTA. When the concentration of OAG was increased to 50 µM the Ca^{2+} current amplitude became reduced by 80 %. Another protein kinase C activator, phorbol ester TPA at concentrations down to 0.4 nM, also mimicked the effect of CCK 8 on the Ca^{2+} current. Such as the effect of CCK 8, both OAG and phorbol ester TPA effects were not reversed by washing. These experiments strongly suggested that protein Kinase C intervened in the CCK 8-induced reduction in Ca^{2+} current in the D2 and F77 neurons.

We therefore studied the effects of intracellular injections of protein kinase C (PKC) in the D2 and F77 neurons. The intracellular injection of PKC in these cells evoked a marked shortening of the duration of the somatic Ca^{2+}-dependent action potential and in voltage-clamped D2 and F77 neurons these enzyme injections markedly reduced the amplitude of the Ca^{2+} current. This effect was similar to those of CCK 8, OAG and phorbol ester TPA and it was also irreversible. Moreover, in another series of experiments we diluted five times the original PKC preparation and injected it intracellularly into the identified neurons. These PKC injections did not evoke any change in the Ca^{2+} current but when the PKC injections preceded the application of a low concentration of CCK 8 (1nM), they induced a marked potentiation of the CCK 8 effect.

These results indicate that the CCK 8-induced decrease in Ca^{2+} current is due the activation of a CCK 8 receptor which activates phospholipase C ; the diacylglycerol thus formed, in the presence of an adequate cytosolic concentration of Ca^{2+}, activates and translocates PKC which would phosphorylates the Ca^{2+} channels or some protein/s associated with them.

We have not yet an explanation for the irreversibility (or slow reversibility) of the CCK 8 effect. One possibility could be that a Ca^{2+}-activated proteases causing a fragmentation of PKC, release the catalytic subunit, which independently from both the cytosolic Ca^{2+} and diacylglycerol (see Murray et al., 1987), would keep the Ca^{2+} channels or some associated protein/s phosphorylated.

There are other recent reports of PKC-induced modulation of

Ca^{2+} channels. On one hand, in aortic muscle cells, the peptides vasopressin as well as bombesin and phorbol esters induce a decrease in the Ca^{2+} influx through L-type channels (Galizzi et al., 1987). On the other hand, PKC activation can also increase the Ca^{2+} current in other molluscan neurons, the bag-cells of Aplysia (DeRiemer et al.,1985). A patch-clamp analysis has shown that the phorbol esters activated a new population of Ca^{2+} channels (Strong et al., 1987).

The dopamine (DA)-induced decrease of voltage dependent calcium current in identified snail neurons . Role of an identified snail G protein . (Paupardin-Tritsch et al, 1985 ; Harris-Warrick et al., 1988).

In contrast to the irreversible effect of CCK 8 on the D2 and F77 snail neurons, the application of DA (1-5 µM) to the same cells evoked a reversible shortening of the duration of their somatic Ca^{2+}-dependent action potential. In voltage-clamped neurons, DA applied at the same concentrations evoked a decrease in Ca^{2+} current that was easily reversed by washing out of the amine (Fig. 3,A). The maximal effect induced by DA consisted in a 30-35 % reduction of the initial Ca^{2+} current amplitude and no desensitization was observed during prolonged applications of DA.

We then investigated the intervention of a second messenger in the DA-induced reduction in Ca^{2+} current. As indicated above, the intracellular injection of either cAMP or cGMP, and the application of forskolin did not affect the Ca^{2+} current in D2 and F77 neurons. At variance with the effect of EGTA on the CCK 8-induced response, a prolonged intracellular injection of this chelating agent did not affect the DA-induced decrease in Ca^{2+} current. This finding strongly suggests that cytosolic Ca^{2+} does not intervene in this DA response. Therefore, an intervention of the phospholipid second messengers resulting from an activation of phospholipase C or of an arachidonic acid metabolite in this DA response appears to be unlikely. Therefore, in collaboration with J. Bockaert, V. Homburger and B. Rouot (Centre

(Centre CNRS-INSERM de Pharmacologie-Endocrinologie, Montpellier), we examined the possible involvement of a G protein in the DA-induced decrease in Ca^{2+} current. As mentioned in the Introduction, the intervention of a PTX-sensitive G protein has been proposed in other neurotransmitter-induced Ca^{2+} current decreases (see Dunlap et al., 1987) and, in particular, G_o as been postulated to be the G protein responsible for the Ca^{2+} current decrease induced by opiate peptides in the tumoral hybrid cell line NG 108-15 (Hescheler et al., 1987).

We first confirmed that a G protein intervened in the DA-induced decrease in Ca^{2+} current by injecting intracellularly non-hydrolysable GTP analogs such as GPT-gamma-S and Gpp $(NH)_p$ in the D2 and F77 neurons. These compounds mimicked the effect of DA and also blocked the DA-induced reduction in the Ca^{2+} current. This latter effect could be due to an uncoupling of the DA receptors from the G protein (Breitweiser and Szabo, 1985). Moreover, the maximal Ca^{2+} current reduction induced by the injection of GTP analogs was similar to that elicited by DA, i.e., they maximally reduced the Ca^{2+} current by 30-35%.

We next examined the effect of PTX on the DA-induced response. The intracellular injection of the toxin previously activated by preincubation with 25 nM dithiothreitol (DTT) blocked the DA-induced decrease in Ca^{2+} current (Fig. 3,B). PTX was injected for less than 20 min and its effect tested 90 minutes after. A similar protocol was followed in a control experiment injecting heat-inactivated DTT-treated PTX. The inactivated PTX had no effect on the Ca^{2+} current or its modulation by DA.

PTX is known to ADP-ribosylate two different G-proteins present in bovine brain : G_o and G_i (see Spiegel, 1987). However, in a membrane fraction of the nervous tissue of the snail _Helix aspersa_ , PTX was found to ADP-ribosylate only one substrate (Homburger et al., 1987): the α subunit of a G protein of 40 KDa, a molecular mass intermediate between those of bovine brain α_o (39 KDa) and α_i (41 KDa).

To explore whether this 40 KDa subunit was related to bovine brain α_o we used an affinity purified polyclonal antibody raised against bovine brain α_o (anti-α_o, Toutant et al., 1987) which did recognize the 40 KDa snail subunit on an immunoblot. The specificity

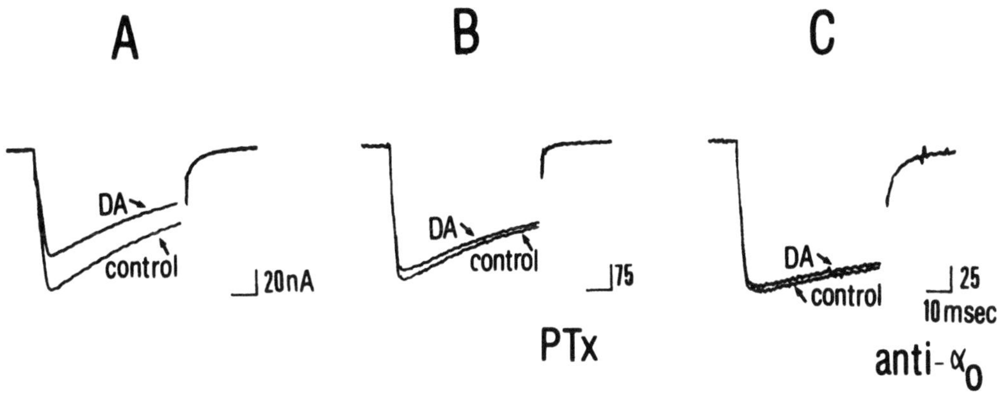

Fig. 3. The dopamine-induced decrease in voltage-dependent calcium current in snail identified neurons is markedly depressed by intracellular injections of either PTX or the anti-α_0 antibody. The voltage clamp protocol is as in Fig. 1.
A, DA (1 uM) decreases the calcium in a D2 neuron.
B, Intracellular injection of PTX blocks the DA-induced decrease in calcium current in a D2 neuron. PTX was previously activated by preincubation in 25 mM dithiothreitol, 0.25 M NaCl and 0.05 M sodium phosphate (pH 7) for 30 minutes at 20^0 and diluted to 10 uM in a medium with final concentrations (in mM): KCl 120, NaCl 5, Na phosphate 0.002, HEPES (pH 7.4) 20. PTX was pressure injected for 30 minutes and the effect of DA was tested after a 90 minutes period allowed for the diffusion and action of PTX.
C, Intracellular injection of anti-α_0 antibody blocks the DA-induced decrease in calcium current in a F77 neuron. The DA response was recorded after pressure injecting the antibody for a 15 minutes period followed by another period of 20 minutes allowed for the antibody diffusion.

of this anti-α_0 antibody was also tested on other G proteins and it was shown that it did not recognized G_s, G_i or the $\beta\gamma$ subunits of transducine. Its specificity was also tested on the α_{40} subunits of bovine brain, C6 glioma cells or chromaffin adrenal granules. It did not recognize any of them (Toutant et al., 1987; Brabet et al., 1988). The intracellular injection of this anti-α_0 antibody in the D2 and F77 neurons evoked a marked depression or a blockade of the DA-induced decrease of Ca^{2+}-current (Fig. 3,C). This effect disappeared if the anti-α_0 antibody was inactivated by heating, prior to its intracellular injection.

In view of the immunological relationship between brain α_0 and the 40 KDa snail subunit, we studied the effect of intracellular injections of the bovine brain α_0 subunit in D2 and F77 neurons. Prior to its injection, α_0 was activated with GTP-gamma-S (10 uM), in the presence of 10 mM $MgCl_2$. The injection of the activated α_0 subunit evoked a gradual and progressive decrease in Ca^{2+} current, beyond the maximal 30-35% reduction induced by DA. Nevertheless, further application of DA was still able to induce a normal decrease in Ca^{2+} current. The explanation of this observation is that the injected α_0 was not limited by the number of DA receptors or α_{40} subunits and thus could modulate more Ca^{2+} channels than DA Moreover, DA could still stimulate the endogenous α_{40} subunit which was neither disconnected as in the case of the GTP-gamma-S injections nor inactivated as in the case of the injections of either PTX or the anti-α_0 antibody.

We interpret these results as being strongly in favor of the view that the 40 KDa α subunit of snail nervous tissue, a protein immunologically and functionally related to mammalian brain α_0, mediates an essential step in the mechanism of the DA-induced decrease of Ca^{2+} current in the D2 and F77 neurons of Helix . We have already detailed the arguments that rule out the intervention of the major known second messengers in the generation of the DA-induced decrease in Ca^{2+} current. However, the existence of a direct coupling between the DA receptor and the Ca^{2+} channels mediated by α_{40} without an intervention of an unknown second messenger remains to be proven.

Concluding remarks

The results from our laboratory summarized in this paper as well as the work of other investigators to which we have referred, show that there is a great diversity of mechanisms underlying the neurotransmitter-induced regulation of Ca^{2+} channels in excitable cells. From the available data about the neurotransmitter-induced increases of voltage-dependent Ca^{2+} current it appears that this kind of modulation generally involves protein phosphorylation of the channels or associated protein/s by one of the following protein kinases: the cAMP-dependent kinase, the cGMP dependent kinase or protein kinase C.

In the case of the neurotransmitter-induced decreases in Ca^{2+} current the mechanisms are more diversified. Phosphorylation by a protein kinase appears to play an important role in two instances in which neurotransmitters induce a Ca^{2+} current reduction, the CCK 8-induced decrease in Ca^{2+} current in molluscan neurones (Hammond et al., 1987) and the vasopressin and bombesin effect on aortic smooth muscle (Galizzi et al., 1987). In another case, the acetylcholine-induced decrease of Ca^{2+} current mediated by muscarinic receptors in cardiac muscle, the mechanism involved appears to be a decrease in cAMP-dependent phosphorylation resulting from the Acetylcholine-induced fall of the intracellular cAMP concentration previously increased by noradrenaline (Fischmeister and Hartzell, 1986). For the other known cases of neurotransmitter-induced decreases in Ca^{2+} current, the most attractive hypothesis at present is that they would be mediated through a direct transmitter receptor-Ca^{2+} channel coupling via a G protein, without intervention of a second messenger system (see Dunlap et al., 1987). The α subunit of mammalian brain G_o protein has been postulated to be involved in some of these modulations (Hescheler et al., 1987) and an α_{40} subunit present in snail nervous tissue , immunologically related to α_o, appears to intervene in the DA-induced decrease in Ca^{2+} current in snail neurons (see above).

Another problem that remains open is whether in neurons, these Ca^{2+} current modulations which have been discovered and analyzed in their cell body also operate at the Ca^{2+} channels of the synaptic endings and whether they play a role in synaptic plasticity. In their pioneer studies, Dunlap and Fischbach (1980) postulated that the different neurotransmitter-induced decreases in Ca^{2+} current that they observed in the soma of dorsal root ganglion cells were probably responsible for presynaptic inhibition at endings of the same cells at the spinal cord. More recently, it has been postulated that presynaptic inhibition at some <u>Aplysia</u> sensory neurons synapses would be mediated by a FMRFamide-induced opening of K^+ channels (S channels) through a mechanism involving phospholipase AII activation and arachidonic acid lipoxygenase metabolites (Piomelli et al., 1987). This hypothesis requires further confirmation since FMRFamide is known to also induce a decrease in Ca^{2+} current in both Helix (Colombaioni et al., 1985) and Aplysia (Brezina et al., 1987) neurons.

References

Barnard EA, Darlison MG, Seeburg P (1987) Molecular biology of the $GABA_A$ receptor/ the receptor/channel superfamily. Trnds Neurosci. 12: 502-509

Bean BP, Nowycky MC, Tsien RW (1984) B-Adrenergic modulation of calcium channels in frog ventricular heart cells. Nature 307: 371-375

Bellardetti F, Kandel ER, Siegelbaum SA (1987) Neuronal inhibition by the peptide FMRFamide involves opening od S K^+ channels. Nature 325: 153-156

Berridge MJ (1986) Inositol phosphates as second messengers. In Phosphoinositides and receptor mechanisms, Putney JW (ed), Alan liss New York pp. 25-39

Brabet P, Dumuis A, Sebben M, Pantaloni C, Bockaert J, Homburger, V (1987) Mol. Pharmac. in press

Brezina V, Eckert R, Erxleben C. (1987) Suppression of calcium current by an endogenous neuropeptide in neurones of Aplysia californica . J. Physiol. Lond. 388: 565-595

Brum G, Flockerzi V, Hofman F, Osterrieder W, Trautwein W (1983) Injection of catalytic subunit of cAMP-dependent protein kinase into isolated myocytes. Pflügers Arch. 398:146-154

Brum G, Osterrieder W, Trautwein W (1984) B-adrenergic increase in the calcium conductance of cardiac myocytes studied with the patch-clamp. Pflügers Arch. 401: 111-118

Changeux JP, Devilliers-Thiery A, Chemouilly P (1984) Acetylcholine receptor: an allosteric protein. Science 225: 1335-1345

Codina J, Yatani A, Grenet D, Brown AM, Birnbaumer L. (1987) The alpha sububit of the GTP binding protein Gk opens atrial potassium channels. Science 236 : 442-445

Colombaioni, L, Paupardin-Tritsch D, Vidal PP, et Gerschenfeld HM (1985) The neuropeptide FMRFamide decreases both the Ca^{2+}-conductance and a cAMP-dependent K^+-conductance in identified snail neurons. J. Neurosci. 6: 2533-2538

DeRiemer SA, Strong JA, Albert KA, Greengard P, Kaczmarek LK (1985) Enhancement of calciul current in Aplysia neurones by phorbol-ester and protein kinase C. Nature 313, 313-316

Dunlap K, Fischbach GD (1980) Neurotransmitters decrease the calcium conductance activated by depolarization of embryonic chick sensory neurones. J. Physiol. Lond. 317: 519-535

Dunlap K, Holz GG, Rane SG (1987) G proteins as regulators of ion channel function. Trends Neurosci. 10: 244

Fesenko EE, Kolesnikov SS Lyubarsky AL (1986) Induction by cGMP of cationic conductance in plasma membrane of retinal rod outer segment. Nature 313: 310-313

Fischmeister R, Harzell HC (1986) Mechanism of action of acetylcholine on calcium current on single cells from frog ventricle. J. Physiol., Lond. 376: 183-202
Galizzi JP, Qar J, Fosset M, Van Renterghem C, Lazdunski M (1987) Regulation of calcium channels in aortic cells by protein kinase activators (diacylglycerol et phorbol esters) and by peptides (vasopressin and bombesin) that stimulate phosphoinositide breakdown. J. Biol. Chem. 15: 6947-6950
Galvan M, Adams PR (1982) Control of calcium current in rat sympathetic neurons by norepinephrine. Brain Research. 244: 135-144
Goldberg ND, Haddox, MK (1977) Cyclic GMP metabolism and involvement in biological regulation. Ann. Rev. Biochem. 46: 823-896
Gray R, Johnston D (1987) Noradrenaline and beta adrenoceptor agonists increase activity of voltage-dependent calcium channels in hippocampal neurons. Nature 620: 621-622
Grenningloh G, Rienitz A, Scmitt B, Methfessel C, Gundelfinger ED, Betz H. (1987) The strychnine-binding subunit of the glycine receptor shows homology with the nicotinic acetylcholine receptors. Nature 328: 215-219
Harris-Warrick R, Hammond C, Paupardin-Tritsch D, Homburger V, Rouot B, Bockaert J, Gerschenfeld HM (1988) An alpha 40 subunit of a GTP-binding protein immunologically related to mammalian alpha o mediates a dopamine-induced decrease of calcium current in snail neurones. Submitted.
Hammond C, Paupardin-Tritsch D, Nairn AC, Greengard P, Gerschenfeld HM (1987) Cholecystokinin induces a decrease in Ca^{2+} current in snail neurons that appears to be mediated by protein kinase C. Nature 325: 809-811.
Hayne L, Yau KW (1985) Cyclic GMP sensitivie conductance in outer segment membrane of catfish cones. Nature 317: 61-64
Hescheler J, Rosenthal W, Trautwein W, Schultz G (1987) The GTP-binding protein, Go, regulates neuronal calcium channels. Nature 325, 445-447
Hoffman F, Nastainczyk W, Röhrkasten A, Schneider T, Sieber M (1987) Regulation of L-type calcium channel. Trends Bioch. Sci. 8: 393-398
Holz GG, Rane SG, Dunlap K (1986) GTP-binding proteins mediate transmitter inhibition of voltage-dependent calcium channels. Nature 319: 670-672
Homburger V, Brabet O, Audigier Y, Pantaloni C, Bockaert J, Rouot B (1987) Immunological localization of the GTP-binding protein G_o in different tissues of vertebrates and invertebrates. Mol. Pharmac. 31: 313-319
Horn JP, McAffee DA (1980) Alpha adrenergic inhibition of calcium-dependent potentials in rat sympathetic neurones. J. Physiol. Lond. 301: 109-204
Levitan, I (1985) Phosphorylation of ion channels. J. Membr. Biol. 87: 177-190
Murray AW, Fournier A, Hardy SJ (1987) Proteolytic activation of protein kinase C: a physiological reaction? Trends Bioch. Sci. 12: 53-54
Nestler EJ, Greengard P (1983) Protein phosphorylation in the brain. Nature 305: 583-588
Nishizuka, N. (1986) Studies and perspectives of protein kinase C. Science 233: 305-312

Noda M, Shimizu S, Tanabe T, Takai T, Kayano T, Ikeda T, Takahashi H, Nakayama H, Kanaoka Y, Miniamino N, Kangawa K, Matsuo H, Raftery MA, Hirose T, Inayama S, Hayashida H, Miyata T, Numa S. (1984) Nature 312: 121-127

Osborne, N.N., Cuello AC, Dockeray GJ (1982) Substance P, cholecystokinin and serotonin in a giant neuron. Science 216: 409-411

Osterrieder, W, Brum G, Hescheler J, W. Trautwein, Flockerzi V, Hofmann F (1982) Injection of subunits of cyclic AMP-dependent protein kinase into cardiac myocytes modulates Ca^{2+} current. Nature 298: 576-578

Paupardin-Tritsch D, Colombaioni L., Deterre P, Gerschenfeld HM (1985) Two different mechanisms of calcium spike modulation by dopamine. J. Neurosci. 5: 2522-2532

Paupardin-Tritsch D, Hammond C, Gerschenfeld HM (1986) Serotonin and cyclic GMP both induce an increase of the calcium current in the same identified molluscan neurones. J. Neurosci. 6: 2715-2723

Paupardin-Tritsch D, Hammond C, Gerschenfeld HM, Nairn AC et Greengard P (1986) cGMP-dependent protein kinase enhances Ca^{2+} current and and potentiates the serotonin-induced increase of Ca^{2+} current in identified snail neurones. Nature 323: 812-814

Pfaffinger PJ, Martin JM, Hunter DD, Nathanson M, Hille B (1985) GTP-binding proteins couple cardiac muscarinic receptors to K channel. Nature 317: 536-538

Piomelli D, Volterra A, Dale N, Siegelbaum SA, Kandel ER, Schwartz JH, Bellardetti, F. (1987) Lipoxygenase metabolites of arachidonic acid as second messengers for presynaptic inhibition of Aplysia sensory cells. Nature 328: 38-43

Rane SG, Dunlap K (1986) Kinase C activator 1,2 oleoylacetylglycerol attenuates voltage dependent calcium current in sensory nuerones. Proc. Natl Acad. Sci. USA 83: 184-188

Reuter H, Cachelin AB, De Peyer JE, Kobuku S (1983) Modulation of calcium channels in cultured cardiac cells by isoproterenol et 8-bromo-cAMP. Cold Spring Harb. Symp. Quant. Biol. 48: 193-200

Schofield PR, Darlison MG, Fujita N, Burt DR, StephensoN FA, Rodriguez H, Rhee L, Ramachandran J, Reale V, Glencorse TA, Seeburg P, Barnard E (1987) Nature 328: 221-227

Spiegel AM (1987) Signal transduction by guanine nucleotide binding proteins. Mol. Cell Endocrinol. 49: 1-16

Strong JA, Fox AP, Tsien RW, Kaczmarek LK (1987) Stimulation of protein kinase C recruits covert calcium channels in Aplysia Cells. Nature 325: 714-717

Toutant M, Aunis D, Bockaert J, Homburger V, Rouot B (1987) Presences of three pertussis toxin substrates and G_o.. immunoreactivity in both plasma and granule membranes of chromaffin cells. FEBS Lett. 215:339-344

Trautwein W, Dudel J (1958) Zum Mechanismus der Membranwirkung des Acetylcholin an der Muskelfasern. Pflügers Arch. 266: 324-330

Yatani A., Codina J, Brown AM, Birnbaumer L. (1987) Direct activation of mammalian atrial muscarinic potassium channels by GTP regulatory protein Gk. Science 235: 207-211

NEURAL AND MOLECULAR MECHANISMS OF SHORT- AND LONG-TERM SENSITIZATION IN APLYSIA

John H. Byrne
Arnold Eskin*
Kenneth P. Scholz
Department of Neurobiology and Anatomy
University of Texas Medical School
P.O. Box 20708
Houston, TX 77225
USA

Introduction

A fundamental problem in neurobiology is to understand the basic processes that endow the nervous system with the ability to store and retrieve long-term memory. A related problem is to understand the relationship between the mechanisms for the induction, storage and retrieval of short-term memory and those for long-term memory. While behavioral studies of vertebrates indicate that short-term memory and long-term memory are distinct entities (Davis and Squire, 1984), there must be some mechanistic link between the two types, since short-term memories can be consolidated into long-term ones.

In order to examine the mechanisms of long-term memory and their relationship to short-term memory, we have used a relatively simple identified neural circuit in the nervous system of Aplysia that controls defensive withdrawal of the tail and siphon. This preparation allows for the analysis of long-term cellular changes that occur in sensory neurons as a result of sensitization, a form of non-associative learning. In addition, with this preparation it is becoming possible to elucidate some of the subcellular processes that are responsible for converting short-term memory into long-term memory.

The Tail-Siphon Withdrawal Reflex

Figure 1 illustrates the basic features of the reflex. Part A shows a dorsal view of an Aplysia in a relaxed state. Application of a stimulus to the tail results in a reflex withdrawal of the tail and the siphon (Fig 1B). The neural circuit that mediates the reflex is illustrated

*Department of Biology, University of Houston, Houston, TX.

NATO ASI Series, Vol. H19
Modulation of Synaptic Transmission and Plasticity in Nervous Systems. Edited by G. Hertting and H.-C. Spatz

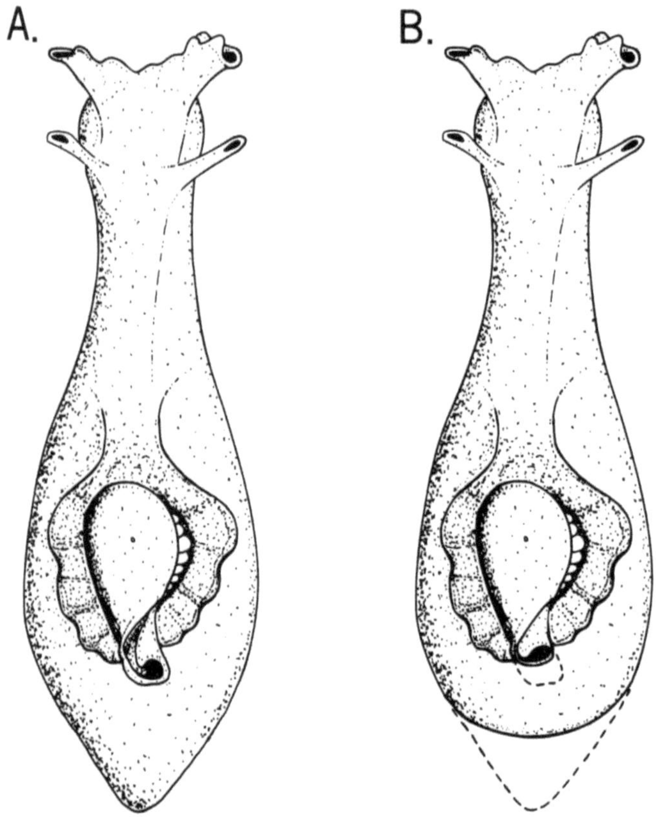

Fig. 1 Dorsal view of Aplysia. (A) relaxed (B) after stimulation of the tail.

schematically in Fig. 2. The cell bodies of the sensory neurons (SN) that innervate the tail are located in symmetric clusters in the left and right pleural ganglia (Walters et al., 1983a). When a test stimulus is applied to the tail, the sensory neurons are excited. The sensory neurons make monosynaptic excitatory connections with motor neurons (MN) that produce withdrawal of the tail. In addition to their connections with the motor neurons, the sensory neurons synapse onto various interneurons (Cleary and Byrne, 1986). Some of these interneurons (IN) activate motor neurons that control reflex withdrawal of the siphon (Fig. 2).

The tail-siphon withdrawal reflex has several unique features that facilitate the analysis of the cellular mechanisms of short- and long-term behavioral changes. First, there are two symmetric and distinct clusters of sensory neurons: one in the left pleural ganglion which innervates the left side of the tail, and one in the right pleural ganglion which innervates the right side of the tail. As we will show, it is possible to sensitize one side of the animal, while the other side can be used as a control (Ocorr et al., 1986; Walters, 1987a; Scholz and Byrne, 1987a). This aids in reducing variability between animals. Second, the clusters of sensory neurons appear to be homogeneous; they can be selectively removed from the animal for

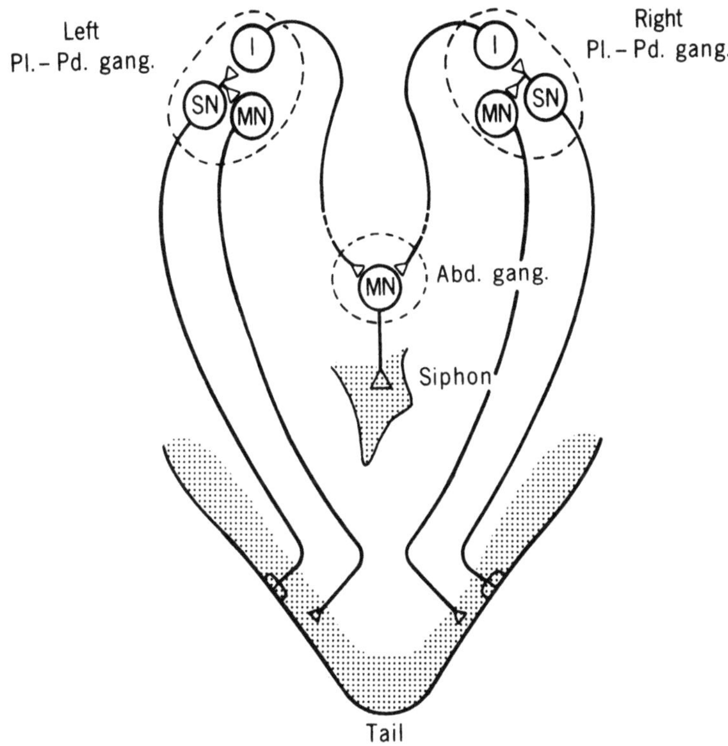

Fig. 2 Schematic diagram of the neural circuit controlling the tail-siphon withdrawal reflex.

biophysical and biochemical analysis (Walters et al., 1983a, Ocorr and Byrne, 1985; Baxter and Byrne, 1986). Third, in addition to the sensory neurons, the motor neurons that control withdrawal of the tail have been identified and are also bilaterally symmetric (Walters et al., 1983a).

Short-Term Sensitization of the Tail Withdrawal Reflex and Identification of a Cellular Locus of Plasticity.

The experimental preparation we used to examine short-term sensitization is illustrated in Fig 3. The animal was split almost in half, leaving only the tail intact. The pleural and pedal ganglia were placed in a small inner chamber so that the periphery and the ganglia could be perfused with separate solutions. Simultaneous intracellular recordings were made from sensory neurons and tail motor neurons. Mechanical or electrical stimuli were delivered to the tail and the resultant reflex contractions were measured with a tension transducer.

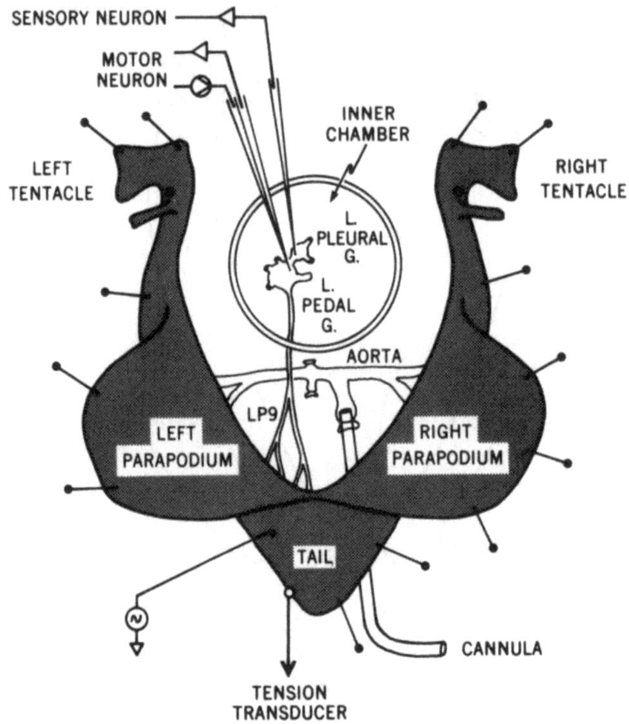

Fig. 3 Schematic diagram of the reduced tail withdrawal preparation. The mantle organs, viscera, and all the ganglia except for the left pleural and pedal ganglia (shown enlarged) were removed and the animal was completely bisected except for the tail, which was connected to a tension transducer. The aorta was cannulated for perfusion of the body and separate perfusion was used in the inner chamber for the central nervous system (CNS). A motor neuron was impaled in the pedal ganglion, while a sensory neuron was recorded from in the pleural ganglion. Mechanical or electrical stimuli were applied to the tail. (From Walters et al., 1983a).

Figure 4 shows an example of short-term sensitization of the tail withdrawal reflex and the correlates of short-term sensitization in the sensory neurons. Part A of Fig. 4 shows the tension and motor neuron recordings. Here, very weak tail shocks were repeatedly delivered to the skin at 5-minute test intervals. These weak test stimuli did not excite the particular sensory neuron being monitored (not shown). The test stimuli produced bursts of synaptic input to the motor neuron, but produced no tail withdrawal because the stimulus intensity was so weak. A sensitizing stimulus was then given to a part of the tail distant from the location where the test stimuli were applied and outside the receptive field of the sensory neuron being monitored. In response to the next test stimulus (trial 5), the tail withdrew briskly. One correlate of this sensitization was increased synaptic input and enhanced spike activity in the tail motor neuron in response to the test stimulus. Part B of Fig. 4 suggests that at least part of the increased synaptic input in the motor neuron is a result of synaptic facilitation of the monosynaptic excitatory postsynaptic potential (EPSP) from the sensory neurons. Prior to the

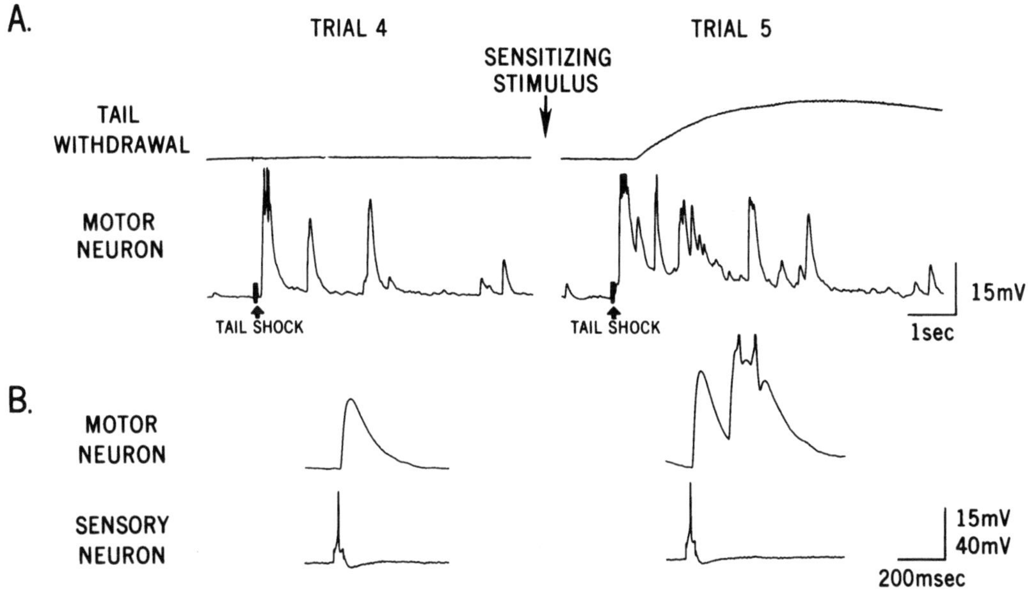

Fig. 4 Sensitization and synaptic facilitation. (A) Test responses of a tail motor neuron 3 min before (Trial 4) and 2 min after (Trial 5) brief application of a sensitizing stimulus to another part of the tail. The test stimuli were delivered at 5 min intervals through an electrode implanted in the tail. The motor neuron was hyperpolarized 30 mV from resting potential throughout. (B) EPSPs recorded in a tail motor neuron produced by a single sensory neuron spike 30 sec before and 90 sec after the sensitizing stimulus. Test EPSPs were elicited every 60 sec by intracellular activation of single sensory neuron action potentials. The short latency component is the monosynaptic EPSP. After sensitizing stimulation the sensory neuron spike evoked a larger monosynaptic EPSP and additional longer latency polysynaptic EPSPs. Spikes in the motor neuron are clipped by the pen recorder. (From Walters et al., 1983b).

sensitizing stimulus, an action potential artificially produced in the sensory neuron resulted in the monosynaptic EPSP in the motor neuron shown to the left of Part B. Immediately after the sensitizing stimulus was delivered, the EPSP from the same sensory neuron was enhanced. This is a form of synaptic plasticity called heterosynaptic facilitation, in which the connection between the two neurons is enhanced as a result of activation of a third independent pathway. Also note that in this example the sensory neuron triggered polysynaptic excitatory components that were not present prior to the sensitizing stimulus. Thus, enhancement of the connections of the sensory neurons may also increase the probability of recruiting interneuronal input to the motor neurons.

As a first step in analyzing the mechanisms underlying short-term sensitization, we attempted to identify the neurotransmitter that produces the heterosynaptic facilitation associated with sensitization. We found that application of the neurotransmitter serotonin (5-HT) to the pleural-pedal ganglia, mimics the effects of the sensitizing stimulus in that it

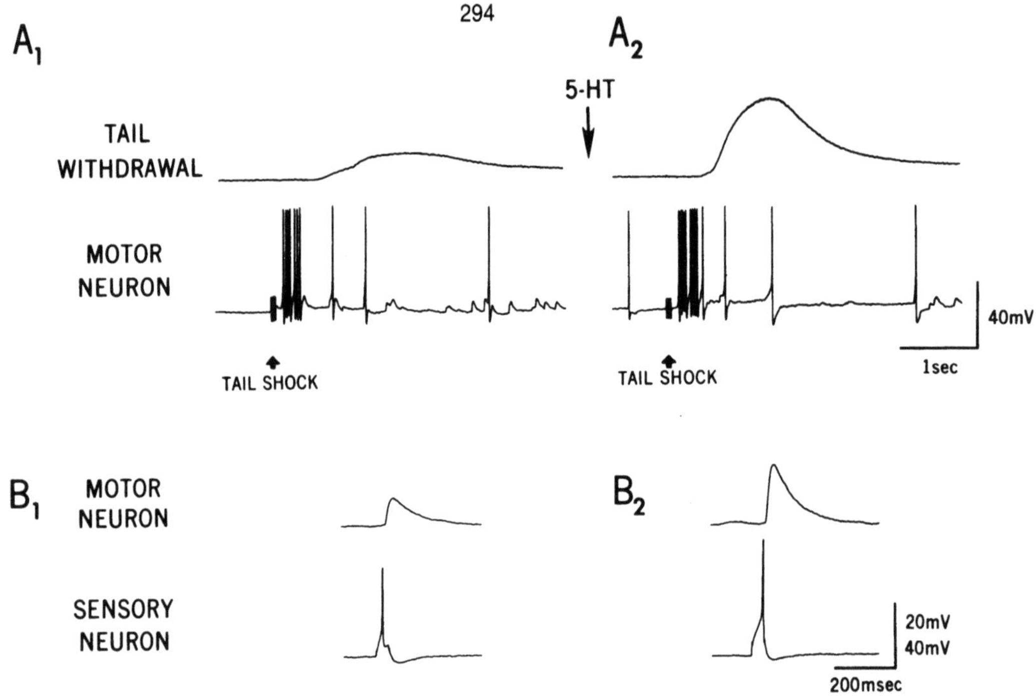

Fig. 5 Sensitization and synaptic facilitation produced by 5-HT. (A) Simultaneous recordings of the responses of the tail and a tail motor neuron to a weak tail shock (delivered outside of the receptive field of the sensory neuron) 4 min before and 12 min after superfusion of the ganglia with 5×10^{-5} M 5-HT. The 5-HT did not reach the peripheral tissues, which were bathed by a separate solution. (B) Facilitation of the monosynaptic EPSP from the sensory neuron to the motor neuron. The EPSP was tested 30 sec before each test shown in Part. A. (From Walters et al., 1983b).

leads to 1) a facilitation of the monosynaptic EPSP from the sensory neuron to the motor neuron, 2) an enhanced discharge in the motor neuron and 3) an enhancement of the tail withdrawal (Fig. 5).

Serotonin is also capable of mimicking the effects of sensitizing stimuli on an analogous reflex system in Aplysia, the gill and siphon withdrawal reflex initiated by stimuli to the siphon. That reflex is controlled by sensory neurons in the abdominal ganglion. Elevation of the intracellular messenger cAMP in the sensory neurons in the abdominal ganglion is critical for short-term synaptic facilitation that contributes to sensitization of the reflex (Castellucci et al., 1976; Brunelli et al., 1976; Klein and Kandel, 1978; Bernier et al, 1982; Klein et al., 1986). If cAMP is involved in synaptic facilitation of the tail sensory neurons, one would expect that sensitizing stimulation of the animal would lead to changes in the levels of cAMP in these cells. Figure 6 illustrates a test of this hypothesis. The animal and its nervous system were completely bisected (Part A). Following delivery of a sensitizing stimulus to one side of the animal, both clusters of sensory neurons were removed surgically and assayed for cAMP levels. Part B illustrates that there is a significant increase in the cAMP content in the sensory neuron cluster from the stimulated side compared to the cluster from the unstimulated control side.

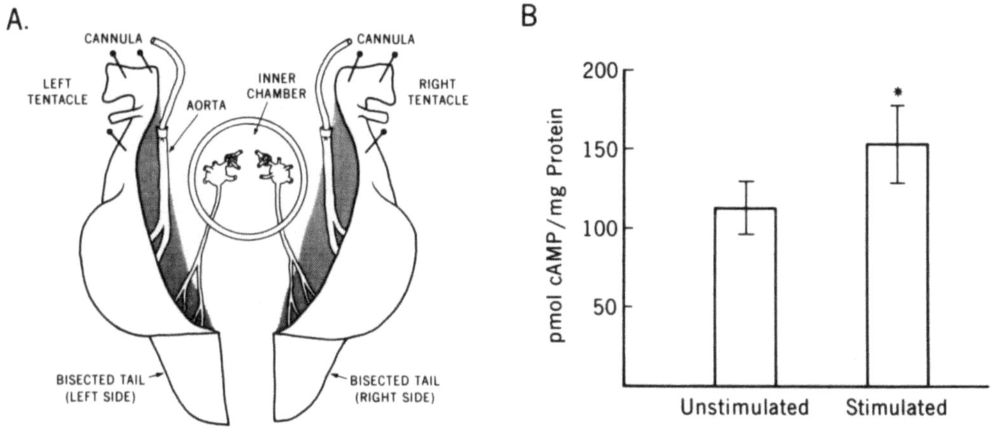

Fig. 6 Effects of sensitization on cAMP levels in the sensory neurons. (A) The preparation used to analyze the effects of sensitization training on the intracellular levels of cAMP in the sensory neurons. The preparation is similar to that shown in Fig. 3 except the animal is completely bisected and the right pleural and pedal ganglia are left intact. (B) One side of the bisected animal was designated as the stimulated side and received four sets of mechanical stimuli, which were delivered to the tail at 30 sec. intervals. Thirty seconds after the last set of stimuli, both sets of pleural-pedal ganglia were simultaneously frozen in a propylene glycol mixture and the sensory neuron clusters removed and analyzed for cAMP content. (From Ocorr et al., 1986).

These results demonstrate that the same stimuli that produce sensitization in Aplysia, also produce a measurable elevation of cAMP levels in the tail sensory neurons (Ocorr et al., 1986).

As mentioned above, 5-HT has been implicated as a facilitatory transmitter in the tail withdrawal reflex (Walters et al., 1983b; Pollock et al., 1985; Walsh and Byrne, 1984). In addition, 5-HT mimics the ability of the sensitizing stimulus to elevate cAMP in the tail sensory neurons (Ocorr and Byrne, 1985). As illustrated in Fig. 7, cAMP levels were measured in surgically isolated sensory neuron clusters. Clusters exposed to 5-HT showed a significant enhancement in cAMP content compared to their contralateral controls.

Some aspects of the mechanisms by which 5-HT exerts its effects are illustrated in Fig. 8. Serotonin has four observable, but probably related, effects on the sensory neurons and their EPSPs. First, 5-HT increases the size of the monosynaptic EPSP from the sensory neurons to the motor neurons. Second, it produces a slow and long-lasting depolarization of the sensory neuron. This depolarization is associated with a decrease in membrane input conductance. Third, there is an increase in the excitability of the sensory neuron, since the same constant current pulse that produced only a single spike prior to 5-HT, produces a burst of 3 action potentials after application of 5-HT. Fourth, 5-HT enhances the amplitude and duration of the sensory neuron action potential. This effect is more clearly illustrated in

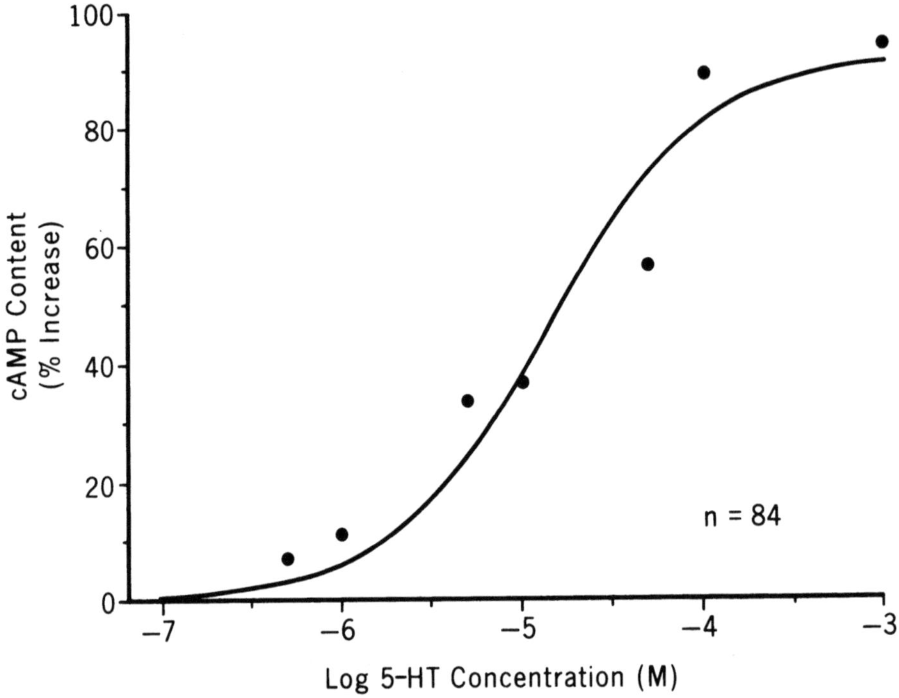

Fig. 7 Dose-response relationship of 5-HT effects on cAMP levels. The effect of 5-HT on the cAMP content of sensory neuron clusters is expressed as a percent of the cAMP content of the contralateral cluster receiving a control treatment. Each point represents the average percent increase determined for 12-13 animals. The estimated ED_{50} is 14 uM. (From Ocorr and Byrne, 1985)

Part B at a faster time base. This increase in the amplitude and duration of the action potential may be critical in enhancing Ca^{2+} influx in the synaptic terminals of the sensory neurons and facilitating transmitter release.

Voltage-clamp and ion substitution experiments indicate that both the slow depolarization and broadening of the spike produced by 5-HT, are due to decreases in K^+ currents of the sensory neurons (Walsh and Byrne, 1984; Pollock et al., 1985; Baxter and Byrne, 1986; see also Klein and Kandel, 1980). A number of experiments indicate that 5-HT alters these membrane currents through elevation of intracellular cAMP in the sensory neurons (Ocorr and Byrne, 1985; Walsh and Byrne, 1984; Pollock et al., 1985; Baxter and Byrne, 1986, 1987; see also Klein and Kandel, 1978, 1980).

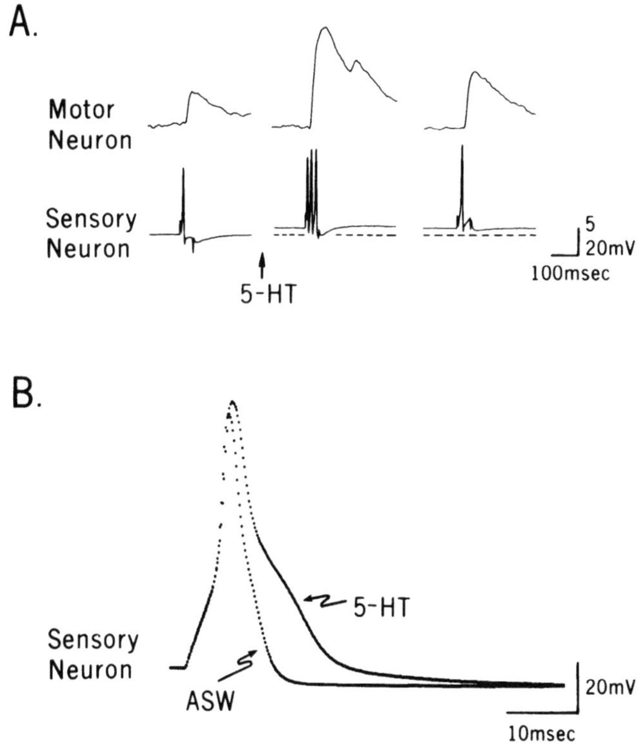

Fig. 8 Serotonin effects on the tail sensory neurons. (A) Effects of 5-HT on membrane potential and amplitude of the monosynaptic EPSP in the semi-intact preparation. Constant current suprathreshold depolarizing pulses (50 msec) were injected into the sensory neuron at 60 sec intervals. After 10^{-4} M 5-HT was perfused into the chamber the brief depolarizing test pulse elicited more spikes. The stimulus current was then decreased so that only single action potentials were elicited. (From Walters et al, 1983b). (B) In an isolated sensory neuron perfused with artificial seawater (ASW), an action potential was initiated by a brief depolarizing current pulse. After application of 30 x 10^{-6} M 5-HT a second action potential was initiated. In the presence of 5-HT the action potential has a greater amplitude and its duration is increased significantly (Baxter and Byrne, unpublished observations).

Long-Term Sensitization

Following a single sensitizing stimulus, synaptic facilitation of the sensory neuron to motor neuron connection lasts about 15-30 minutes. Multiple training sessions increase the magnitude of the facilitation as well as its duration. Long-term sensitization (lasting at least 24 hrs) of the siphon withdrawal reflex has also been demonstrated (Pinsker et al., 1973). Although long-term sensitization of the siphon withdrawal reflex was found to be associated with synaptic facilitation (Frost et al., 1985), until recently, little was known about the cellular changes that accompany long-term sensitization. We have begun to study long term sensitization in the tail-siphon withdrawal reflex in an effort to elucidate further the cellular and molecular mechanisms that underly simple forms of long-term memory.

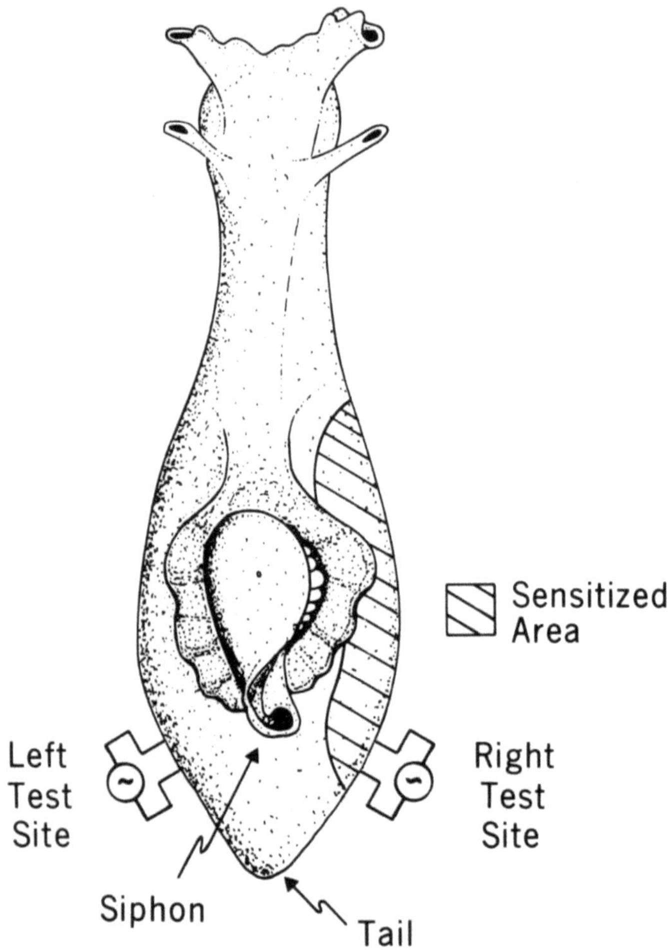

Fig. 9 Preparation used for behavioral analysis of long-term sensitization. (See text for details).

The behavioral effects of long-term sensitization were examined by observing withdrawal of the siphon elicited by stimulation of the tail. The experimental procedure is illustrated schematically in figure 9. For behavioral analysis, both sides of each animal were initially tested by the delivery of weak electric shock to the posterior part of the body wall. The test stimuli produced a coordinated set of defensive responses, including reflex withdrawal of the tail and siphon (see also Fig. 1). We used the siphon component of this reflex as a measure of the sensitivity of the animal to the test stimuli and changes in the siphon response, relative to the pretest responses, as a measure of sensitization.

Immediately after initial testing, one randomly chosen side of the animal was subjected to sensitization training which consisted of 4 trains of shocks delivered over a 2 hour period.

The other side of the animal served as the untrained control. Twenty-four hours after sensitization training, the animal was again tested.

For each animal, the averages of the test scores before and after training were calculated, and the percent change in the average test score after training relative to the average test score before training was determined for each side. As a result of sensitization training, there is a significant enhancement of the reflex that persists for at least 24 hours.

Since the behavioral effects of long-term sensitization training were lateralized, we were able to examine biophysical correlates of long-term sensitization using each animal as its own control. Specifically, we compared the membrane properties of neurons mediating the response induced by test stimuli to one side, with the membrane properties of the neurons mediating the response to the opposite side. Since the tail sensory neurons had been shown to be a cellular locus for the changes that accompany short-term sensitization, we began by examining these same sensory neurons to determine their role in the cellular changes underlying long-term sensitization. In addition, the sensory neurons serve as the first central relay of the response pathway (as shown in Fig. 2), making these cells a logical place to begin.

Immediately after the behavioral analysis described above (24 hours after sensitization training), the left and right pleural-pedal ganglia were removed from the animal so that the tail sensory neurons could be used for a biophysical analysis. In all experiments, sensory neurons were voltage-clamped at a holding potential of -50 mV, and the current-voltage relationship of each cell examined. Fig. 10 illustrates averaged families of voltage-clamp responses recorded from sensory neurons innervating the control side (A) and the sensitized side (B). These measurements were made 24 hours after the animal had been sensitized. The consequences of long-term sensitization in the sensory neurons include a reduction in net outward membrane current. Thus, long-term sensitization, like short-term sensitization, is correlated with a reduction of net outward membrane current in the sensory neurons. These results indicate that the sensory neurons are not only a cellular locus for short-term sensitization as described above, but also serve as a cellular locus for long-term sensitization.

Further insight into the nature of the net membrane current that is reduced in a long-term fashion as a consequence of sensitization training, was obtained by subtraction of the response families of cells innervating the sensitized side, from the response families of cells innervating the control side. This procedure yielded the sensitization difference current (Fig 10C). There is some similarity between the currents modulated by 5-HT in the short-term (Klein and Kandel, 1980; Baxter and Byrne, 1986) and the currents altered in a long-term fashion by sensitization. The presence of the long-term sensitization difference current at membrane potentials near the resting potential of the cell, indicate that the functional consequences of this reduction in net membrane current may include enhancement of

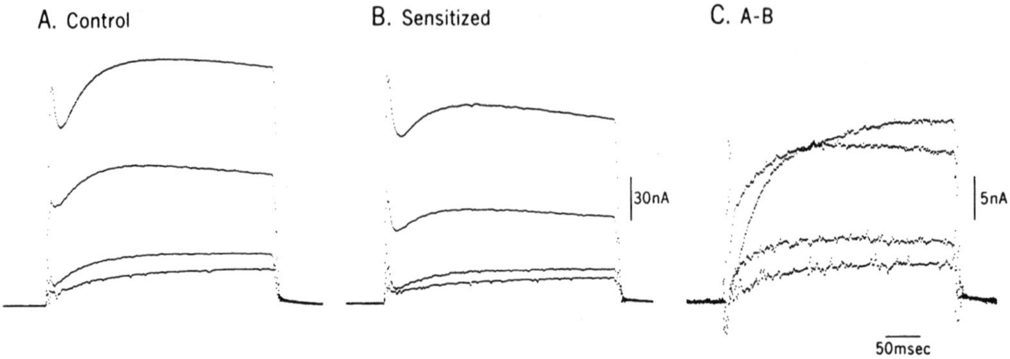

Fig. 10 Long-term sensitization training alters membrane properties of sensory neurons. Membrane currents in response to a series of different voltage pulses were recorded and digitized. Thus, for each cell, a response family was generated. The response families of the sensory neurons in each cluster were then averaged by computer. The overall average of control (A) and sensitized (B) sides were then obtained by averaging the characteristic response families of the respective clusters. The data in (C) were obtained by subtracting the response family of the sensitized side of each animal from its corresponding control family. The results for all of the animals were then averaged. Thus, the family of traces in (C) represent the net outward currents that are reduced 24 hours after sensitization training. (From Scholz and Byrne, 1987a).

excitability of the sensory neurons and an increase in the probability of repetitive firing. These conclusions are supported by the findings of Walters (1987b) and Dale et al. (1987).

Subcellular Processes Contributing to Long-Term Sensitization.

The results presented above indicate that the sensory neurons are a locus for both short- and long-term memory for sensitization. While many of the cellular changes that occur in the sensory neurons and that contribute to short-term sensitization are triggered by an elevation of cAMP, a major unresolved question is the identitiy of the intracellular signal that triggers the long-lasting cellular changes. A number of studies indicate that cAMP, in addition to promoting protein phosphorylation through the activation of a protein kinase, may also participate in the regulation of protein synthesis (Jungmann and Kranias, 1977; Murdoch et al., 1982; Mangiarotti et al., 1983; Waterman et al., 1985). Thus, one hypothesis for the induction of long-term sensitization is that the transient elevation of cAMP during sensitization training, may trigger other cellular processes that produce the long-lasting changes in the sensory neurons. To test this hypothesis, we directly injected cAMP into sensory neurons in order to mimic the effects of sensitization training at the level of single sensory neurons. The long-lasting effects of cAMP were assessed by analyzing the membrane currents of the sensory neurons 24 hours after injection of cAMP.

The results of these experiments indicate that a transient elevation of intracellular cAMP in the sensory neurons, can induce a long-term reduction in K^+ currents in those cells (Scholz and Byrne, 1987b). Furthermore, the membrane current that was reduced in a long-term fashion by cAMP, was very similar to that previously found to be reduced after 24 hours as a consequence of sensitization training. These findings suggest that cAMP is one of the critical intracellular signals that triggers the long-lasting cellular changes that underly long-term sensitization.

A fundamental question that remains, however, is how cAMP mediates these long-term effects. Studies by Montarolo et al. (1986), Dale et al. (1987) and Schacher et al. (1987) point to a critical role for protein synthesis in producing long-term synaptic facilitation and increases in excitability of sensory neurons in culture. To further assess the role of protein synthesis in long-term sensitization, we have begun to examine patterns of protein synthesis in the sensory neurons and how they are altered by treatments that mimic sensitization training.

Using two-dimensional polyacrylamide gel electrophoresis (2D-PAGE), we found that prolonged (2 to 3 hr) exposure to the adenylate cyclase activator forskolin, leads to changes in incorporation of labeled amino acid into several proteins from isolated clusters of sensory neurons. Furthermore, the same proteins that are affected by treatment with forskolin are also altered by 5-HT. These results are consistent with a role for protein synthesis in the long-term effects of cAMP and 5-HT. However, they do not rule out the possibility that other mechanisms are involved.

Conclusion

A model that summarizes some of the neural and molecular mechanisms contributing to short-and long-term sensitization is illustrated in Fig. 11. Sensitizing stimuli lead to the activation of facilitatory neurons that release a neurotransmitter with properties similar to, or the same as, serotonin. This transmitter acts on the sensory neurons to increase the levels of cAMP, which in turn leads to closure of membrane K^+ channels through protein phosphorylation. Closure of these K^+ channels leads to membrane depolarization and an enhancement of the excitability of the sensory neurons. An additional consequence of reduced K^+ channel activity is a reduction of the repolarizing current during an action potential. Thus, action potentials initiated in the sensory neurons are broader. This may serve to enhance Ca^{2+} influx in the presynaptic terminal and increase transmitter release from the sensory neurons.

The mechanisms for the long-term changes are not yet fully understood. However, it appears that the same intracellular messenger that leads to the expression of the short-term changes, also triggers cellular processes that lead to the long-term changes. One possible mechanism for this action is through the regulation of protein synthesis by cAMP. Our results

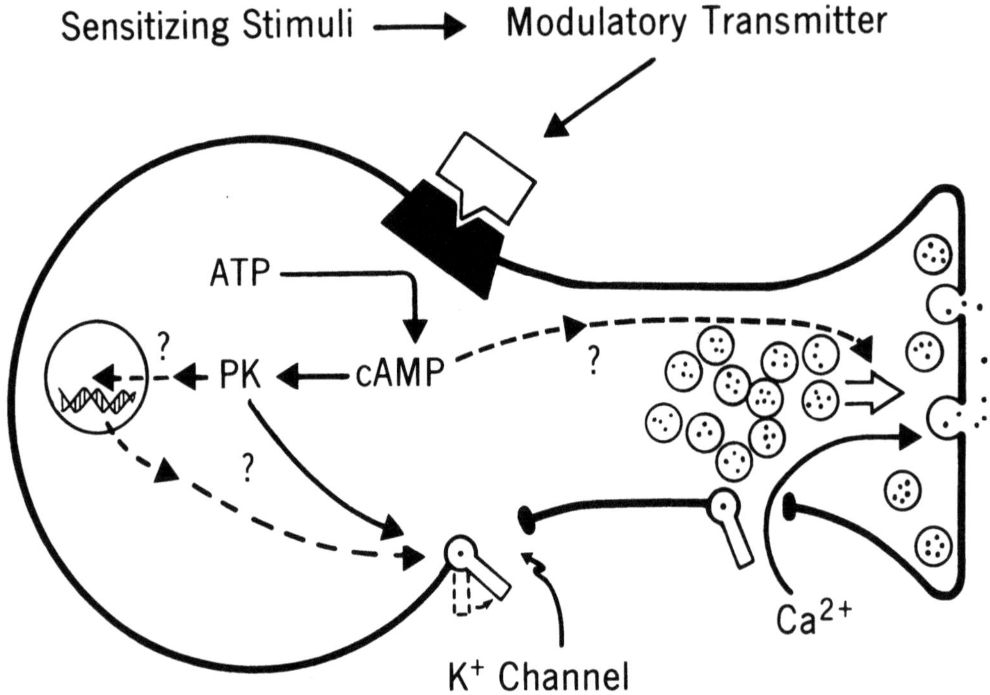

Fig. 11 Model of heterosynaptic facilitation that contributes to sensitization. (See text for details).

with protein analysis using 2D-PAGE, as well as experiments using protein synthesis inhibitors (Montarolo et al., 1986) support a role for this mechanism. It is also possible that the long-term changes in membrane currents could be due at least in part to changes in the ratio of subunits of the cAMP-dependent protein kinase (Greenberg et al., 1987).

In addition to the long-term biochemical and biophysical changes that we have described, morphological changes have been observed in the synaptic terminals of the sensory neurons as a consequence of long-term sensitization (Bailey and Chen, 1983). This may contribute to the synaptic facilitation that has been observed under similar conditions (Frost et al., 1985; Montarolo et al., 1986). Thus, both short- and long-term sensitization are associated with a number of cellular alterations that appear to act in concert to increase the excitability of the sensory neurons and enhance transmitter release, thereby producing an enhanced withdrawal response.

Acknowledgements

Supported by National Institute of Mental Health Award K02 MH00649, National Institutes of Health Research Grant NS 19895 and National Institute of Mental Health Grant MH 41979.

REFERENCES

Bailey C. H., Chen M. (1983) Morphological basis of long-term habituation and sensitization in Aplysia. Science 220: 91-93.

Baxter, D. A., Byrne, J. H. (1986) Serotonin- modulated membrane currents in Aplysia tail sensory neurons. Soc. Neurosci. Abstr. 12: 765.

Baxter, D. A., Byrne, J. H. (1987) Modulation of membrane currents and excitability by serotonin and cAMP in pleural sensory neurons of Aplysia. Soc. Neurosci. Abstr. 13: 1440.

Bernier L., Castellucci V. F., Kandel E. R., Schwartz, J. H. (1982) Facilitatory transmitter causes a selective and prolonged increase in adenosine 3':5'-monophosphate in sensory neurons mediating the gill and siphon withdrawal reflex in Aplysia. J. Neurosci. 2: 1682-1691.

Brunelli M., Castellucci V. F., Kandel E. R., (1976) Synaptic facilitation and behavioral sensitization in Aplysia: Possible role of serotonin and cyclic AMP. Science 194: 1178-1181.

Byrne J. H. (1987) Cellular analysis of associative learning. Physiol. Rev. 67: 32- 439.

Castellucci V. F., Kandel E. R. (1976) Presynaptic facilitation as a mechanism for behavioral sensitization in Aplysia. Science 194: 1176-1178.

Cleary, L. J., Byrne, J. H. (1986) Associative learning of the gill and siphon withdrawal reflex in Aplysia: Interneurons mediating the unconditioned response. Soc. Neurosci. Abstr. 12: 397.

Dale N., Kandel E. R., Schacher S., (1987) Serotonin produces long-term changes in the excitability of Aplysia sensory neurons in culture that depend on new protein synthesis. J. Neurosci. 7: 2232-2238.

Davis H. P., Squire L. R., (1984) Protein synthesis and memory: A review. Psychol. Bull. 96: 518-559.

Frost W. N., Castellucci V. F., Hawkins R. D., Kandel, E. R. (1985) Monosynaptic connections made by the sensory neurons of the gill- and siphon-withdrawal reflex in Aplysia participate in the storage of long-term memory for sensitization. Proc. Natl. Acad. Sci. USA 82: 8266-8269.

Greenberg S. M., Castellucci V. F., Bayley H., and Schwartz, J. H. (1987) A molecular mechanism for long-term sensitization in Aplysia. Nature 329: 62-65.

Jungmann, R. A., Kranias, E. G. (1977) Nuclear phosphoprotein kinases and the regulation of gene expression. Int. J. Biochem. 8: 819-830.

Klein M., Hochner B., Kandel E. R. (1986) Facilitatory transmitters and cAMP can modulate accommodation as well as transmitter release in Aplysia sensory neurons: Evidence for parallel processing in a single cell. Proc. Natl. Acad. Sci. USA. 83: 7994-7998.

Klein M., Kandel E. R. (1978) Presynaptic modulation of voltage-dependent Ca^{2+} current: Mechanism for behavioral sensitization in Aplysia californica. Proc. Natl. Acad. Sci. USA 75: 3512-3516.

Klein M., Kandel E. R. (1980) Mechanism of calcium current modulation underlying presynaptic facilitation and behavioral sensitization in Aplysia. Proc. Natl. Acad. Sci. USA 77: 6912-6916.

Mangiarotti G., Ceccarelli A., Lodish H. (1983) Cyclic AMP stabilizes a class of developmentally regulated Dictostelium discoideum mRNAs. Nature 301: 616-618.

Maurer R. A. (1981) Transcriptional regulation of the prolactin gene by ergocryptine and cyclic AMP. Nature 294: 94-97.

Montarolo P. G., Goelet P., Castellucci V. F., Morgan, J., Kandel, E. R., Schacher, S. (1986) A critical period of macromolecular synthesis in long-term heterosynaptic facilitation in Aplysia. Science 234: 1249-1254.

Murdoch G. H., Rosenfeld M. G., Evans R. M. (1982) Eukaryotic transcriptional regulation and chromatin-associated protein phosphorylation by cyclic AMP. Science 218: 1315-1317.

Ocorr K. A., Byrne J. H. (1985) Membrane responses and changes in cAMP levels in Aplysia sensory neurons produced by 5-HT, tryptamine, FMRFamide and SCP_B. Neurosci. Letts. 55: 113-118.

Ocorr K. A., Tabata M., Byrne J. H. (1986) Stimuli that produce sensitization lead to elevation of cyclic AMP levels in tail sensory neurons of Aplysia. Brain Res. 371: 190-192.
Pinsker H. M., Hening W. A., Carew T. J., Kandel, E. R. (1973) Long-term sensitization of a defensive withdrawal reflex in Aplysia Science 182: 1039-1042.
Pollock, J. D., Bernier, L., Camardo, J. S. (1985) Serotonin and cyclic adenosine 3':5'-monophosphate modulate the S potassium current in tail sensory neurons in the pleural ganglion of Aplysia. J. Neurosci. 5: 1862-1871.
Schacher S., Castellucci, V. F., Kandel, E. R. (1987) Exogenous cyclic AMP analog produces long-term facilitation of Aplysia sensorimotor synapses in culture that is blocked by the protein synthesis inhibitor anisomycin. Soc. Neurosci. Abstr. 13: 390. ,
Scholz K. P., Byrne J. H. (1987a) Long-term sensitization in Aplysia: Biophysical correlates in tail sensory neurons. Science 235: 685-687.
Scholz K. P., Byrne J. H., (1987b) cAMP leads to long-term (24 hr) changes in membrane current in Aplysia tail sensory neurons: A possible mechanism for long-term sensitization. Soc. Neurosci. Abstr. 13: 390.
Thompson R. F., Berger T. W., Madden J. (1983) Cellular processes of learning and memory in the mammalian CNS. Annu. Rev. Neurosci. 6: 447-491.
Walsh, J. P., Byrne, J. H. (1984) Forskolin mimics and blocks a serotonin-sensitive decreased K^+ conductance in tail sensory neurons of Aplysia. Neurosci. Lett. 52: 7-11.
Walters E. T. (1987a) Site-specific sensitization of defensive reflexes in Aplysia: A simple model of long-term hyperalgesia. J. Neurosci. 7: 400-407.
Walters E. T. (1987b) Multiple sensory neuronal correlates of site-specific sensitization in Aplysia. J. Neurosci. 7: 408-417.
Walters E. T., Byrne J. H., Carew T. J., Kandel, E. R. (1983a) Mechanoafferent neurons innervating tail of Aplysia. I. Response properties and synaptic connections. J. Neurophysiol. 50: 1522-1542.
Walters E. T., Byrne J. H., Carew T. J., Kandel, E. R. (1983b) Mechanoafferent neurons innervating tail of Aplysia. II. Modulation by sensitizing stimulation. J. Neurophysiol. 50: 1543-1559.
Waterman M., Murdoch G. H., Evans R. M., Rosenfeld, M. G. (1985) Cyclic AMP regulation of eukaryotic gene transcription by two distinct molecular mechanisms. Science 229: 267-269.

MODULATION OF NEUROMUSCULAR TRANSMISSION IN THE LOCUST BY FMRFAMIDE-LIKE PEPTIDES

Peter D. Evans, Sandra Robb and Brian A. Cuthbert
AFRC Unit
Department of Zoology
University of Cambridge
Downing Street
Cambridge CB2 3EJ
Great Britain

The levels of modulatory substances present in the local hormonal environment of a synapse can clearly affect the efficacy of synaptic transmission between the presynaptic and postsynaptic neurone. Modulatory compounds may arrive at their target sites either as circulating neurohormones or as locally released hormones, but may also be released as cotransmitters from anatomically defined synaptic connections. Many of the chemical effectors present in the nervous system, such as particular peptides or biogenic amines, may function as either neurotransmitters, neuromodulators or neurohormones in different contexts. Thus for any specific chemical effector, its functional role must be studied individually in relation to the particular synaptic system under consideration and, wherever possible, in relation to the effects of uniquely identifiable neurones.

Invertebrate nervous systems have many advantages for the study of basic synaptic mechanisms because of the large size of many of their neurones and the ease with which individual neurones can be identified from one preparation to the next. Insect skeletal muscles, for instance, are frequently innervated by small numbers of motor neurones, all of which can be uniquely identified physiologically (Hoyle,1983). They, therefore, present excellent model systems in which to study the basic principles of neuromuscular transmission and the modulation of these effects by biogenic amines and peptides (Evans, 1980; O'Shea, 1985;O'Shea and Schaffer, 1985; Evans and Myers,1986a). These modulatory effects can be studied in detail all the way from the pharmacology of the receptors mediating the effects to the molecular details of their modes of action.

NATO ASI Series, Vol. H19
Modulation of Synaptic Transmission and Plasticity
in Nervous Systems. Edited by G. Hertting and H.-C. Spatz

Table 1. Invertebrate FMRFamide Peptide Family

Met-Enkephalin	Tyr-Gly-Gly-Phe-Met		
	Tyr-Gly-Gly-Phe-Met-Arg-Phe-NH_2	Octopus	1
FMRFamide	Phe-Met-Arg-Phe-NH_2		2
FLRFamide	Phe-Leu-Arg-Phe-NH_2	Pomacea	3
	pGlu-Asp-Pro-Phe-Leu-Arg-Phe-NH_2	Helix	3
	Ser-Asp-Pro-Phe-Leu-Arg-Phe-NH_2	Helix, Lymnea	3
	Gly-Asp-Pro-Phe-Leu-Arg-Phe-NH_2	Lymnea	3
SCP_B	Met-Asn-Tyr-Leu-Ala-Phe-Pro-Arg-Met-NH_2	Aplysia	4
	pGlu-Asp-Val-Asp-His-Val-Phe-Leu-Arg-Phe-NH_2	Leucophaea	5
Antho RFamide	pGlu-Gly-Arg-Phe-NH_2	Anemone	6
L5 peptide	...?..Arg-Pro-Gln-Gly-Arg-Phe-NH_2	Aplysia	7

(1) Voigt and Martin (1986) (2) Price and Greenberg (1977) (3) Price et al (1987) (4) Morris et al (1982) (5) Holman, Cook and Nachman (1986) (6) Grimmelikhuijzen and Graff (1986) (7) Shyamala, Fisher and Scheller (1986).

Table 2. Vertebrate FMRFamide Peptide Family

Tyr-Gly-Gly-Phe-Met	Met-Enkephalin	
Tyr-Gly-Gly-Phe-Met-Arg-Phe-OH	Striatum, Adrenals	1
Phe-Met-Arg-Phe-NH_2	FMRFamide	
Phe-Leu-Arg-Phe-NH_2	FLRFamide	
Leu-Pro-Leu-Arg-Phe-NH_2	Chicken Brain	2
Phe-Leu-Phe-Gln-Pro-Gln-Arg-Phe-NH_2	Bovine Brain	3
...(+10) ...Ser-Leu-Ala-Ala-Pro-Gln-Arg-Phe-NH_2	Bovine Brain	3
..................-Phe-Arg-Trp-Asp-Arg-Phe-NH_2	MSH	
..................-Met-Gly-Trp-Met-Asp-Phe-NH_2	CCK octapeptide	
..................-Ile-Thr Arg- Glu -Arg-Tyr-NH_2	NPY	
..................-Leu-Thr-Arg-Pro-Arg-Tyr-NH_2	BPP	

(1) Stern et al. (1979) (2) Dockray et al (1983) (3) Yang, Fratta, Majane and Costa (1985)

In the present paper we will review the modulatory actions of the neuropeptide FMRFamide (Phe-Met-Arg-Phe-NH_2) and related peptides on the functioning of skeletal muscle in the locust Schistocerca gregaria. FMRFamide was originally isolated and identified from the nervous system of the clam Macrocallista nimbosa (Price and Greenberg,1977). Since that time FMRFamide, and a family of structurally related peptides,have been shown to be present in nervous tissue from a wide range of invertebrate species (see Table 1). Structurally related peptides have also been reported to be present in vertebrate nervous tissue (see Table 2).

In the locust nervous system immunocytochemical evidence has been presented for the presence of peptides related to FMRFamide in specific subsets of neurones in the ganglia of the ventral nerve cord (see Fig 1) (Myers and Evans, 1985a,b) and brain (Myers and Evans, 1987). No immunoreactive processes could be detected leaving the ganglia in the paired lateral nerves, but an extensive series of immunoreactive processes was detected in the neurohaemal organs of the median nerves from the thoracic ganglia. Recently extracts of the neurohaemal organs from the locust metathoracic ganglia have been shown to mimick the modulatory actions of FMRFamide-like peptides on locust skeletal muscle (Walther and Schiebe, 1987). Analysis of serial sections of locust thoracic ganglia stained with antibodies raised against FMRFamide revealed that axons from a paired group of posterior lateral cells in each ganglion could be traced into the median nerve, where some of their processes end as neurosecretory terminals in the median neurohaemal organs. Other processes continue further along the median nerve towards the spiracle muscle giving off branches that ramify over the surface of identified skeletal muscles (Myers and Evans, 1985a,b). The presence of FMRFamide-like peptides in the locust nervous system has now been confirmed by radioimmunoassay (see Table 3) and the data agrees well with our previous immunocytochemical findings (Robb and Evans, in preparation). At present we do not know the chemical structure of the endogenous ligands responsible for the FMRFamide immunoreactivity in the locust nervous system but this is currently under investigation.

The immunocytochemical evidence suggested to us that locust skeletal muscle might be an important target site for the actions of FMRFamide-like peptides in the locust (Myers and Evans, 1985a,b). However, the muscles

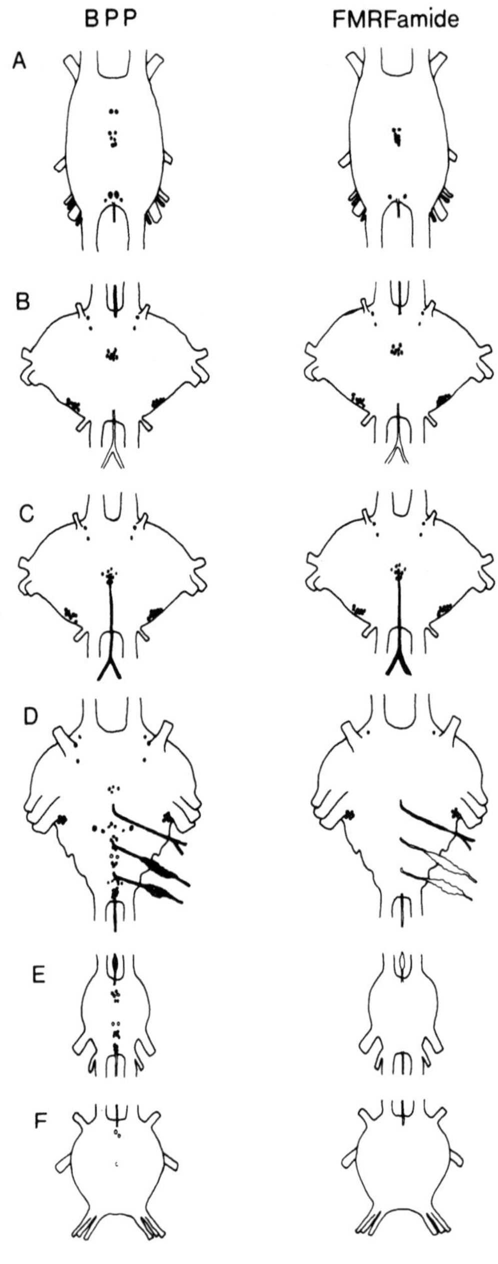

Fig.1. A-F. Diagram of two ventral nerve cords in dorsal view comparing the distributions of cell bodies immunoreactive with BPP antiserum, on the left, and with FMRFamide antiserum, on the right. A, suboesophageal ganglion. B, prothoracic ganglion, C, mesothoracic ganglion, D, metathoracic ganglion, E, 2nd abdominal ganglion; the 3rd and 4th ganglia are all similar. F, terminal 5th abdominal ganglion. Filled (●) circles represent ventral cell bodies; open (o) circles represent dorsal cell bodies. (From Myers and Evans, 1985b).

Table 3. FMRFamide-like Immunoreactivity in ganglia from the nervous system of the locust (2-7 day old adult males) expressed as fmoles/structure.

TISSUE	MEAN	S.E.	n
Brain	421.5	86.3	8
Optic lobe	132.4	32.9	8
Suboesophageal	71.2	14.7	9
Prothoracic	134.0	25.5	9
Mesothoracic	127.9	21.2	9
Metathoracic	117.7	20.7	9
Abd.N.C.	51.0	9.9	6
Terminal Abd.	39.8	13.9	5
Heart	151.8	44.9	5

directly displaying immunoreactivity, such as the spinopleural muscle, the lateral oblique intersegmental muscle and the pleural subalar muscle, are small muscles that degenerate after afew days of adult life. Thus we decided to investigate the modulatory actions of FMRFamide- like peptides on the extensor tibiae muscle of the locust hindleg,since this muscle had been used extensively for the study of the modulatory actions of octopamine in the locust (Evans, 1985). Some of the modulatory effects of FMRFamide-like peptides on this preparation (Walther, Schiebe and Voigt, 1984; Evans and Myers, 1986a,b) are described below. In vivo these effects are likely to be mediated via FMRFamide-like peptides released into the circulating haemolymph of the locust, since none of the four neurones known to innervate this muscle are immunoreactive to antibodies raised against FMRFamide (see Myers and Evans, 1985a,b).

In the locust extensor tibiae muscle preparation FMRFamide-like peptides can modulate twitch tension induced by the firing of the slow motor neurone (SETi) to this muscle (Walther et al 1984; Evans and Myers, 1986a,b) and also alter basal tension in the muscle, but they do not alter fast motor neurone generated twitch tension or affect the frequency of a myogenic rhythm of contraction and relaxation found in a proximal bundle of muscle fibres (Evans and Myers, 1986a,b). Thus the extensor muscle displays differential effects in different regions of the muscle with the largest effects of these peptides occurring in the slow muscle fibres located in the most distal regions of the muscle. Fig 2 shows that a 5-min pulse of either FMRFamide, or the structurally related peptide, YGGFMRFamide, at a concentration of 10^{-6}M increases the amplitude, contraction rate and relaxation rate of SETi-induced twitch tension. The responses to YGGFMRFamide develop more rapidly than those due to FMRFamide itself and saturate after a 2-min exposure. Shorter pulses in the region of 30s produce effects that persist for several minutes after the removal of the peptide from the superfusate. Dose- response curves indicate that YGGFMRFamide has a threshold between 10^{-9} and 10^{-10}M for a 5-min exposure for all the three parameters measured and that the dose-response curves for all three parameters are superimposible with a roughly equal maximal effect of 200% increases occurring at 10^{-6}M. FMRFamide is almost one order of magnitude less potent than YGGFMRFamide. Increasing the frequency of stimulation of the slow motor neurone indicates that just as for the effects of octopamine (Evans and Siegler, 1982) and proctolin (Evans, 1982) on this muscle, FMRFamide has a frequency-dependent effect on maintained tension with maximal increases in slow contractions occurring between 5 and 6 Hz (Fig 2B) (Evans and Myers, 1986a,b). Thus FMRFamide has a very similar effect to proctolin on basal maintained tension with both peptides increasing it rather than decreasing it as is the case with octopamine. In addition FMRFamide increases the amplitude, contraction rate and relaxation rate of twitch tension as does octopamine. FMRFamide, however, increases all three parameters to the same extent, whilst octopamine produces a much more marked increase in relaxation rate than in the other two parameters. In addition the effects of FMRFamide appear to follow a much slower time course and are not blocked by phentolamine (Evans and Myers, 1986b). FMRFamide and YGGFMRFamide produce some of their effects presynaptically on the terminals of SETi since YGGFMRFamide does not increase the depolarizations caused by glutamate, the putative rapidly acting neurotransmitter of SETi (Walther et al 1984) but both peptides increase the frequency, but not the amplitude of spontaneous miniature end-plate potentials from SETi (Evans and Myers, 1986b). A presynaptic site of action for FMRFamide would be consistent with its increasing EJP amplitude,twitch tension amplitude and the rate of contraction of twitch tension. However, it is difficult to reconcile this with an increase in the rate of relaxation of twitch tension, since this is presumably related to the rate at which Ca^{++} ions are taken up by the muscle sarcoplasmic reticulum

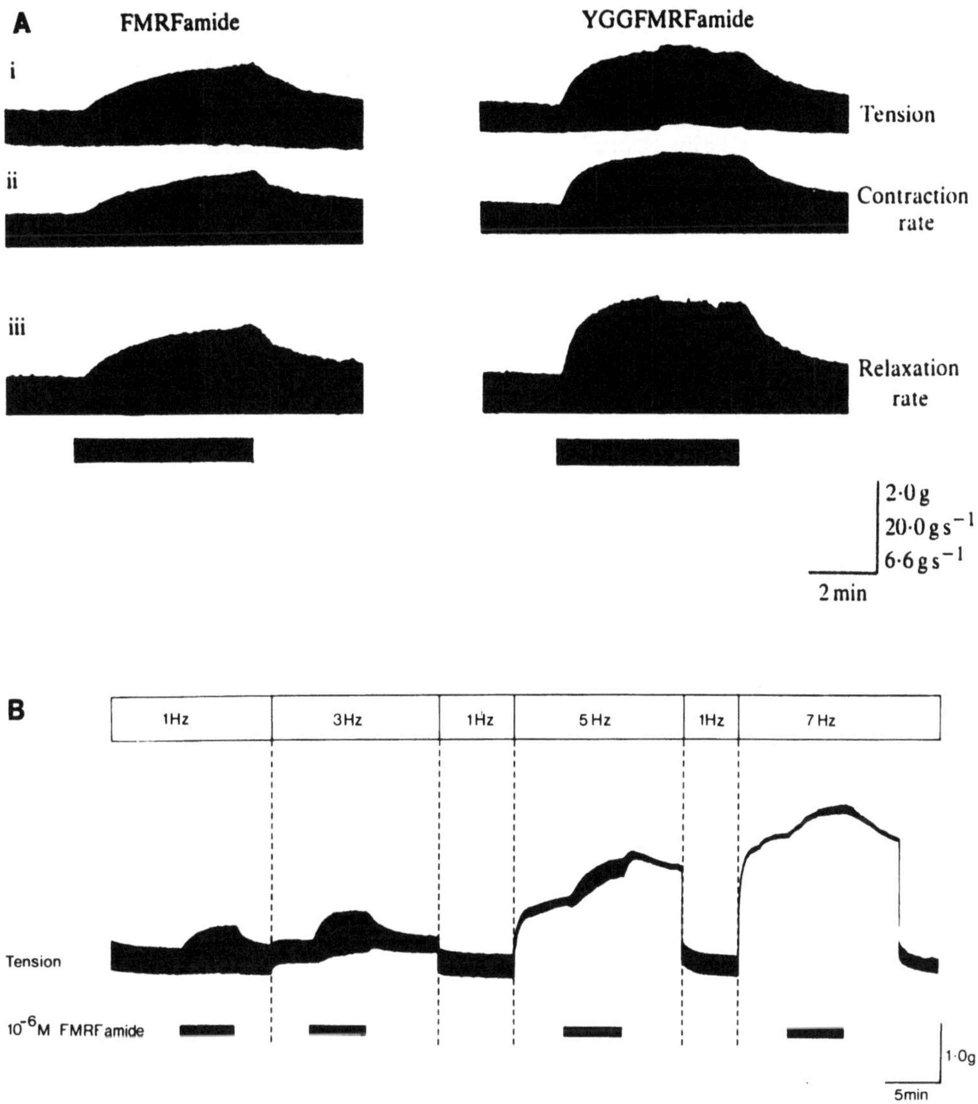

Fig.2. The effect of FMRFamide-like peptides on SETi-induced tension in the extensor-tibiae muscle. (A) The effect of 5-min pulses of 10^{-6}M FMRFamide (left hand side) and YGG-FMRFamide (right-hand side) (black bars) introduced into the muscle superfusate on twitch tension response elicited by stimulating SETi at a frequency of 1HZ. Trace i shows the response of twitch tension, and traces ii and iii show the effects on contraction and relaxation rates respectively (B) The frequency dependence of FMRFamide effects on maintained tension. A continuous recording of the tension profile from a metathoracic extensor-tibiae muscle produced by firing SETi at different frequencies is shown. 5 min pulses of 10^{-6}M FMRFamide (bars) were introduced into the superfusate. The preparation was returned to 1Hz stimulation between the 3- and 5-Hz stimulation and also between the 5- and 7-Hz stimulation (from Evans and Myers, 1986a).

(Evans and Myers, 1986b) and also with fact that YGGFMRFamide potentiates potassium induced contractures in this muscle (Walther et al 1984). We are therefore forced to postulate that FMRFamide also has direct postsynaptic or extrajunctional actions on the slow muscle fibres of this muscle. Since FMRFamide did not affect the relaxation rate of fast twitch tension it seems likely that the intermediate muscle fibres in this muscle, that are innervated by both the fast and slow excitatory motor neurones, lack receptors for FMRFamide-like peptides and that they are confined to the slow muscle fibres that are only innervated by SETi. At present the exact contributions of pre- and post-synaptic effects to the observed actions of FMRFamide and related peptides on EJP size, twitch tension amplitude and contraction rate remain to be resolved.

Immunocytochemical blocking experiments on the locust nervous system (Myers and Evans, 1985a,b,1987) and radioimmunoassay analysis of HPLC fractions of locust nervous tissue extracts (Robb and Evans, in preparation) suggest that there is more than one endogenous FMRFamide-like peptide present in the locust nervous system. At present we do not know the exact amino acid sequences of these peptides but they are currently being isolated and sequenced. In the locust extensor tibiae muscle specificity studies on various FMRFamide analogues (Table 4) indicate that NH_2-terminal extensions of the peptide can increase its potency by an order of magnitude. Thus YGGFMRFamide is more potent than FMRFamide itself. In addition, substitution of leucine for methionine does not decrease activity. The C-terminal Arg-Phe-NH_2 sequence is essential for activity in the locust and the preceding neutral amino acid (Met or Leu) may also be important since γ_1-MSH, in which the neutral amino acid is replaced by aspartate is inactive. Since LPLRFamide, the chicken brain peptide (Dockray et al 1983) is an order of magnitude less effective than FMRFamide it suggests that the Phe residue at the beginning of the FMRFamide sequence may also be important for the actions of these peptides on locust muscle. Some cells in the locust nervous system appear to stain with antibodies raised against bovine pancreatic polypeptide, but not with those raised against FMRFamide (Myers and Evans, 1985a,b,1987) (see Fig 1), thus we tested the actions of various pancreatic polypeptides ending with the Arg-Tyr-NH_2 sequence. None of these were active on the extensor tibiae muscle, confirming that the C-terminal -Arg- Phe-NH_2 sequence is essential for activity in this muscle.

Table 4. Structure-activity relations for effects of FMRFamide peptides on locust extensor-tibiae muscle.

Peptide	Amino-acid sequence	Relative potency	Reference
YGGFMRFamide	Tyr-Gly-Gly-Phe-Met-Arg-Phe-NH_2	1	1, 2
Met^5-Enk-Arg^6-Phe^7	Tyr-Gly-Gly-Phe-Met-Arg-Phe	$0–3\times10^{-5}$	1
	Tyr-Gly-Gly-Phe-Leu-Arg-Phe-NH_2	1	1
Met-Enkephalin	Tyr-Gly-Gly-Phe-Met	0	1, 2
FMRFamide	Phe-Met-Arg-Phe-NH_2	$10^{-1}–10^{-2}$	1, 2
	Phe-Met-Arg-Phe	0	1
FLRFamide	Phe-Leu-Arg-Phe-NH_2	10^{-2}	1
LPLRFamide	Leu-Pro-Leu-Arg-Phe-NH_2	5×10^{-2}	2
RFamide	Arg-Phe-NH_2	0	2
γ_1-MSH	.... Phe-Arg-Trp-Asp-Arg-Phe-NH_2	approx. 3×10^{-5}	1
CCK-Octapeptide	.......Met-Gly-Trp-Met-Asp-Phe-NH_2	$0–3\times10^{-5}$	1
Neuropeptide Y	.... Ile-Thr-Arg-Glu-Arg-Tyr-NH_2	0	1
BPP	.... Leu-Thr-Arg-Pro-Arg-Tyr-NH_2	0	2
APP	........Val-Thr-Arg-His-Arg-Tyr-NH_2	0	2
HPP	........Leu-Thr-Arg-Pro-Arg-Tyr-NH_2	0	2

The potencies of the peptides are expressed relative to that of YGGFMRFamide and the table combines data from Walther, Schiebe & Voigt (1984) (1) and from P.D. Evans & C.M. Myers (1986b) (2)

γ_1-MSH, melanocyte stimulating hormone; CCK, cholecystokinin; BPP, bovine pancreatic polypeptide; APP, avian pancreatic polypeptide; HPP, human pancreatic polypeptide.

The mode of action of FMRFamide-like peptides on locust muscle has not yet been determined. However, it appears not to involve cyclic nucleotides, since the effects are not potentiated by phosphodiesterase inhibitors such as IBMX and changes in cyclic AMP levels in this muscle were not observed after 10-min incubations of the muscle with concentrations of FMRFamide ranging from 10^{-11} to 10^{-4}M in the presence of 10^{-4}M IBMX (Evans and Myers, 1986b). In other preparations, such as some molluscan neurones, FMRFamide can induce changes in potassium and calcium conductances, and some of the effects are mediated by cyclic nucleotides in some cells but not in others (Colombaioni, Paupardin-Tritsch, Vidal and Gerschenfeld, 1985). Recently, it has been suggested that lipoxygenase metabolites of arachidonic acid may act as second messengers for the inhibitory responses of FMRFamide on Aplysia sensory neurones (Piomelli et al 1987). However, in our hands exogenously applied arachidonic acid (at estimated concentrations up to 3.3×10^{-4}M) did not appear to mimick the actions of FMRFamide on the locust extensor tibiae muscle. In addition the actions of FMRFamide (10^{-6}M) are not blocked in the presence of 4-bromophenacyl bromide (10^{-5}M) which inhibits the enzymes responsible for the breakdown of phospholipid to arachidonic acid or by nordihydroguaiaretic acid (10^{-6}M) which blocks the action of lipoxygenases. In addition the actions of FMRFamide (10^{-6}M) on locust muscle are not blocked by indomethacin (10^{-5}M) which blocks the production of prostoglandins by cyclo-oxygenase. At present it is not clear if these different effects observed in locust muscle represent differences in the susceptibility of the corresponding enzymes in insect tissue, different rates of penetration for the drugs used compared with Aplysia nervous tissue or if the receptors for FMRFamide-like peptides in locust muscle are linked to a different transduction mechanism. Further studies are currently in progress to elucidate the mode of action of FMRFamide-like peptides in locust skeletal muscle.

Besides modulating neuromuscular transmission and muscle contraction in locust skeletal muscles, FMRFamide and related peptides also modulate the spontaneous contractions of locust visceral muscles, such as the heart and oviducts (Cuthbert and Evans, in preparation). FMRFamide (10^{-11} to 10^{-5}M) and low doses of YGGFMRFamide (up to 10^{-7}M) (Fig 3A) increase both the frequency and the amplitude of spontaneous heart contractions and also regularise the beating of the heart. However, at higher concentrations (10^{-6}M and above) YGGFMRFamide exhibits a biphasic effect on the spontaneous contractions of the heart. Initially it speeds up the frequency of the heart beat but this is followed by a reduction in frequency for as long as the peptide remains in the superfusate. When the peptide is washed of the preparation a second pronounced period with an increased frequency of contractions occurs (Fig 3B). This suggests that there is more than one receptor mediating the actions of FMRFamide-like peptides in locust heart. One receptor can increase the frequency of contractions and is activated by FMRFamide and low doses of the extended peptide YGGFMRFamide and a second receptor can reduce the frequency of contractions and can be activated by high doses of YGGFMRFamide. The modes of action of these different receptor types is currently being investigated.

Recent immunocytochemical studies (Evans and Myers, 1986a) have demonstrated that the lateral posterior group of neurones in the locust thoracic ganglia that are immunoreactive to antibodies raised against

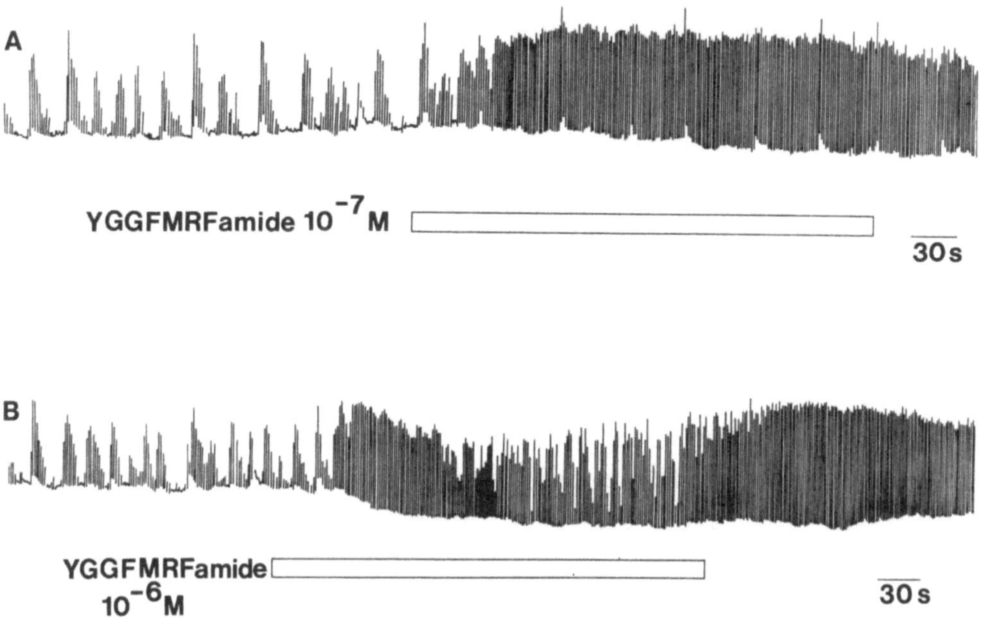

Fig.3 A The effect of a 5 min pulse of 10^{-7}M YGGFMRFamide on a heart with bursting activity. B. The effect of 5 min pulse of 10^{-6}M YGGFMRFamide on a heart with bursting activity. Note the nature of the biphasic response in the latter case.

FMRFamide are also immunoreactive to a monoclonal antibody raised against another molluscan peptide, namely small cardioactive peptide (SCP_B) (Morris et al 1982). SCP_B does not have any modulatory actions on neuromuscular transmission in the locust nor does it block the actions of FMRFamide as it does in a variety of molluscan preparations (Murphy, Lukowiak and Stell, 1985; P.E.Lloyd, personal communication). However, SCP_B does appear able to increase the frequency of contraction of the myogenic bundle of muscle fibres at the proximal end of the extensor tibiae muscle (Evans and Myers, 1986a).

Thus the locust extensor-tibiae muscle is modulated by a variety of biogenic amines and peptides (see Evans and Myers, 1986a). It is likely that the primary function of octopaminergic modulation is to change the response of the muscle from one that favours the maintenance of posture to one that favours rapid changes in joint position or force, such as might occur during locomotion. Under such conditions the muscle's output would be more closely tuned to its neuronal input, since it would have an increased frequency-following capability and any residual maintained or catch-like tension would be removed. In addition any interference from the myogenic rhythm would also be removed making the responses of the muscle ideally suited to deal with a stressful situation. In contrast the muscles responses to proctolin (and also to adipokinetic hormone-like peptides) (O'Shea, 1985) would seem more suited to a postural mode rather than a dynamic mode of action. The induction of long-lasting contractures of the muscle by these peptides would help maintain tension independently of the neuronal input. This would be important to the locust when the extensor muscle is being used to oppose gravity in a postural state or perhaps even during flight when the hind legs are held in a fixed extended positon close to the body (Kutsch, 1971). The physiological roles of the locust FMRFamide-like and SCP_B like peptides are at present difficult to forecast since nothing is known of the conditiond under which they are released or of the variability of their levels in the haemolymph. It seems likely, however, that specific times may exist when it would be of use to speed up the myogenic rhythm without modulating any of the other contractile properties of the muscle; a situation for which the SCP_B -like peptide would be ideally suited. Equally FMRFamide-like peptides appear to combine some of the properties such as the potentiation of twitch tension and increase in relaxation rate which are similar to those produced by octopamine, together with the effects on stimulation-frequency dependence and basal tension which are similar to those produced by proctolin. The exact behavioural situation when such a combination of properties would be of an advantage to the locust is unclear at present. However, it seems likely that the modulatory amines and peptides referred to above will not be released singly in response to a particular behavioural stimulus, but that rather the important physiological consequence to the locust will be changes in the relative amounts of each of the modulators present at any one time.

References

Colombaioni L, Paupardin-Tritsch D, Vidal PP, Gerschenfeld HM (1985) The neuropeptide FMRFamide decreases both the Ca^{2+} conductance and a cyclic 3',5'-adenosine monophosphate-dependent K^+ conductance in identified molluscan neurons. J. Neurosci. 5: 2533-2538.

Dockray GJ, Reeve JR, Shirely J, Gayton RJ, Barnard CS (1983) A novel active pentapeptide from chicken brain identified by antibodies to FMRFamide. Nature (Lond.) 305: 328-330.

Evans PD (1980) Biogenic amines in the insect nervous system. Adv. Insect Physiol. 15: 317-473.

Evans PD (1982) Properties of modulatory octopamine receptors in the locust. Ciba Fdn. Symp. 88: 48-69.

Evans PD (1985) Octopamine. In Eds. G.A. Kerkut and L. Gilbert Comprehensive Insect Biochemistry, Physiology and Pharmacology. pp499-530, Pergamon Press, Oxford.

Evans PD, Myers CM (1986a) Peptidergic and aminergic modulation of insect skeletal muscle. J. exp. Biol. 124: 143-176.

Evans PD, Myers CM (1986b) The modulatory action of FMRFamide and related peptides on locust skeletal muscle. J. exp. Biol. 126: 403-422.

Evans PD, Siegler MVS (1982) Octopamine mediated relaxation of maintained and catch tension in locust skeletal muscle. J. Physiol. 324: 93-112.

Grimmelikhuijzen CJP, Graff D (1986) Isolation of <Glu-Gly-Arg-Phe-NH_2 (Antho-RFamide) a neuropeptide from sea anemones. Proc. Natl. Acad. Sci. U.S.A. 83: 9817-9821.

Holman GM, Cook BJ, Nachman RJ (1986) Isolation, primary structure and synthesis of leucomyosuppressin, an insect neuropeptide that inhibits spontaneous contractions of the cockroach hindgut. Comp. Biochem. Physiol. 85C: 329-333.

Hoyle G (1983) Muscles and their neural control. Wiley-Interscience, New York.

Kutsch W (1971) The development of the flight pattern in the desert locust, Schistocerca gregaria. Z. vergl. Physiol. 74: 156-168.

Morris HR, Panico M, Karplus A, Lloyd PE, Riniker B (1982) Elucidation by FAB-MS of the structure of a new cardioactive peptide from Aplysia. Nature (Lond.) 300: 643-645.

Murphy AD, Lukowiak K, Stell WK (1985) Peptidergic modulation of patterned motor activity in identified neurons of Helisoma. Proc. Natl. Acad. Sci. U.S.A. 82: 7140-7144.

Myers CM, Evans PD (1985a) The distribution of bovine pancreatic polypeptide/FMRFamide-like immunoreactivity in the ventral nervous system of the locust. J. Comp. Neurol. 234: 1-16.

Myers CM, Evans PD (1985b) A FMRFamide antiserum differentiates between populations of antigens in the ventral nervous system of the locust, Schistocerca gregaria. Cell and Tissue Res. 242: 109-114.

Myers CM, Evans PD (1987) An FMRFamide antiserum differentiates between populations of antigens in the brain and retrocerebral complex of the locust, Schistocercia gregaria. Cell and Tissue Res (in press).

O'Shea M (1985) Are skeletal motoneurons in arthropods peptidergic? In Ed. A.I. Selverston Model Neural Networks and Behavior pp401-413, Plenum Press, New York.

O'Shea M, Schaffer M (1985) Neuropeptide function: The invertebrate contribution. Ann. Rev. Neurosci. 8: 171-198.

Piomelli D, Volterra A, Dale N, Siegelbaum SA, Kandel ER, Schwartz JH and Belardetti F (1987) Lipoxygenase metabolites of arachidonic acid as second messengers for presynaptic inhibition of Aplysia sensory cells. Nature (Lond.) 328: 38-43.

Price DA, Davies NW, Doble KE, Greenberg MJ (1987) The variety and distribution of the FMRFamide-related peptides in molluscs. Zoological Sci. 4: 395-410.

Price DA, Greenberg MJ (1977) Structure of a molluscan cardioexcitatory neuropeptide. Science 197: 670-671.

Shyamala M, Fisher JM, Scheller RH (1986) A neuropeptide precursor expressed in Aplysia neuron L5. DNA 5: 203-208.

Stern AS, Lewis RV, Kimira S, Rossier J, Gerber LD, Brink L, Stein S, Udenfriend S (1979) Isolation of the opioid heptapeptide Met-enkephalin [Arg^6,Phe^7] from bovine adrenal medullary granules and striatum Proc. Natl. Acad. Sci. U.S.A. 76: 6680-6683.

Voigt K-H, Martin R (1986) In Ed. G.B. Stefano 'Handbook of Comparative Opioid and related neuropeptide mechanisms' Vol.1: pp 127-138, C.R.C. Press, Boca Raton.

Walther C, Schiebe M, Voigt KH (1984) Synaptic and non-synaptic effects of molluscan cardioexcitatory neuropeptides on locust skeletal muscle. Neurosci. Letters. 45: 99-104.

Walther C, Schiebe M (1987) FMRFamide-NH_2-like factor from neurohaemal organ modulates neuromuscular transmission in the locust. Neurosci. Letts. 77: 209-214.

Yang H-YT, Fratta W, Majane EA, Costa E (1985) Isolation, sequencing, synthesis and pharmacological characterization of two brain neuropeptides that modulate the action of morphine. Proc. Natl. Acad. Sci. U.S.A. 82: 7757-7761.

CONSTANCY AND PLASTICITY IN THE OPERATION OF THE MOTOR CONTROL SYSTEM FOR LOCUST FLIGHT

Heinrich Reichert
Department of Animal Biology
University of Geneva
Rue de l'Ecole-de-Medecine 20
CH 1211 Geneve 4
Switzerland

INTRODUCTION

At a basic level, the ability to move can be regarded as one of the most important characteristics of animal life. It is thus not surprising that locomotor control systems have evolved into complex and interactive neuronal networks which are capable of producing coordinated and adaptive motor acts (Stein, 1978; Henneman, 1980). However, in order to be adaptive, locomotor control systems constantly have to deal with an apparent paradox. On the one hand they must assure a stabilized and repetitive motor output and be able to produce stereotyped rhythmic activity for extended time periods (Grillner and Wallen, 1985). They must thus be resistant to fluctuations and perturbations in internal operation. On the other hand, they must be able to react to changes in external environment and behavioral context. Thus such systems must also be accessible to short-term and to long-term functional modification (Gallistel, 1980). Some of this plasticity occurs at the level of synaptic transmission, a fact which is the subject of many

NATO ASI Series, Vol. H19
Modulation of Synaptic Transmission and Plasticity
in Nervous Systems. Edited by G. Hertting and H.-C. Spatz

of the articles in this volume. In addition to this, motor systems have developed a number of network solutions to the need for constancy and plasticity. These include switches between modes of operation, central gating of sensory information and network reconfiguration during different behavioral states. These properties are part of what Sherrington (1947) referred to as the integrative action of the nervous system. Here I shall discuss some of these features by focusing on a cellular analysis of one specific model motor system, the flight control system of the locust.

The locust flight system is a classical system for studies of locomotor control. It is one of the systems in which the concept of central neuronal oscillators was established (Wilson, 1961, 1968) and the role of sensory feedback in locomotor control was demonstrated (Wendler, 1974). Moreover, in the last few years much has been learned about the neuronal circuits responsible for generating and modulating the flight motor program. This is due to the fact that the flight producing neuromuscular system is simple. The number of muscles involved in powering the wings is limited to about 20 bilaterally symmetrical pairs. Moreover, the approximately 80 motoneurons controlling them are found in the accessible ganglia of the thorax (Snodgrass, 1935; Tyrer and Altman, 1974). The motoneurons and interneurons in these ganglia are identifiable as individuals and can be studied with intracellular recording and staining techniques (e.g. Burrows, 1973; Robertson and Pearson, 1982; Reichert and Rowell, 1985).

CONSTANCY OF RHYTHM GENERATION

It is generally agreed, that the basic motor activity underlying locust flight is produced within the ganglia of the CNS by a system of interconnected interneurons. The connectivity of some of these has been characterized by intracellular recording (Robertson and Pearson, 1983, 1985 a, b). This interneuron network is referred to as the neuronal oscillator for flight and is capable of producing rhythmic activity and channeling this rhythmic drive to the elevator and depressor motoneurons of the wing thus contributing to a coordinated wingbeat. The neuronal oscillator is remarkably resistant to perturbations in its internal operations. Examples of this internal stability are the persistance of (somewhat abnormal) oscillatory activity following connective transsection or ganglion ablation (Wolf and Pearson, 1987; Stevenson and Kutsch, 1987). Even more striking is the fact that an entirely normal flight rhythm is observed after hemisection of the mesothoracic ganglion, a lesion which functionally impairs about one half of the known interneuron types in the flight oscillator (Ronacher and Reichert, in preparation).

There are several factors which contribute to the constancy in operation of the rhythm generating circuitry even when individual elements of the circuitry are malfunctioning or disabled. First, the operation of this neural oscillator is reinforced by stabilizing sensory feedback circuits involving wing proprioceptors. Second, many of the oscillator interneuron types are represented within each hemiganglion in multiple copies, which are virtually identical in their physiology and morphology. Third, for many of the interneurons there exist segemental homologs in the six neuromeres (three thoracic and first three abdominal) of the thoracic ganglia. Thus the neuronal oscillator is organized in a distributed and apparently redundant manner.

The fact that this multisegmental interneuron network is functionally resistant to massive loss of neuronal connectivity suggests that selection pressure for redundant organization occurred during network evolution. Although this may have been the case, other evolutionary processes were probably involved as well. For example, the segmentally homologous organization of the flight interneurons in the abdominal neuromeres might have no apparent functional necessity (wings are thoracic structures). Yet, it may reflect a prior evolutionary stage in which prowing-like moveable appendages were serially repeated along all segments of thorax and abdomen (Robertson, Pearson and Reichert, 1982). This would be in accordance with data from the fossil record which support the pleural appendage theory for the origin of the insect wing (see Dumont and Robertson, 1986).

ADAPTIVE VARIABILITY AND THE FLIGHT AUTOPILOT

Whatever the evolutionary history of the neuronal oscillator for flight in the locust, its current multiply-reinforced and stabilized neuronal organization allows production of stereotyped rhythmic premotor activity. However, in nature this stereotyped motor program is insufficient. In flight the locust must do more than rhythmically beat its wings. It must also detect course deviations caused for example by turbulence or by faulty motor performance and it must correct these deviations. For this it is equipped with the equivalent of an autopilot circuitry. This course correction circuitry which makes modification of locomotor output possible has two features which are noteworthy here. The first is that the correction circuitry is operative only when the animal is flying; it is context dependent. The second is that this circuitry provides adaptive variability in motor output without necessarily influencing the stabilized rhythm-generating neuronal oscillator. In order to understand these two features it is necessary to consider the organization of the autopilot circuitry in the context of the overall flight control system. A simplified schematic diagram of this is shown in figure 1.

Deviations from course are exteroceptively detected by three sensory systems. These are the compound eyes, the ocelli and the cephalic windhairs. Each of these systems is capable of detecting different consequences of a deviation but all three converge in a highly interactive manner onto the same

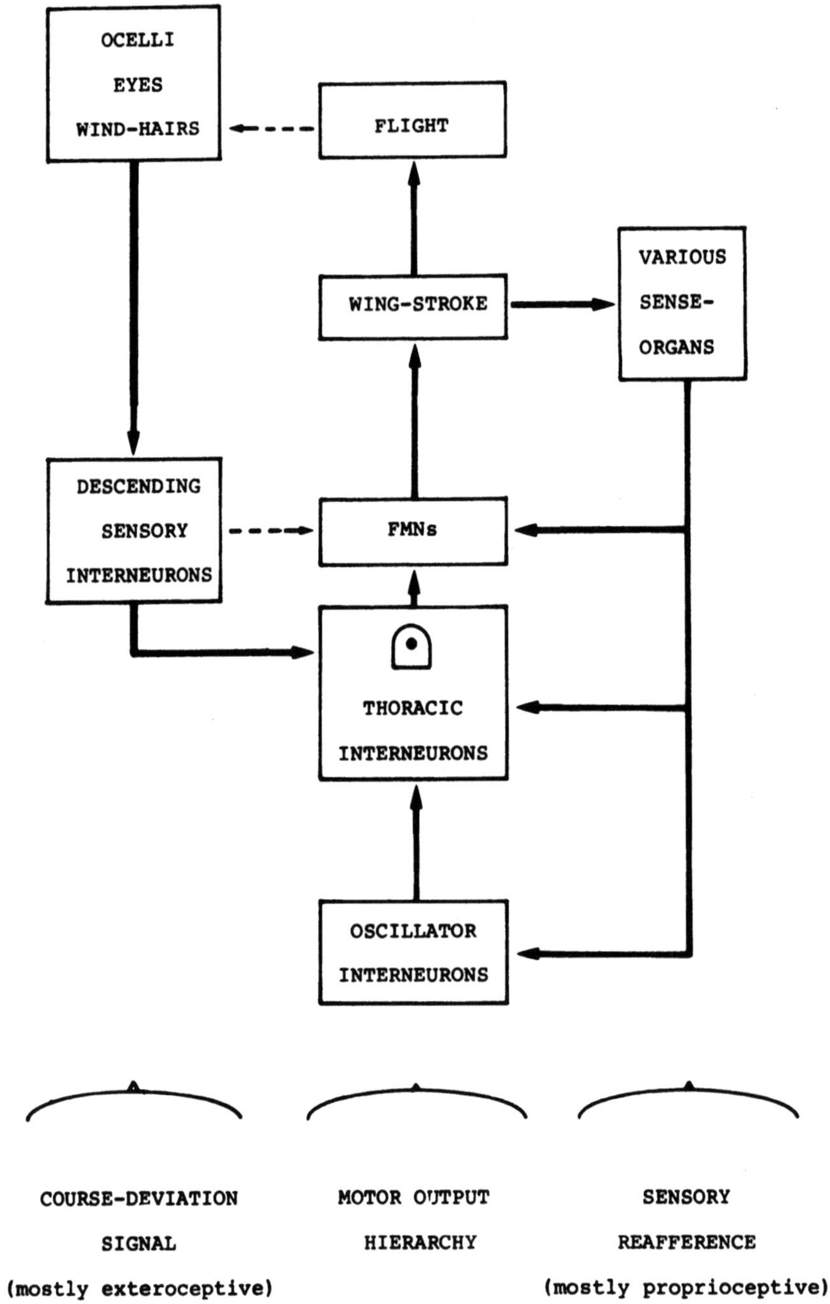

Figure 1. Schematic representation of the locust flight control system. For description see tex.

population of descending deviation detector neurons (Rowell and Reichert, 1986). These descending neurons carry processed information concerning specific types of course deviation to the flight motor centers of the thoracic ganglia where they make synapses with a population of thoracic interneurons (Reichert, 1985). In a non-flying animal this error detection part of the autopilot system is operative. However, the error correction part is not. This is because the thoracic interneurons function like gating elements that distribute the deviation information to postsynaptic flight motoneurons if and only if the central neuronal oscillator for flight is active (Reichert and Rowell, 1985). Only then can the subthreshold PSPs caused in the thoracic interneurons by descending detector neurons summate with rhythmic drive from the oscillator. Thus, only then can changes in the firing pattern of the thoracic interneurons and consequently modulation of the rhythmic drive to the flight motoneurons result.

This is one of many examples in which a given polymorphic neural network is reconfigured into different functional circuits depending on the behavioral state of the animal (see Getting and Dekin, 1985). When the animal is not flying, the autopilot circuitry is turned off (literally short circuited at the thoracic interneuron level). When the animal is flying, that is when the neuronal oscillator for flight is active, the autopilot circuitry is active and deviation information is channeled to the appropriate flight motoneurons (Reichert, Rowell and Griss, 1985). Furthermore, since sensory gating via the neuronal oscillator phasically alternates at the flight rhythm, the processed deviation information is automatically phase coupled to the flight rhythm before it is routed to the appropriate flight motoneurons. In a sense, a copy of the centrally generated motor rhythm is used to guide the flow of exteroceptive information from deviation detector neurons to motoneurons during steering reactions. This is an important process since it adapts the unpredictable and variable exteroceptive information from the outside world to the constant rhythmic activity generated within the nervous system (Reichert and Rowell, 1986).

Adaptive variability in the flight motor program is made possible through these interactions. Moreover, constancy of operation of the central oscillator and adaptive variability of motoneuron firing are compatible in this type of circuit configuration. This can be directly observed by recording simultaneously from an oscillator interneuron and from a flight motoneuron involved in steering. Figure 2 shows an example of this kind of recording. As part of the steering response to a simulated roll deviation, the regular rhythmic firing of the flight motoneuron is essentially turned off. Yet the stereotyped firing pattern of the oscillator interneuron continues unperturbed throughout the steering manoeuvre.

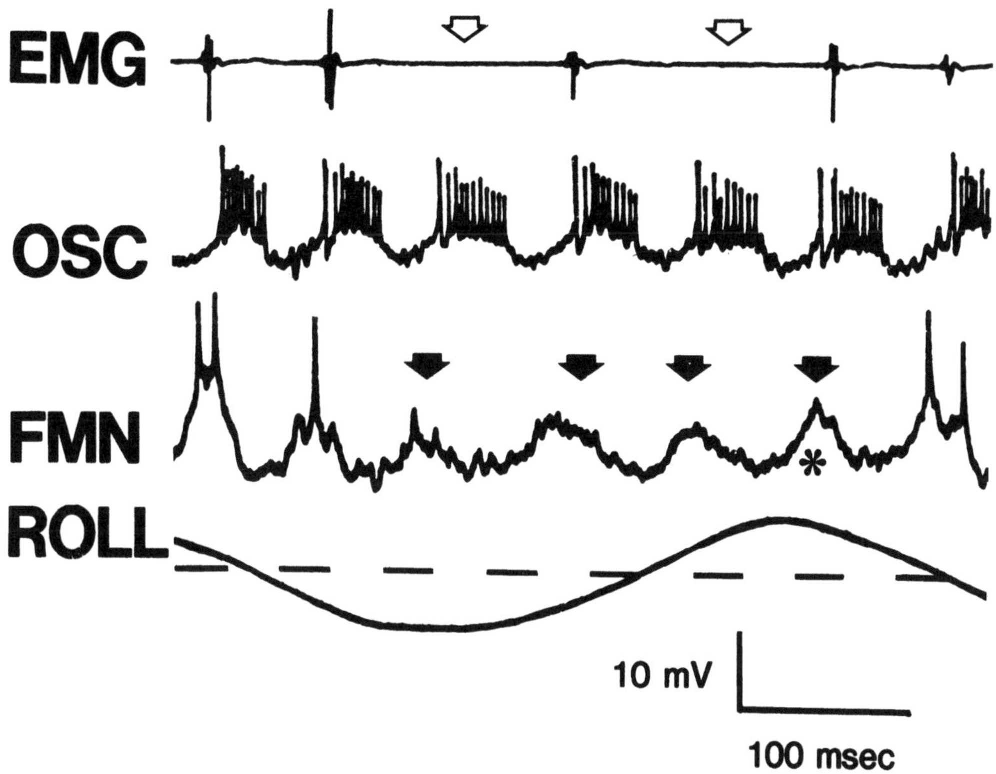

Figure 2. Variability in motor output and constancy in oscillator activity are compatible. Simultaneous intracellular recordings from an identified metathoracic oscillator interneuron (511) "OSC" and an identified mesothoracic flight motoneuron (98, wing depressor) "FMN". An electromyogram recording from a dorsolongitudinal flight muscle "EMG" is presented as well. Throughout the recording an artificial horizon was moved sinusoidally through 20° about the animal's longitudinal axis at 2.5 Hz ("ROLL") and thus used to simulate roll deviations. In response to a given directionality of the simulated roll deviation, the firing pattern of the flight motoneuron and of the flight muscle are profoundly altered; no spikes are produced during several wingbeat cycles (arrows). In contrast, the regular, rhythmic activity pattern of the oscillator interneuron continues virtually unperturbed throughout the steering manoeuvre.

SYSTEMS ORCHESTRATION THROUGH NEUROACTIVE PEPTIDES

During long-term flight in the locust, a behavior which can continue nonstop for well over 24 hours, hormonal signals are released which alter metabolism and neuronal function. The whole animal including its nervous system is orchestrated into its new behavioral role. Among the controlling factors with widespread integrating effects are the so called adipokinetic hormones AKH I and AKH II (for reviews see O'Shea, 1986; Orchard, 1987; Schaffer and Noyes, 1987). These are small neuropeptides that belong to the general family of arthropod RPCH/AKH-like peptides. In the locust, AKH is synthesized in the corpora cardiaca, a special neurosecretory organ associated with the brain, as well as in numerous neurons in the ventral nerve cord (Schooneveld et al., 1983; Hekimi and O'Shea, 1987; Siegert, Fischer-Lougheed, Hekimi and O'Shea, in preparation). The primary stimulus for release of AKH from the corpora cardiaca into the haemolymph is prolonged flight. Elevated titers of AKH can be detected 15 minutes after flight initiation and approximately 1 picomol of AKH can be recovered from the haemolymph after 30 minutes (Cheeseman and Goldsworthy, 1979; Orchard and Lange, 1983).

AKH has numerous integrating effects which are important for long-term flight. As a circulating hormone it serves to mobilize lipid stores, hence its name. It acts on muscle to suppress carbohydrate oxidation and stimulate fatty acid oxidation. It also has a heart acceleratory effect. As is the case for many other peptide hormones, AKH has neurogenic effects as well. We are just now beginning to explore the roles of AKH in the nervous system but it is already clear that it will prove to be a typical neuropeptide with a diverse set of neural actions. The observation that AKH can stimulate contraction of skeletal muscle (O'Shea, Witten and Schaffer, 1984) suggests a role at neuromuscular junctions. Furthermore AKH affects identified neuronal elements of the flight control system.

Two targets for AKH action in the flight control system are currently being studied. These are the tegula (Pearson, personal communication) and the strech receptor (Reichert, in preparation). Both are wing proprioceptors and both are part of sensory feedback circuits which stabilize the operation of the neuronal oscillator for flight and compensate for errors in the effector system. It is via its action on these proprioceptors that AKH appears to exert its acceleratory effects on the flight rhythm. No direct effects of AKH on central neurons of the flight oscillator have been described. Thus, it seems that the neuropeptide alters the operation of the flight oscillator indirectly by modulating sensory structures which have been "predesigned" to interact with that oscillator.

CONCLUSIONS

The flight control system of the locust deals with the need for both constancy and plasticity in its operation through the integration of multiple neuronal network properties. In this report, I have described some of these properties and the cellular and molecular components that give rise to them. Many of these features are likely to be generally applicable and significant. For example, in vertebrate motor systems, central neuronal oscillators in the spinal cord (Grillner and Wallen, 1985; Roberts, Soffe and Dale, 1986), sensory gating process at the level of the segmental interneurons (Forssberg, Grillner and Rossignol, 1975; Sillar and Roberts, in preparation) and orchestrating peptidergic neuroendocrine systems (Kupfermann, 1985) are being studied which are directly comparable with the organization of the flight control system in the locust.

Acknowledgements: I thank Elisabeth Jeantin for help in preparing the manuscript and Drs. Michael O'Shea and Siegfried Hekimi for discussions. The research presented in this report was supported in part by the Schweizerische Nationalfonds.

References:

Burrows M (1973) The morphology of an elevator and a depressor motoneuron of the hindwing of a locust. J Comp Physiol 83: 165-178

Cheeseman P, Goldsworthy GJ (1979) The release of adipokinetic hormone during flight and starvation in Locusta. Gen Comp Endocr 37:35-43
Dumont JPC, Robertson RM (1986) Neuronal circuits: an evolutionary perspective. Science 233:849-853
Forssberg H, Grillner S, Rossignol S (1975) Phase dependent reflex reversal during walking in chronic spinal cats. Brain Res 85:103-107
Gallistel CR (1980) The organization of action: a new synthesis. Erlbaum, Hillsdale
Getting PA, Dekin MS (1985) Tritonia swimming: a model system for integration within rhythmic motor systems. In: Selverston Al (ed) Model neural networks and behavior. Plenum, New York, London, pp 21-35
Grillner S (1975) Locomotion in vertebrates: central mechanisms and reflex interaction. Physiol Rev 55:247-304
Grillner S, Wallen P (1985) Central pattern generators for locomotion, with special reference to vertebrates. Ann Rev Neurosci 8:233-262
Hekimi S, O'Shea M (1987) Identification and purification of precursors of the insect neuropeptide adipokinetic hormone. J Neurosci (in press)
Henneman E (1980) Motor functions of the brain stem and basal ganglia in Mountcastele VB (ed) Medical physiology, Vol 1. Mosby, St Louis, pp 787-812
Kupfermann I (1985) Hypothalamus and limbic system I: peptidergic neurons, homeostasis, and emotional behavior. In: Kandel ER, Schwartz JH (eds) Principles of neural science. Elsevier, New York, pp. 611-625
Orchard I (1987) Adipokinetic hormones - an update. J Insect Physiol 33:451-463
Orchard I, Lange AB (1983) Release of identified adipokinetic hormones during flight and following neural stimulation in Locusta migratoria. J Insect Physiol 29:425-429
O'Shea M (1986) Insect neuropeptides - pure and applied. In: Borkovec AB, Gelman DB (eds) Insect neurochemistry and neurophysiology. Humana, New Jersey, pp 3-27
O'Shea M, Witten J, Schaffer MH (1984) Isolation and characterization of two myoactive neuropeptides: further evidence of an invertebrate family. J Neurosci 4:521-529
Reichert H (1985) The cellular basis of sensorimotor coordination in the flight control system of the locust. In: Bush BMH, Clarac F (eds) Coordination of motor behavior. University Press, Cambridge, pp 121-140
Reichert H, Rowell CHF (1985) Integration of nonphaselocked exteroceptive information in the control of rhythmic flight in the locust. J Neurophysiol 53:1201-1218
Reichert H, Rowell CHF (1986) Neuronal circuits controlling flight in the locust: how sensory information is processed for motor control. Trends Neurosci 9:281-283
Reichert H, Rowell CHF, Griss C (1985) Course correction circuitry translates feature detection into behavioural action in locusts. Nature, London 315:142-144

Rowell CHF, Reichert H (1986) Three descending interneurons reporting deviation from course in the locust. II Physiology. J Comp Physiol 158:775-794
Roberts A, Soffe SR, Dale N (1986) Spinal interneurons and swimming in frog embryos. In: Grillner S, Herman R, Stein PSG, Stuart D (eds) Neurobiology of vertebrate locomotion. Macmillan, London
Robertson RM, Pearson KG (1982) A preparation for the intracellular analysis of neuronal activity during flight in the locust. J Comp Physiol 146:311-320
Robertson RM, Pearson KG (1983) Interneurons in the flight system of the locust: distribution, connections and resetting properties. J Comp Neurol 215:33-50
Robertson RM, Pearson KG (1985a) Neural circuits in the flight system of the locust. J Neurophysiol 53:110-128
Robertson RM, Pearson KG (1985b) Neural networks controlling locomotion in locusts. In: Selverston Al (ed) Model neural networks and behavior. Plenum, New York, London, pp 21-35
Robertson RM, Pearson KG, Reichert H (1982) Flight interneurons in the locust and the origin of insect wings. Science 217:177-179
Schaffer MH, Noyes BE (1987) The potential of adipokinetic hormone to teach us about neuropeptides. BioEssays 7:67-71
Schooneveld H, Tesser GI, Veenstra JA, Romberg-Privee HM (1983) Adipokinetic hormone and AKH-like peptide demonstrated in the corpora cardiaca and nervous system of Locusta migratoria by immunocytochemistry. Cell Tissue Res. 230:67-76
Sherrington CS (1947) The integrative action of the nervous system, 2nd ed. Yale University Press, New Haven
Snodgrass RE (1935) Principles of insect morphology. McGraw Hill, New York
Stein PSG (1978) Motor systems, with specific reference to the control of locomotion. Ann Rev Neruosci 1:61-82
Stevenson PA, Kutsch W (1987) A reconsideration of the central pattern generator concept for locust flight. J Comp Physiol 161:115-129
Tyrer NM, Altman JS (1974) Motor and sensory flight neurones in a locust demonstrated using cobalt chloride. J Comp Neurol 157:117-138
Wendler G (1974) The influence of proprioceptive feedback on locust flight coordination. J Comp Physiol 88:173-200
Wilson DM (1961) The central nerovus control of flight in a locust. J Exp Biol 38:471-490
Wilson DM (1968) The nervous control of insect flight and related behaviour. Adv Insect Physiol 5:289-338
Wolf H, Pearson KG (1987) Flight motor patterns recorded in surgically isolated sections of the ventral nerve cord of Locusta migratoria. J Comp Physiol 161:103-114

NEUROPHARMACOLOGY OF LEARNING AND MEMORY IN HONEY BEES

R. Menzel, B. Michelsen, P. Rüffer and M. Sugawa

Institut für Tierphysiologie - Neurobiologie
Freie Universität Berlin
Königin-Luise-Str. 28/30
D-1000 Berlin 33

Local injections of small quantities ($\leqslant$ 10 nl) of drugs reveal a selective facilitatory action of octopamine and noradrenaline, and an inhibitory action of 5-HT and dopamine on particular associative and non-associative components of olfactory learning and memory in bees. It is concluded that the mushroom bodies are the neuropiles in which the antagonistic effects of octopamine and 5-HT act on olfactory memory. A single conditioning trial proceeding or following drug injection enables the distinction of whether a drug affects memory formation or memory recall. Dopamine reduces selectively memory recall, octopamine facilitates both memory recall and formation, and 5-HT also inhibits both memory recall and formation.

INTRODUCTION

The modulation of behavioral patterns by the biogenic amines serotonin (5-HT), octopamine (OA), dopamine (DA), and noradrenaline (NA), is a common feature of vertebrate and invertebrate nervous systems. In invertebrates, the action of certain amines can frequently be addressed to a set of certain well-defined neurons or even single identified neurons. It appears that these neurons are specifically involved in the modulation of stereotyped behaviors and often act in a complimentary fashion (Menzel and Bicker, 1987; Bicker and Menzel, subm). For example, serotonergic cells in molluscs and annelids modulate biting, feeding and swimming (Glover and Kramer, 1982; Kupfermann and Weiss, 1981; Lent and Dickinson, 1984), whilst 5-HT and OA act in an opposing manner to determine the posture of the lobster (Harris-Warrick and Krawitz, 1984; Livingstone et al., 1980). In insects the eliciting or enhancement of motor programs such as flying and jumping is controlled by octopaminergic cells (Evans, 1980; Sombati and Hoyle, 1984; Stevenson and Kutsch, 1987). Furthermore, associative learning in insects is under the control of aminergic cells and circuits (Mercer and Menzel, 1982; Mercer, 1987; Menzel and Michelsen, 1986; Michelsen, in press). In this paper, we report the results of experiments in which pharmacological substances are injected into well-defined regions of the honey bee brain during olfactory conditioning. These substances have a selective affect on associative learning in bees, the nature of the effect depending on the substance injected and the location of the injection into the brain.

Olfactory conditioning in honey bees is a particularly fast and effective associative learning process (v. Frisch, 1967; Menzel, 1983). A single learning trial with a floral odor changes the behavior of free flying bees in such a way that more than 90 % of

NATO ASI Series, Vol. H19
Modulation of Synaptic Transmission and Plasticity
in Nervous Systems. Edited by G. Hertting and H.-C. Spatz

the choices are directed only to this odor. Likewise, the conditioning of the proboscis extension reflex to an odor is very fast; one conditioning trial changes the behavior from not responding to more than 70 % responses to the conditioned odor alone (Menzel et al., 1974; Bittermann et al., 1983). Even if a capillary is inserted into the exposed brain, the bees continue to learn well. Thus, the odor conditioning of fixed (i.e. not free flying) bees (Kuwabara, 1957; Vareschi, 1971) offers favourable access to questions concerning the role of neuromodulators in the associative processes of memory formation and recall, as well as giving an insight into which areas of the brain these neuromodulators function.

In the following experiments, the learning trial lasts only 3 seconds and constitutes a single pairing of a conditioned stimulus (CS = odor) with an unconditioned stimulus (US = sucrose solution to the antennae and the proboscis). To determine whether a drug affects the learning processes (the phenomena associated with the formation of memory) or those involved in the retrieval or recall from the memory, the drug injection is carried out before or after the learning trial. The first type of experiments are called "storage tests" and the latter "recall tests".

RESULTS

(1) Experimental Design and Representative Examples for the Storage and Recall Test

The standard procedure of conditioning bees to an odor stimulus under restrained conditions has been described earlier (Menzel et al., 1974; Bitterman et al., 1983). The motivation to learn is high, if the bees are fed for the last time in the evening before the experiment. If a bee responds to the odor before conditioning ("spontaneous response") it is excluded from the conditioning experiment. In other words the response rate in all experiments before conditioning is 0 %. The rate of spontaneous responses varies from 0 % to 30 %, and on average is approximately 10 %. Special test procedures are used to determine if drug injection alters the spontaneous response rate.

Two experimental protocols are applied, the storage test and the recall test. In both cases, the bees are conditioned once only by pairing CS and US. The two protocols differ with respect to the time of injection of the drug. In the recall test, the injection follows the conditioning 30-90 min later, whilst in the storage test the drug is injected 5-20 min prior to conditioning i.e. at a time when the drug action is optimal. In each experiment, the control group of bees receives an injection of saline solution which is of the same volume as that received by the experimental group, and which is applied at the same time and location. Fig. 1a and b give examples for the

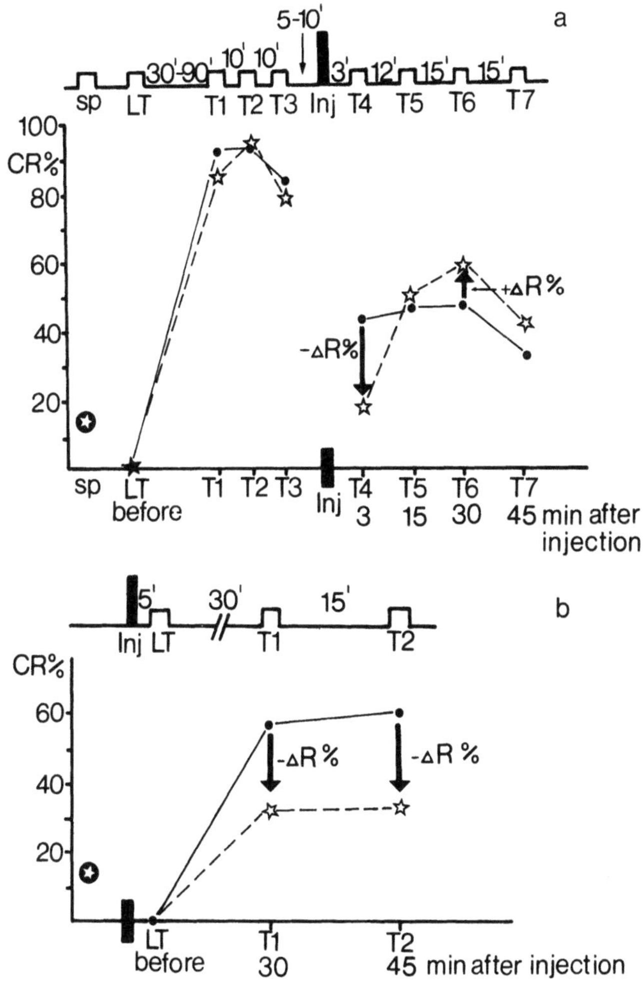

Fig. 1 Experimental design for the two test procedures: (a) Recall test, (b) Storage test. A group of bees (30-40 in number) are individually fixed to a holder the day before the experiment. Each bee is then exposed to the odor (carnation) which is also later used as the CS (spontaneous response test (Sp)). The few bees responding to the odor (less than 20 %, ✪) are excluded from the experiment. In the recall test (upper figure, a) the bees are then conditioned by one learning trial (LT) and tested 3 times, 30-90 min later, by exposure to the CS without US (T1, T2, T3). Afterwards, the head capsule is opened, the frontal brain surface exposed and the drug injected. In this particular example, half of the bees are injected with 5 nl of 5-HT (10^{-8}M) in each alpha-lobe (experimental group, ☆) or 5 nl of saline (control group, •). 3 min after injection, the unreinforced tests start again (T4-T7). The dissection and injection causes an unspecific reduction of the conditioned response rate (CR in % at the ordinate). The effect caused by 5-HT is the difference between the control group and the experimental group and is expressed as ΔR %. In the case - Δ R % the drug effect is inhibitory, and in the case of + Δ R % the effect is facilitatory (see text).
(b) Storage test. The bees are dissected and injected in the same way as described for the recall test. In this example, the bees are injected with 5 nl of 5-HT (10^{-4}M) in each alpha-lobe (experimental group, ☆) or 5 nl of saline (control group, •). After a period of 5 min, each animal is conditioned with one learning trial (LT). Tests begin 30 min later, when most of the immediate effects of the drug injection have disappeared. Statisitcs: The Kolmogoroff-Smirnoff χ^2-test is used to test the homogeneity of the experimental and control group. In these examples, only - ΔR % values are statistically significant.

two experimental conditions. In Fig. 1a (recall test), 5 nl of 5-HT are injected into each alpha-lobe of the mushroom bodies. The experimental dissection that permits injection (i.e. removal of the frontal cuticle, the exposure of the brain, removal of the front trachea and injection of the saline) reduces the high rate of conditioned response to about half. Control experiments show that this reduction is due to the dissection procedure and that very little is due to additional injection of saline, as long as the injection volume is kept below 10 nl. We consider this reduction in the response rate as an unspecific and unavoidable effect in our experimental procedure, but which nevertheless bares no relation to the test procedure (recall or storage test) or drug injection. Since we are mainly interested in the specific effects of drugs, we shall express the response rate of each experimental group relative to its saline control (ΔR % = CR % of saline control - CR % of experimental group). Fig. 1a and b show that ΔR % can be positive or negative. Repeated presentations of CS without US, as in T1 - T7 in the recall test and T1 + T2 (sometimes T3) in the storage test, may reduce CR % by experimental extinction. This is indeed the case as can be seen in Fig. 1a. However, the effect is minimal and this would imply that the bees are relatively resistant to extinction, even after a single learning trial as long as the CS presentations are spaced by 10 min or more (Bitterman et al., 1983).

(2) Time course of the Drug Effects

Recall test: The recall test is optimal for studying the time course of the effect of a drug on the performance of a learned response. Since the bee's memory is established before drug injection, the consequent response level is solely dependent on the manner in which the drug interferes with the memory retrieval process and/or the sensory and motor components of the conditioned response. A few examples are shown in Fig. 2. In most cases, and particularly after local injections, the behavioral effect of the drug appears within the first 5 min after injection and then slowly disappears within 30-40 min. In all cases, the effect is completely reversible within 45-60 min. Similar time courses were found by Mercer and Menzel (1982), Menzel and Michelsen (1986), and MacMillan and Mercer (1987). Mercer, Menzel and Michelsen injected large quantities (several hundred nl's) of different substances through the median ocellar tract into the brain ("global injections"), whereas MacMillan and Mercer flushed the antennal lobe with approximately 250 nl of DA (10^{-4}-10^{-6}M). More complicated time courses are found if 5-HT is injected in small quantities (5 nl) into the two alpha-lobes of the mushroom body (see Fig. 1a). An initial inhibitory action, which is particularly obvious for a concentration of 10^{-8}M, is followed by a facilitatory effect. Once again, the effect of 5-HT is fully reversible after 45 min. The effects of some drugs develop very slowly. An extreme example is the time course of phentolamine (Phent) when it is injected into the pedunculus of each mushroom body (Fig. 2).

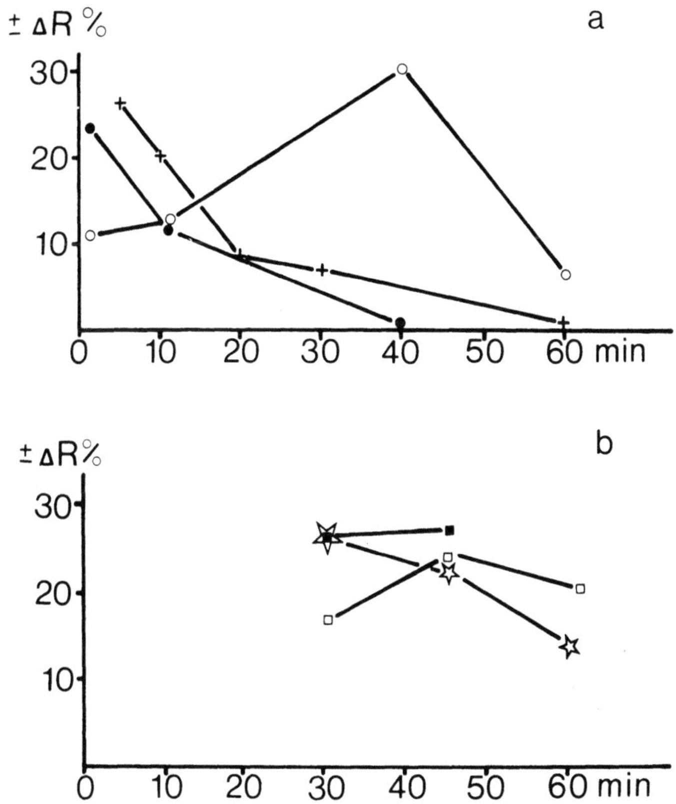

Fig. 2 (a) Three examples of the time course of drug action as determined in the recall test (see Fig. 1a). The ordinate gives either + ΔR % or - ΔR %, where ΔR is the difference in the conditioned response rate between the experimental and control group. + ΔR % indicates an improvement in response due to the action of the drug, and - ΔR % indicates a reduction in response. +NA (10^{-6} M, ca. 100 nl) is injected globally into the protocerebrum through the median ocellus. The ordinate gives - ΔR %. • Synephrin is injected (10^{-5} M, 5 nl) into each pedunculus of the mushroom bodies. The ordinate gives + ΔR %. ○ Phentolamine is also injected (10^{-3} M, 5 nl) into the pedunculus. The ordinate gives - ΔR %.
(b) Three examples of the time course of drug action as determined in the storage test (see Fig. 1b). There are no tests during the 30 min after injection.■5-HT (10^{-4} M, 5 nl) was injected into each alpha-lobe; the ordinate gives - ΔR %. □ NA (10^{-8} M, 2 nl) was injected into each alpha-lobe; the ordinate gives + ΔR %.☆OA (10^{-6} M, 2 nl) was injected into the pedunculus; the ordinate gives + ΔR %.

In the recall test, the effects of the injected drug may be the consequence of a variety of different modes of action. For example, the odor stimulus may no longer be perceived due to a blocking, degrading or changing of the sensory processes, or the motor program of the proboscis extension may be disturbed to such an extent that it is unable to function properly. Alternatively, the specific pattern of excitation elicited by the odor stimulus may be changed to such an extent that it can no-longer be matched with the stored image in the memory, or the retrieved and properly activated memory image may be unable to control the conditioned response. The possibility of the improper functioning of the motor program of the proboscis extension may be disregarded, since we tested the sucrose reflex for all conditions of drug injection and found it to be unchanged. That the sensory coding properties remain unaltered, is demonstrated in the storage test.

Storage test: The aim of this experiment is to determine the ability of bees to learn a stimulus under conditions of optimal drug action. If the bees are treated shortly before the single learning trial, and conditioned at the time of optimal drug action, persistant changes are found in the conditioned response for certain substances and localizations of injections (Fig. 2b). In most storage tests, the animals are tested 30 min and later after drug injection i.e. at a time when the drug should have no or little effect on performance. Therefore, any persistant change in CR % is indicative of an anterograde effect of the drug on the processes of memory formation.

Memory formation requires time, and any interference with this process of fomation would be expected to result in anterograde and/or retrograde effects, the nature of these effects depending on the time at which such interference takes place. In bees, the time course of memory consolidation after a single learning trial is within the range of several minutes (3-5 min half time)(Menzel et al., 1974; Erber et al., 1980; Sugawa, 1986; Sugawa and Menzel, 1986). A drug injected before the learning trial will, therefore, not only act on the learning process alone, but will also affect the consolidation period following the learning trial. The procedure of the storage test, as we have applied it, cannot distinguish between anterograde and retrograde effects of the drug since the injected drugs were effective throughout the learning trial and the consolidation period. Therefore, the main emphasis of the experiments reported here is the separation of those processes involved in memory formation (anterograde and retrograde components) and those involved in the recall from the memory store.

(3) Pharmacology of the Proboscis Extension Reflex

(a) Sensory components:

Stimulation of the antennae with sucrose solution leads to a reflexive extension of the proboscis. The proportion of animals responding is directly dependent on their hunger status. Hungry animals respond with 100 % reliability, and satiated animals with a very low rate or not at all. If OA (10^{-6}M) is injected into the hemolymph, or into the protocerebrum through the median ocellar nerve, even satiated animals respond at a very high rate ($\geq$90 %). The same is also true for OA agonists (synephrine (Syn), chlordimeform (CDM)), whereas the antagonist phentolamine reduces the response rate of hungry animals. Other amines do not change the reflex when they are injected into the protocerebrum.

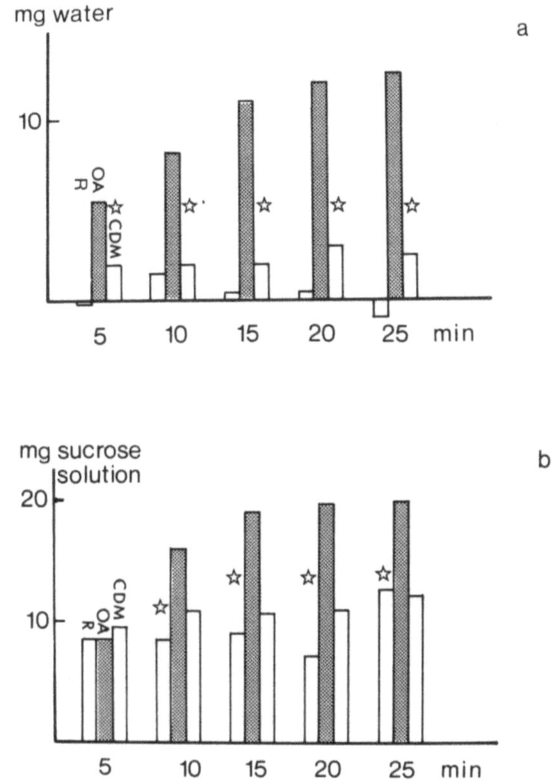

Fig. 3 The effect of OA (10^{-5}M) and chlordimephorm (CDM, 10^{-4}M) on the uptake of water (upper figure, a) and sucrose solution (lower figure, b) when injected into the hemolymph. The first column (marked R) is the control group, which received the same volume (5 nl) of Ringer solution. Statistically significant differences are indicated by stars. In a) each column represents the average of 18 animals, in b) that of 25 animals. OA stimulates the sucking behavior of both water and sugar water.

OA also enhances another component of the reflex. It has been previously reported (Mercer and Menzel, 1982) that sucrose stimulation of the antennae arouses the animals and increases the probability that animals perform the reflex to a water stimulus at the antennae. Similar observations were reported for the fly, and Dethier (1976) interpreted this effect as an indication of an increase in the "central excitatory state". OA injected into the hemolymph, or protocerebrum, mimicks the sucrose induced sensitization (Fig. 3). The same effect is observed after local injections of small amounts of OA (2 nl, 10^{-6}M) into the deutocerebrum, especially the dorsal lobes, whereas injections into other regions (alpha-lobe, pedunculus) have either no effect or even an inhibitory effect on the reflex. The only other amine found to interfere with the sensory components of the reflex is DA, which blocks olfactory coding when flushed over the antennal lobes (Macmillan and Mercer, 1987).

(b) motor components:

In hungry animals, the proboscis starts to suck and lick when extended. Satiated animals only imbibe very little or no sucrose solution at all. OA injected into the hemolymph, or protocerebrum, stimulates sucrose uptake in even fully satiated animals (Fig. 3b). Local injections reveal that OA acts upon the antennal lobes, the dorsal lobes and the alpha-lobes of the mushroom bodies. The latter is particularly interesting because the sensory components of the reflex are not enhanced when OA is injected into the alpha-lobes (see above). As Fig. 3b shows, the amount of sucrose solution imbibed can be twice as much as that under control conditions. These animals have an extremely swollen abdomen and the pressure in the crob is so large that the solution drops out through the proboscis when they stop sucking. Similar observations are reported for the fly (Long and Murdoch, 1983), where OA injections into the hemolymph also result in the animals becoming hyperphagic. As in the case of the fly, the pharmacological profile for induced hyperphagia indicates OA-2 receptors as the major receptor system. In the bee, noradrenaline (NA) and amphetamine (Amph) are also found to stimulate hyperphagia when injected into the antennal and alpha-lobes. Since OA-2 and alpha-adrenergic receptors have partially overlapping pharmacological profiles (Evans, 1980; Evans, 1982; Robertson et al., 1976), it is as yet unknown whether adrenergic receptors are involved in addition to the OA-2 receptors.

So far, no antagonistically acting amine has been found for the motor components of the proboscis reflex and sucrose uptake. However, very little attention has been paid to the role of 5-HT or DA in the amount of sucrose uptake in hungry animals.

(4) Pharmacology of Memory Recall

Global injections into the mid-protocerebrum after conditioning reveal that DA, 5-HT, NA and Amph significantly reduce the conditioned response to the CS (Fig. 5, upper line). As we shall see later, none of these drugs reduce the sensory or motor components of the reflex. One can conclude, therefore, that the injected transmitters interfere with a process related to memory recall, as opposed to inhibition of the sensory coding or motor performance.

OA injected globally has no effect. However, enhancement due to OA may not have been detected because of the already very high level of conditioned responses in the control group (90 % CR). Furthermore, local injections of OA analogs indicate a facilitatory action. Syn, a strong OA receptor agonist, improves CR % if it is injected in the pedunculus, but not when it is injected into the alpha-lobes (Fig. 5, line 2 and 3). Phen, an alpha-adrenergic blocker which also antagonises the OA-2 receptors (Evans, 1980), inhibits CR % when injected into the pedunculi and/or in the alpha-

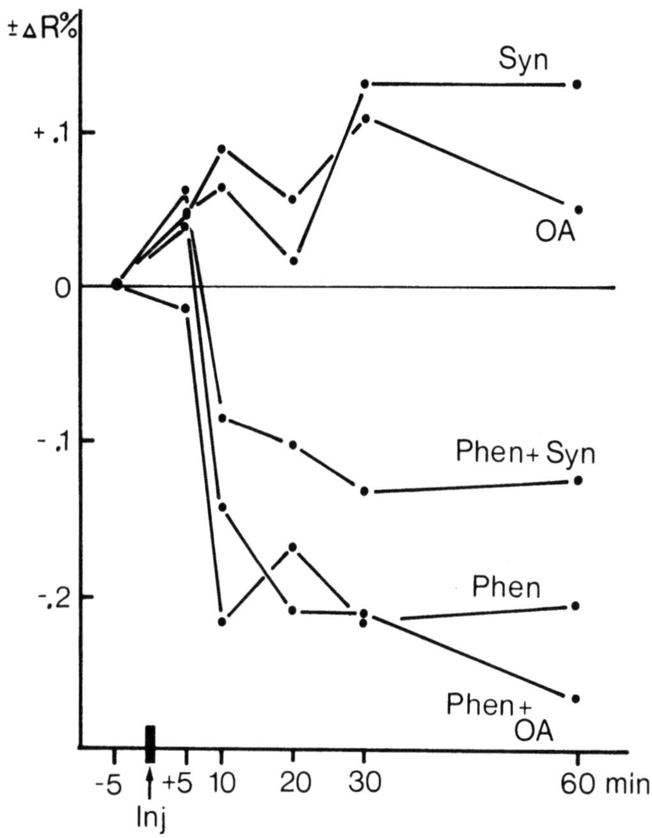

Fig. 4 The effect of OA (10^{-6}M), Syn (10^{-5}M), Phen (10^{-5}M), or equal mixtures thereof, on the response to water stimuli at the antennae when injected into the mid-protocerebrum. 5 min before the injection, the 20 animals of each group are tested for their response (proboscis extension) to a stimulation of the antennae with pure water. The average response level is 23 %. After the injection each animal is tested 5 times - 5, 10, 20, 30 and 60 min after injection. The change of the response level is calculated as ΔR (see text). OA and Syn increase the response level, Phen or equal mixtures of Phen with OA or Syn decrease the response.

lobes. The time course for the inhibitory effect of Phen is exceptionally slow (see Fig. 2a), and the time of optimal action is approximately 45 min after injection. This is true regardless of whether the drug is injected into the alpha-lobes or the pedunculi. This observation rules out the possibility that it is necessary for the drug to diffuse from the alpha-lobe to the pedunculus (or vice versa) to become active. Instead, it would seem that Phen acts in an inhibitory fashion at the site of injection, and since it is an antagonist of both OA-2 and alpha-adrenergic receptors, it would appear that it interferes with both receptor systems. If one compares the action of Phen with the selective effect of Syn in the pedunculi, then these results argue in favor of a facilitatory mechanism localized in the mushroom bodies, an alpha-adrenergic mechanism in the alpha-lobes and an octopaminergic mechanism in the pedunculi.

5-HT has so far only been tested with local injections into the alpha-lobes. The time course (Fig. 1) and the dose dependence indicate a complex effect with both facilitatory and inhibitory components, but only the early inhibitory component is statistically significant.

DA reduces memory recall when injected globally into the protocerebrum (Fig. 2a, Fig. 5 upper line). As has been pointed out above (see Fig. 2b), bees are successfully conditioned at the time of optimal action of DA (see also Fig. 6, upper line). This would seem to indicate that DA selectively inhibits neural processes involved in the retrieval from memory, as opposed to sensory or motor processes connected with the conditioned or unconditioned reflex. Local injections of DA have yet to be performed, but we can nevertheless conclude that the mushroom bodies and/or the central complex (central body, protocerebral bridge) are the likely targets of the DA injections, and not the antennal lobes. This is because (1) the injection site is further away from the antennal lobes, and (2) Macmillan and Mercer (1987) found a pronounced inhibitory effect of DA on olfactory coding in the antennal lobe, whereas we did not observe any interference with olfactory coding.

(5) Pharmacology of Memory Formation

If a drug affects the recall of formation of memory selectively, it should not interfere with the sensory perception of the CS and US, or the motor pattern of the UR or CR. Furthermore, in the case of a selective effect on memory formation, the memory should be permanently changed and outlast the immediate action of the drug, whereas a selective change of memory recall should be limited to the action of the drug and be fully reversible. These conditions are fulfilled for the biogenic amines and their analogs when tested with the storage test. The results obtained with DA are a representative example (see above, also Fig. 5 and 6, upper line)(Michelsen, in press).

When the same kind of experiment is carried out with global injections of NA or Amph a permanent depression of CR % is recorded (Fig. 6, upper line). This indicates that an adrenergic component is involved in the formation of memory. Indeed, local injections of NA into the alpha-lobe, pedunculus and antennae have been found to affect memory formation, but in these cases memory is enhanced rather than reduced. This effect is only statistically significant for alpha-lobe injections and only if the bees have a very low rate of learning.

The significance of the reversal of the effects caused by NA is not presently understood. Nevertheless, they clearly show the importance of an adrenergic (or closely related) mechanism in memory formation. A much stronger facilitatory effect

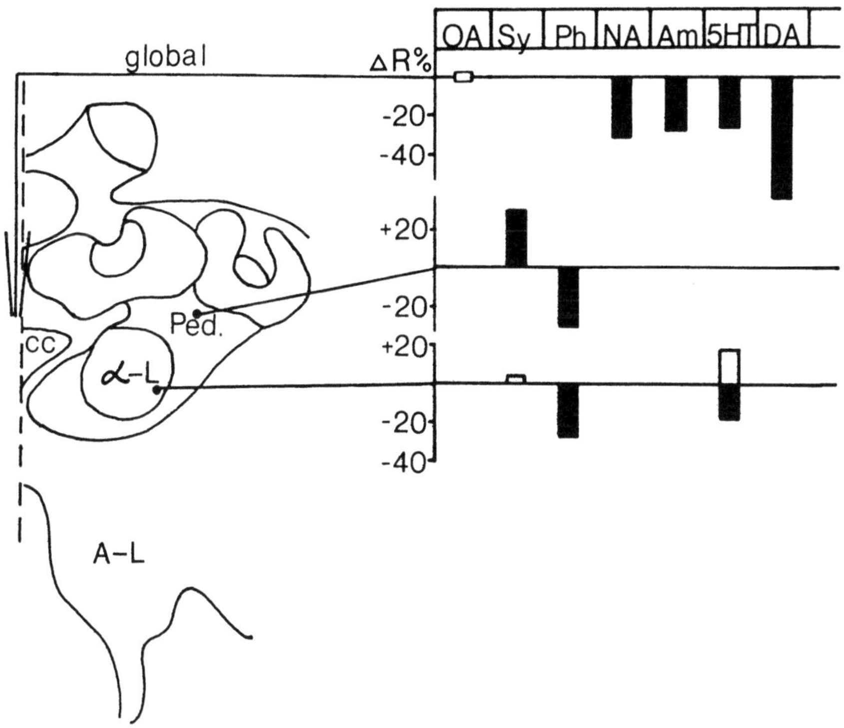

Fig. 5 Pharmacology of odor conditioning: recall test.
The effect of global (upper line) and local injections (lower 2 lines, pedunculus and alpha-lobe) on the conditioned response rate as determined in the recall test. The ordinates (ΔR %) of the three graphs give + ΔR % with columns pointing upwards, and - ΔR % with columns pointing downwards. Statistically significant differences between the experimental and control groups are indicated by filled columns; open columns indicate no statistically significant difference (see text). OA: Octopamine (10^{-6} M), Sy: Synephrine (10^{-5} M), Ph: Phentolamine (10^{-5} M), NA: Noradrenaline (10^{-5} M), Am: Amphetamine (10^{-5} M), 5-HT: Serotonin (10^{-8} M), DA: Dopamine (10^{-6} M). Two columns are given for the injection of 5-HT into the alpha-lobe. The dark column pointing downwards indicates the effect shortly after injection, the open column pointing upwards represents the late facilitatory effect (compare with Fig. 1a).

on memory formation is found if Syn, an OA-agonist, is injected into the pedunculi (Fig. 6), whilst Phen, an OA-2 receptor and alpha-adrenergic antagonist, inhibits memory formation. Since clonidine and yohimbine have no significant effects, it is concluded that an OA-2 receptor mechanism in the region of the upper mushroom bodies, rather than an alpha-adrenergic or OA-1 receptor mechanism, is involved in memory formation. The supposed OA-2 receptor mechanism in the upper mushroom body is not specific for memory storage, because Syn and Phen also facilitate and inhibit respectively the retrieval from memory (see Fig. 5). However, the suspected OA-2 receptor mechanism is unlikely to be a general arousing system, since OA, Syn and Phen, have no significant effect on the sensory or motor components of the reflex if the drugs are injected into the pedunculus.

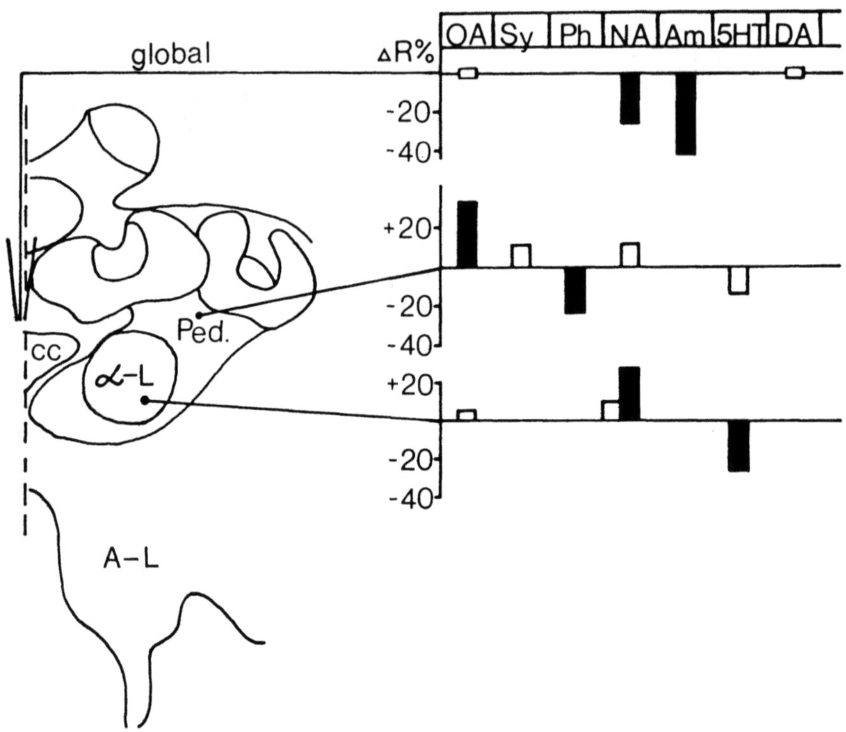

Fig. 6 Pharmacology of odor conditioning: storage test.
The effect of global (upper line) and local injections (lower 2 lines, pedunculus and alpha-lobe) on the conditioned response rate as determined in the storage test. The ordinates (ΔR %) of the three graphs give + ΔR % with columns pointing upwards, and - ΔR % with columns pointing downwards. Statistically significant differences between the experimental and control groups are indicated by filled columns; open columns indicate no statistically significant difference. Local injection of NA into the alpha-lobe was carried out for two different groups of bees - those with a high learning rate (left, open column) and those with a low learning rate (right, dark column) (see text). OA: Octopamine (10^{-6} M), Sy: Synephrine (10^{-5} M), Ph: Phentolamine (10^{-5} M), NA: Noradrenaline (10^{-5} M), Am: Amphetamine (10^{-5} M), 5-HT: Serotinin (10^{-4} M), DA: Dopamine (10^{-6} M).

It is more than likely that drugs injected into the pedunculi act upon the synapses of the closely attached calyces in each mushroom body. The different drug profiles found for the pedunculi and the alpha-lobes, indicate that there is a separation between these two regions of the mushroom bodies. For example, NA enhances the spontaneous response rate to the odor stimulus (Sp %) when injected into the pedunculi but does not change Sp % when injected into the alpha-lobe. OA inhibits SP % in the alpha-lobe but has no effect on Sp % in the pedunculi. Syn enhances memory recall in the pedunculi, but does not change recall in the alpha-lobes. OA enhances storage in the pedunculi but does not effect storage when injected into the alpha-lobes. The inhibitory action of Phen in the alpha-lobes on recall is not inconsistent with the emerging picture since Phen is also an adrenergic agonist acting on an adrenergic mechanism in the alpha-lobes.

The effect of 5-HT in the storage test is predominantly inhibitory (Fig. 6). If bees learn relatively well (46 % CR after one-trial in the control group) injection of 5-HT into either both pedunculi or both alpha-lobes reduces the CR % (The inhibitory effect resulting from injection into the pedunculus is only just not statistically significant). 5-HT injected into the antennal lobes does not alter the CR % in the storage test.

DISCUSSION

The aim of this study is to determine which neurotransmitter systems in the bee brain are important for learning and memory. The emphasis is on the biogenic amines because:

1) The amines DA, NA, OA, and 5-HT are found in considerable amounts in the mid-protocerebrum (Mercer et al., 1983), the brain region which was found to contain the processes that underly the consolidation of an olfactory memory into a stable and long lasting form of memory (Menzel et al., 1974; Erber et al., 1980; Sugawa, 1986), and

2) Biogenic amines are found to be essential modulators for learning and memory in vertebrates (Dunn, 1986) and invertebrates (Menzel and Bicker, 1987).

Since learning and memory is highly integrated into the overall function of an animal's nervous system, it is to be expected that any disturbance in the functioning of any neural component will alter learning and memory. We have, therefore, taken several steps to ensure a more specific pharmacological analysis:

1) Drugs are directly injected into the brain in either relatively large quantities (several hundred nl) through the median ocellar tract (called "global injection"), or in small quantities (2-10 nl) into specific brain regions ("local injection"). This procedure prevents any peripheral drug action.
2) The motor pattern of the unconditioned or conditioned response (UR, CR), the effectiveness of the US and the CS before conditioning, and a variety of other behavioral measures are monitored and compared after drug treatment with a control group of animals.
3) The animals are conditioned by a single CS-US pairing that lasts for only a few seconds. This procedure allows accurate control over the time course, and in turn enables a clear distinction to be made with respect to whether a drug affects memory retrieval ("recall test") or, alternatively, memory formation ("storage test"). Such a

distinction is not possible if the training procedure includes several conditioning trials, since retrieval and storage would then be compounded in the learning trials that follow.

The "interference" technique used in this study is still open to criticism despite the precautions we have taken. For example, no single neurotransmitter system works in isolation and thus the actions of one system must necessarily have ramifications on many other transmitter systems (Dunn, 1986; Martinez, 1986). Furthermore, an excess amount of transmitter or modulatory substance resulting from an artificial injection does not solely act on its own receptors but probably also affects other receptor types. Additional complications arise from the possibility that such receptors may be located both pre- and post synaptically. In other words, a sudden rise of a transmitter or modulator in a dense neuropile would certainly result in a whole range of competing effects.

Despite these complications, our survey has nevertheless uncovered some specific modulatory actions of the biogenic amines OA, 5-HT, NA and DA, as observed in the context of olfactory learning in bees. An overall summary of these results is given in Fig. 7. The amines are not likely to be transmitters in the sensory or motor pathways, but rather appear to regulate excitability in the pathways. OA and NA are synergists for facilitatory actions. OA stimulates sensory and motor components of the reflex particularly in the deutocerebrum (antennal lobe, dorsal lobe). In the mushroom bodies, OA facilitates both components of learning, memory formation and retrieval from memory, particularly in the synaptic input region of the mushroom bodies. NA enhances memory formation when injected into the alpha-lobes of bees with a low learning rate. This effect is much weaker in normal learning bees. Small facilitatory effects are also seen when NA is applied to the antennal lobes and the pedunculi of the mushroom bodies. It is still unclear, however, why global injection of NA (and Amph) reduces memory formation (Fig. 5, upper line). A similar controversial effect is seen with the action of OA on the sensory component of the reflex. Global injection of OA enhances the spontaneous response to the odor and the response to water (Fig. 3a), but local injection into the alpha-lobes reduces these responses.

Memory formation is reduced when 5-HT is injected into the alpha-lobe, and memory retrieval is inhibited after global injection or local injection into both parts of the mushroom bodies (pedunculus and alpha-lobe). In contrast, the sensory and motor components of the reflex are not changed after treatment with 5-HT, even when 5-HT is injected into the antennal lobes or the dorsal lobes. 5-HT appears to oppose the effects of OA/NA on learning and memory retrieval. We also tested

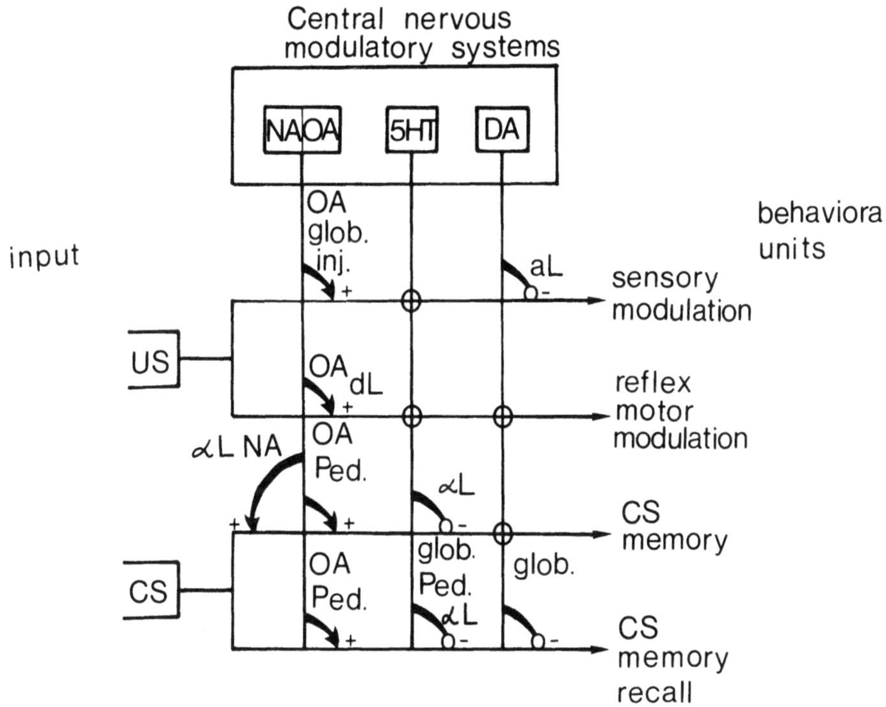

Fig. 7 Summary diagram and interpretation
The inputs (US, CS) are connected to the outputs (behavioral units) through direct pathways (sensory and motor components of the US pathway) and indirect pathways (memory components of the CS pathway). The central nervous modulatory systems are interpreted as circuits which change the connections between the inputs and outputs. Arrows indicate a facilitatory action of the particular modulator, curved lines with open circle indicate an inhibitory action. Large circles around the cross-points symbolise no action of the particular neuromodulator. Abbreviations: glob: global injection; dL: dorsal lobe (the motor center in the deutocerebrum); αL: alpha-lobe of the m.b; ped: pedunculus of the m.b; aL: antennal lobe. The DA inhibitory effect in the antennal lobe (upper right, cross point) was observed by Macmillan and Mercer (1987), (see text).

whether injections of 5-HT into the antennal lobe affect the extinction of the conditioned response after repeated presentations of an unreinforced CS. This is not the case. These results are somewhat surprising since each antennal lobe is densely invaded by the ramifications of a serotonergic cell, the deutocerebral giant cell (Rehder et al., 1987). The type of modulation originating from the serotonergic deutocerebral giant cell remains unknown.

Macmillan and Mercer (1987) observed that application of DA to the antennal lobes blocks olfactory coding. This blocking effect is less pronounced if the odor is conditioned more strongly. It is tempting to speculate that a strong memory trace can even be activated with a weakened sensory input. In this case it is likely that the memory trace resides outside the antennal lobes. Indeed, this would support our conclusion (Erber et al., 1980) that the mushroom bodies are the loci of the permanent memory trace. In our experiments, global injections of DA (Fig. 5 and 6) inhibit memory retrieval but not learning. Since we have not performed local injections of DA, it is unknown which neuropile in the protocerebrum participates, via a DA mechanism, on memory retrieval. Immunocytochemical studies reveal dense DA-immunoreactivity in the central body and neurons closely attached to the alpha-lobes (Schäfer, pers. communication). Further experiments will demonstrate whether or not these neurons are involved in the retrieval process.

It appears from these studies that olfactory learning in honey bees is modulated by two antagonising aminergic systems, the facilitatory OA/NA system and the depressing 5-HT system. DA is unlikely to be a modulatory substance but may instead be a transmitter which is involved in integrative processes at the sensory and motor command level. The intricate pharmacology of the OA and NA receptors makes it impossible at the present time to separate the roles of the two amines. In any case, our results support the notion that the facilitatory modulatory system(s) acts in parallel on many levels. This modulatory system is likely to serve several processes during learning: enhancement of sensory processing and attention, facilitation of the motor response, and transmission of the US-excitation to an associative network for CS-specific synaptic changes. The antagonising 5-HT system appears to be more specific.

The electrophysiological analysis of neuropiles at the single cell level is extremely difficult due to a dense packing of the small neurons. The local injection technique, however, may help unravel the complexities of the pharmacological subsytems operating in such neuropiles.

REFERENCES

Bicker, G., and Menzel, R. 1987. Nature (subm).

Bitterman, M.E.,; Menzel, R.,; Fietz, A.,; and Schäfer, S. 1983. Classical conditioning of proboscis-extension in honey bees. J. Comp. Physiol. Psychol. 97: 107-119.

Dethier, V.G. 1976. The hungry fly. Cambridge, Mass. Harvard University Press.

Dunn, A. 1986. Neurochemistry of learning and memory: An evaluation of recent data. J. Ann. Rev. Psychol. 31: 343-390.

Erber, J.; Masuhr, Th.; and Menzel, R. 1980. Localization of short-term memory in the brain of the honey bee. Physiol. Entomol. 5: 343-358.

Evans, P.D. 1980. Biogenic amines in the insect nervous system. Adv. Insect 15: 318-473.

Evans, P.D. 1982. Porperties of modulatory octopamine receptors in the locust In: Neuropharmacology of insects. Ciba Foundation Symposium 88: pp. 48-69. London: Pitman.

Frisch, K. von. 1967. The Dance Language and Orientation of Bees. Cambridge: University Press.

Glover, J.C., and Kramer, A.P. 1982. Serotonin analogue selectively ablates identified neurones in the leech embryo. Science 216: 317-319.

Harris-Warrick, R.M., and Krawitz, E.A. 1984. Cellular mechanisms for modulation of posture by octopamine and serotonin in the lobster. J. Neurosci. 4: 1976-1993.

Kupfermann, I, and Weiss, K.R 1981 The role of serotonin in arousal of feeding behavior of Aplysia In: Serotonin Neurotransmission and Behavior, eds. D.L. Jacobs and A.Gelperin, Cambridge, MA, London: MIT Press.

Kuwabara, M. 1957. J. Fac. Sci. Hokkaido Univ. Series 6, Zoology 13, 458-464.

Lent, C.M., and Dickinson, M.H. 1984. Serotonin integrates the feeding behavior of the medicinal leech. J. Comp. Physl. A 154: 457-471.

Livingstone, M.S.; Harris-Warrick, R.M.; and Kravitz, E.A. 1980. Serotonin and octopamine produce opposite postures in lobster. Science (NY) 208: 76-79.

Long, T.F., and Murdock, L.L. 1983. Stimulation of blowfly by octopaminergic drugs. Proc. Natl. Acad. Sci. 80: 4159-4163.

MacMillan, C.S. and Mercer, A.R. 1987. An investigation of the role of dopamine in the antennal lobes of the honeybee, Apis mellifera. J. Comp. Physiol. 160: 359-366.

Martinez, J.L. 1986. Memory: drugs and hormones. In: Learning and Memory. A biological view, eds. J.L. Martinez and R.P. Kesner, pp. 127-164. Academic Press Orlando.

Menzel, R. 1983. Neurobiology of learning and memory: the honey bee as a model system. Naturwiss. 70: 504-511.

Menzel, R.,; Erber, J.,; and Masuhr, Th. 1974. Learning and memory in the honeybee. In: Experimental Analysis of Insect Behaviour, eds. L. Barton-Browne, pp. 195-217. Berlin-Heidelberg-New York: Springer Verlag.

Menzel, R., and B. Michelsen. 1986. Neuropharmacology of associative learning in bees. In: Learning and Memory: Mechanisms of Information storage in the Nervous System, eds. M. Matthies, pp. 273-275. Oxford, Pergamon Press.

Menzel, R., and Bicker, G. 1987. Plasticity in neuronal circuits and assemblies of invertebrates. In: The Neural and Molecular Basis of Learning, eds. J. P. Changeux and M. Konishi, pp. 433-472. J. Wiley & Sons Ltd.

Mercer, A.R. 1987. Biogenic Amines and the Bee Brain In: Neurobiology and Behavior of Honeybees, eds. R. Menzel and A. Mercer, pp. 244-252. Berlin-Heidelberg-New York-London-Paris-Tokyo: Springer.

Mercer, A.R., and Menzel, R. 1982. The effect of biogenic amines on conditioned and unconditioned responses to olfactory stimuli in the honey bee, Apis mellifica. J. Comp. Physiol. 145: 363-368.

Mercer, A.R.; Mobbs, P.G.; Davenport, A.P.; and Evans, P.D. 1983. Biogenic amines in the brain of the honeybee, Apis mellifera. Cell Tissue Res. 234: 655-677.

Michelsen, D.B. In press. Catecholamines affect storage and retrieval fo conditioned odour stimuli in honey bees. J. Biochem. Physiol. C.

Rehder, V.; Bicker, G.; and Hammer, M. 1987. Serotonin-immunoreactive neurons in the antennal lobes and suboesophageal ganglion of the honeybee. Cell Tissue Res. 247: 59-66.

Robertson, H.A., and Juorio, A.V. 1976. Octopamine and some related noncatecholic amines in invertebrate nervous systems. Inter. Rev. Neurobiol. 19: 173-224.

Sombati, S., and Hoyle, G. 1984. Central nervous sensitization and dishabituation of reflex action in an insect by the neuromodulator octopamine. J. Neurobiol. 15: 455-480.

Stevenson, P.A., and Kutsch, W. 1987. A reconsideration of the central pattern generator concept for locust flight. J. Comp. Physiol. A 161: 115-129.

Sugawa, M. 1986. Zeitliche Dynamik und Lokalisation des Engramms im Honigbienengehirn (Apis mellifera). Diplomarbeit Technische Universität Berlin.

Sugawa, M., and Menzel, R. 1986. Zeitliche Dynamik und Lokalisation der Gedächtnis-konsolidierung bei der Biene. In: Sensomotorik identifizierter Neuronen, eds. N. Elsner und W. Rathmeyer, p. 174. Thieme Verlag, Stuttgart.

Vareschi, D.M. 1971. Duftunterscheidung bei der Honigbiene. Einzellableitungen und Verhaltensreaktionen. Z. vergl. Physiol. 75: 143-173.

HABITUATION OF THE LANDING RESPONSE OF *DROSOPHILA*

W.C. Wittekind and H.Ch. Spatz
Institut für Biologie III
Universität Freiburg
Germany

A device was developed to automatically record the landing response of *Drosophila*. Tethered *Drosophilae* fly in front of a viewing screen onto which moving pattern can be projected. The movement of the front legs during a landing response can be monitored by two sensitive light gates: one records the onset of the forward thrust, the other the completion of the response as defined by a full forward extension of the legs.

If the visual pattern consists of a grey bar moving upwards on the screen, the fly reacts with a response time of 200 msec. Upon repeated stimulation, this time increases to 300 msec and the response occurs less frequently. The behavioral analysis shows that this is not due to muscle fatigue nor to receptor adaptation but has to be attributed to a habituation of the fly to the landing stimulus.

Wildtypes *Berlin* and *C-S* and the mutants *dunce*, *rutabaga* and *amnesiac* have been characterized by their timecourse of habituation. It is found that the mutants habituate faster than the corresponding wildtype. The effects of neuropharmaca on the habituation are studied after injection of nanoliter quantities into the head capsule in situ. Caffeine or theophylline almost eliminate habituation. This is also found for dopa-mine.

NATO ASI Series, Vol. H19
Modulation of Synaptic Transmission and Plasticity
in Nervous Systems. Edited by G. Hertting and H.-C. Spatz

INTRODUCTION

"The simplest, most universal form of learning is habituation. It may be as fundamental a characteristic of life as DNA. It differs from other forms of learning in that it is a loss of habitual responses rather than the acquisition of new ones." (Dethier, 1976). The term habituation is used for a stimulus induced decrease of response to this stimulus, not attributable to the receptor or the effector peripherie of the nervous system.

Thorpe (1939) first described "pre-imaginal conditioning" in Drosophila, a phenomenon which after some controversy was reinterpreted by Manning (1967) as due to habituation. Habituation and sensitization of flies to chemical stimuli have been extensively studied by Dethier et. al. (1965) and Getting (1971). Duerr and Quinn (1982) could show that three mutations blocking olfactory learning also affect habituation and sensitization to chemosensory stimuli.

Habituation of Drosophila to visual stimuli was first reported by Dürrwächter (1957). He uses the term "Ermüdung" (weariness) but excludes muscle fatigue. Also in his paradigm, adaptation of the visual receptor cells seems an unlikely cause. Thon (1980) has reported habituation of an optocardiac response in Calliphora vomitoria. Perez de Talens and Taddei-Ferretti (1970) noted a decrease of the landing response of Musca domestica upon repeated presentation of a visual landing stimulus. The first systematic investigation of habituation and sensitization of the landing response used Drosophila melanogaster (Fischbach, 1981; see also this volume) in particular neuroanatomically defined mutants as well as mutants affected in olfactory learning (Fischbach, 1983).

In this communication an automated device to monitor the landing response of Drosophila and to measure characteristic times of the response is described. Pilot studies of the effects of neuropharmaca on landing activity and habituation of the response are reported.

MATERIALS AND METHODS

Neuropharmaca used: caffeine (Sigma), theophylline (Roth), theobromine (Roth), dopamine hydrochloride (Sigma), serotonine hydrochloride (Sigma), forskolin (Sigma).

Stocks and culture methods: *Drosophila melanogaster* wildtype Berlin are used for most experiments. Only where mutant strains *dunce*^{2}, *amnesiac* and *rutabaga* are tested, wildtype C-S is used as control. All of these strains were kindly provided by Dr. W.G. Quinn, M.I.T. Cambridge, Mass., U.S.A.

The stocks are maintained on standard cornmeal medium in transparent plastic bottles. The flies are reared at 24°C in a temperature controlled room which is illuminated by two 60 W bulbs with a 12h light and 12h dark cycle. On the day of eclosion flies are transferred to a fresh culture bottle. Experiments are carried out on the next day. Only male flies were tested.

Mounting of flies: Individual male flies are anaesthetized by low temperatures and placed on a copper block cooled to 0°C. A mount consisting of a small steel tube supporting a wire loop is approached towards the dorsal side of the thorax of the fly. The wire loop holds a tiny drop of a wax-collophonium mixture, which can be melted via a pulse of electrical current through the loop. This way head and thorax can be fastened to the mount. The tethered flies are tested for good flight activity and transferred to an illuminated box with high humidity. In the time ($>$30 min) before the actual test for landing responses a small piece of soft paper soaked with a sucrose solution is given to the flies.

Landing stimuli: For the test an individual tethered fly is placed in front of a viewing screen. A cardanic suspension and a micromanipulator allows fine adjustment of the fly's position in all 6 degrees of freedom. In all experiments the equator of the eyes is horizontal . The viewing screen covers a range of -42° to +42° in the vertical direction and -39° to +39° in the horizontal dircction as viewed by the fly. Landing is elicited by the movement of a grey stripe over the otherwise homogenously illuminated viewing screen (Leitz pradovit projector, 250 W halogen bulb,

1 W/m^2 on the side facing the fly). The grey stripe covers $7,5^0$ in the vertical direction and the entire breadth in the horizontal position. Unless otherwise stated the light intensity on the screen within the stripe is reduced by 15%. The stripe moves upwards over the entire screen at a speed of 170^0/sec with respect to the fly's viewing field. In the following we will refer to this as the landing stimulus. In the test the landing stimulus is repeated 70 times at 2 sec intervalls.

Automatic recording of the landing response: In the normal flight position all legs of Drosophila are extended backwards (Fig. 1)

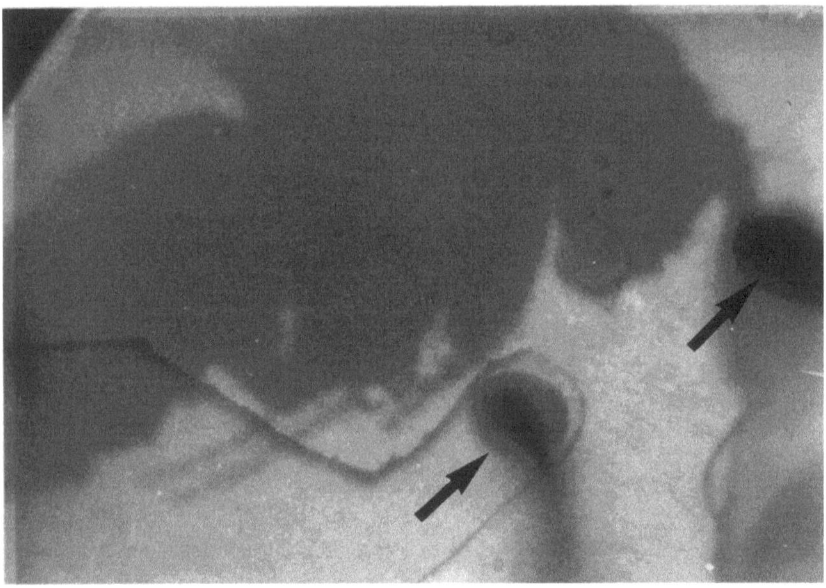

Figure 1. The image of a flying Drosophila projected through a microscope onto a focussing screen. The arrows point to the endings of two light guides positioned to record the onset and the completion of the landing reaction. Movement of the front legs reduces the light flux through the light guides by approximately 20 % for a period of a few milliseconds. This change of the light flux is detected by sensitive optoelectronics.

A landing response is defined as a forward thrust of the forelegs until they are streched out approximately at the height of the antennae, almost in a horizontal position. The optoelectronic device described here is designed to automatically monitor the landing response and to measure the time of the onset of the response and its performance speed (Wittekind, 1987a). For this the image of the fly is projected with red light (λ>630 nm) through a microscope onto a screen. After appropriate adjustment the body and the head of the fly are shielded from the red light. Two adjustable light guides are placed on the screen, one directly in front of the image of the forelegs, the other just below the antennae, in order to monitor beginning and completion of the landing response. Special care has to be taken to separate the signal due to leg movement from the periodic signal of approximately 200 Hz due to the wing-movement.

The criterion for a landing response is a signal from light gate 1 followed by a signal from light gate 2 within 500 msec. The automatic recording is reliable: in less than 5% of the cases the automated recording differed from the recording by eye. (Habituation on the side of the observer is likely to account for an error of this order of magnitude). Also the influence of the remaining red light on body and head of the fly proves to have no significant influence on the landing behavior.

The signals from the light gates are also used to determine characteristic times for the response. For this an electronic clock is started when the visual signal appeares on the viewing screen. The clock is stopped by the signal from light gate 1, that is at the onset of the landing response. At the same time a second electronic clock is started and stopped by the signal from light gate 2. This clock measures the time from onset to near completion of the response. The data are processed and stored by computer.

Application of neuropharmaca: Neuropharmaca are injected into the head capsule in situ, i.e. directly in the test apparatus. For this 10 ± 2 nanoliter of Ringer solution with the neuropharmaca dissolved at appropriate concentrations is sucked into a microcapillary drawn out to an external diameter of about 15 micrometer by a Narashige electrode puller. With a micromanipulator the microcapillary is inserted into the head capsule in the ocellar region. The solution is injected into the head by air pressure. Within one minute the solution is distributed over the entire head capsule. A small amount is also found in the thorax. Since the volume of the head capsule is approximately 100 nl, the overall concentration of the neuropharmaca is approximately a 1 : 10 dilution from the injected concentrations.

RESULTS AND DISCUSSION

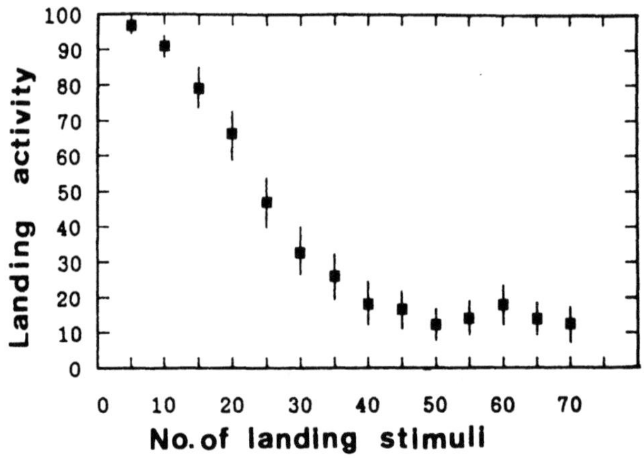

Figure 2a. The landing activity in % against the number of stimuli presented. The activity is measured as the number of responses to 5 consecutive stimuli and averaged over all flies, which maintained steady flight during the 140 sec of the test. The error bars denote the standard deviation of the mean. 32 flies, wildtype Berlin.

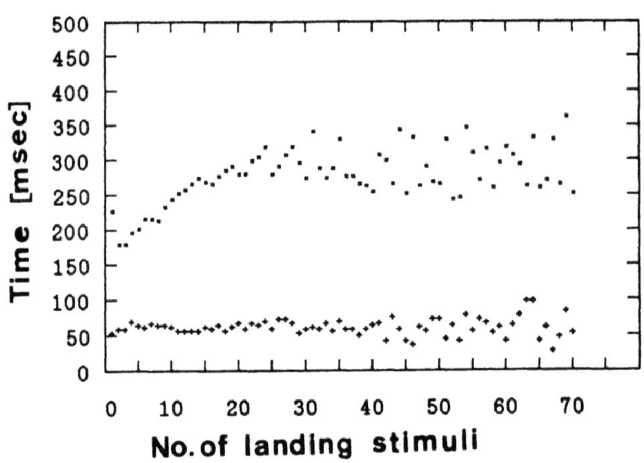

Figure 2b. The time till the onset of the landing response (squares) and the time between onset and completion of the landing response (crosses) are plotted against the number of stimuli. The response times are plotted for each stimulus and are averaged over all flies which responded to this stimulus. The data result from the same experiment as those of figure 2a.

Basic observations:: In Drosophila melanogaster the landing response can easily be elicited by grey stripes moving upwards through the visual field. (Fischbach, this volume). Upon repeated stimulation with weakly contrasting stripes the flies show a pronounced decrease of the landing activity with the number of stimuli (Fig. 2a). Furthermore the time till the onset of the response, after an initial decrease of about 50 msec, raises from 180 msec to 300 msec (Fig. 2b).

The decrease in landing activity might be attributed to muscle fatigue, adaptation of visual receptors and/or habituation.

Two experiments to rule out muscle fatigue: The time between onset and near completion of the landing reaction, i.e. the actual time for the movement of the front legs is represented by the lower trace in Fig. 2b. Contrary to expectation on the basis of muscle fatigue this time remains remarkably constant at approx. 50 msec.

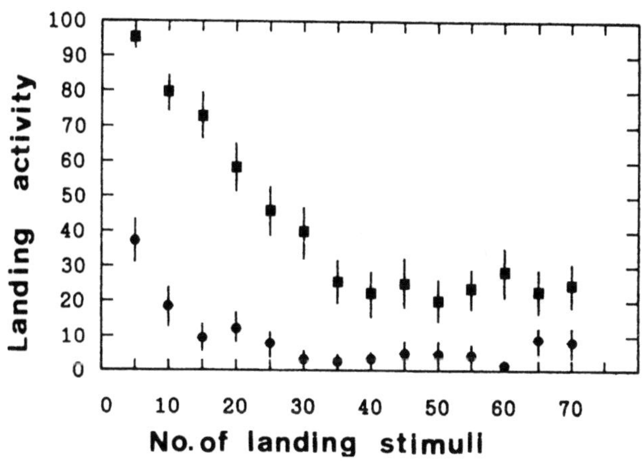

Figure 3. The landing activity against the number of stimuli for a group of flies, which receive prestimulation with 120 upward moving stripes (filled circles), and a control, which is presented with the illuminated viewing screen without stripes (squares) for the same period of time. 33 flies, wildtype Berlin, are tested in each group. For this and all other comparisons the controls are tested within the same hour as the experimental flies.

Another experiment uses the fact that tarsal contact to a substrate, e.g. to a small piece of paper, completely suppresses flight behaviour in Drosophila as well as any landing response. Flies mounted in the test apparatus holding a piece of paper with their tarsae can be prestimulated with landing stimuli. Upon removal of the paper Drosophilae start flight immediately. Although no reaction is observable during prestimulation, landing activity is strongly reduced in comparison to a control, in which no upward moving stripes are presented to the flies during the time of prestimulation (Fig. 3).

Two experiments to rule out visual adaptation: Prestimulation of Drosophilae with tarsal contact can also be used to test, whether visual adaptation is responsible for the decrease in landing activity. Two groups of flies were compared. One group received prestimulation with upward moving stripes the other prestimulation with downward moving stripes. After initiating of flight both groups were tested with normal (upward) landing stimuli. The effect of prestimulation is clearly direction sensitive, which can not be explained by adaptation of visual receptor cells (Fig. 4).

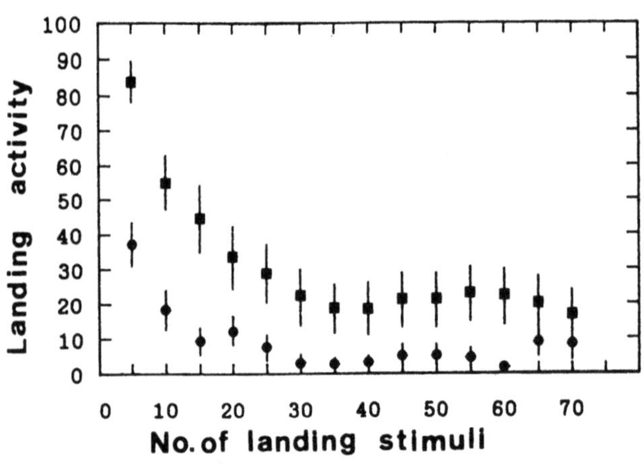

Figure 4. The landing activity against the number of stimuli for a group of 33 flies which receive prestimulation with 120 upward moving stripes (filled circles, same data as in figure 3), and a group of 23 flies which receive prestimulation with 120 downward moving stripes (squares). Wildtype Berlin.

In a similar experiment flying Drosophilae were exposed to 120 short periods of darkening, equivalent to the darkening by 120 landing stimuli. This does not elicit a landing response in Drosophila. Upon the following exposure to landing stimuli, the flies show landing activity even slightly higher than the controls which had not experienced darkening during the 240 sec period prior to the test (Fig. 5). If simply darkening and not movement of the grey stripe were effective, a lower activity than in the control should have been found.

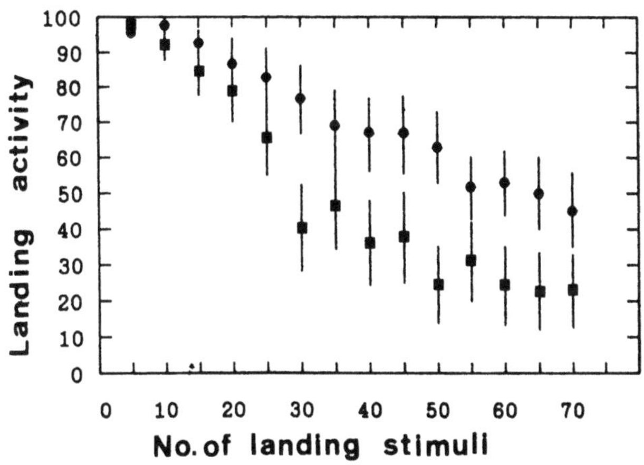

Figure 5. The landing activity against the number of stimuli for a group, which prior to the test receives 120 periods of darkening by 15 % for 24 msec every 2 seconds (filled circles), and a control, which is flying 4 min in front of the illuminated viewing screen without experiencing darkening (squares). 20 flies, wildtype Berlin, in each group.

Two experiments supporting habituation: Thompson and Spencer (1966) have listed a number of behavioural criteria for habituation. Here we present data concerning dishabituation and spontaneous recovery. A more complete treatment is given by Wittekind (1987a).

Dishabituation can be induced by exposing a fly, which had been habituated to a repetitive stimulus, to another stimulus of preferentially different modality. This should lead to recover full or at least partial responsiveness to the repetitive stimulus. Fig. 6 shows an experiment in which Drosophilae are tested with repeated landing stimuli. Between landing stimulus 35 and 36 they are shortly vibrated in situ. This suffices to increase the landing activity to over 50% (Fig. 6).

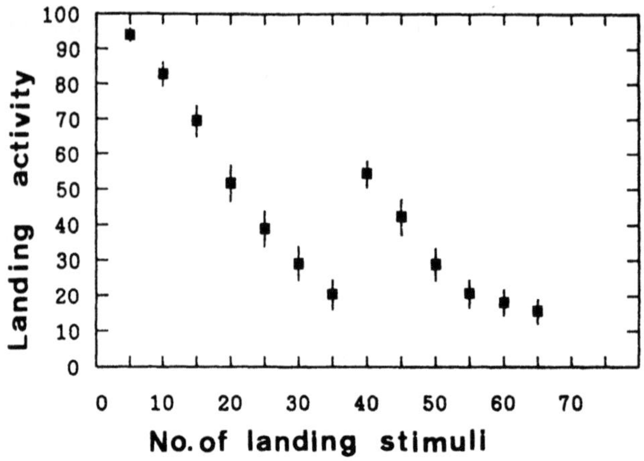

Figure 6. The landing activity against the number of stimuli. In the 2 seconds between stimulus 35 and 36 the flies are shortly vibrated by operating a loudspeaker, which is mechanically coupled to the mount of the flies. 64 flies, wildtype Berlin.

Spontaneous recovery can be demonstrated by placing a time interval in which Drosophilae continue to fly undisturbed between two periods of exposure to landing stimuli. Fig. 7 shows that recovery is nearly

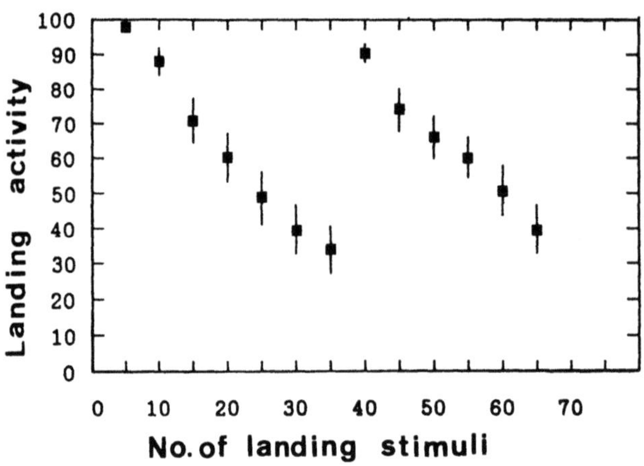

Figure 7. The landing activity against the number of stimuli. Presentation of 35 stimuli is followed by a period of 70 sec without stimulus and another period of 30 stimuli. The flies maintain flight during the entire time. 38 flies, wildtype Berlin.

complete after a 70 sec interval without landing stimuli. The time for 50% recovery is of the order of 20 sec. (A. Wachenfeld, unpublished), demonstrating that *habituation of the landing response* is a rather short term process in contrast to visual learning in Drosophila (Folkers and Spatz, 1981).

Interpretation of the time until the onset of the landing response: If very short but effective expanding stimuli are presented, *Drosophila* is able to start the landing response within 50 msec (Wittekind, 1987b). Less direct evidence (Wittekind, 1987a) leads to the conclusion that the delay time of the response to upward moving grey stripes is also of the order of 50 msec, independent of the degree of habituation. Wittekind (1987a) has shown, that the time till the onset of the landing response is mainly determined by the requirement for temporal and spatial integration of the landing stimulation (compare Fischbach, this volume). In order to elicit the response the stimulus has to be presented over a retinal angle increasing with the degree of habituation, i.e., at a given velocity of the grey stripe for an increasing period of time. This interpretation is supported by a strong negative correlation between landing activity and time till the onset of the response in all experiments using wildtype *Berlin* not treated with neuropharmaca.

Effects of neuropharmaca on the landing response and its habituation: A major incentive to develop a single fly test is the possibility to study the effects of neuropharmaca, directly injected into the head capsule (Mercer and Menzel, 1982), on plasticity of behaviour of *Drosophila*. This was not feasible in our earlier population studies (Folkers and Spatz, 1984). The key role of cAMP in plasticity of behaviour has been pointed out by Kandel and Schwarz (1982) for *Aplysia* and summarized by Quinn (1984) for *Drosophila*. Caffeine and theophylline, inhibitors of phosphodiesterases, are known to impair olfactory learning (Byers, personal communication) as well as visual learning (Folkers and Spatz, 1984) in *Drosophila*. Upon injection of 10 nl 10^{-2} M theophylline the decrease of landing activity is only small as compared to the control, which received an injection of 10 nl Ringer solution (Fig. 8a), or to untreated flies (Fig. 2a). As expected the increase in the time till the onset of the response is much less pronounced in theophylline treated flies than in the control (Fig. 8b). Very similar results are obtained if caffeine at a concentration of 10^{-2} M is injected into the head capsule of the flies. Similar but smaller effects are obtained injecting 5×10^{-3} M theobromine.

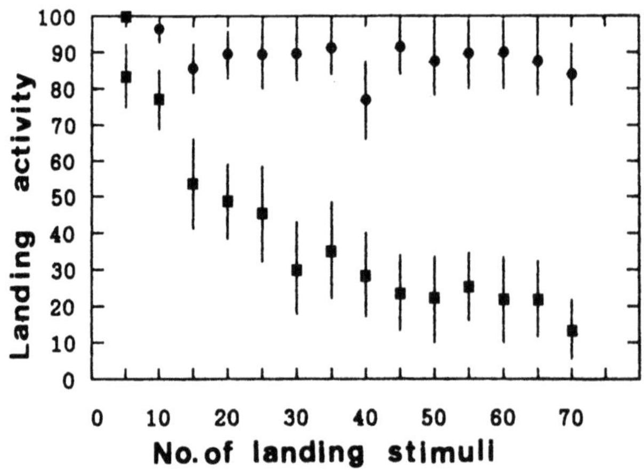

Figure 8a The landing activity against the number of stimuli. A group of 11 flies received injections of 10 nl 10^{-2} theophylline (filled circles), the control group of 12 flies injections of 10 nl Ringer solution (squares). Wildtype Berlin.

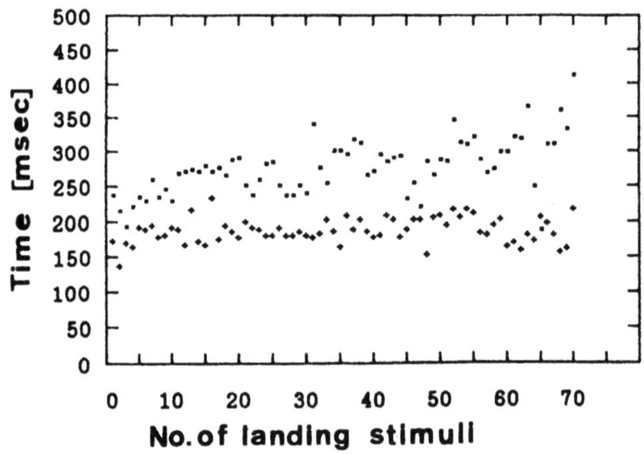

Figure 8b. The time until the onset of the landing response for the theophylline treated flies (crosses) and the Ringer control (squares). The data result from the same experiment as those of figure 8a.

In another attempt to influence the cAMP metabolism, forskolin known to stimulate adenylate cyclase (Seamon et. al., 1981) was injected into the head capsule. At 10^{-4}M concentration, the limit of solubility of forskolin in Ringer with 1 % DMSO, no significant effect on the landing response was observed.

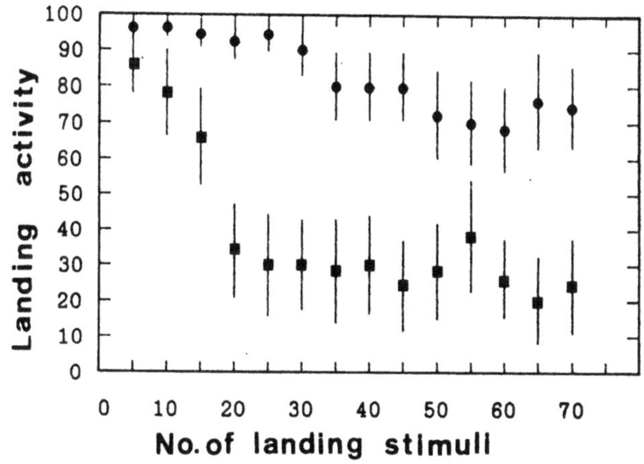

Figure 9a. The landing activity against the number of stimuli for a group of flies, which received injections of 10 nl 5 x 10^{-2} M dopamine (filled circles), and a Ringer control (squares). 10 flies, wildtype Berlin, in each group.

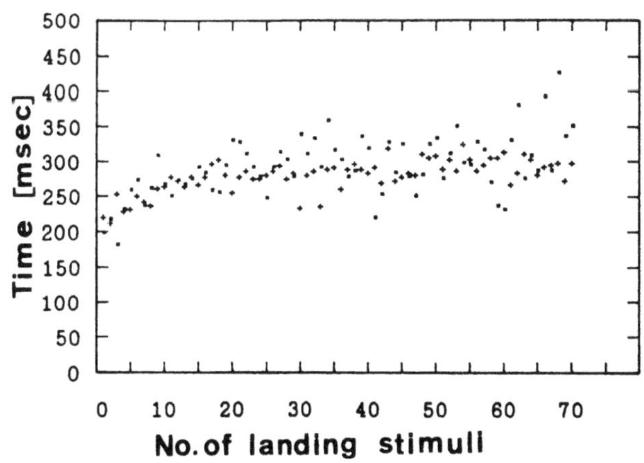

Figure 9b. The time until the onset of the landing response for the dopamine treated flies (crosses) and the Ringer control (squares). The data result from the same experiment as those of figure 9a.

Dopamine reduces conditioned responses in bees (Mercer and Menzel, 1982). The effects of an injection of 10 nl 5 x 10^{-2} M dopamine on the habituation of the landing response in Drosophila is shown in Fig. 9a. Again habituation of the landing response is weaker as in the control. However, no significant difference in the times till the onset of the response are observed Fig. 9b. Theophylline and dopamine have a similar effect on the landing activity but influence the time till the onset of the response differently. This shows that the two behavioural traits should not be considered equivalent.

Galissie (1986) has pointed out the importance of reduced serotonine levels in the chemosensory conditioning paradigm developed by Medioni and Vaysse (1975). In our test injection of 10 nl 5 x 10^{-2} M serotonine had no significant effect on the landing response.

Landing activity of mutants of olfactory learning: A complementary approach towards an understanding of neural plasticity in molecular terms is the study of mutants, preferentially with well known biochemical defects. Several Drosophila mutants of olfactory learning and memory are available (Hall, 1982). dunce is defective in the phosphodiesterase II (Byers et. al., 1981) rutabaga in adenylate cyclase (Livingstone et. al., 1984; Dudai and Zvi, 1985). The biochemical defect in amnesiac is not yet known.

Wildtype C-S serves as control for the observations of habituation of the landing response. C-S is much less sensitive to landing stimuli than wildtype Berlin. In order to elicit landing behaviour similar to that reported in the previous sections, the contrast between moving stripe and viewing screen has to be increased to 80 %.

For all three mutants the decrease of the landing activity is more pronounced than in the wildtype C-S (Fig. 10). Particularly for mutant dunce this seems to contradict the findings for Drosophilae treated with inhibitors of phosphodiesterases (Fig. 9a). However, corresponding discrepancies in the neuropharmacology of associative visual learning have been found earlier (Folkers and Spatz, 1984). They may possibly be reconciled, if the role of cGMP in visual learning and habituation is understood.

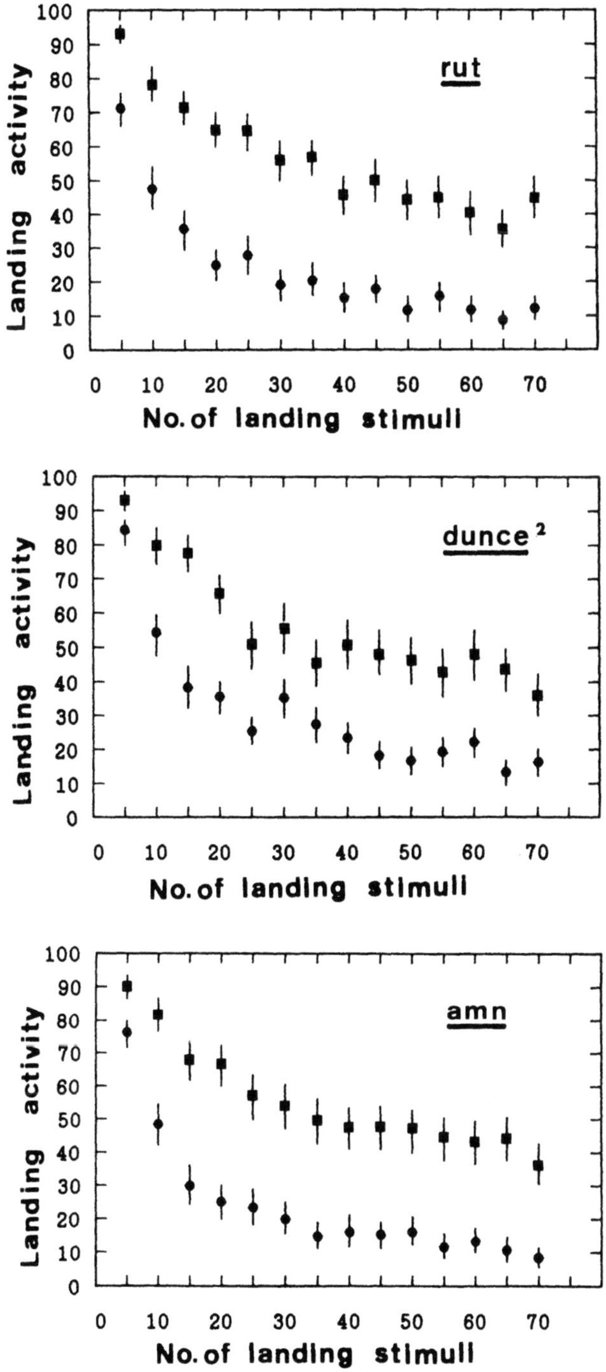

Figure 10. The landing activity against the number of stimuli for mutants defect in olfactory learning (filled circles) and wildtype C-S (squares). 34 to 45 flies in each group.

Although the interpretation of behavioural differences between strains always bear the uncertainties accompanied with unknown differences in the genetic background, we feel that the mutants dunce, rutabaga and amnesiac try to tell us a different story about habituation to visual stimuli than the mutants dunce, turnip and amnesiac about habituation to chemosensory stimuli. However, we don't understand them yet.

The landing activity in rutabaga decreases faster than in the wildtype C-S, yet the times till the onset of the response are indistinguishable in these strains. This is also true for mutant $dunce^2$ with the exception of the time for the onset of the very first landing response, which is significantly longer than in the wildtype. This lends support to an interpretation of the initial decrease of the time till onset (Fig. 2b) as an indication for a sensitization process. In amnesiac this initial decrease is not larger than in the wildtype but the times till the onset of the response are in the whole longer than in C-S. It seems likely that these times bear more information than will become accessible through studying the landing activity.

This work was supported by the Deutsche Forschungsgemeinschaft. We are grateful for stimulating discussion with Drs. K.F. Fischbach, R. Willmund R. Hertel and W. Michalke.

REFERENCES

Byers D, Davis RL, Kiger JA (1981) Defect in cyclic AMP phosphodiesterase due to the dunce mutation of learning in Drosophila melanogaster. Nature 289:79-81

Dethier VG (1976) The hungry fly. Harvard University Press.

Dethier VG, Solomon RL, Turner LH (1965) Sensory input and central excitation and inhibition in the blowfly. J. Comp, Physiol. Psychol. 60:303-313

Dudai Y, Zvi S (1985) Multiple defects in the activity of adenylate cyclase from the Drosophila memory mutant rutabaga. J. of Neurochem. 45:355-364

Duerr JS, Quinn WG (1982) Three Drosophila mutations which block associative learning also affect habituation and sensitization. Proc. Natl. Acad. Sci. USA 79:3646-3650

Dürrwächter G (1957) Untersuchungen über Phototaxis und Geotaxis einiger Drosophilamutanten nach Aufzucht in verschiedenen Lichtbedingungen. Z. Tierpsychologie 14:1-28

Fischbach KF (1981) Habituation and sensitization of the landing response of Drosophila melanogaster. Naturwiss. 68:332

Fischbach KF (1983) Neurogenetik am Beispiel des visuellen Systems von Drosophila melanogaster. Habilitationschrift, Universität Würzburg

Folkers E, Spatz HCH (1981) Visual learning behaviour in Drosophila melanogaster wildtype AS. J. Insect Physiol. 27:615-622

Galissie M (1986) La serotonine chez Drosophila melanogaster aspects neuroanatomiques et comportementaux. These, Universite Toulouse

Getting PA (1971) The sensory control of motor output in fly proboscis extension. Zeit. vergl. Physiol. 74:103-120

Hall JC (1982) Genetics of the nervous system in Drosophila. Quart. Rev. of Biophys. 15:223-479

Kandel ER, Schwartz JH (1982) Molecular biology of learning: modulation of transmitter release. Science 218:433-443

Livingstone MS, Sziber PP, Quinn WG (1984) Loss of calcium/calmodulin responsiveness in adenylate cyclase of rutabaga, a Drosophila learning mutant. Cell 37:205-215

Manning A (1967) "Pre-imaginal conditioning" in Drosophila. Nature 216:338-340

Medioni J, Vaysse G (1975) Suppression conditionelle d'un reflexe chez la Drosophile (Drosophila melanogaster): aquisition et extinction. C.R. Soc. Biol. 5:1386-1391

Mercer AR, Menzel R (1982) The effects of biogenic amines on conditioned and unconditioned responses to olfactory stimuli in the honeybee Apis mellifera. J.Comp. Physiol. A 145:363-368

Perez de Talens AF, Taddei-Ferretti C (1970) Landing reaction of Musca domestica: Dependence on dimensions and position of the stimulus. J. Exp. Biol. 52:233-256

Quinn WG (1984) Work in invertebrates on mechanisms underlying learning in: The biology of learning eds. Marler P, Terrace HS, Dahlem Konferenzen 1984, Springer Verlag

Seamon BS, Padgett W, Daly JW (1981) Forskolin: unique diterpene activator of adenylate cyclase in membranes and in intact cells. Proc. Natl. Acad. Sci. USA 78:3363-3367

Thomson RF, Spencer WA (1966) Habituation: A model phenomenon for the study of neuronal substrates of behavior. Psychological Review 73:16-43

Thon B (1980) Habituation of cardiac and motor responses to a moving visual stimulus in the blow fly, Calliphora vomitoria. J. Comp. Physiol. Psychol. 94:886-893

Thorpe WH (1939) Further experiments on pre-imaginal conditioning in insects. Proc. Roy. Soc. B 127:424-433

Wittekind W (1987a) Analyse des stereotypen und insbesondere plastischen Landeverhaltens von Drosophila melanogaster. Thesis, Universität Freiburg

Wittekind W (1987b) The landing response of tethered flying Drosophila is induced at a critical object angle. J. of exp. Biol. in press

HABITUATION AND SENSITIZATION OF THE LANDING RESPONSE OF *DROSOPHILA MELANOGASTER*: II. RECEPTIVE FIELD SIZE OF HABITUATING UNITS

Karl-Friedrich Fischbach[1] and Bernhard Bausenwein[2]

[1]Institut für Biologie III der Universität
Schänzlestr.1
7800 Freiburg i. Brsg.
W.-Germany
[2]Institut für Genetik und Mikrobiologie der Universität
Röntgenring 11
8700 Würzburg
W.-Germany

The landing response of *Drosophila* can easily be elicited by visual stimuli. The response probability decreases with repeated stimulus presentation. The present paper shows that this habituation of the response occurs near to the level of the sensory input. The neurons habituating have relatively small receptive fields.

For upward movement the horizontal extent of the habituating units corresponds to the receptive field width of one visual element. In the vertical direction their extent is also rather small and may coincide with the extent of elementary detectors for upward movement. Under the experimental conditions a global component of habituation for upward movement is measured in addition which is about 20% of the total habituation.

Different receptive fields are found with movement from front to back. While no evidence for a global habituation component has been obtained in this case, the receptive field of the local habituating units extends vertically and horizontally about 30°. Therefore, habituation to progressive movement seems to occur after pooling of several elementary movement detectors.

INTRODUCTION

The visual system of *Drosophila melanogaster* has become a favoured object for the study of the genetic control of development and for the genetic dissection of behaviour (Fischbach and Heisenberg, 1984). Unfortunately, however, the functional analysis of the visual system of the fruitfly is hampered by the small size of the neurons constituting the optic lobe neuropile. This has so far prevented electrophysiological studies at the single cell level. This gap is only partially filled by the

NATO ASI Series, Vol. H19
Modulation of Synaptic Transmission and Plasticity
in Nervous Systems. Edited by G. Hertting and H.-C. Spatz

successful use of deoxyglucose activity staining (Buchner et al., 1984). Therefore, all other tools for a dissection of functional pathways should be applied. Selective habituation of visual channels is the one which is presented in this and the accompanying papers.

The visually released landing response of *Drosophila melanogaster* shows habituation and sensitization (Fischbach, 1981). These basic plastic properties are of interest in connection with the genetic dissection of learning behavior (Duerr and Quinn, 1982) and qualify as convenient tools for the analysis of the visual system in general. Here we evaluate the transfer of habituation between different sectors of the visual field. The results obtained will be discussed in connection with the following questions:

1. What is the extent of the receptive field of individual habituating units?
2. At what level of the information processing are the habituating synapses situated?
3. At what stage is the landing response pathway separated from other movement-dependent functions?

MATERIALS AND METHODS

1. Stocks and preparation of flies

If not otherwise stated experiments were performed with 3 days old females of the *Drosophila melanogaster* wild-type strain "Berlin" (Jacob et al., 1977). In some control experiments wild-type strain "Kapelle" (Heisenberg and Buchner, 1977) was used. Flies were prepared for stationary flight one day before testing as described by Götz (1964).

2. Visual stimulation

Flies were kept suspended in front of an oscilloscope screen (distance fly-screen = 8 cm) on which the visual stimuli were presented (Fig.1). A pattern generator (200 images/s) allowed the independent control of two vertically or horizontally oriented dark stripes which were moved either alone or together in the horizontal (front to back) or in the vertical (upward) direction. The width of the stripes in front of the fly was chosen to be 7^0. Their length was variable and depended on the nature of the experiment. If not otherwise stated stimuli were automatically given in batteries of 24. Such a sequence of stimuli constituted one test. The frequency of stimulus presentation was 0.37/s. The pause between consecutive tests lasted 7 s. Normally a long sequence of tests was displayed (see below).

3. Recording of landing response

The landing response was visually recorded by inspection of the fly under the stereomicroscope. The landing response of *Drosophila* is a graded response (own observations). However, it was recorded as an all or none reaction by only counting a response as positive, if both frontlegs were at least lifted above the lower boun-

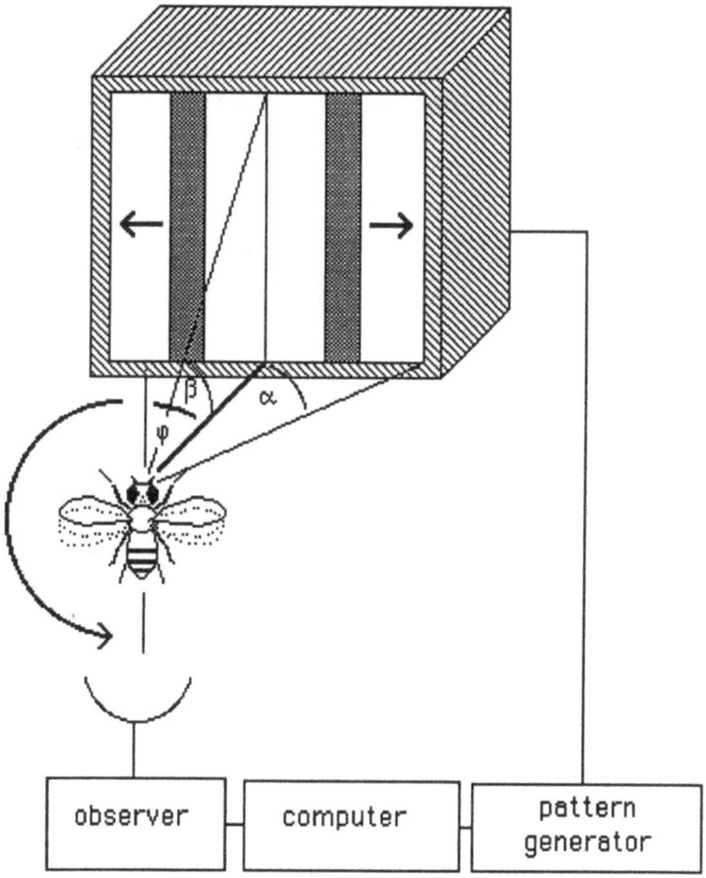

Fig.1: Experimental apparatus. Flies were kept in stationary flight in front of an oscillo - scope screen. A pattern generator (200 images/s) allowed the independent control of two vertically or horizontally oriented dark stripes. These were moved either alone or to - gether in the horizontal (front to back) or in the vertical (upward) direction. The width of the stripes in front of the fly was 7 degree. Their length is given for each experiment separately. The fly could manually be turned between tests around its vertical axis by the angle φ. α = 39 degree; β = 64 degree.

dary of the wing envelopes as observed from behind.

4. Determining habituation and transfer of habituation in single fly experiments

In principle the habituation index should measure the decrease in the response frequency caused by a single stimulus. For single fly experiments, however, it proved to be a good strategy to compare the response frequencies between consecutive tests of 24 stimuli each. The definition of the *habituation index* was thus chosen as:

$$H = (R_i - R_{i-1})/(R_i + R_{i-1}) \quad (1)$$

with R_{i-1} and R_i denoting the number of landing responses in the consecutive tests i-1 and i. This definition of the habituation index takes into account that habituation curves can be described by exponential functions of the form described in Fig.2. In contrast to the difference between two consecutive tests (Duerr and Quinn, 1982) the habituation index is thus in theory independent of the state of habituation. For this reason a sequence of tests can be used to calculate the characteristic habituation index of a single fly. As defined, the index offers the further advantage of being equally well suited to measure sensitization instead of habituation. Habituation and sensitization have the same scale (habituation $0>H>-1$, sensitization $0<H<1$). Sensitization of the landing response will be characterized in detail in another paper (Waldvogel and Fischbach, in preparation).

In principle, the *cross-habituation index* H_{xy} is a measure of the influence of a test with stimulus x on the response frequency in a consecutive test with stimulus y averaged with the value obtained when x and y are exchanged. We measured the habituation indices H_{xx} and H_{yy} together with the cross-habituation index H_{xy} for single flies by evaluating testsequences of the following nature:

$$T_{x,1}\ T_{x,2}\ T_{y,3}\ T_{y,4}\ T_{x,5}\ T_{x,6}\ T_{y,7}\ T_{y,8}\ T_{x,9}\ T_{x,10}\ T_{y,11}\ T_{y,12}$$

$T_{x,i}$ and $T_{y,j}$ denote here test numbers i and j using stimulus x or y, respectively. Between consecutive tests of the same kind the habituation index can be calculated and for each fly an averaged H_{xx} and H_{yy} can be given. The cross-habituation index was calculated in several steps: First, the influence of stimulus y on the response to stimulus x is assessed by relating the response frequency R in test $T_{x,i}$ to that in $T_{x,i+3}$:

$$R_{x,i+3} = R_{x,i}\, F_{xx}\, (F_{yx})^2 \qquad (2)$$

$$\text{where } F_{xx} = R_{x,i}/R_{x,i-1} = (1+H_{xx})/(1-H_{xx}) \qquad \text{using (1)} \qquad (3)$$

As habituation curves can be described by exponential functions, F_{xx} equals as well the factor by which test $T_{x,i}$ decreases the response in test $T_{x,i+3}$. Accordingly F_{yx} is the factor by which tests $T_{y,i+1}$ or $T_{y,i+2}$ decrease the response in test $T_{x,i+3}$.

Equation 2 is resolved for F_{yx}. It only depends on the measured landing response frequencies in successive tests. We prefer to use the normalized coefficient H_{yx} which is derived from F_{yx} by analogy to formula (3):

$$F_{yx} = (1+H_{yx})/(1-H_{yx}) \qquad (4)$$

$$\text{yielding} \qquad H_{yx} = (F_{yx}-1)/(F_{yx}+1) \qquad (5).$$

The *cross-habituation index* H_{xy} is defined as the mean of H_{yx} and H_{xy}. H_{xy} is a measure of the mutual transfer of habituation between tests for stimulus x and y. This is a simplification and assumes the existence of symmetrical interactions.

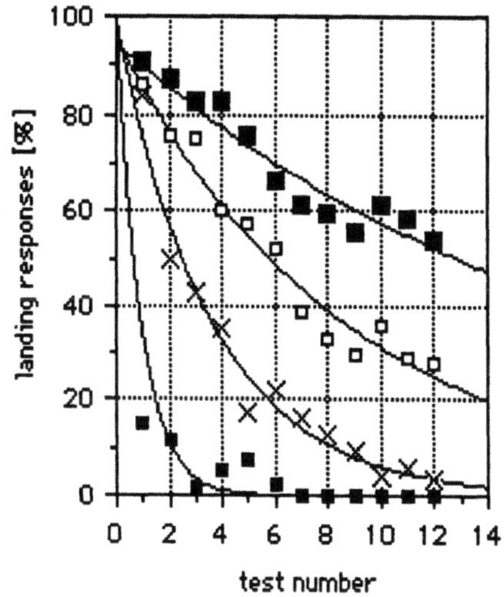

Fig. 2: Frequency of landing response elicited by bilateral progressive motion of two vertical bars on the screen of an oscilloscope plotted as a function of test number (16 stimuli /test). Female flies of wild-type "Berlin" were stimulated with different mean angular velocities of stripe movement (120 degree/s large black squares; 60 degree/s open squares; 30 degree/s small black squares). The data show that habituation to weak stimuli (i.e. low angular velocities) is faster than habituation to strong stimuli. Furthermore, female flies of wild-type "Kapelle" tested at 120 degree/s (X) show under these stimulus conditions faster habituation than females of wild-type "Berlin". Using 10-23 flies per experiment, individual data points are calculated from the responses to 160-368 stimuli. Exponential functions of the form exp(-k*testnumber) have been fitted to the time course of habituation (continuous lines) yielding correlation coefficients of 0.95, 0.97, 0.98, 0.89 (from top).

For identity between x and y the habituation index H_{xx} should equal the cross-habituation index H_{xy}. In fact this has been used as a test of the procedure. Negative values of H_{xy} indicate that habituation to stimulus x also implies partial or complete habituation to stimulus y and vice versa. Positive values of H_{xy} result, if there is mutual sensitization of the landing response.

Interpretation of H_{xy}-values is complicated, if H_{xx} and H_{yy} greatly differ from each other. Therefore it is useful to compare H_{xy} with the mean of H_{xx} and H_{yy} and to define a *habituation transfer function* as:

$$HT_{xy} = 200\, H_{xy}/(H_{xx}+H_{yy})\ [\%] \qquad (6).$$

The purpose of the above outlined, somewhat complicated operational procedures is that H_{xy} or HT_{xy} can be evaluated together with H_{xx} and H_{yy} in a single sequence of tests for one fly.

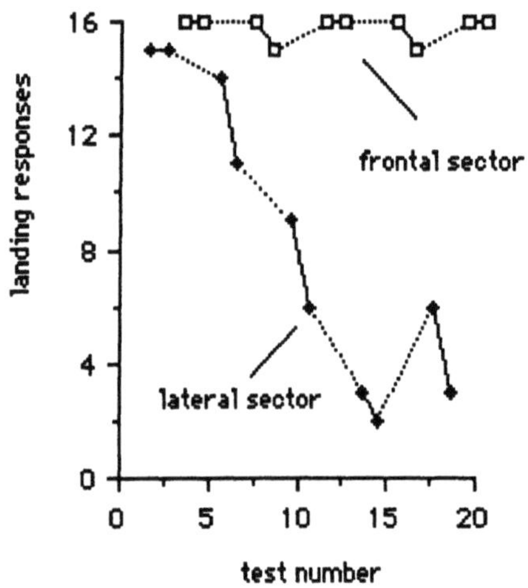

Fig.3: Response of a single female fly (wild type "Kapelle") to alternating tests in a frontal and a lateral sector (centered at 0 and 40 degrees relative to the fly's midline) to upward movement of a horizontal bar of 29 degree length. In this experiment a test consisted of 16 stimuli. Note that all the data points are derived from one single sequence of tests where the responses to frontal stimuli (open squares) and to lateral stimuli (black squares) were measured alternately.

RESULTS

1. Habituation to upward movement

a) The horizontal extent of habituating units for vertical movement

The landing response of *Drosophila* can be elicited by a single dark, horizontal stripe which is moving upward on the screen of an oscilloscope (Fig.1). In the experiments of this chapter the horizontal extent of the stripe was 29^0, the width 7^0. The mean velocity of stripe movement was 180^0/s. By turning the fly around its vertical axis the stimulus could be presented in vertically oriented sectors at horizontally displaced positions of the visual field. The test sectors extended vertically from -32^0 to $+32^0$ relative to the equator of the compound eye. The horizontal position of these testsectors is given in table I together with the frequency of the landing response in the first test (R_1) and the habituation index for wild-type "Berlin" females.

The frequency of the landing response decreases laterally while the habituation index becomes more negative. This result and the data shown in Fig.2 demonstrate that habituation to weak stimuli is faster than habituation to strong stimuli. Fig.3 gives an extreme single fly example. This wild-type female fly ("Kapelle") did not show habituation in the frontal test sector at all, while it simultaneously displayed strong habituation in a lateral test sector.

The cross-habituation indices obtained between the different sectors shown in table I are plotted in Figs.4A-F. The most clear cut result is that habituation transfer is maximal if the first and second test sectors are identical, i.e. habituation has a strong local component. A further observation is that partial overlap between the test sectors also secures a high amount of habituation transfer. The third observation is that the size of the gap between non-overlapping test sectors does not seem to play an important role. This is best seen in Fig.4E, but is supported by the super-

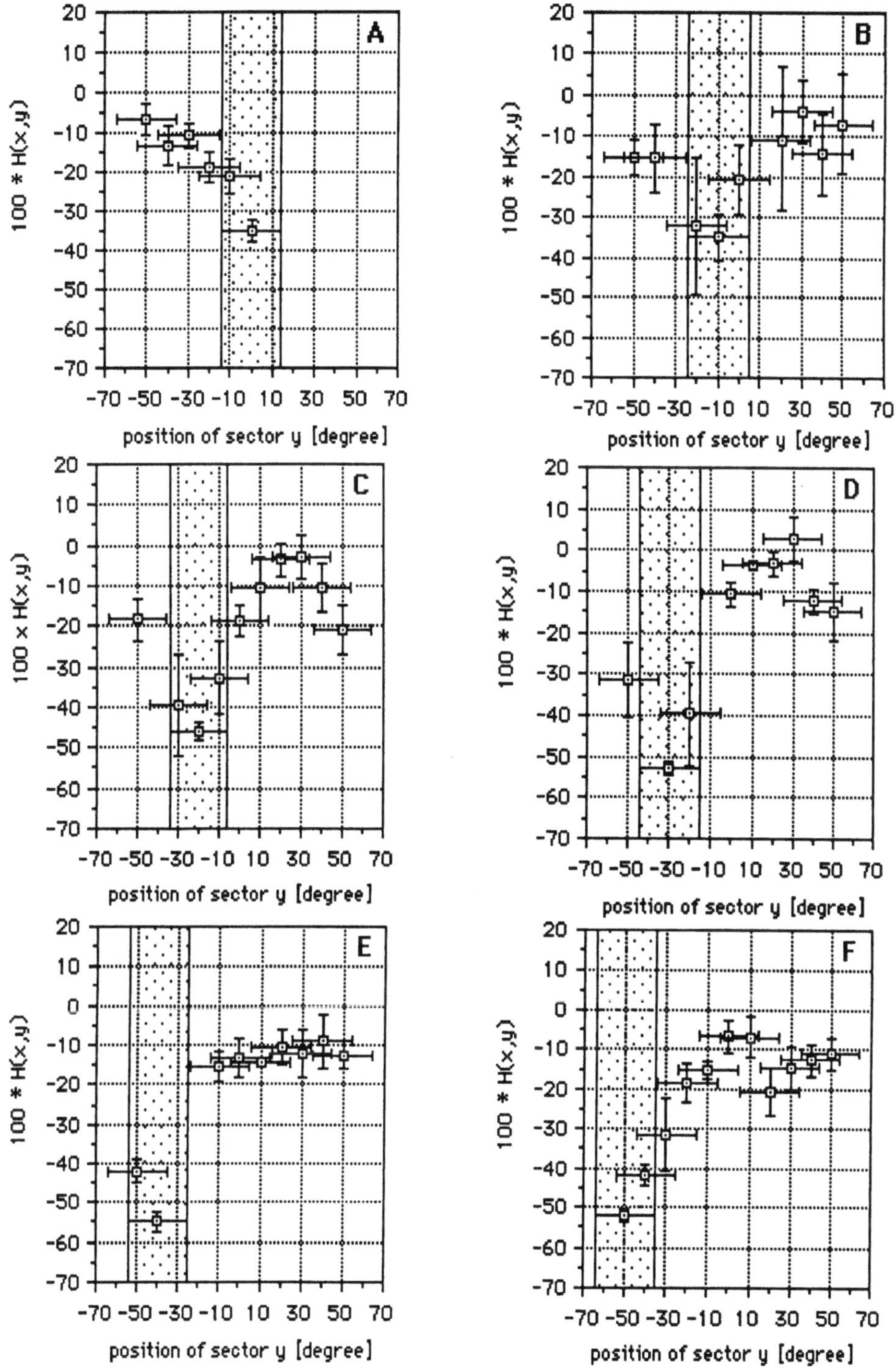

Figs.4 A-F: Plot of the cross-habituation index between horizontally displaced vertical sectors of the visual field stimulated by an upward moving horizontal stripe against the position of test sector y. The varied parameter in A-F is the position of sector x (emphasized by shading). The lateral extent of the test sectors is shown by the horizontal bars. Vertical bars are standard errors of the mean. 318 female flies (wild-type "Berlin") were tested.

position of the other data as well. If all data are added together an increasing distance between non-overlapping test sectors is not correlated with a further decrease of habituation transfer. This is best explained by assuming a minimal lateral extent of the habituating units, i.e. a receptive field width of one visual element.

The mean habituation transfer between non-overlapping test sectors is -0.105 ± 0.014. This we call the global habituation component for upward movement.

Table I: Response to an upward moving horizontal bar of 29° width in different vertical sectors of the visual field extending from - 32° to + 32° relative to the equator of the compound eye (wild type "Berlin")

Test sector	horizontal extent [°]	R_1	H_{xx}
I	-14.5 to 14.5	60.0±3.1	-0.35±0.045
II	-4.5 to 24.5	58.0±3.0	-0.35±0.03
III	5.5 to 34.5	52.0±3.2	-0.46±0.02
IV	15.5 to 44.5	42.5±3.5	-0.53±0.02
V	25.5 to 54.5	47.8±4.1	-0.55±0.025
VI	35.5 to 64.5	41.6±4.0	-0.52±0.015

b) On the vertical extent of the habituating units for upward movement

The investigation of the vertical extent of the habituating units was severely hampered by the fact that the stripe had to move a certain minimal distance in order to be an adequate stimulus for the landing response. Therefore we first consider the elicitation of the landing response as a function of the distance a horizontal stripe moves upwards on the screen. Using the data summarized in Table II and Fig.5 the dependence is plotted in Fig.6 for wild type Berlin, and is compared with that of wild type Kapelle in a different experiment. It is seen that the appearance of the stripe on the screen is not an efficient stimulus. The stripe has to be moved over about 30-35° to elicit a half maximal reaction. Comparison of the responses to stimuli in sectors I and II in table II shows that this cannot be explained by a more frequent elicitation of the landing response above the equator. Therefore, the spatial extent (or duration) of movement and not positional effects seem to be responsible for the shape of the curves in Fig.6. Similar results have recently been obtained by Wittekind (1987).

The next question concerns the vertical extent of the habituating units for upward movement. For this purpose combinations of the sectors shown in Fig.5 have been stimulated (see table II).

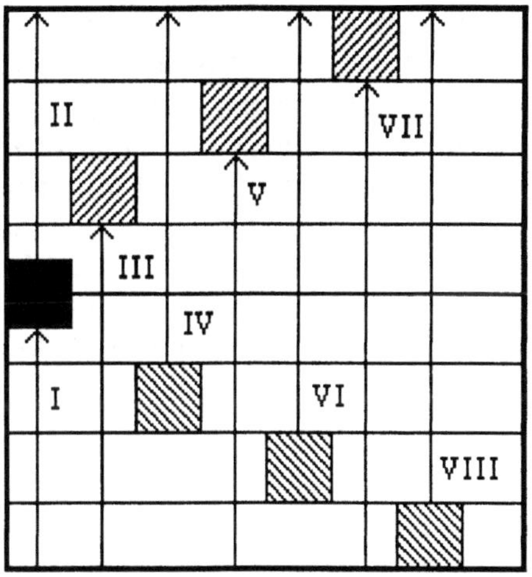

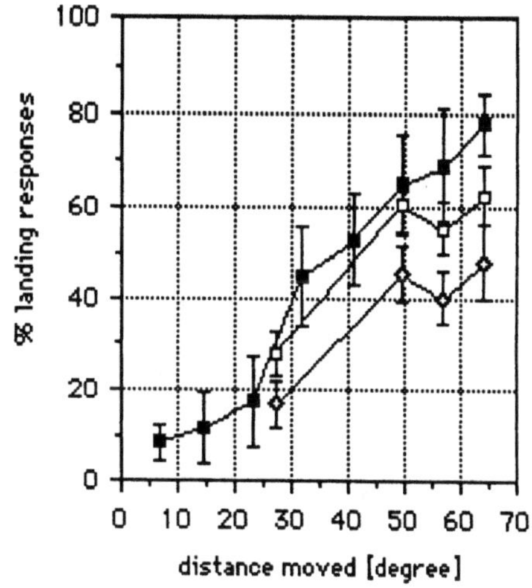

Fig.5 (left): Vertically overlapping frontal test sectors (I-VIII) used for the experiments described in table II. On the screen of an oscilloscope a dark stripe of 29 degree horizontal extent was moved upwards. In the sectors I, III, V, VII the stripe appears on the screen and stops at the position indicated (hatched area). In the sectors II, IV, VI, and VIII the stripe starts moving on the screen and disappears at the top of it. To secure a complete separation of test sectors I and II, the end (starting) position of the stripe has been masked (symbolized by black area). The travel distances given in table II refer to the visible distance moved by the lower edge of the stripe in the even sectors and to the visible distance moved by the upper edge in the uneven sectors.

Fig.6 (right): Dependence of the frequency of the landing response on the distance travelled by a horizontal bar moving upwards on the screen of an oszilloscope in front of female flies from wild type "Kapelle" (black squares) or "Berlin" (open symbols). The latter data points are derived from table II and represent responses to stimulation in the test sectors depicted in Fig.5. The open squares represent the uneven sectors and the diamonds the even ones.

Table II : Habituation transfer between frontal sectors of different vertical extent (see Figs.5,6).

Test sector	length [°]	R_1±sm	H_{xx}±sm	% overlap	H_{xy}	HT_{xy}
I	27.5	27.8±5	-.49±.05	0	-.16±.06	28±11
II	27.5	16.8±5.1	-.62±.06			
III	49.5	60.4±6	-.27±.05	44	-.20±.07	51.6±18
IV	49.5	45.5±6	-.52±.06			
V	57	55.4±6	-.34±.07	70	-.21±.07	51.4±18
VI	57	40.2±5.7	-.47±.07			
VII	64	62.5±6.3	-.22±.06	88	-.16±.07	64±28
VIII	64	48.1±8.4	-.27±.08			

The habituation of the landing response to an upward moving horizontal stripe is not mediated by units with a large vertical extent of their receptive fields. This is shown by the rather low cross-habituation index $H_{I,II}$ between the non-overlapping test sectors I and II. Cross-habituation is significantly lower than the mean of the habituation indices $H_{I,I}$ and $H_{II,II}$. It is, however, also significantly different from zero. This remaining habituation transfer is not significantly different from the global habituation component (see above). With increasing overlap between the test sectors the cross-habituation increases as expected. We therefore cannot exclude the possibility that habituation to vertical movement occurs at the output level of local movement detectors for upward movement.

2. Habituation to movement from front to back

a) Vertical extent of the habituating units for front to back motion

Two vertical stripes moving from the middle of the screen to both edges simulate the horizontal component of expansion and are efficient stimuli for the elicitation of the landing response even if their vertical extent is rather small. This allows to stimulate stationary flying flies in five non-overlapping frontal sectors of their visual field. The frequency of landing responses and the habituation indices are rather similar in these sectors (see table III). Cross-habituation experiments show a significant amount of habituation transfer between all sectors. However, self-habituation is strongest and the amount of transfer of habituation is a function of the gap size between the non-overlapping sectors (see Fig.7). The data can satisfactorily be described by the linear regression which reads "position of second sector = -10.3 - 95.7 H_{xy}" yielding a coefficient of correlation of 0.93.

Table III : Response to bilateral progressive stimulation in vertically stacked frontal test sectors extending horizontally from -40° to +40° (wild type "Berlin")

Test sector	vertical extent°	R_1	H_{xx}
I	32-25	57.6±3.6	-.43±0.04
II	25-17.5	62.9±7.0	-.37±0.08
III	17.5-9	62.8±7.0	-.38±0.06
IV	9-0	46.9±6.3	-.54±0.06
V	0-(-9)	51.8±4.6	-.42±0.06

This implies that zero habituation transfer (H_{xy}=0) is achieved when the second test sector is centered at -10°. This is 39° apart from the center of the reference sector. Using 9° as the vertical extent of test sectors (see table III), we conclude that habituation transfer occurs only when the gap between the test sectors is < 30°. A global habituation component for movement from front to back does not seem to exist (see below).

b) Horizontal extent of the habituating units for front to back motion

Fig.8 shows the decrease of the landing response frequency to progressive movement when the test sector is displaced laterally. The horizontal extent of the test sectors is indicated by the horizontal bars. The result clearly distinguishes the landing response from the optomotor yaw response which in walking or flying flies is elicited by front to back movement in all parts of the visual field (Götz and Wenking, 1973). It can be seen in table IV that the habituation in the test sectors differs greatly. Therefore, we use the normalized habituation transfer function in Figs. 9A,B.

The habituation transfer between the most frontal test sector I and lateral ones is shown in Fig.9A. Fig.9B describes the habituation transfer between test sector V and more frontal sectors. In both cases it is obvious that the amount of habituation transfer decreases with the distance between the sectors. There is no habituation transfer between sectors I and VII. This is evidence against a global habituation component for progressive motion.

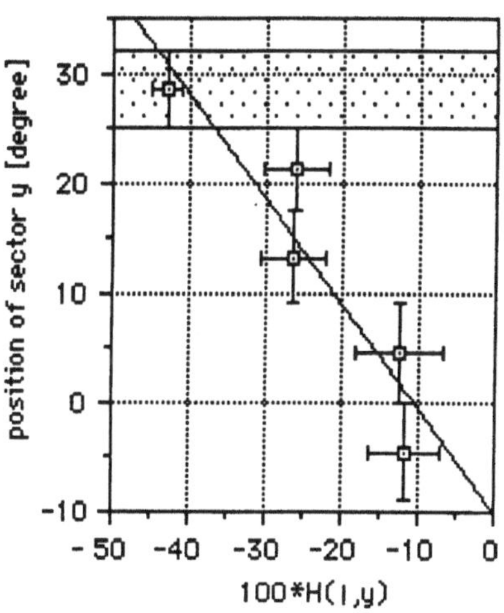

Fig.7: Bilateral progressive stripe movement in vertically stagged horizontal sectors of the visual field (see table III): Plotted is the vertical position of test sectors y against their cross-habituation index with sector I (shaded area). The vertical width of the test sectors is indicated by the vertical bars. The horizontal bars are standard errors of the mean. Zero degree corresponds to the equator of the compound eye. Using the linear regression (straight line) it is possible to give an estimate of the vertical extent of the habituating units (see text).

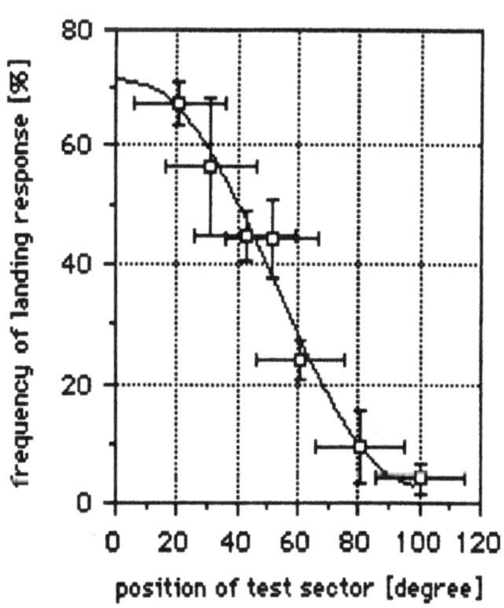

Fig.8 : Frequency of the landing response to unilateral progressive motion of a vertical bar of 45 degree length (extending from -15 degree to +30 degree relative to the equator of the compound eye) in horizontally displaced sectors of the visual field. Zero degree corresponds to the elongation of the saggital midline of the fly. The horizontal extension of the test sectors is indicated by the horizontal bars. The mean angular velocity of stripe movement was 140 degree/s.

The theoretical curves in Figs.9A,B (straight lines) describe the data under the assumption that the horizontal extent of the receptive fields of the habituating neurons is 30 degrees, i.e., if the gap size between two test sectors reaches this value, habituation transfer is zero.

Table IV : Response to unilateral progressive stimulation in horizontally displaced test sectors extending vertically from -15° to +30° relative to the equator of the compound eye

sector	horizontal extent[1] [°]	R_1	H_{xx}
I	6-36	67.3± 3.7	-.16±.03
II	16-46	56.7±11.8	-.09±.04
III	26-59	44.7± 4.3	-.27±.05
IV	36-67	44.3± 6.5	-.26±.05
V	46-75	24.1± 3.2	-.44±.06
VI	66-95	9.4± 6.0	-.23±.17
VII	86-115	4.2± 2.6	-.39±.17

[1] Given is the distance moved by the leading edge of the vertical stripe.

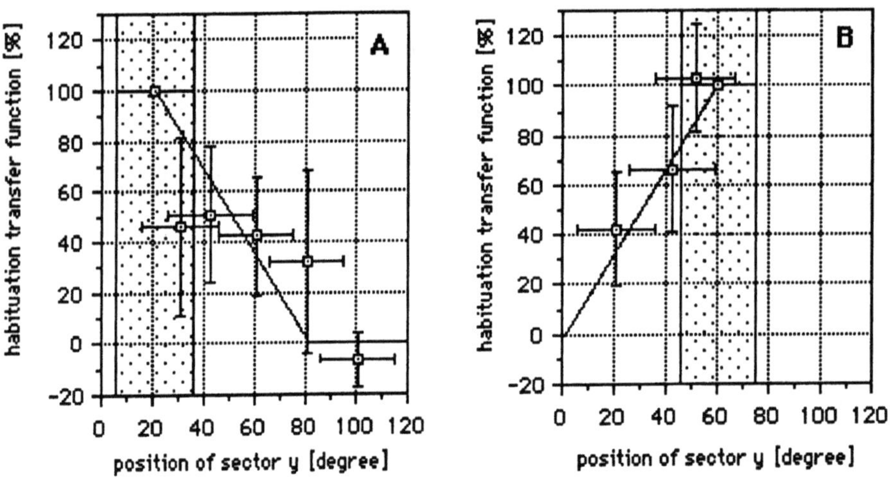

Fig.9: Unilateral stimulation with a progressively moving vertical stripe: A) The normalized habituation transfer function between the most frontal test sector (shaded) and five laterally displaced test sectors is plotted versus the position of the center of these test sectors. B) The habituation transfer function between a lateral test sector (shaded) and three frontally displaced test sectors is plotted versus the position of the center of these test sectors. The horizontal bars indicate their horizontal extent. Error bars are vertical. Straight lines describe the data under the assumption that the horizontal extent of the receptive fields of the habituating neurons is 30 degrees.

DISCUSSION

The evaluation of the presented data has to be done on the background of what is already known about the neuronal and functional organization of the insect optic lobe. Especially relevant in the context of this paper is the retinotopic and layered organization of the optic neuropils (Campos-Ortega and Strausfeld, 1972; Fischbach and Dittrich, in preparation)

According to the classical concept of a neuron, the receptive field of a presynaptic site coincides with that of the whole neuron. Although exceptions to this rule seem to exist (see e.g. discussion of lamina amacrine cells in Shaw, 1981), we will use this assumption as a working hypothesis. Following the results of Kandel obtained with *Aplysia* (1976, 1981), habituation is a plastic change at the presynaptic site. Therefore, if habituation of a response is mediated by a symmetrical binocular neuron, habituation induced by unilateral stimulation should be indistinguishable from habituation induced by stimulation of the contralateral side. In this case, habituation should be transferable from one side to another. Such a result has been obtained by Duerr and Quinn (1982) for the habituation of the proboscis extension reflex elicited by tarsal sugar contact. In this paradigm habituation is - at least partially - transferable from one tarsus to the contralateral one. Unilateral habituation of the landing response to front to back movement is different. It is not transferable to the other side (Fischbach, 1981). Therefore, the neurons habituating do not seem to possess binocular visual fields.

In the present paper the same logic has been applied to investigate the extent of the monocular receptive fields of habituating neurons. For this purpose small parts of the visual field were stimulated and subsequently different parts were probed for their state of habituation. The most important result is that habituation is mainly a local phenomenon, i.e. it has to be assumed that the receptive fields of the main habituating units are rather small.

1. Visual field size of habituating units

a) Upward movement

The elicitation of the landing response in *Drosophila melanogaster* depends on directional motion. In the case of vertical (upward) motion the receptive field of the habituating units is minimal concerning its horizontal extent. This is most clearly demonstrated by the data of Fig.4E. Superposition of all data of Figs.4A-F yields the same result. Non-overlapping vertical test sectors display only a more or less constant global habituation transfer. Transfer of the stronger local habituation component is directly correlated with the degree of sector overlap. This can only be explained if one assumes that the local habituating units use a vertical row of visual elements.

How many visual elements in such a vertical row feed into a habituating unit? The experiments depicted in Fig.5 and in table II show that some habituation jumps over a gap of 9^0. However, a significant amount of this habituation transfer is due to the global habituation component. Therefore, we have to conclude that the vertical extent of the habituating units is small as well.

Do the habituating units have the same field size as the vertically oriented movement detectors? Unfortunately, no direct measurements on the vertical motion detector system feeding the landing response have been done. Data on vertical motion detection as measured by thrust responses (Buchner et. al., 1978) show the dominant detectors to be of nearest neighbour interactions. Buchner et al. also found a further detector with a larger sampling base of about 12^0. This value would also be compatible with the vertical size of the habituating units for vertical movement. Habituation thus could take place directly at the level of the movement detectors.

b) Front-to-back movement

In case of front-to-back movement the receptive field of the habituating units was determined to be fairly large (about $30^0 \times 30^0$). Therefore, the habituating units for front-to-back movement are different from those for upward movement. The horizontal extent is by far larger than the minimal sampling base for horizontal motion detectors which is twice the interommatidial angle (Buchner, 1976).

In a study on the movement detectors mediating the landing response of *Musca domestica* to binocular front-to-back motion, detectors with large sampling bases (up to the 8fold interommatidial angle) were shown to contribute equally strong as the ones with the minimal sampling base (Borst and Bahde, 1987). Since habituation is direction-sensitive, the horizontal extent of the receptive fields should be at least as large as that of movement detectors. The estimate of 30^0 obtained from our data is compatible with this notion.

The large vertical extent of the habituating units, however, cannot simply be explained by the vertical extent of elementary front-to-back motion detectors. It rather suggests that habituation occurs after pooling of neighbouring movement detectors.

2. Towards a location of the habituating neurons

By use of the deoxyglucose activity labelling computation of directional motion could not be demonstrated to occur peripheral to the lobula complex (Buchner, 1984). However, as the deoxyglucose labelling in the lobula plate of the mutant *omb* ^{H31} which lacks the giant HS and VS neurons (Heisenberg et al., 1978) is very similar to wild type (Bülthoff et al., 1985), the direction-sensitive label seen at the level of the lobula plate seems to reside in small field input neurons from the proximal medulla. Golgi studies (Fischbach and Dittrich, in preparation) suggest that the respective cells are T4 neurons. It looks as if in the lobula plate fine grained small field input is directly fed into large tangential neurons which scan most parts of the visual field.

Indeed, certain optomotor responses, the large field responses (Heisenberg and Wolf, 1984), very likely mediated by the HS-cells, are thought to get input from front-to-back movement detectors with minimal sampling bases distributed over the whole visual field (Buchner, 1976). Our data (e.g. Fig.8) therefore exclude that the

HS-cells are part of the landing response network. Supporting evidence derives again from studies with the mutant *omb* ^{H31}. Flies of this strain show normal landing responses to front to back motion (Heisenberg et al., 1978), while they do not land in response to upward motion (Fischbach, 1984).

In summary, our data do not favour the participation of the lobula plate in the landing response. We rather suggest that the landing response is mediated by the lobula. This is supported by landing-stimuli-specific deoxyglucose labelling in this neuropil (Bausenwein, in preparation). Unfortunately, up to now nearly nothing is known about the physiological properties of lobula neurons.

3. Plasticity of visual neurons in other insects

In larger flies the giant tangential cells of the lobula plate have been electrophysiologically studied in great detail (e.g. Hausen 1984, Hengstenberg, 1982). Most cells have extensive receptive fields and respond best to large field motion. In most cases, however, not much attention has been paid to dynamic changes of response properties, possibly due to relatively short test times. A decrease of responsiveness with repeated stimulation has been measured at the level of the H1 neuron (Eriksson, 1984, Maddess and Laughlin, 1985, de Ruyter van Steveninck et. al, 1986). H1, which is a unique whole-field neuron, responds to back-to-front stimulation by increasing its spike rate. Here, "adaptation" seems to occur in presynaptic small field units which are retinotopically distributed and could thus coincide with elementary motion detectors. Some evidence for adaptational effects in other giant cells have also been reported (Maddess and Laughlin, 1985). However, no data are available on the locus of their "adaptation".

In locusts, habituative effects have been visualized electrophysiologically at the level of giant interneurons. LGMD which connects directly to a giant descending neuron (DCMD) gives an example for habituation at the level of a feature detecting unit immediately prior to the descending level. It responds to moving small field stimuli irrespective of their position in the visual field (O`Shea and Rowell, 1976). Both response and response decrement (habituation) are inhibited by large field stimuli. Again in the locust, 3 descending ´course deviation receptor´ neurons (DNI, DNM and DNS), which respond to rotations of the animal with different spatial and temporal preferences, display habituational effects (Rowell and Reichert, 1978). Here habituation is seen at a high level of spatial and multimodal (wind, ocellar, visual) convergence.

4. Habituation can be used to distinguish between parallel visual channels

Considering the high number of neuronal cell types inside each column of the visual neuropils and knowing e.g. of the coexistence of different systems contributing to optomotor behaviour (Bausenwein et. al., 1986), one should not be surprised to find a high level of parallel organization throughout the visual system (Fischbach and Heisenberg, 1984). Characterizing plastic properties of visual channels can be useful as a tool for separating the function of cells which respond to similar or identical stimuli in parallel systems.

At what stage is the landing response which shows habituation separated from other directionally motion dependent systems which do not habituate? It is clear that habituation has to occur after the separation of the respective neural pathways. For example, we can ask: Is there a special set of movement detectors feeding into the landing response? Borst and Bahde (1987) compared the movement detection system underlying the large field optomotor yaw response and the landing response in *Musca domestica* and concluded that both responses are mediated by movement detectors with very similar properties. Our inference that habituation of the landing response to front to back movement occurs after pooling of several movement detectors is compatible with the notion that the pathways for the optomotor yaw response and the landing response are separated after elementary movement detection.

ACKNOWLEDGEMENT

We thank J. Hohmann, R. Zierz, G. Kimmerle, J. Winter, and D. Unger for participation in the experiments. H. Mitschulat, H.Ch. Spatz, and W. Wittekind read the manuscript and made useful suggestions. We are especially thankful to A. Borst who helped us to improve the interpretation of our data. We profited from the discussions with M. Heisenberg and R. Wolf. The work was supported by the DFG (grant Fi336/1-5 to K.F.F.).

LITERATURE

Bausenwein B, Wolf R and Heisenberg M (1986) Genetic Dissection of Optomotor Behavior in *Drosophila melanogaster*. Studies on Wildtype and the Mutant *optomotor-blind*^{H31}. J. Neurogenetics 3, 87-109

Borst A and Bahde S (1987) Comparison between the Movement Detection System Underlying the Optomotor and the Landing Response in the Housefly. Biol.Cybern. 56, 217-224

Buchner E (1976) Elementary movement detectors in an insect visual system. Biol. Cybernet. 24, 85-101.

Buchner E, Buchner S, Bülthoff I (1984) Deoxyglucose mapping of nervous activity induced in *Drosophila* brain by visual movement Part I. Wildtype. J.Comp.Physiol. (A) 155, 471-483

Buchner E, Götz K G, Straub C (1978) Elementary Movement Detectors for Vertical Movement in the Visual System of *Drosophila*. Biol. Cybern. 31, 235-242

Bülthoff I and Buchner E (1985) Deoxyglucose mapping of nervous activity induced in *Drosophila* brain by visual movement. II. *optomotor blind*^{H31} and *lobula plate-less*^{N684}, visual mutants. J. Comp. Physiol (A) 156, 25-34

Campos-Ortega, J and Strausfeld, N J (1972) Columns and layers in the second synaptic region of the fly's visual system. The case for two superimposed neuronal architectures. *In* Information processing in the visual system of arthropods. Ed. R Wehner. Springer Verlag. Berlin

Duerr J S and Quinn W G (1982) Three *Drosophila* mutations that block associative learning also affect habituation and sensitization. Proc.Natl.Acad.Sci. USA 79, 3646-3650

Eriksson E S (1984) Vector analysis in a neural network. J. Insect Physiol. 30, 363-368

Fischbach K F (1981) Habituation and sensitization of the landing response of *Drosophila melanogaster*. Naturwissenschaften 68, 332

Fischbach, KF (1983) Neurogenetik am Beispiel des visuellen Systems von *Drosophila melanogaster*. Habilitation Thesis, Würzburg

Fischbach, K F and Heisenberg, M (1984) Neurogenetics and Behaviour in Insects. J. Exp.Biol. 112, 65-93

Götz K G and Wenking H (1973) Visual control of locomotion in the walking fruitfly *Drosophila*. J. Comp. Physiol. 85, 235-266

Hausen K (1984) The Lobula Complex of the Fly : Structure, Function and Significance in Visual Behaviour. *in* : Photoreception and Vision in Invertebrates Ed. MA Ali Plenum,New York

Heisenberg M and Buchner E (1977) The role of retinula cell types in visual behavior of *Drosophila melanogaster*. J. Comp. Physiol. 117, 127-162

Heisenberg M and Wolf R (1984) Vision in Drosophila. Springer Verlag Berlin

Heisenberg M, Wonneberger R, and Wolf R (1978) *Optomotorblind*^{H31} - a *Drosophila* mutant of the lobula plate giant neurons. J. Comp. Physiol. 124, 287-296

Hengstenberg R (1982) Common Visual Response Properties of Giant Vertical Cells in the Lobula Plate of the Blowfly *Calliphora*. J. Comp.Physiol. (A) 149, 179-193

Jakob K G, Willmund R, Folkers E, Fischbach KF, and Spatz HCh (1977) T-maze phototaxis of *Drosophila melanogaster* and several mutants in the visual systems. J. Comp. Physiol. 116, 209-225

Kandel E R (1976) Cellular basis of behavior. An introduction to behavioral neurobiology. WH Freeman and Company San Francisco

Kandel E R (1981) Behavioral Biology of Aplysia. W.H. Freeman and Co.. San Francisco

Maddess T and Laughlin SB (1985) Adaptation of the motion-sensitive neuron H1 is generated locally and governed by contrast frequency. Proc.R.Soc.Lond B 225, 251-275

Rowell C H F and H Reichert (1986) Three descending interneurons reporting deviation from course in the locust II. Physiology. J. Comp. Physiol. (A) 158, 775-794

Ruyter van Steveninck R R de, Zaagman W H and Mastebroek H A K (1986) Adaptation of transient responses of a movement sensitive neuron in the visual system of the blowfly *Calliphora erythrocephala.* Biol. Cybern. 53, 451-463

Shaw S R (1981) Anatomy and physiology of identified non-spiking cells in the photoreceptor-lamina complex of the compound eye of insects especially Diptera. *In* Neurones without impulses (Ed. A Roberts and BMH Bush). Cambridge University Press.

O'Shea M and Rowell C H F (1976) The neuronal basis of a sensory Analyser, the Acridic Movement Detector System II. Response Decrement, Convergence, and the Nature of the Excitatory Afferents to the Fan-like Dendrites of the LGMD. J. Exp. Biol. 65, 289-308

Wittekind W (1987) Analyse des stereotypen und insbesondere plastischen Landeverhaltens von *Drosophila melanogaster*. Dissertation. Universität Freiburg.

MOLECULAR ASPECTS OF PLASTICITY IN PHOTOTAXIS

Mitschulat, H. and Willmund, R.
Institut für Biologie III
D-7800 Freiburg

Prolonged adaptation of <u>Drosophila</u> <u>melanogaster</u> with blue light induces a change in light/dark choice behavior that lasts for hours.
It has been demonstrated recently, that this visual adaptation induces long-term modulation of the in vitro autophosphorylation of a Ca^{2+}/calmodulin-dependent protein kinase. Here evidence is shown that an 86 kDa phospho-protein is a substrate of the kinase in <u>Drosophila</u>, and its molecular weight and phospho-peptide composition are similar to those of rat synapsin I. Furthermore it shows immunological cross-reactivity with antibodies against rat synapsin I. We conclude that 86 kDa protein is homologous to the vertebrate synapsin I.

INTRODUCTION

<u>Drosophila</u> <u>melanogaster</u> is a suitable object for analyzing behavioral plasticity and the molecular mechanisms this plasticity is based on.
The present paper deals with biochemical changes in behavior of wildtype flies, which can be induced by sensory stimulation of the living animals.
Herein a visual stimulation is used which causes an artificial light/dark choice behavior: in <u>Drosophila</u> prolonged illumination with blue light converts the photopigment in

NATO ASI Series, Vol. H19
Modulation of Synaptic Transmission and Plasticity
in Nervous Systems. Edited by G. Hertting and H.-C. Spatz

the peripheral retinula cells R1-R6 into the photostable metarhodopsin and induces a prolonged depolarizing afterpotential (PDA) in these cells (1). This prolonged bleaching of R1-R6 is shown to be involved in processes causing a long-lasting modification in a phototactic behavior (1-6). This long-lasting modification in the behavior can be observed for up to 3 hours even if the photopigment is reconverted by yellow light, restoring the function of the retinula cells R1-R6. Thus this behavior can be seen as a simple form of learning.
Many authors have postulated that protein phosphorylation mediates neural plasticity during learning processes in vertebrates and invertebrates (for a review see 7). Several ways of regulating phosphorylation in the nervous system are known; for example, so-called first messengers (extracellular signals) directly cause changes in protein kinase and/or phosphatase activity: enzymes are coupled to membrane-bound receptor molecules. Another way is the change of intracellular concentrations of second messengers such as cyclic AMP, cyclic GMP or Ca^{2+}.

In Aplysia both cAMP-dependent phosphorylation of K^+-channel proteins (8,9) and Ca^{2+} as an intracellular signal have been shown to be involved in the mechanism of associative learning (9,10).

In Aplysia a Ca^{2+}/calmodulin-dependent protein kinase is activated after in vitro application of serotonin to the nerve cell. Recent studies demonstrated that this membrane-bound enzyme can be released after phosphorylation and subsequently exhibits increased activity. In vitro added vertebrate synapsin I is an excellent substrate for this enzyme (11).

Synapsin I is a phosphoprotein found in the central and peripheral nervous systems (7). It is described as composed

of two polypeptides of 86 kDa (synapsin Ia) and 80 kDa (synapsin Ib) which exhibit similar properties including immunological cross-reactivity (12,13). Synapsin is associated with the membrane of synaptic vesicles (14-16), and it is speculated that it is involved in the translocation of the synaptic vesicles or their prefusion with the plasma membrane (17).
Bähler and Greengard (18) described an actin bundling activity associated with synapsin I. The dephospho form of the enzyme is shown to be the active one. In their working model synapsin serves to anchor the vesicles to a spectrin/F-actin network in the cytoplasm and prevents prefusion of the vesicles with the pre-synaptic membrane; thereby no transmitter release occurs.

In agreement to the findings in Aplysia both cAMP and Ca^{2+} were shown to be important in the learning mechanisms of Drosophila (19-23).

It could be demonstrated by one of us recently that the phosphorylation of a Ca^{2+}/calmodulin-dependent protein kinase parallels the time course of the plasticity of the phototactic behavior described above in Drosophila (24). Here we report evidence that a synapsin-like protein which is possibly a substrate of this protein kinase exists in Drosophila brain. Some of its properties are similar to those of vertebrate synapsin I.

MATERIAL AND METHODS

(1) Phosphorylation assay

Heads of 15 instantly frozen flies were cut off in liquid nitrogen (- 196 °C) and quickly homogenized (10 sec) in 50 ul $K-PO_4$ buffer (0.1 M, pH 6.8) containing 2 M glycerol.

For some experiments this crude head extract is separated in a membrane and a cytosolic fraction by centrifugation either at 40.000 x g for 20 min or at 400.000 x g for 10 min at 4 ^oC.

Protein phosphorylation was assayed in $K-PO_4$ buffer (50 mM, pH 6.8) containing 5 mM $MgCl_2$, 2 uM ATP and γ-^{32}P-ATP (300 - 600 Ci/mM). Other additions (1 mM $CaCl_2$, 50 uM cAMP or 10 mM EGTA) are indicated.

The reaction was carried out at room temperature for 2.5 min and stopped by addition of a half volume of Tris-HCl buffer (0.5 M, pH 6.8) containing 10 % 2-mercaptoethanol and 6 % SDS.

(2) "Translocation"

Carefully washed membrane fractions were phosphorylated in $K-PO_4$ buffer (50 mM, pH 6.8) containing 5 mM $MgCl_2$, 1 mM $CaCl_2$ and ATP as indicated. After subsequent centrifugation, a half volume of Tris-HCl buffer (see above) was added to the supernatant and the remaining pellet was resuspended in the same buffer. Proteins were separated by gel electrophoresis (see below).

In one set of experiments exogenous synapsin I from rat was added to the phosphorylation assay.

(3) SDS-PAGE

Using the method of Laemmli (25), proteins were separated on 12.5 % gels. The gels were stained with Coomassie blue R 250, destained, dried and exposed to Kodak X-Omat AR film. Autoradiograms were scanned with a Kontron Uvikon 810 Spectrophotometer.

(4) Immunoblots

Immunoblotting was carried out as described by Towbin (26): Samples were run on 12.5 % gels and electroblotted onto nitrocellulose filters, presaturated with TBS-buffer (10 mM Tris, pH 7.4, 0.9 % NaCl, and 0.02 % NaN_3) containing 2 % BSA overnight.
Incubation with antibodies against rat synapsin I was carried out at room temperature for one hour. After washing with TBS containing additional 0.05 % SDS and 0.1 % Triton-X-100 for 1.5 hours, the filters were incubated with the second antibody (anti-mouse IgG) for 1 hour and washed again. Filters were stained using second antibody-linked alkaline phosphatase. The reaction was carried out in Tris-HCl buffer (0.1 M, pH 8.8) containing 0.1 M NaCl, 0.005 M $MgCl_2$, 5-bromo-4-chloro-3-indolyl phosphate and nitro blue tetrazolium.

(5) Peptide mapping

Using the method of Cleveland et al. (27) the bands from dried SDS-gels were cut out and placed in the sample wells of a second SDS-gel. Digestion with Staphylococcus aureus V8 protease was carried out in the stacking gel for 20 min. The gels were stained with Coomassie blue R 250, destained, dried and autoradiographed by exposing to Kodak X-Omat AR film.

RESULTS

(1) Release of Ca^{2+}/calmodulin-dependent protein kinase from the membrane

After phosphorylation, the Ca^{2+}/calmodulin-dependent protein kinase is partially released from the membrane (Fig. 1).

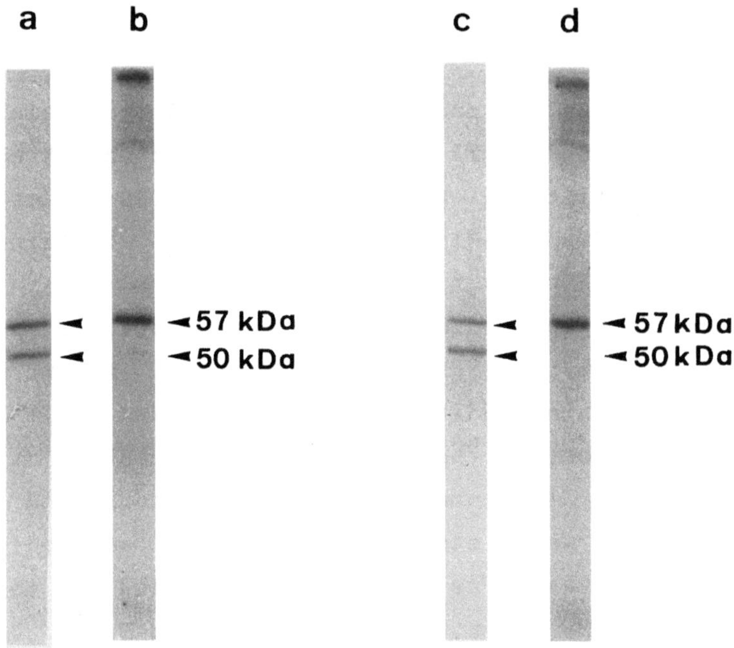

Fig. 1: ATP-dependent release of Ca^{2+}/calmodulin-dependent protein kinase from the membrane.
Carefully washed membrane fractions were incubated with K-PO_4 buffer containing $MgCl_2$ and $CaCl_2$ and ^{32}P-ATP or EGTA and ^{32}P-ATP, and subsequently centrifugated: here the released proteins in the supernatant are shown. Lanes a and c show the released protein after phosphorylation in the presence (a) or absence (c) of Ca^{2+}.
In lanes b and d the incubation is carried out in buffer without ATP, proteins were phosphorylated after centrifugation with labeled ATP: No 50 kDa protein kinase is released in the absence of ATP.

This release of the phosphorylated form does not depend on Ca^{2+}, because no difference can be detected between phosphorylation in the presence (Fig. 1, lane a) or absence (lane c) of Ca^{2+}. But in the absence of ATP no enzyme release occurs, as shown in lanes b and d in Fig. 1.

The phosphorylation state of the released protein seems to be very stable; neither dephosphorylation nor further phosphorylation can be detected for at least 30 min (data not shown).

In contrast to the release of the protein the activity in phosphorylating another substrate depends on Ca^{2+} (Fig. 2):

When rat synapsin is added to a phosphorylation assay with 50 kD kinase released from the membrane fraction, it is phosphorylated as shown in Fig. 2. The 50 kDa is released from the membrane fraction by phosphorylation with unlabeled ATP, so it cannot be seen. But the Ca^{2+}-dependency of the phosphorylation of rat synapsin is obvious.

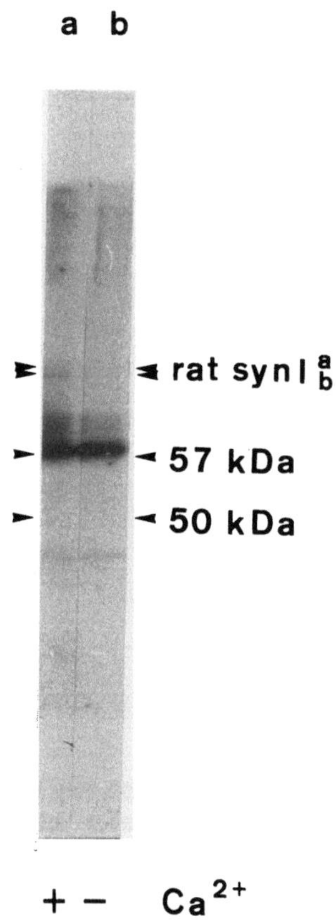

Fig. 2: In vitro phosphorylation of vertebrate synapsin I. Synapsin I from rat brain is added to the phosphorylation assay containing Ca^{2+}/calmodulin-dependent protein kinase, which is released from the membrane after phosphorylation with unlabeled ATP. Phosphorylation was carried out in the presence (lane a) or absence of Ca^{2+} (lane b). Enzyme activity depends on Ca^{2+}, since synapsin I is phosphorylated in the presence of Ca^{2+}, only (lane a).

(2) 86 kDa phospho-protein in <u>Drosophila</u>

Looking for endogenous substrates of the Ca^{2+}/calmodulin-dependent protein kinase in <u>Drosophila</u>, we detected a phospho-protein with the apparent molecular weight of 86 kDa, which showed phosphorylation states strongly correlated to the phosphorylation of the 50 kDa protein kinase (Fig. 3).

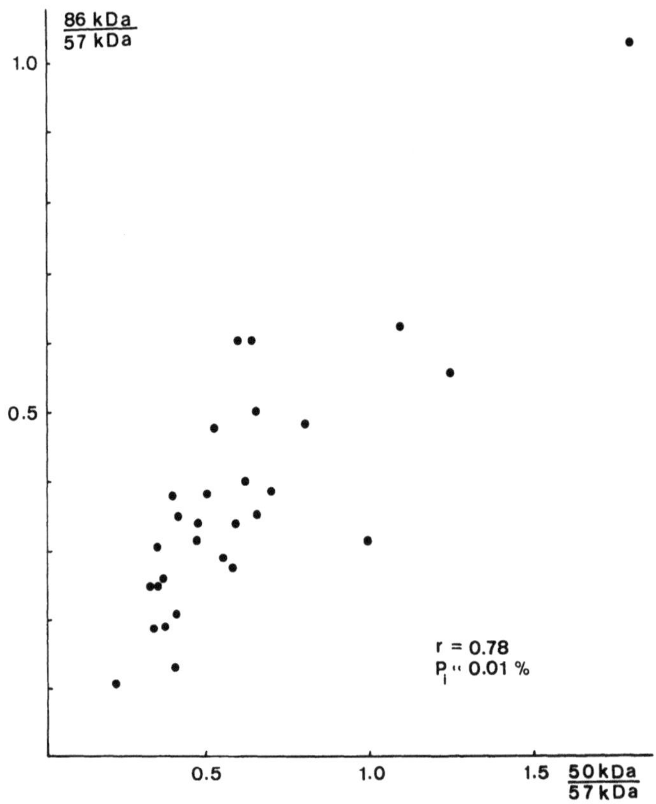

Fig. 3: Correlation of phosphorylation of the 50 kDa Ca^{2+}/calmodulin-dependent protein kinase to phosphorylation of the 86 kDa phospho-protein.

(3) Cross-reactivity of the 86 kDa protein with antibodies against rat synapsin I

In Fig. 4 a Western blot with monoclonal antibodies against rat synapsin I bound to Drosophila protein is shown. Antibody binding to the 86 kDa phosphoprotein can be seen. The

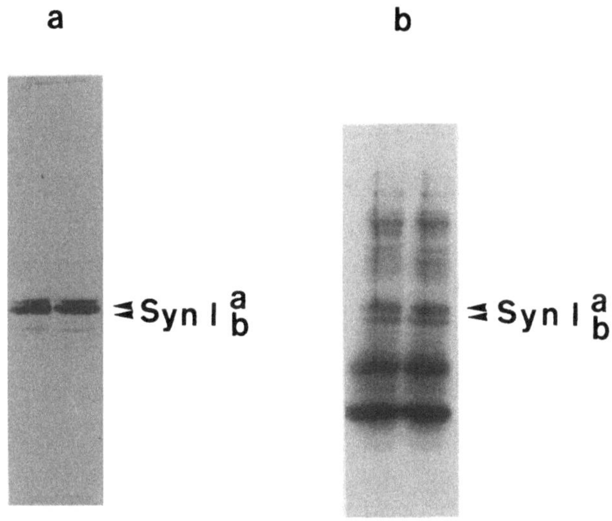

Fig. 4: Western blot of antibodies against rat synapsin I. Lanes a show the specificity of the antibodies: The two phosphopeptides of synapsin I with apparent molecular weights of 80 and 86 kDa show strong antibody-binding. Lanes b show the corresponding autoradiograph of phosphorylated proteins from crude head homogenate of rat.

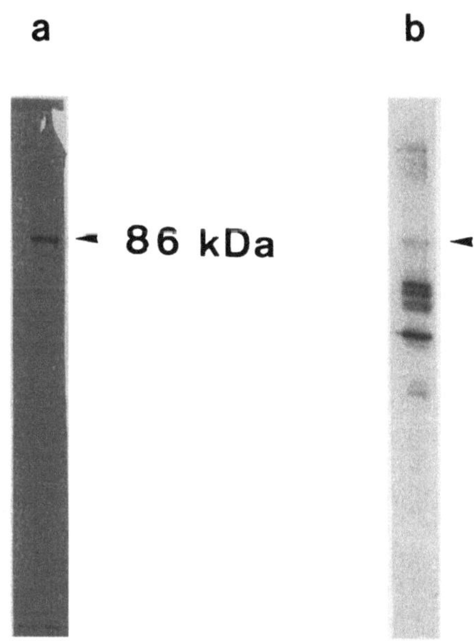

Fig. 5: Western blot of antibodies against rat synapsin I bound to Drosophila protein.
Lane a: binding of the antibodies against the vertebrate synapsin I to 86 kDa phospho-protein in Drosophila. Lane b: corresponding autoradiogram of the in vitro phosphorylated membrane-fraction of the fly.

specificity and purity of the antibodies against rat synapsin is shown in Fig. 5: strong antibody-binding to the 86 and 80 kDa phosphoproteins of rat brain is shown in lanes a. Lanes b show the corresponding autoradiography of in vitro labeled phosphoproteins from rat brain. However, binding sometimes is not as specific as shown in Fig. 5. There are several weak signals. However, only one prominent band with the molecular weight of 86 kDa can be seen.

(4) Peptide mapping of Drosophila and rat brain protein

Protein digestion with V8 protease should show whether the vertebrate and invertebrate proteins are similar with respect to their phospho-peptide composition.

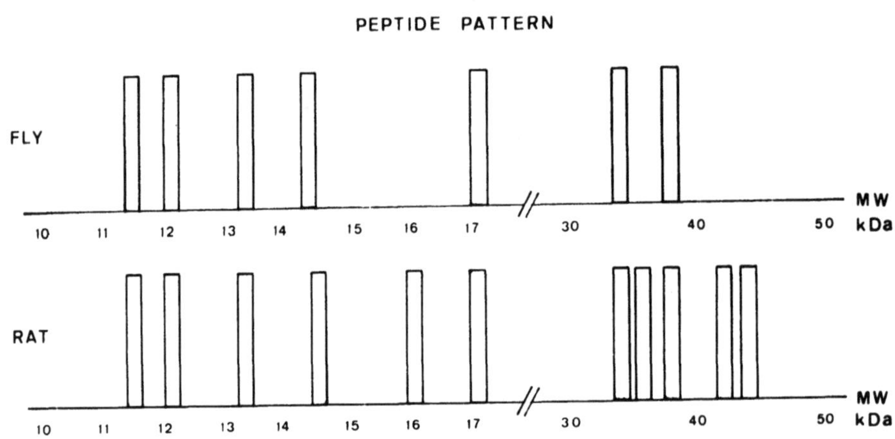

Fig. 6: Comparison of the phospho-peptide composition of 86 kDa protein of the fly and rat synapsin I.
The bars indicate the positions of the phospho-peptides on the SDS-gel. These patterns look very similar: In the low molecular weight range only two differences can be found; the 16.1 kDa peptide of the rat protein is missing in the pattern of the fly, and the 14.5 kDa peptide in rat protein is at 14.3 kDa in the fly. The other 4 peptides are identical with respect to their molecular weight. In the high molecular weight range there are two identical peptides at 38 and 34 kDa, the others are missing in the pattern of the fly. But these differences may be trivial, since too much protein from rat compared to protein of Drosophila was digested with the same amount of protease in a limited time causing only partially digested high molecular weight fragments.

In Fig. 6 the patterns of phospho-peptides of rat synapsin I and Drosophila 86 kDa protein are compared. These patterns look very similar. Only few differences can be found; in the low molecular weight range the 16.1 kDa peptide in rat synapsin is missing in the pattern of the fly, and the 14.5 kDa peptide in rat protein is at 14.3 kDa in fly protein. The molecular weights of the other 4 peptides of this range are identical.
The apparent differences in the high molecular weight range may be trivial, since too much protein from rat compared to protein from Drosophila was digested with the same amount of protease in a limited time, leading to an amount of only partially digested high molecular weight fragments.

DISCUSSION

Parallel to findings in Aplysia, a Ca^{2+}/calmodulin-dependent protein kinase is involved in long-term adaptational processes in neural tissues. As in the snail, the 50 kDa subunit of the protein kinase is associated to the membrane and it is released after Ca^{2+}-dependent autophosphorylation (11). After dissociation from the membrane the phosphorylation state remains stable for at least 30 min. Enzyme activity can be released from the membrane fraction into the supernatant: In vitro added vertebrate synapsin I is phosphorylated by the released enzyme in the presence of Ca^{2+}.

A phosphoprotein with the apparent molecular weight of 86 kDa seems to be an endogenous substrate of the Ca^{2+}/calmodulin-dependent protein kinase in the in vitro assay of the fly. This phosphoprotein exhibits properties similar to the synapsin I of rat: Its molecular weight and composition of phosphopeptides are similar. Antibodies against rat synapsin I do bind to 86 kDa protein of the fly.

We conclude that the 86 kDa protein in Drosophila is homologous to synapsin I.

Since the 86 kD protein seems to be a substrate of the 50 kD protein kinase in vitro, it is very likely that this protein is a substrate of the protein kinase in vivo as well. This leads to the following working hypothesis: Ca^{2+}-influx as a result of stimulation of the nerve cell stimulates the autophosphorylation of the Ca^{2+}/calmodulin-dependent protein kinase. This causes the protein kinase to dissociate at least partially from the membrane and to phosphorylate the synapsin I-homologous 86 kDa protein. According to the working modell of Nestler and Greengard (7) the phosphoform of this protein dissociates from the membrane of the synaptic vesicles, which subsequently prefuse with the pre-synaptic plasma membrane and so transmitter molecules are released.

This plastic pre-synaptic process, which can be analyzed easily in the fly, might be involved in neural information storage.

Acknowledgements: The authors wish to thank Dr. L.J. DeGennaro for kindly providing rat synapsin I and antibodies against this protein, and Prof. K.F. Fischbach for critical reading of the manuscript. The research was supported by research grant from the Deutsche Forschungsgesellschaft.

LITERATURE

1. WILLMUND, R. and FISCHBACH, K.F. (1977) Light induced modification of phototactic behavior of Drosophila melanogaster wildtype and some mutants in the visual system. J. Comp. Physiol.118, 261-271
2. COSENS, D., BRISCOE, D. (1972) A swith phenomenon in the compound eye of the white-eyed mutant of Drosophila melanogaster. J. Insect Physiol. 18, 627-632
3. STARK, W.S. (1975) Spectral selectivity of visual response alterations mediated by interconversions of native and intermediate photopigments in Drosophila. J. Comp. Physiol. 96, 343-356
4. MINKE, B., WU, C.F., PAK, W.L. (1975) Isolation of light induced response of the central retinula cells from the electroretinogram of Drosophila. J. Comp. Physiol. 98 345-355
5. COSENS, D., WRIGHT, R. (1975) Light elicited isolation of the complementary visual input systems in white-eyed Drosophila. J. Insect Physiol. 21, 1111-1120

6. BRODA, H., WRIGHT, R. (1978) prolonged depolarising afterpotentials in red-eyed Drosophila. J. Insect Physiol. 24, 681-684
7. NESTLER, E.J., GREENGARD, P. (1984) Protein phosphorylation in the nervous system. John Wiley & Sons, New York
8. KANDEL, E.R., SCHWARTZ, J.H. (1982) Molecular biology of learning: modulation of transmitter release. Science 218, 433-443
9. KANDEL, E.R., ABRAMS, T. BERNIER, L., CAREW, T.J. HAWKINS, R.D. and SCHWARTZ, J.H. (1983) Classical conditioning and sensitization share aspects of the same molecular cascade in Aplysia. Cold Spring Harbor Symp. Quant. Biol. 48, 821-830
10. WALTERS, E.T., BYRNE, J.H., CAREW, T.J. and KANDEL, E.R. (1983) Mechanoefferent neurons innervating the tail of Aplysia: II. modulation by sensitizing stimulation. J. Neurophysiol. 50, 1543-1559
11. SAITOH, T. and SCHWARTZ, J.H. (1985) Phosphorylation-dependent subcellular translocation of a Ca^{2+}/calmodulin-dependent protein kinase produces an autonomous enzyme in Aplysia neurons. J. Cell Biol. 100, 835-842
12. UEDA, T. and GREENGARD, P. (1977) Adenosine 3'-5'-monophosphate-regulated phosphoprotein system of neuronal membranes. J. Biol. Chem. 252, 5155-5163
13. BROWNING, M.D., HUANG, C.-K. and GREENGARD, P. (1987) Similarities between protein IIIa and protein IIIb, two prominent synaptic vesicle-associated phosphoproteins. J. Neurosci. 7, 847-853
14. DeCAMILLI, P., CAMERON, R. and GREENGARD, P. (1983a) Synapsin I (Protein I), a nerve terminal specific phosphoprotein. I. Its general distribution in synapses of the central and peripheral nervous system demonstrated by immunoflourescence in frozen and plastic sections. J. Cell Biol., 96, 1337-1354
15. DeCAMILLI, P., HARRIS, S.M. Jr., HUTTNER, W.B. and GREENGARD, P. (1983b) Synapsin I (Protein I), a nerve terminal specific phosphoprotein. II. Its specific association with synaptic vesicles demonstrated by immunocytochemistry in agarose-embedded synaptosomes. J. Cell Biol. 96, 1355-1373
16. HUTTNER, W.B., SCHIEBLER, W., GREENGARD, P. and DeCAMILLI, P. (1983) Synapsin I (Protein I), a nerve terminal specific phosphoprotein. III. Its association with synaptic vesicles studies in a highly purified synaptic vesicle preparation. J. Cell Biol. 96, 1374-1388
17. LLINAS, R., McGUINNES, T.L., LEONARD, C.S., SUGIMORI, M., GREENGARD, P. (1985) Intraterminal injection of synapsin I or Ca^{2+}/calmodulin-dependent protein kinase II alters neurotransmitter release at the squid giant synapse. Proc.Natl. Acad. Sci. (USA) 82, 3035-3039
18. BAEHLER, M. and GREENGARD, P. (1987) Synapsin I bundles F-actin in a phosphorylation dependent manner. Nature 326, 704-707

19. DUDAI, Y., UZZAN, A., Zvi, S. (1983) Abnormal activity of adenylate cyclase ibn Drosophila memory mutant rutabaga. Neurosci. Lett. 42, 207-212
20. QUINN, W.G., SZIBER, P.P. and BOOKER, R. (1979) The Drosophila memory mutant amnesiac. Nature 277, 212-214
21. ACEVES-PINA, E.O., BOOKER, R., DUERR, J.S., LIVINGSTONE, M.S., QUINN, W.G., SMITH, R.F., SZIBER, P.P., TEMPEL, B.L. and TULLY, T.P. (1983) Learning and memory in Drosophila, studied with mutants. Cold Spring Harbor Symp. Quant. Biol. 48, 831-840
22. LIVINGSTONE, M.S., SZIBER, P.P. and QUINN, W.G. (1984) Loss of calcium/calmodulin responsiveness in adenylate cyclase of rutabaga, a Drosophila learning mutant. Cell 37, 205-215
23. WILLMUND, R., MITSCHULAT, H. SCHNEIDER, K. (1986) Long-term modulation of Ca^{2+}-stimulated autophosphorylation and subcellular distribution of the Ca^{2+}/calmodulin-dependent protein kinase in the brain of Drosophila. Proc. Natl. Acad. Sci. (USA) 83, 9789-9793
24. WILLMUND, R. (1986) Longlasting modulation of protein kinase activity in the brain of Drosophila melanogaster induced by visual adaptation. J. Insect Physiol. 32, 1 - 8
25. LAEMMLI, U.K. (1970) Cleavage of structural proteins during the assembly of the head of bacteriophage T4. Nature 227, 680-685
26. TOWBIN, H. STAEHLIN, T. and GORDON, J. (1979) Electrophoretic transfer of protein from polyacrylamide gels to nitrocellulose sheets: Procedure and some applications. Proc. Natl. Acad. Sci. (USA) 76, 4350-4354
27. CLEVELAND, D.W., FISCHER, S.G., KIRSCHNER, M.W. and LAEMMLI, U.K. (1977) Peptide mapping by limited proteolysis in sodium dodecyl sulfate and analysis by gel electrophoresis. J. Biol. Chem. 252, 1102-1106

ON THE ROAD TO A BETTER UNDERSTANDING OF LEARNING AND MEMORY IN *DROSOPHILA MELANOGASTER*

T. Tully
Department of Biology
Brandeis University
Waltham MA 02254
U.S.A.

In the decade or so since S. Benzer and coworkers at CalTech began to isolate mutations that affect learning and memory in *Drosophila*, much work has revealed interesting, if not exciting, biochemical and behavioral details about normal and mutant flies. Below, I will summarize our current understanding of the biochemistry underlying associative learning. Then, I will discuss interesting behavioral aspects of normal and mutant memory formation, which is produced by a classical conditioning procedure.

THE CYCLIC AMP CELL SIGNALLING SYSTEM AND BEYOND

Four years ago, the biochemistry underlying associative learning in fruit flies seemed rather simple (Quinn, 1984). Four mutations were reported to interfere with different components of the monoamine-stimulated adenylate cyclase pathway -- *dunce*, *rutabaga*, *turnip* and *Ddc* (dopa decarboxylase deficient). These results were considered to support strongly a model of associative learning in *Aplysia* proposed by Kandel and coworkers (Kandel et al, 1987). More recent biochemical work on the learning mutants suggests that associative learning in *Drosophila* may not be quite so simple. Below, I will update the reader on these new developments.

dunce

Recent molecular cloning experiments have provided direct proof that the *dunce*+ gene encodes a phosphodiesterase (Chen et al, 1986). Substantial sequence homology exists between the (predicted) translation product of *dnc*+ and the amino

NATO ASI Series, Vol. H19
Modulation of Synaptic Transmission and Plasticity
in Nervous Systems. Edited by G. Hertting and H.-C. Spatz

acid sequences of bovine Ca^{2+}/calmodulin-dependent PDE and of yeast cAMP-dependent PDE. Moreover, the dnc+ gene product and the RII subunit of cAMP-dependent protein kinase share a short but perfect sequence homology, which may correspond to the cAMP binding domain. Importantly, the dnc+ protein shows (weak) homology to the precursor of Aplysia egg-laying hormone (ELH). The ELH precursor homology is confined to its own exon, suggesting that alternative RNA splicing of dnc+ may produce a protein involved with egg-laying in Drosophila. RNA transcript analysis also suggests that the dnc+ gene encodes more than one function (Davis & Davidson, 1986). In both dnc^{M11} and dnc^{M14} flies, alterations in the size and number of these transcripts can be detected. Thus, dunce mutations may cause female sterility by altering the sequence or splicing of RNA transcripts containing the ELH exon.

rutabaga

Direct molecular evidence that rutabaga is the structural gene for a Ca^{2+}-sensitive adenylate cyclase (Livingstone et al, 1984) does not yet exist, although several labs currently are trying to clone the gene. Biochemical and genetic data continue to support such a hypothesis, however. Stimulation of adenylate cyclase (AC) by octopamine, GppNHp (a GTP analogue) and NaF is normal in rutabaga flies, suggesting that receptor and G_s components of the AC membrane complex are normal. Moreover, GppNHp inhibits manganese-stimulated AC activity normally in rutabaga flies, indicating that G_i is not affected by the mutation. In contrast, differences in AC activity between wild-type and mutant flies can be detected when ligands thought to interact directly with the catalytic subunit are used. Calcium, forskolin, manganese, free Mg^{2+} and the Mg^{2+}-ATP substrate complex all produce abnormal activation of AC in rutabaga flies (Livingstone et al, 1984; Livingstone, 1985; Dudai & Zvi, 1985).

Genetic experiments have co-mapped rutabaga's defects in Ca^{2+} and forskolin stimulation of AC and learning to chromomeres 12F5-7 on the X chromosome (Dudai et al, 1985; Livingstone et al, 1984; Livingstone, 1985). Furthermore, Ca^{2+}-stimulated AC activity is proportional to the number of rut+ gene copies in the genome (Livingstone, 1985), which is a result similar to those of other genes that encode enzymes in Drosophila (see Stewart & Merriam, 1980).

Ddc

Flies homozygous for either of two temperature-sensitive alleles of the Ddc (dopa decarboxylase deficient) gene do not synthesize dopamine or serotonin in restrictive temperatures (Livingstone & Tempel, 1983), and one study claims that they learn poorly in several associative tasks (Tempel et al, 1984). These results suggested that one or both of these putative neurotransmitters may be necessary for associative learning in fruit flies. Unfortunately, neither of these Ddc strains show temperature-sensitive effects in classical conditioning experiments (Tully, unpublished data; see below). Recent attempts to replicate the learning deficit, originally measured in a shock-avoidance conditioning procedure (Quinn et al, 1974), have failed (Tully, Chan & Quinn, unpublished data).

At first glance, these failures to produce a learning deficit in Ddc mutants (under appropriate conditions) might lead one to conclude that the original behavioral results of Tempel et al (1984) cannot be replicated (as is the case with many behavioral experiments). However, the situation is not that simple. Dopa decarboxylase (DDC) probably is not the rate-limiting enzyme in the synthetic pathways of dopamine and serotonin; as little as 3-5% of normal levels of DDC activity may be sufficient to produce normal levels of transmitters (Wright, 1987). In restrictive temperatures, Ddc^{ts2} flies show 1-2% normal DDC activity. Possibly, our recently tested Ddc^{ts} strains have accumulated genetic modifiers, which have increased DDC activity to the 3-5% level or which have increased the concentrations of precursor pools. In either case, near-normal levels of transmitters might result. In support of this hypothesis, HPLC measurements reveal that endogenous levels of dopamine and serotonin are no less than 50% of wild-type levels in those Ddc strains that now learn normally.

New data actually support the alternate hypothesis -- that dopamine and serotonin are not necessary for associative learning in flies (Morgan, Tully & Hirsh, in preparation). The Ddc gene has been cloned, and study of its gene structure has led to the discovery of two tissue-specific forms of dopa decarboxylase, one expressed in the CNS and one in the hypoderm (Morgan et al, 1986). Genetic engineering of a frame-shift mutation in the CNS-specific exon, followed by germ-line transformation, has produced transformant strains (on a Ddc null background), which express little or no DDC in the CNS and normal levels in hypoderm. These transformants show less than 1% normal levels of serotonin and 30% normal levels of dopamine in dissected larval CNSs. Significantly, these flies do not show severe learning deficits. These results suggest that serotonin, at least, is not necessary for associative learning. Since dopamine levels still are relatively high, evidence that this transmitter is not necessary for learning remains weak. However, enzymatic studies never have revealed significant differences in substrate affinities for DDC. Therefore, the most likely reason that HPLC-quantified dopamine levels are high is that dopamine leaked into CNS tissue from hypodermal tissue during dissection.

turnip

In early experiments on turnip flies, membrane extracts from heads showed reduced monoamine receptor binding, lower GppNHp-stimulated adenylate cyclase activity and abnormal GTPase activity (Aceves-Pina et al, 1983). All three biochemical defects are not severe, however, and have not yet been shown to co-map genetically with turnip's learning defect. Nevertheless, these results are consistent with the idea that the turnip mutation in some way affects G-protein function, again implicating the cAMP cell signalling system with associative learning.

More recent biochemical experiments have revealed little or no EGTA-extractable protein kinase C (PKC) activity in turnip head membranes (Smith et al, 1986). Genetic experiments, however, indicate that turnip is not the structural gene for PKC (also see Rosenthal et al, 1987). Flies heterozygous for turnip and the wild-type allele have normal, rather than intermediate, levels of PKC activity (in contrast to similar experiments on dunce and rutabaga above). Furthermore, bovine PKC cDNA probes do not hybridize in situ to the turnip chromosomal region (Smith et al, 1986). Taken together, these biochemical results lead us to suspect that 1) turnip may affect a membrane-bound regulator of PKC and 2) PKC may "crosstalk" with the cAMP second messenger system in an as yet unidentified manner (cf. Katada et al, 1985; Yoshimasa et al, 1987).

Such profound biochemical effects in turnip flies only emphasize the need for formal genetic proof that the biochemical defects co-map genetically with the learning defect. To this end, genetic recombination experiments have indicated that the turnip gene is located on the X-chromosome proximal to forked. Moreover, several recombinants that show high or low PKC activity also show corresponding high or low learning scores. Preliminary deletion mapping experiments, however, do not support the conclusion that defects in PKC activity and learning co-map to the same locus in turnip flies. Flies heterozygous for turnip and an X-chromosome deletion do not show any PKC activity, yet they show normal classical conditioning (Choi, Smith, Tully & Quinn, unpublished). Currently, we are repeating these experiments, using several different turnip strains, to determine whether genetic background effects may have confounded our deletion mapping results.

Does the cAMP second messenger system underlie associative learning in Drosophila?

Figure 1 depicts the convergence between Kandel's model of associative learning in Aplysia and the biochemical abnormalities in Drosophila mutants. The facts that

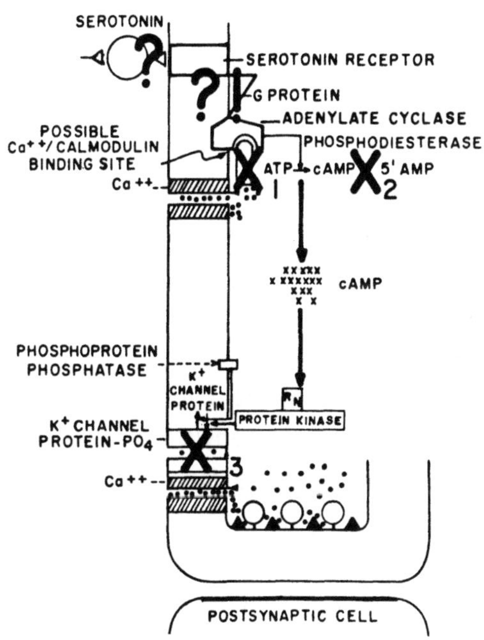

Figure 1. Convergence between the Aplysia biochemical model of associative learning and mutations affecting learning in Drosophila.

the dunce (X_2), rutabaga (X_1) and, maybe, Ddc (?) and turnip (?!) mutations all affect components of the monoamine-stimulated adenylate cyclase pathway cannot be coincidental. Three of these mutations were isolated independently because they disrupted shock-avoidance learning (Aceves-Pina & Quinn, 1979; Dudai et al,

1976; Quinn et al, 1979). The fourth mutation was known to affect biogenic amine metabolism and, therefore, was suspected to affect learning.

In fact, this latter approach -- to assay learning in flies carrying mutations known to affect specific biochemical or physiological systems -- has provided further evidence consistent with Kandel's model. Classical conditioning experiments on flies homozygous for the Shaker mutation (X_3 in Fig. 1) show only moderate learning and a rapid short-term memory loss, a phenotype similar to the other learning mutants (see below). In contrast, learning and memory in nap^{ts} flies are essentially normal (Cowan & Siegel, 1986). Importantly, the Shaker mutation affects a voltage-sensitive K^+ channel (Tempel et al, 1987), while nap disrupts Na^+ channel function (Wu & Ganetsky, 1980). Although work on Aplysia suggests that a voltage-insensitive K^+ channel is phosphorylated and inactivated during associative learning, results with Drosophila ion channel mutants again are in general agreement with the Aplysia model.

In spite of the convergence of results from Aplysia and Drosophila, three observations suggest that the cAMP cell signalling system is not the only biochemical response underlying associative learning in flies: 1) According to the Kandel model, the molecular point of convergence between two stimuli [the conditioned stimulus (CS) and the unconditioned stimulus (US)], which must be presented to the organism in close temporal proximity to produce associative learning, is a Ca^{2+}-sensitive adenylate cyclase. Interestingly, the rutabaga mutation produces a complete loss of adenylate cyclase activity in vitro. Therefore, the cAMP model predicts that rutabaga flies would be incapable of associative learning. Nevertheless, rutabaga flies are capable of moderate levels of associative learning (see below). 2) The turnip mutation affects PKC, which normally can be activated in vivo via the phosphatidylinositol (PI) pathway. Although PKC may interact with the cAMP pathway (see above), the possibility still exists that turnip's effects on learning and memory derive from disruption of other biochemical response systems. 3) An exhaustive biochemical analysis of monoamine-stimulated adenylate cyclase activity in amnesiac flies, which have a behavioral phenotype similar to dunce, rutabaga and turnip (see below), has not uncovered any consistent effects (Tully & Brandes, unpublished).

The discrepant biochemical data from fruit flies has provoked some to shy away from the Kandel model of associative learning. It seems that learning and in flies is more complicated than learning in a monosynaptic circuit. Furthermore, PKC has been implicated with long-term potentiation in rat hippocampal slices (Akers et al, 1986) and with learning in Hermissenda (Farley & Auerbach, 1986) and in Aplysia (Hochner et al, 1986; Sacktor et al, 1986). In light of these developments, we are forced to consider two possibilities: 1) Associative learning in simple (monosynaptic) neuronal systems may induce multiple biochemical responses via more than one second messenger system. 2) Associative learning in complex neuronal circuits may induce different subsets of neurons, which respond with different biochemical systems. Thus, associative learning in whole-organisms may result from the temporal and spatial summation of multiple underlying biochemical response systems. Moreover, different constellations of biochemical responses may be responsible for different associative tasks in a given species, or in different species. In short, comlex biochemistries may act in concert to produce whole-organism learning. Accordingly, the cAMP cell signalling system may not account for all of fly learning, but such a mechanism most likely is responsible, in part, for the appearance of conditioned responses.

ASSOCIATIVE LEARNING AND MEMORY IN NORMAL AND MUTANT FLIES: A PAVLOVIAN PERSPECTIVE

Four years ago, our understanding of learning and memory in normal and mutant flies was confused at best. Two mutations (turnip and Ddc) appeared to disrupt learning acquisition (Tempel et al, 1983; Quinn et al, 1979), one mutation (dunce) disrupted memory (Dudai, 1979, 1983) and three mutations (rutabaga, cabbage and amnesiac) seemed to interfere with memory retrieval (Quinn et al, 1979; Dudai, 1983). The puzzling fact was that mutations from each of these phenotypic classes were thought to affect the cAMP cell signalling system.

One possible explanation for such confusion rests on the fact that the behavioral assays for associative learning were instrumental conditioning procedures. Flies received a reinforcing stimulus in the presence of a conditioned stimulus only if they first elicited a particular behavioral response. When exposed to an odor, for instance, flies were either shocked, or rewarded with food, only if they walked toward the odor. Thus, wild-type and mutant flies, which differ in their capacities to associate events, may have received different amounts of reinforcement during training. These different amounts of training then may have interacted with acquisition and memory processes in strain-specific ways to produce unpredictable effects on performance in any particular behavioral assay.

This rationale provided Tully & Quinn (1985) with the motivation to design a T-maze choice chamber so that carefully controlled air currents could be drawn through it. In this manner, the original shock-avoidance paradigm (Quinn et al, 1974) was adapted to a classical (Pavlovian) conditioning procedure. About 150 flies are sequestered in a closed chamber and are trained by exposing them sequentially to two odors delivered in air currents. Flies receive twelve 1.25-s 60V (DC) electric shock pulses every five seconds during a 60 s presentation of the first odor (CS+) but not during a 60 s presentation of the second odor (CS-). To test for conditioned avoidance, flies are tapped gently into a sliding compartment, which transports them (elevator fashion) to the T-maze choice point between converging currents of the two odors. Typically, 93% of wild-type flies avoid the CS+, 4% avoid the CS- and 3% remain in the middle of the T-maze choice point, yielding a learning index of 0.89 (see Tully & Quinn, 1985, for more experimental details).

This learning score for wild-type flies is significantly higher than that of 0.33 produced by the original shock-avoidance procedure (Quinn et al, 1974). Two changes most likely account for the improved results: 1) During training with the classical conditioning procedure, flies cannot escape electric shock. All flies receive equal, maximal amounts of shock reinforcement. 2) Testing for conditioned responses after classical conditioning is done in dim red light, thereby eliciting little or no phototaxis. To perform correctly, flies simply have to learn to avoid a specific shock-paired odor. In contrast, to perform correctly after shock-avoidance conditioning, flies were required to elicit two conditioned responses -- avoidance of a specific odor and inhibition of phototaxis. The latter response may have been too difficult for most flies to learn or to remember.

During the classical conditioning procedure, presentations of CS+, CS- and US could be controlled more precisely, which permitted Tully & Quinn (1985) to characterize several behavioral properties of classically conditioned avoidance responses such as acquisition, extinction, and CS and US saliencies. Using standard control procedures to assess the effects of nonassociative learning, Tully & Quinn also were able to demonstrate empirically that the learning index is a measure of associative learning unbiased by nonassociative factors (which are present during classical conditioning). More importantly, Tully & Quinn showed that conditioned responses 1) are not produced when the US precedes the CS and 2) wane as the interval of time between CS and US presentations increases. Thus, associative learning in Drosophila shows order dependence and temporal specificity -- two properties common to associative learning across species.

Given such robust effects of classical conditioning, Tully & Quinn (1985) went on to compare learning and memory in wild-type (Can-S), amnesiac (amn), rutabaga (rut), dunce (dnc) and turnip (tur) flies. From results depicted in Figure 2, three general conclusions can be drawn: 1) The mutants all show moderate levels of

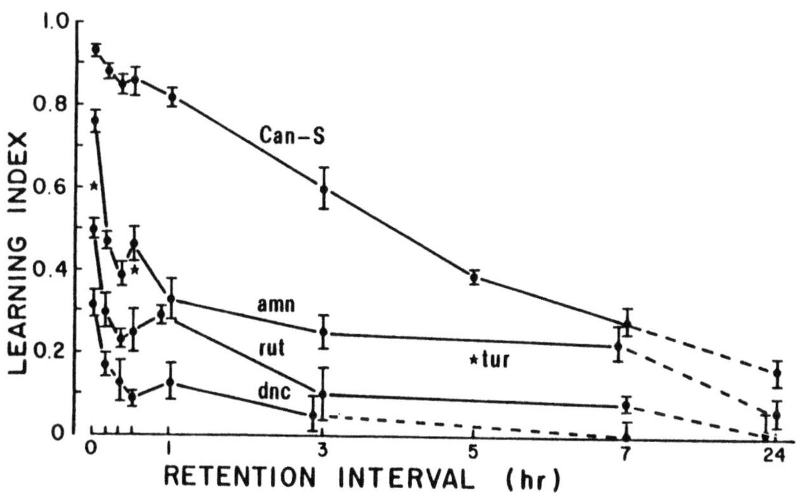

Figure 2. Memory retention in normal and mutant flies after classical conditioning.

learning. 2) Mutant memory curves look similar. They differ quantitatively rather than qualitatively. 3) Mutant memory curves appear to be composed of at least two components. Their memories decay rapidly at first, but then memory loss slows dramatically.

Earlier work on wild-type Drosophila revealed that an anesthesia-resistant phase of memory begins to appear within an hour after training (Quinn & Dudai, 1976; Tempel et al, 1983). Tully & Block (Fig. 3; unpublished data) have confirmed these findings using the classical conditioning procedure. Wild-type flies, trained in the usual way, were anesthetized at various Intervals afterwards by placing them in a test tube and submerging the tube in 4° C water for 2 min. Conditioned avoidance

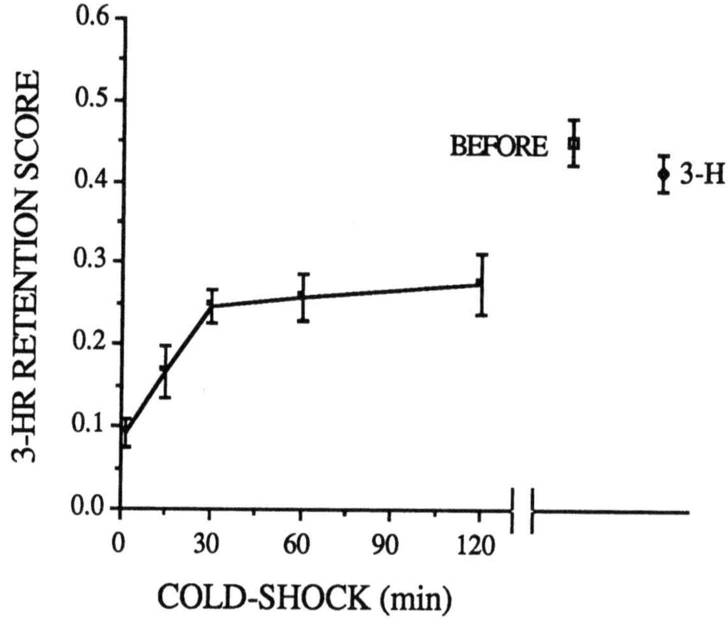

Figure 3. Memory consolidation after classical conditioning in wild-type (Can-S) flies.

then was tested three hours after training. Anesthetizing flies during the first 30 min after training produces relatively large deficits in 3-hr memory. In contrast, cold-shock 30 to 120 min after training affects 3-hr scores minimally.

These results indicate that long-term memory begins to form during training and is maximal 30 to 60 min later. Interestingly, maximum anesthesia-resistant memory levels for wild-type flies in the cold-shock experiments are equivalent to those for amnesiac flies one or more hours after training in the standard classical conditioning procedure (see Fig. 2). Taken together, these observations suggest that the amnesiac, rutabaga, turnip and dunce mutations affect acquisition and short-term memory, leaving long-term memory formation substantially intact. In fact, Dudai et al (1987) recently have confirmed that rutabaga flies can form anesthesia-resistant memory.

The appearance of an anesthesia-resistant form of memory traditionally has been considered an indication that a short-term, labile phase of memory has been "consolidated" into a more stable, long-term phase (Lynch et al, 1984; McGaugh & Herz, 1972). Given the existence of an anesthesia-resistant form of memory in flies, Tully & Kruse (unpublished data) decided to see just how stable long-term memory might be. Third-instar larvae were trained to avoid an odor paired with shock (Aceves-Pina & Quinn, 1979). Afterwards, half of the trained larvae was tested immediately for conditioned avoidance; the other half was tested for conditioned avoidance as adults. Results shown in Figure 4 indicate that larvae can learn to

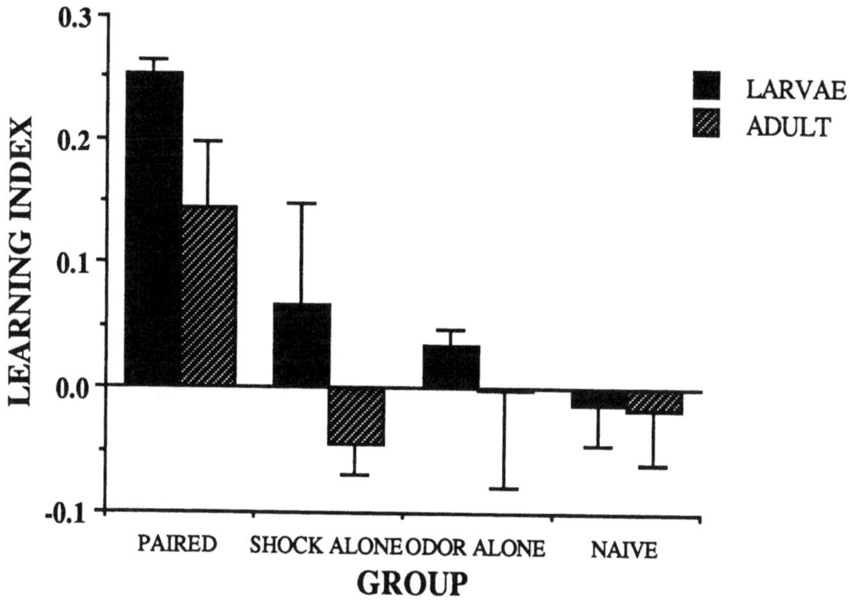

Figure 4. Memory through metamorphosis in wild-type (Can-S) flies.

avoid a shock-associated odor and, more interestingly, that they can remember this event as adults seven days later. Thus, wild-type flies can form long-term memory that is stable even through metamorphosis. Experiments are underway to see whether similar memory is formed in the learning mutants.

Work on memory in Aplysia indicates that the decay rate of sensitization-induced increases in cAMP levels parallels the decay rate of short-term memory of the behavioral response, which lasts about an hour (Schwartz et al, 1983). In Drosophila, mutations that affect components of the cAMP cell signalling system also produce short-term memory deficits. We do not know yet whether associative learning produces a transient increase in neuronal cAMP levels in flies. However, lower-than-normal learning-induced increases in cAMP in the mutants could explain the observed deficits in acquisition and short-term memory. Moreover, if a learning-induced increase in cAMP ultimately triggers the formation of long-term memory (Goelet et al, 1986), then lower levels of long-term memory in the mutants (see Fig. 2) would be expected -- weaker learning produces less long-term memory, perhaps because cAMP (and, therefore, short-term memory) levels do not reach a critical threshold.

These ideas can be combined with a healthy amount of speculation to advance a preliminary model of memory formation in fruit flies (Fig. 5). Acquisition and short-term memory (STM) are maximal immediately after training, but STM decays within the first hour of retention. Long-term memory (LTM) begins to appear shortly after training but does not reach maximal levels for one hour. Interestingly, if STM and LTM phases act additively (which certainly may be an oversimplification) and no other memory phase exists, then the STM+LTM curve should equal observed retention levels. In fact, this is not the case. One must postulate the existence of a middle-term memory (MTM) phase, which also acts additively with STM and LTM, to account for the observe memory curve. Currently, no evidence for MTM exists in flies. However, behavioral and pharmacological experiments on day-old chicks have detected a MTM phase with kinetics similar to our hypothetical MTM (Gibbs & Ng, 1976; Patterson et al, 1985).

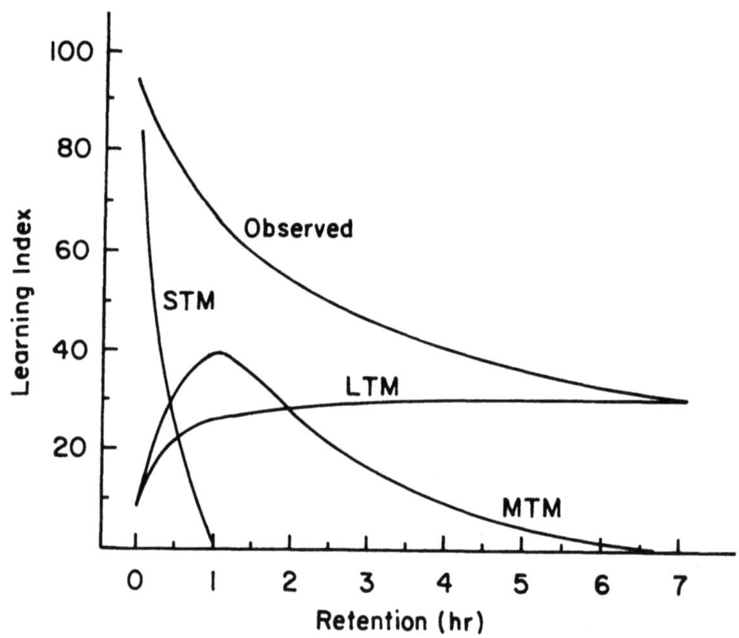

Figure 5. Hypothetical model of memory formation in Drosophila.

This model of memory formation in Drosophila is generally consistent with a biochemical model of memory formation (Goelet et al, 1986). Associative learning may cause a rapid increase in cAMP levels in relevant neurons during training.

Elevated cAMP levels activate a cAMP-dependent protein kinase (Foster et al, 1984), which leads to two independent biochemical events : 1) The catalytic subunit of the kinase phosphorylates K^+ channels (and other intracellular proteins), leading to an influx of Ca^{2+} in presynaptic terminals and increased transmitter release, which corresponds to STM (Schwartz et al, 1983). As cAMP levels return to baseline within minutes, phosphorylated proteins are dephosphorylated by extant (and most likely up-regulated) phosphatases. 2) A regulatory subunit of kinase derepresses the expression of specific genes (Constantinou et al, 1985; Waterman et al, 1985)-- some that promote biochemical changes in neurons lasting minutes to hours, which corresponds to MTM (Gibbs & Ng, 1976). Other newly expressed proteins produce more stable biochemical (Lisman, 1985) or structural changes (Bailey & Chen, 1983) lasting hours to days, which corresponds to LTM.

CONCLUSIONS

Our understanding of the biochemistry of learning and memory in Drosophila is far from complete. Strong biochemical and genetic evidence implicates the cAMP cell signalling system with assoiciative learning. However, behavioral and biochemical data suggest that other biochemical systems also must be involved. Nevertheless, these new clues do not disprove or minimize the involvement of the cAMP second messenger system. On the contrary, we might have expected a more complex biochemistry to underlie associative learning in whole organisms, when compared to the cellular biochemistry of an identified monosynaptic circuit.

On the behavioral level of analysis, the development of a classical conditioning procudure for Drosophila has clarified our understanding of how extant learning mutations affect conditioned behavior. Furthermore, classical conditioning produces long-lasting memory, which permits a more detailed study of memory formation. Initial results from experiments on the learning mutants suggest that short-term and long-term memory are genetically distinct processes.

REFERENCES

Aceves-Pina EO, Quinn WG (1979) Learning in normal and mutant Drosophila larvae. Science 206: 93-96

Aceves-Pina EO, Booker R, Duerr JS, Livingstone MS, Quinn WG, Smith RF, Sziber PP, Tempel BL, Tully TP (1983) Learning and memory in Drosophila, studied with mutants. Cold Spring Harbor Symp. Quant. Biol. 48: 831-839

Akers RF, Lovinger DM, Colley PA, Linden DJ, Routtenberg A (1986) Translocation of protein kinase C activity may mediate hippocampal long-term potentiation. Science 231: 587-589

Bailey CH, Chen M (1983) Morphological basis of long-term habituation and sensitization in Aplysia. Science 220: 91-93

Chen C-N, Denome S, Davis RL (1986) Molecular analysis of cDNA clones and the corresponding genomic coding sequences of the Drosophila dunce+ gene, the structural gene for cAMP-dependent phosphodiesterase. Proc. Natl. Acad. Sci., U.S.A. 83: 9313-9317

Constantinou AI, Squintto SP, Jungmann RA (1985) The phosphoform of the regulatory subunit RII of cyclic AMP-dependent protein kinase possesses intrinsic topoisomerase activity. Cell 42: 429-437

Cowan T, Siegel RW (1986) Drosophila mutations that alter ionic conduction disrupt acquisition and retention of a conditioned odor avoidance response. J. Neurogenet. 3: 187-201

Davis RL, Davidson N (1986) The memory gene dunce+ encodes a remarkable set of RNAs with internal heterogeneity. Mol. Cell. Biol. 6: 1464-1470

Dudai Y (1979) Behavioral plasticity in a Drosophila mutant, dunce^{DB276}. J. Comp. Physiol. 130: 271-275

Dudai Y (1983) Mutations affect storage and use of memory differentially in Drosophila. Proc. Natl. Acad. Sci., U.S.A. 80: 5445-5448

Dudai Y, Zvi S (1985) Multiple defects in the activity of adenylate cyclase from the Drosophila memory mutant rutabaga. J. Neurochem. 45: 355-364

Dudai Y, Corfas G, Hazvi S (1987) What is the possible contribution of Ca^{2+}-stimulated adenylate cyclase to acquisition, consolidation and retention of an associative olfactory memory in Drosophila. J. Comp. Physiol. (in press)

Dudai Y, Jan Y-N, Byers D, Quinn WG, Benzer S (1976) dunce, a mutant of Drosophila deficient in learning. Proc. Natl. Acad. Sci., U.S.A. 73: 1684-1688

Dudai Y, Sher B, Segal D, Yovell Y (1985) Defective responsiveness of adenylate cyclase to forskolin in the Drosophila memory mutant rutabaga. J. Neurogenet. 2: 365-380

Farley J, Auerbach S (1986) Protein kinase C activation induces conductance changes in Hermissenda photoreceptors like those seen in associative learning. Nature 319: 220-223

Foster JL, Guttman JJ, Hall LM, Rosen OM (1984) Drosophila cAMP-dependent protein kinase. J. Biol. Chem. 259: 13049-13055

Gibbs ME, Ng KT (1976) Memory formation: a new three-phase model. Neurosci. Lett. 2: 165-169

Goelet P, Castellucci VF, Schacher S, Kandel ER (1986) The long and the short of long-term memory - a molecular framework. Nature 322: 419-422

Hochner B, Braha O, Klein M, Kandel ER (1986) Distinct processes in presynaptic facilitation contribute to sensitization and dishabituation in Aplysia: possible role of C kinase in dishabituation. Soc. Neurosci. Abstr. 12: 1340

Kandel ER, Klein M, Hochner B, Shuster M, Siegelbaum SA, Hawkins RD, Glanzman DL, Castellucci VF, Abrams TW (1987) Synaptic modulation and learning: new insights into synaptic transmission from the study of behavior. In Edelman, GM, Gall, WE, Cowan, WM (eds) Synaptic Function. John Wiley & Sons New York pp. 471-518

Katada T, Gilman AG, Watanabe Y, Bauer S, Jakobs KH (1985) Protein kinase C phosphorylates the inhibitory guanine nucleotide-binding regulatory component and apparently suppresses its function in hormonal inhibition of adenylate cyclase. FEBS Lett. 151: 431-437

Lisman JE (1985) A mechanism for memory storage insensitive to molecular turnover: a bistable autophosphorylating kinase. Proc. Natl. Acad. Sci., U.S.A. 82: 3055-3057

Livingstone MS (1985) Genetic dissection of Drosophila adenylate cyclase. Proc. Natl. Acad. Sci., U.S.A. 82: 5992-5996

Livingstone MS, Tempel BL (1983) Genetic dissection of monoamine transmitter synthesis in Drosophila. Nature 303: 67-70

Livingstone MS, Sziber PP, Quinn WG (1984) Loss of calcium/calmodulin responsiveness in adenylate cyclase of rutabaga, a Drosophila learning mutant. Cell 37: 205-215

Lynch G, McGaugh JL, Weinberger NM (eds) (1984) Neurobiology of Learning and Memory. Guilford London

McGaugh JL, Herz MJ (1972) Contemporary issues in psychobiology: Memory consolidation. Albion San Francisco

Morgan BA, Johnson WA, Hirsh J (1986) Regulated splicing produces different forms of dopa decarboxylase in the central nervous system and hypoderm of Drosophila melanogaster. EMBO J. 5: 3335-3342

Patterson TA, Rosenzweig MR, Bennett EL, Lucey SE (1985) Evidence for multiple stages of memory formation in the chick. Soc. Neurosci. Abstr. 11: 534.

Quinn WG (1984) Work in invertebrates on the mechanisms underlying learning. In Marler, P, Terrace, HS (eds) The Biology of Learning. Dahlem Konferenzen Springer-Verlag Berlin pp. 197-246

Quinn WG, Dudai Y (1976) Memory phases in Drosophila. Nature 262: 576-577

Quinn WG, Harris WA, Benzer S (1974) Conditioned behavior in Drosophila melanogaster. Proc. Natl. Acad. Sci., U.S.A. 71: 708-712

Quinn WG, Sziber PP, Booker R (1979) The Drosophila memory mutant amnesiac. Nature 277: 212-214

Rosenthal AR, Rhee L, Yadegari R, Paro R, Ullrich A, Goeddel DV (1987) Structure and nucleotide sequence of a Drosophila melanogaster protein kinase C gene. EMBO J. 6: 433-441

Sacktor TC, O'Brian CA, Weinstein IB, Schwartz JH (1986) Translocation from cytosol to membrane of protein kinase C after stimulation of Aplysia neurons with serotonin. Soc. Neurosci. Abstr. 12: 1340

Schwartz JH, Bernier L, Castellucci VF, Palazzolo M, Saitoh T, Stapleton A, Kandel ER (1983) What molecular steps determine the time course of the memory for short-term sensitization in Aplysia? Cold Spring Harbor Symp. Quant. Biol. 48: 811-819

Smith RF, Choi K-W, Tully T, Quinn WG (1986) Deficient protein kinase C activity in turnip, a Drosophila learning mutant. Soc. Neurosci. Abstr. 12: 399

Stewart B, Merriam JR (1980) Dosage compensation. In Ashburner M, Wright T The Genetics and Biology of Drosophila Vol. 2d Academic Press New York pp. 107-140

Tempel BL, Bonini N, Dawson DR, Quinn WG (1983) Reward learning in normal and mutant Drosophila. Proc. Natl. Acad. Sci., U.S.A. 80: 1482-1486

Tempel BL, Livingstone MS, Quinn WG (1984) Mutations in the dopa decarboxylase gene affect learning in Drosophila. Proc. Natl. Acad. Sci., U.S.A. 81: 3577-3581

Tempel BL, Papazian DM, Schwarz TL, Jan YN, Jan LY (1987) Sequence of a probable potassium channel component encoded at Shaker locus of Drosophila. Science 237: 770-775

Tully T, Quinn WG (1985) Classical conditioning and retention in normal and mutant Drosophila melanogaster. J. Comp. Physiol. 157: 263-277

Waterman M, Murdoch GH, Evans RM, Rosenfeld MG (1985) Cyclic AMP regulation of eukaryotic gene transcription by two discrete molecular mechanisms. Science 229: 267-269

Wright TRF (1987) The genetics of biogenic amine metabolism, sclerotization and melanization in Drosophila melanogaster. Adv. Genet. 25: (in press)

Wu CF, Ganetzky B (1980) Genetic alteration of nerve membrane excitability in temperature-sensitive paralytic mutants of Drosophila melanogaster. Nature 286: 814-816

Yoshimasa T, Sibley DR, Bouvier M, Lefkowitz RJ, Caron MG (1987) Cross-talk between cellular signalling pathways suggested by phorbol-ester-induced adenylate cyclase phosphorylation. Nature 327: 67-70

SHORT-TERM PERSISTENCE OF SEXUAL AROUSAL IN *DROSOPHILA MELANOGASTER* MALES : CONTRASTING BEHAVIOURAL EXPRESSION IN TWO WILD-TYPE STRAINS.

J. Médioni
Département de Biologie du Comportement
Université de Toulouse-3
118 route de Narbonne
31062 Toulouse
France

A. Manning
Department of Zoology
University of Edinburgh
West Mains Road
Edinburgh EH9 3JT
United Kingdom

Interruption of the courtship induces in *Drosophila* males a persistent sexual arousal, lasting up to 1 hour. Odour trails are not implied in this effect : for a clean space offered to the male's post-courtship display induces a longer persistence than the meeting place does (possible enhancement by novelty). Food deprivation may also modulate this effect, according to the duration of fasting. Finally, the dominant behavioural expression of sexual arousal is quite different in two wild-type strains. These results are discussed in terms of non-specific processes of central activation.

Many years ago (Médioni 1959 *a, b*) the possible operation of some non-specific processes of central activation was considered in insects, as a modulator of various behavioural phenotypes : phototaxis and geotaxis (Médioni 1959, Campan 1966), take-off rapidity (Romaneli 1968), sexual efficiency of males (Vaysse 1973). All these studies, performed on *Drosophila melanogaster,* were inspired by the model of non-specific arousal proposed by Moruzzi and Magoun (1959) for mammals, relating wakefulness to an aspecific processing of various sources of sensory stimulations by the central nervous system. A convergent hypothesis was formulated by Dethier, Solomon and Turner (1965), accounting for some aspects of the food-research behaviour in the blowfly *Phormia regina,* by inference of a "central excitatory state" (C.E.S.) induced by sugar stimulation ; the importance of this factor was established, as a source of artefact in associative learning, by Nelson (1971) and Hirsch and McCauley (1977) on *Phormia,* as well as by Vargo and Hirsch (1982) on *Drosophila melanogaster.*

In the present article, we shall describe and analyse an effect of persistence of "sexual arousal" in male *D. melanogaster* ; in particular, we shall investigate whether this effect may be enhanced by some non-specific, external or internal, factors of central activation. We shall also stress the differences in the behavioural expression of sexual arousal between two wild-type laboratory strains of D. *melanogaster.*

NATO ASI Series, Vol. H19
Modulation of Synaptic Transmission and Plasticity
in Nervous Systems. Edited by G. Hertting and H.-C. Spatz

Persistent sexual arousal was discovered in A. Manning's laboratory by himself and a doctoral student, P. Thorpe, in 1971 : they looked at what is happening of the male partner, if the courtship of an isolated pair of fruit-flies is interrupted before the copulation, by letting the female flee away : typically, the "forsaken" male becomes then able to emit an important flow of wing-display : lateral vibrations and mainly "scissoring" movements. In the case of *melanogaster,* scissoring is rarely observed in presence of the female, whereas this peculiar display is much more frequent during the courtship of *simulans* (Manning, 1959).
These observations had remained unpublished. We recently tried to analyse further the persistence of sexual arousal, especially to investigate whether the remnant sexual arousal was specifically linked to a sexual motivation ; or, alternatively, was dependent on a more general, non-specific, activatory process. An important part of this research was performed on some mutants (especially *dunce^2* and *Dopadecarboxylase-ts^2* : Médioni and Manning 1987, *in press*). In this paper we shall concentrate on some striking differences observed between the males of two wild-type strains, all confronted with young, still unreceptive females.

Materials and procedure

In these observations, the encounter between one male and one female fly was controlled and watched under a stereomicroscope, with the help of a two-parts cell, fitted with a slide door (Figure 1) ; each "room" was 5 mm high and measured 25 mm in diameter.
Three minutes after introduction of the two potential partners in their respective room, a systematic observation and quantification of some characteristic behaviour patterns of the male was started and lasted five more minutes : in particular, we recorded the number of quadrants entered by the male with all his six legs, minute after minute. This index of initial activity was labelled A1.
Then the slide-door was open and the observer had to wait that one of the flies came to visit the other ; the door was then gently closed again and the observation of the female-male interaction could start. Once the courtship had been initiated, it was *interrupted* after a given number of "bouts" of typical wing-displays (4-5 bouts). This interruption was easy to perform : for the male was so engaged in the courtship that his tendency to fly away was then non-existent ; whereas a very young, unreceptive, female still tended to avoid the male and could be allowed to fly away in a few seconds, by discarding the ceiling-slide.
After a variable time of seeming inertia, *sometimes exceeding one minute,* the male typically began to display a high level of locomotor activity, with a plenty of directional changes, as well as an impressive flow of wing-display (mainly scissoring, and sometimes lateral

vibrations) which could in some cases exceed 30 per minute.- This stage of the observation was performed during 5 minutes at least ... and much longer, if necessary : the levels of locomotor activity (number of quadrants "A2") and of wing-display ("Sc") were rated minute after minute, during the whole post-courtship display, which in some cases could last up to one hour ! - In this paper, we shall only take into account the first five minutes following the female's departure, which may be compared with the five minutes of initial activity of the male, previous to meeting the female.

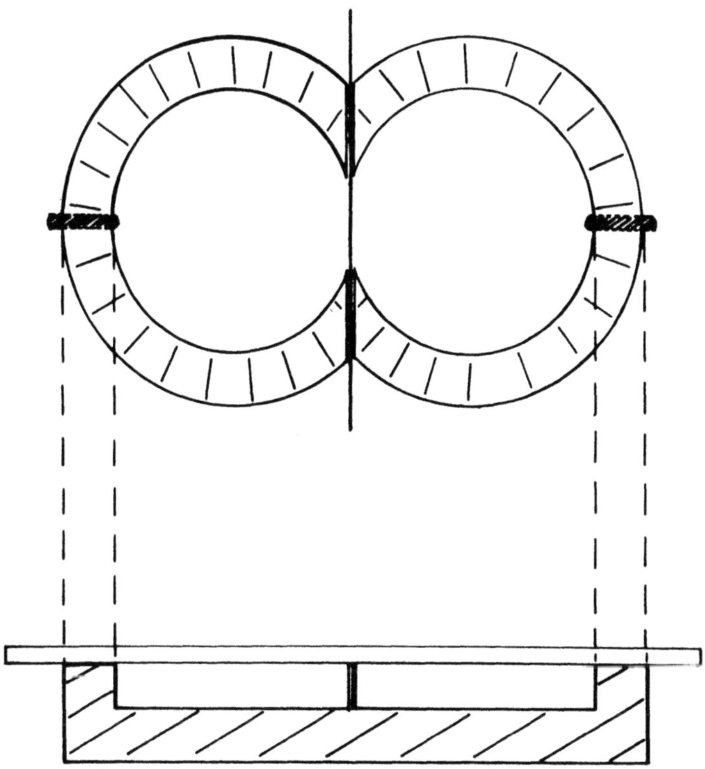

Figure 1 : Schematic drawing of the two-cell enclosure designed for observing the post-courtship sexual arousal. The *upper part* of the figure gives an overview of the apparatus, showing in particular the slide-door, as well as the plugs designed for introducing the flies.- The lower part shows a vertical section ; the "ceiling" is a transparent object-slide, allowing steromicroscopic observation of the animals.

All these observations - as well as rearing the flies - were performed in a room maintained at 21±1°C, under a fixed, artificial photoperiod (L:D=12:12 ; light on at 8 a.m.). Males of two wild-type laboratory strains (Canton-S 231, or C-S, vs. Dahomey, or Dh) have been observed as 3-4 days old and devoid of any previous sexual experience. They all met virgin, still unreceptive females from the Dh strain, never older than 24 hours. The observations took place between 8^{15} a.m. and 5 p.m. (starting-time of a record).

Statistical processing of data mainly resorted to various forms of variance analysis (Winer 1962). All considered variables were submitted to a square-root transformation.

We shall describe the results of four experiments, respectively designed for :

1) controlling the eventual role of odour trails let by the female, after leaving the cell where she had met the male ;

2) looking for an eventual arousing role of novelty ;

3) investigating the possible modulation of sexual arousal by the circadian rhythms of activity and reactivity ;

4) looking for the modulatory role of hunger.

The first two experiments only used males from the Dh strain. In the 3rd and 4th operations, we made a systematic comparison between males of the two strains Dh and C-S.

Experiment 1 : Control for arousal by odour trails

For this control, we designed a peculiar procedure in which the female was directly introduced in the male's own room and never allowed to enter the other cell, which kept "clean".

Two samples of 12 males, all from the Dh wild-type strain, were compared in this situation :

- the "migrant animals (m), who were given the opportunity to enter the other cell and to express their post-courtship arousal in this space, free from any odour trails ;
- the "sedentary" male (s), who had to remain in their own cell after departure of the female partner.

Would chemical trails let by the female be critical for the post-courtship arousal, then one should expect it to be more pronounced in sedentary than in migrant males.

As shown by Figure 2, this is *not* the case : initial, pre-courtship locomotor activity A1 does not differ significantly between *s* and *m* ; hence these two samples compare well, as regards their general activity level. This is also the case for the level of remnant wing-display (post-scissoring Sc) ; although, in this case, the dispersal of individual values is considerable, so that the two samples may not be satisfactorily differentiated. The only variable which significantly ($P < 0.05$) discriminates the two groups of males is the post-courtship level of locomotor activity A2.

But the slight difference observed is in favour of the *migrant* males, expressing their sexual arousal in the clean cell. Hence the persistence of sexual excitation can*not* be reduced to a trivial effect of some chemical trails let by the female.

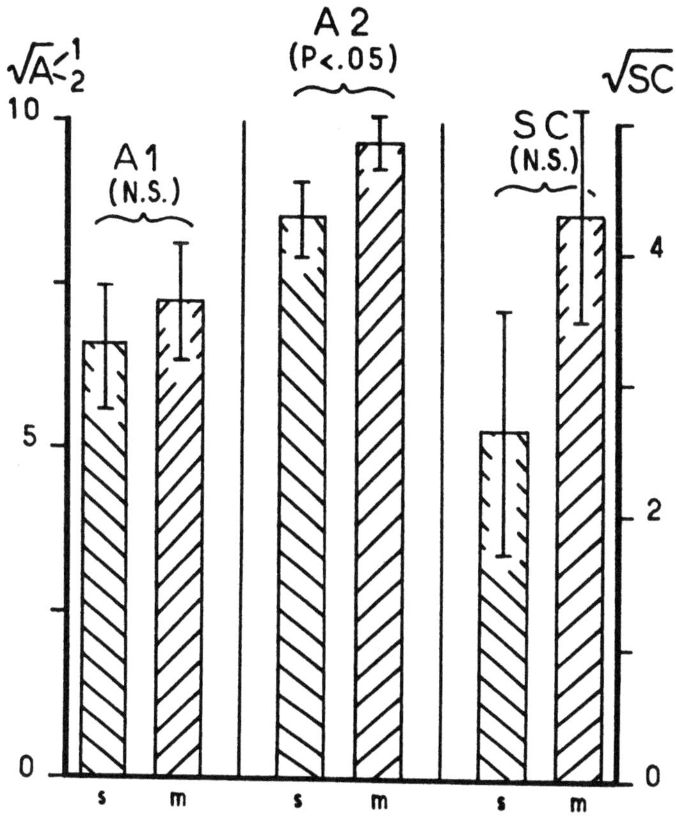

Figure 2 : Main results of the control experiment, showing the initial activity level (A1) as well as the locomotor reactivity of the male to the interruption of his courtship (A2) and the related specific wing-displays (Sc). All these variables were submitted to a square-root transformation, for the sake of variance homogeneity : *s* vs *m* mean "sedentary" vs. "migrant" males (see text). Dispersion indices are ± 1 standard deviation of the mean.

Expermiment 2 : The possible role of novelty.

Although the discrimination between migrant and sedentary male was not very clear in the previous experiment, all indices tended to suggest that the expression of sexual arousal could be enhanced by an unknown environment. If we also look at the maximal durations of persistence obtained, we find 26 and 46 min for the migrant males, against 12 and 20 min for the sedentary ones.

Therefore we imagined an experiment specifically designed to evidence the possibly arousing role of novelty for the males submitted to the interrupted courtship procedure. In this design, each male was initially familiarised with one of the two cells (C1) during a much longer period than in Experiment 1 (at least 20 min, instead of 3). Among the 24 males under study, 12 were then put in the other, unknown, cell (C2) where they met the female until interruption of their courtship ; finally, they were gently put back into C1 and could express their sexual perseveration in a comparatively well-known environment. Another sample of 12 males had to initiate the courtship in C1, after a similar period of familiarisation. Then they were introduced in the novel cell C2 for the post-courtship phase.

The main results of this experiment are shown in Figure 3.

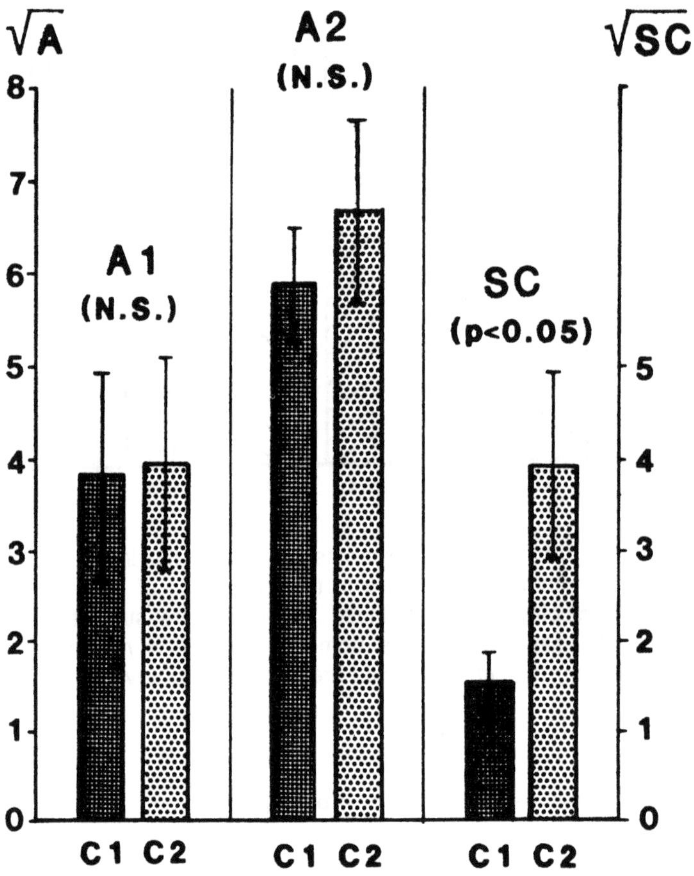

Figure 3 : Main results of Experiment 2 : possible role of an unknown environment, as a factor of non-specific activation, enhancing the post-courtship sexual arousal. Variables and transformates, cf. Figure 2. C1, post-courtship in the familiar environment ; C2, post-courtship in the unfamiliar cell.

The average levels of locomotor activity A1 and A2 do not significantly differ between the two groups. But contrasting with these results, the mean level of post-courtship wing-display Sc is significantly higher ($P < 0.05$) in males moved toward the new, clean, environment C2, than in males moved toward the familiar cell C1. Similarly, the maximal duration of sexual persistence averages to 9.58 ± 3.36 min in the new environment (highest individual values :

42 and 22 min) against only 3.42 ± 1.31 min in the familiar cell (highest individual values : 11 and 6 min).

Although this experiment is only preliminary, the results tend to confirm the suggestion of an *arousing effect of novelty,* already emerging from the above-mentioned control experiment.

Experiment 3 : Influence of day-time.

In this experiment (as well as in the next one), we compared the two wild-type strains Dahomey (Dh) and Canton-S (C-S). Among 24 males of each strain, 8 were studied during the early morning (AM1, from 8^{15} to 9^{39} a.m., starting time of the observation), classically known (Dürrwächter 1957 ; Médioni 1967, 1968) as giving rise in *melanogaster* to a peak of locomotor acitivy ; 8 during the second part of the morning (AM2, from 9^{40} a.m. to 12^{59}) ; 8 during the afternoon, before the evening peak (P.M., between 1 and 5 p.m.).

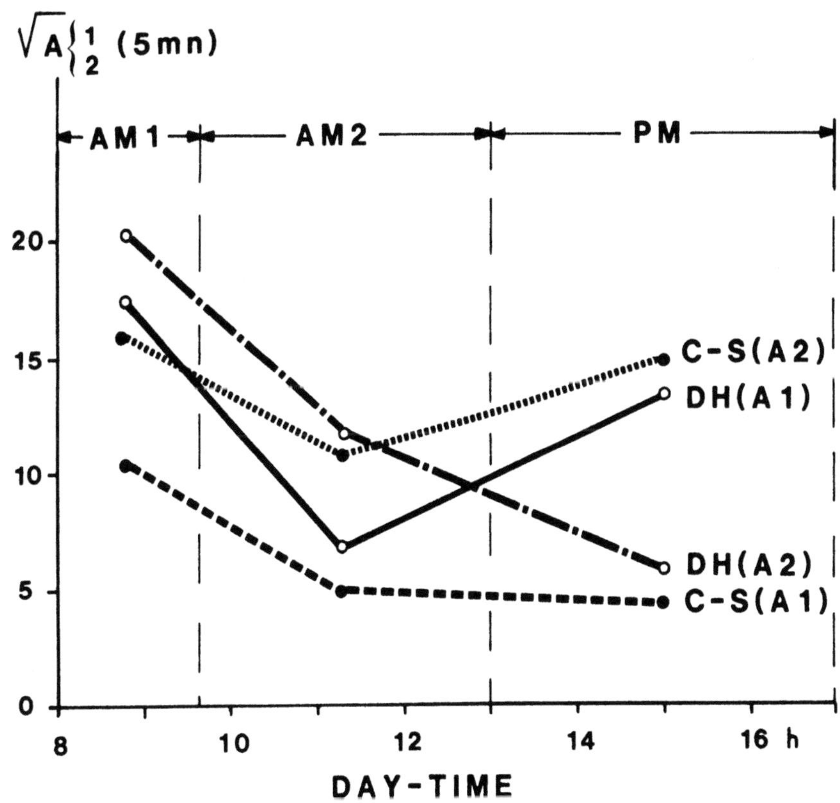

Figure 4 : Variations in initial activity A1, vs. locomotor reactivity to the female's departure A2 (ordinates), as a function of day-time (abcissae) in the two wild-type strains Dh (Dahomey) and Canton-S (C-S).

The temporal variation of the activity variables : A1 (initial activity of the male), vs. A2 (locomotor reactivity to the interrupted courtship) is shown by Figure 4. - For A1, the average level is higher in Dh than in C-S flies ($P < 0.01$). An analysis of variance also reveals a significant variation ($P < 0.01$) among the levels of the day-time factor, with a minimum observed during AM2. However the time-evolution of A1 does not differ significantly between Dh and C-S (strain x day-time interaction, N.S.).

As regards A2, no significant difference between the strains can be detected ; whereas the time-linked variation is highly significant ($P < 0.01$), as well as the strain x day-time interaction : in the Dh strain, there is a monotonous decrease of A2 level from the early morning to the afternoon ; whereas in C-S, the level of A2 is minimal during AM2.

However a clearer view of this evolution may be gained by combining the two activity variables : the *difference* between their square-root transformates ($\sqrt{A2}$-$\sqrt{A1}$) measures the increase in locomotor activity (in 5 min) possibly associated with the post-courtship arousal. Figure 5 reveals a remarkable contrast between the two strains, as regards the time-evolution of this difference : during the afternoon, the level of A2 by reference to A1 increases in C-S, whereas it decreases in Dh flies. This statistical strain x day-time interaction is significant at $P < 0.01$.

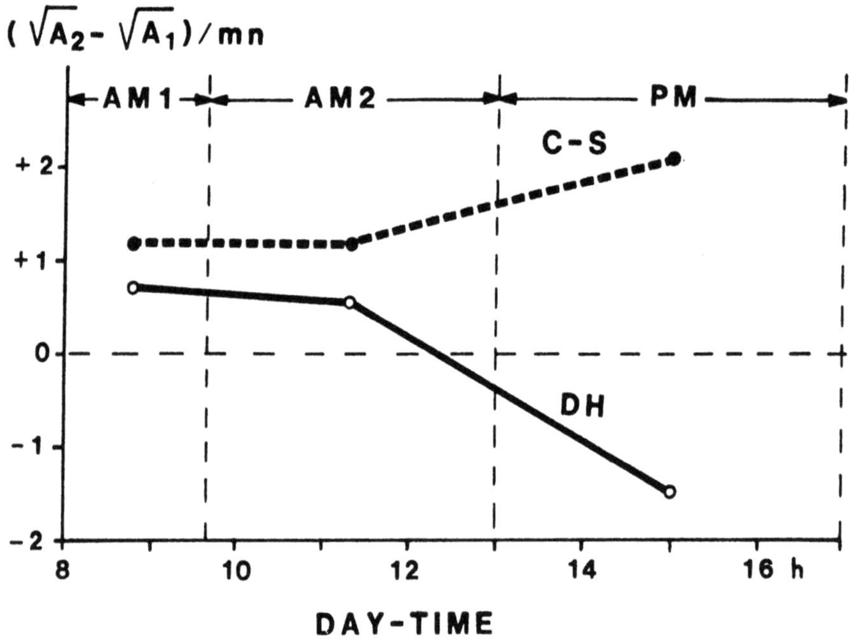

Figure 5 : Variations in the level of locomotor activation - rated by the difference between A2 and A1 transformates (ordinates) - as a function of day-time (abscissae), in the two wild-type strains Dahomey (Dh) and Canton-S (C-S).

This result does not provide any direct evidence about the possible contribution of a general arousal process to the post-courtship arousal effect. However it suggests that a similar change in the internal state, linked to the recent experience of an uncompleted sexual contact, has a quite different behavioural expression in males of the two strains under study.
This view is confirmed by the results concerning the level of post-courtship wing-display (Figure 6) : the flow of scissoring follows a different evolution along the day in the two strains (strain x day-time interaction significant at P = 0.01). Also to be stressed is the fact that average level of wing-display, as a whole, is much higher in Dh than in C-S males.

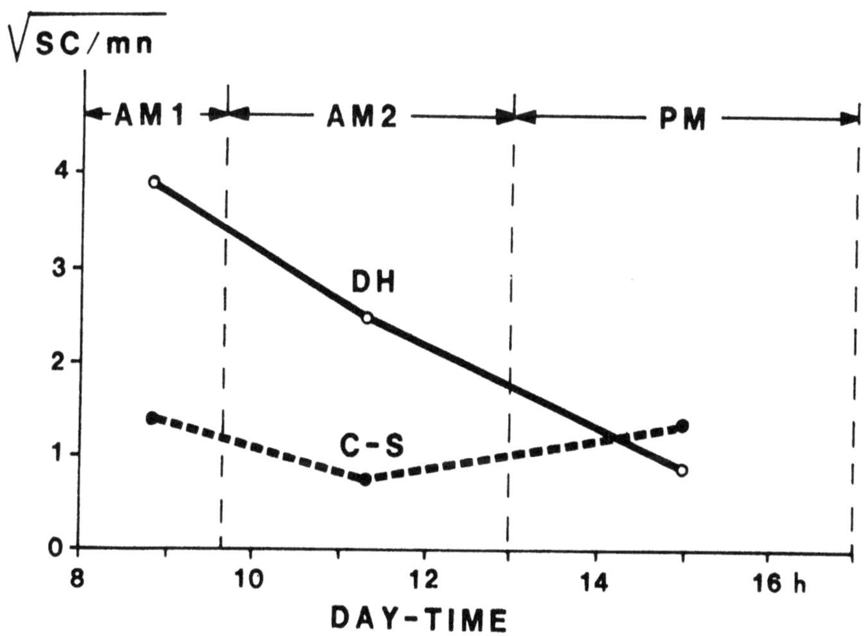

Figure 6 : Variations in the level of wing-displays induced by sexual frustration, √Sc (ordinates), as a function of day-time (abscissae), in the two wild-type strains Dahomey (Dh) and Canton-S (C-S).

To sum up, this experiment reveals a general trend to decreasing *from early morning until mid-afternoon* for the levels of the two major variables of sexual arousal A2 and Sc, as it is the case for the level of "general activity", as rated here by the variable A1. This might orient us toward an interpretation of "sexual arousal" in terms of *non-specific processes of central activation,* perhaps analogous to the "Central Excitatory State" (Dethier, Salomon and Turner 1965).
However, the qualitative contrast found between the two strains, as regards the dominant behavioural expression of this sexual arousal (majoration of locomotor activity in C-S, *vs.* emergence of wing-display in Dh) has now to be investigated further.

Experience 4 : Effect of hunger.

To start handling this possibly activating factor (see Connolly 1966), we first controlled that comparable times of hydric fasting had similar effects on survival in the two strains under study. At the same temperature (21°C), young (3 days-old) males, held in isolation on an agar substrate devoid of any nutriments, had an average survival duration of 86 h in the Dh strain, against 83 $^{1}/2$ h in the C-S strain (100 males were studied in each case) ; the longest survivals recorded exceeded 5 days in the two strains.
The experimental design was then performed with 6 groups of 16 males in each strain (N = 192), respectively subjected to fasting durations of zero - $1^{1}/2$ h (± 15 min) - 6 h (± 30 min) - 24 h (± 1h) - 48 h (± 3 h) and 72 h (± 4 h). The animals were observed and rated for their sexual arousal at day-times between 10 a.m. and 5 p.m. (time-intervals AM2 and PM, as defined earlier) ; no subjects were tested during the early morning (AM1), to avoid possible correlates of the peak of locomotor activation already found at that time.

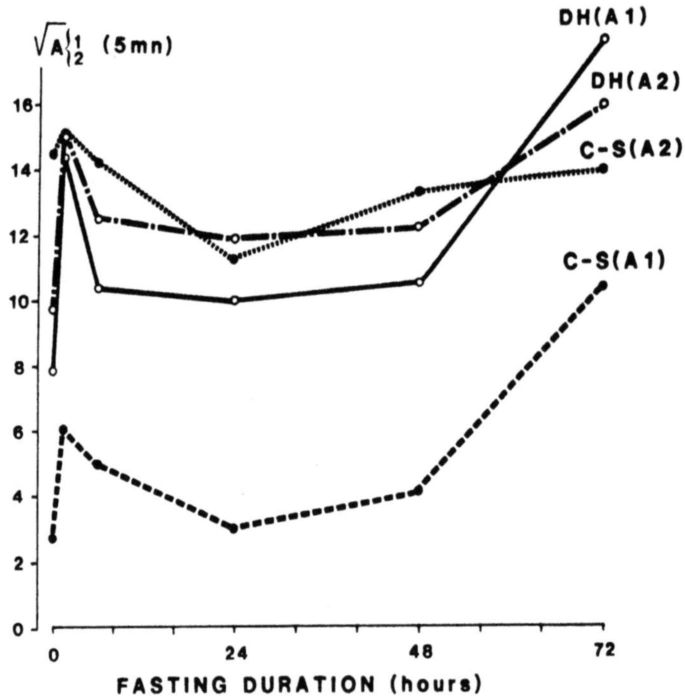

Figure 7 : Variations in initial activity A1, vs. locomotor reactivity to the female's departure A2 (ordinates), as a function of the level of starvation (abscissae) in the two wild-type strains Dahomey (Dh) and Canton-S (C-S).

Figure 7 shows the results obtained for the level of initial (pre-courtship) locomotor activity (√A1/min, recorded during 5 min). In the two strains, a maximum was found for the shortest fasting duration ($1^{1}/2$ h), then a decline around 24 h of starvation, as well as a marked increase for the longest period of fasting (72 h). This evolution is highly significant, as shown by a variance analysis (P < 0.001), and nearly parallel for the two strains (no significant strain x fasting duration interaction). However the most striking feature is the mean level of A1, much higher for Dh than for C-S males, at any degree of starvation (P < 0.001).

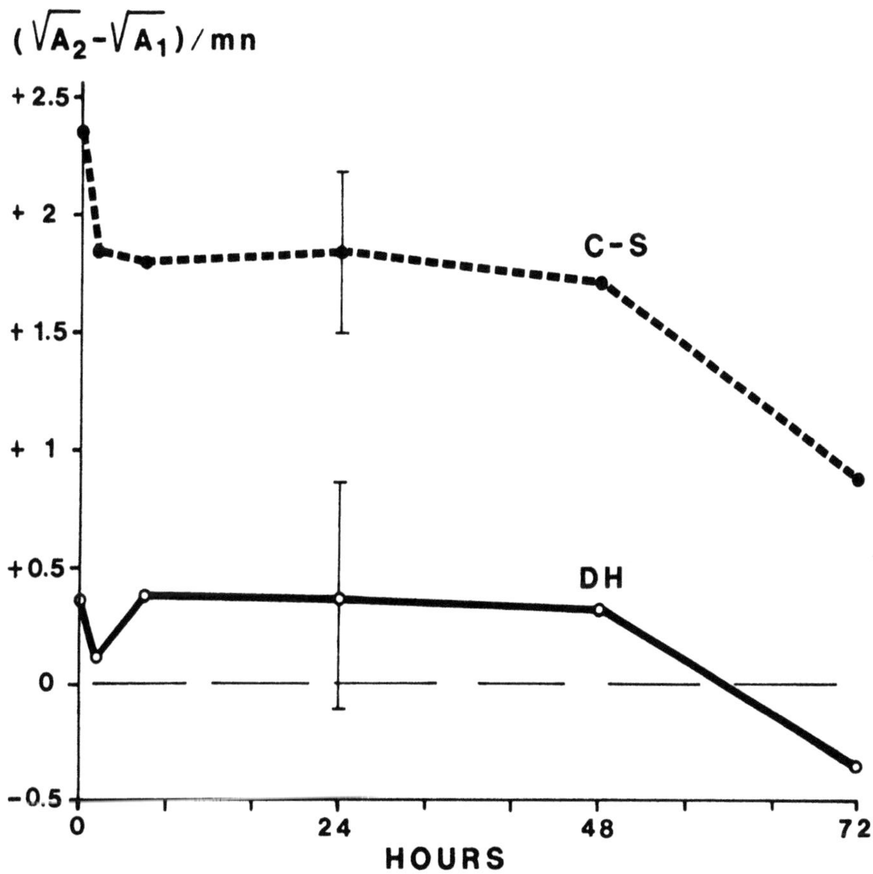

Figure 8 : Variations in the level of locomotor activation - rated by the difference between A2 and A1 transformates (ordinates) - as a function of the level of starvation (abscissae) in the two wild-type strains Dahomey (Dh) and Canton-S (C-S).

As shown by Figure 8, the level of the post-courtship sexual arousal, when rated by the difference ($\sqrt{A2}-\sqrt{A1}$)/min, appears much higher in the C-S than in the Dh males, which fully confirms the results obtained on unstarved flies, tested at differentiating day-times (see Figure 5). This difference (P < 0.001) simply expresses the fact that, in Dh males, the locomotor activity (A1) is already at a high level before any encounter of a female ; whereas, in C-S males, this initial actiivty is, on an average, much more limited, so that it can be strongly increased during the post-courtship period.

As regards the level of the characteristic wing-displays recorded after interruption of the courtship, Figure 9 shows a clear, statistically significant ($P < 0.01$), strain x fasting duration interaction : for the shortest fasting-duration (1 $^1/_2$ h), the flow of these wing-displays strongly increases in Dh males (as compared with unstarved animals, group 0 h), whereas it remains nearly unchanged in the case of C-S.

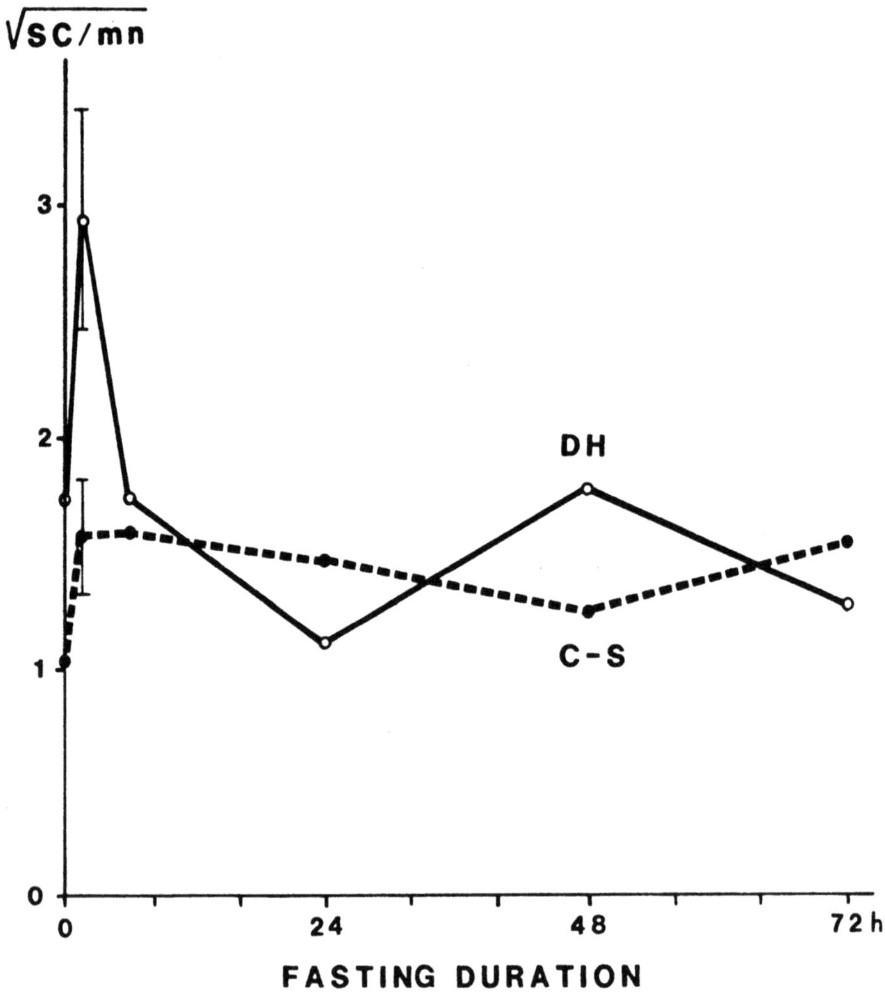

Figure 9 : Variations in the level of wing display induced by sexual frustration, Sc (ordinates), as a function of the level of starvation (abscissae), in the two wild-type strains Dahomey (Dh) and Canton-S (C-S).

An overview of the between-strains differentiation - all starvation times being put together - is given by Figure 10. Whereas the level of delayed wing-display is significantly higher in Dh than in C-S males, the post- courtship locomotory enhancement, negligible in the case of Dh, is very high in the C-S strain.

ALL FASTING DURATIONS CUMULATED

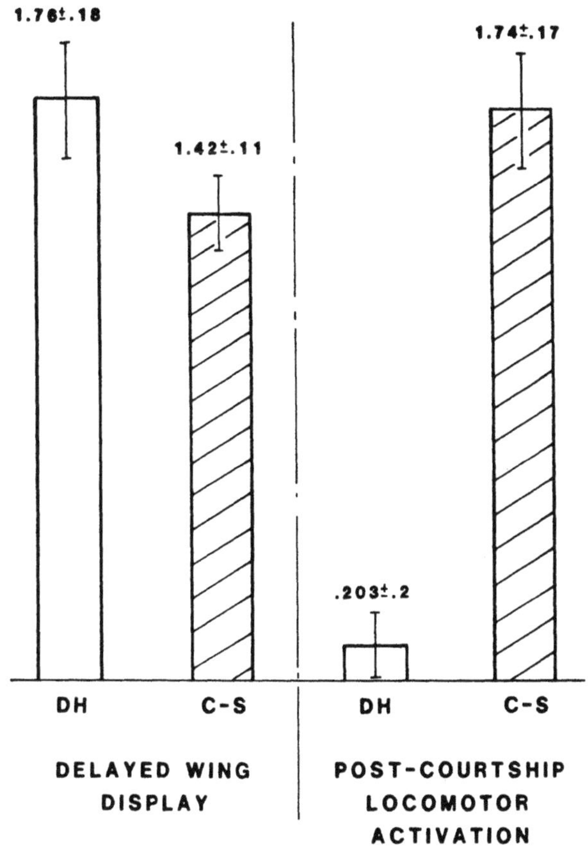

Figure 10 : Overview on the between-strains differentiation in the starvation experiment. The results *pooled for all fasting durations* show that responding to courtship interruption is expressed more clearly by specific wing-displays in the Dahomey males (Dh) ; and by increase in locomotory activity (A2-A1) in the Canton-S males. Although more marked in this latter case, the difference is significant for these two indices of sexual arousal.

The main issue of this last experiment therefore appears as fully converging with the results of Experiment 3, comparing unstarved animals, observed at more differentiated day-times. The two wild-type strains under study present some remarkably contrasting properties :

- their average level of locomotor activity (A1, vs. A2-A1) constitutes the basis of a *quantitative* between-strains discrimination, the expression of a "general arousal" function by this channel being much more evident in the Dh strain ;

- as a consequence of this important quantitative difference emerges a *qualitative* differentiation in the behavioural expression in response to the same situation of sexual frustration.

The fact that internal modifications induced by a similar psychological experience may result in different behavioural issues, for different populations of the same animal species, is a point of great methodological interest which should also be taken into consideration in a more neurobiological reference frame. The commonly accepted view of a "behavioural expression" of internal states, either psychological, or neural, or else ... has to be used most carefully.

Litterature cited

Campan M (1966) Etude quantitative de la facilitation exercée par des stimulations vibratoires sur la géocinèse et sur la photocinèse de *Drosophila melanogaster* Meigen. C.R. Séances Soc. de Biol. 160: 672-674.

Connolly KJ (1966) Locomotor activity in *Drosophila* as a function of food deprivation. Nature, London, 209: 224.

Dethier VG, Solomon RL, Turner LH (1965) Sensory input and central excitation and inhibition in the blowfly. J. compar. physiol. Psychol. 60: 303-313.

Dürrwächter G (1957) Untersuchungen über die Phototaxis und Geotaxis einiger *Drosophila* - Mutanten nach Aufzucht in verschiedenen Lichtbedingungen. Zschr. Tierpsychol. 14: 1-28.

Hirsch J, McCauley L (1977) Successful replication of, and selective breeding for, classical conditioning in the blowfly *Phormia regina.* Animal Behav. 25: 784-785.

Manning A (1959) The sexual behaviour of two sibling *Drosophila* species. Behaviour 15: 123-145.

Médioni J (1959 a) Mise en évidence et évaluation d'un "effet de stimulation" dû aux ocelles frontaux dans le phototropisme de *Drosophila melanogaster* Meigen. C.R. Séances Soc. de Biol. 153: 1587-1590.

Médioni J (1959 b) Les organes de stimulation photique chez *Drosophila melanogaster.* Non-spécificité de cette stimulation relativement aux réactions à la lumière. C.R. Séances Soc. de Biol. 153: 1845-1848.

Médioni J (1967) Nachweis einer Tagesrhythmik der lokomotorischen Aktivität bei *Drosophila melanogaster.* in "Biologische Rhythmen" G Birukow and L Rensing Eds. Nachrichten der Akad. der Wissenschaften in Göttingen : 117.

Médioni J (1968) Premières données sur le rythme journalier de l'activité locomotrice chez *Drosophila melanogaster* (Dipt. Drosophilidae). Annales des Epiphyties 19: 193-194.

Médioni J, Manning A (1987) Age-related perseveration of the precopulatroy behaviour in male *Drosophila melanogaster.* Experientia, *in press.*

Moruzzi G, Magoun HW (1949) Brain stem reticular formation activation of the EEG. EEG Clin. Neurophysiol. 1: 455-473.

Nelson M (1971) Classical conditioning in the blowfly *(Phormia regina)* Associative and escitatory factors. J. compar. physiol. Psychol. 77: 353-368.

Romaneli E (1968) Etude expérimentale de l'influence de quelques facteurs sur les réactions d'envol chez un insecte : *Drosophila melanogaster* Meig. Thèse Doctorat Univ. de Toulouse, p. 87.

Vargo M, Hirsch J (1982) Central excitation in the fruit fly *(Drosophila melanogaster).* J. compar. physiol. Psychol. 96: 452-459.

Vaysse G (1973) Sur le rôle activateur des ocelles frontaux dans la compétition sexuelle entre les mâles de *Drosophila melanogaster.* C.R. Séances Soc. de Biol. 167: 771-776.

Winer BJ (1962) Statistical Principles in Experimental Design. Mc Graw-Hill, New York: 907pp.

GENE STRATEGY IN THE MODULATION OF SYNAPTIC ACTIVITY

A. Ferrús; S. LLamazares and F. Gaunitz
Instituto Cajal. C.S.I.C.
Velazquez,144
Madrid-28006
Spain

Introduction

When the genetic foundations of a cellular system are studied the most relevant question becomes, how to account for so much functional and structural diversity on the basis of so scant a genetic material. This problem seems to be more tangled the more sophisticated the cellular system is and often, the nervous system is used to illustrate a paramount example of biological complexity.

Two different attitudes are found among the investigators that have tackled this problem. Many, consider the coordination of biological activities as epigenetic phenomena determined by the developmental conditions of the system itself. Others, envision a strict genetic control of this concert. Nowadays, although most experts admit that the genetic material is the only recipient of the information required to build and put to work an organism, few would admit that there is information in the genome to specify, for instance, the fine structural details of neuronal morphology or the tight coupling between phosphorilation and modulation of certain ionic currents.

In this review we will analyze this biological problem, and the attitudes towards it, through the filter of our own experience with a genetic system that plays a key role in K^+ conductance and other functions that act in physiological harmony.

A definition of terms

Diversity and specificity are two key concepts in all this argument. Both are quite subjective and dependent on the level of observation. Two homologous neurons from two organisms of the same species that show different architecture are said to be diverse. When these two organisms are members of a clone, and assumed to be isogenic, then it becomes an evidence for the lack of genetic control of fine cellular morphology[12,20].

NATO ASI Series, Vol. H19
Modulation of Synaptic Transmission and Plasticity
in Nervous Systems. Edited by G. Hertting and H.-C. Spatz

Similarly, within a single nervous system neurons show an apparent enormous diversity of forms, biophysical properties, etc.

The apparent diversity within a single nervous system has already been reconsidered to its real dimension by investigators that have analyzed this cellular system comprehensively. Thus, on morphological criteria the estimated 10^{12} cells of a human brain can be grouped in 7 classes of neurons and 6 classes of glial cells or the estimated 10^{5} cells of Musca's CNS can be grouped in 5 categories[3,5,31]. The physiological properties are not yet known in sufficient number of neurons as to attempt a similar classification. It could be argued that the main issue remains, namely that within a given category of neurons the diversity (more properly: variability) of forms is too great to be under strict genetic control. We think that the question should be stated in a different way: how much of the biologically significant diversity is under genetic control?. That is, when two homologous cells of two supposedly isogenic organisms are found to be different in architecture, one would have to correlate this observation with a change in behaviour[14]. Otherwise the observed variability can not be said biologically significant. It is critical to distinguish diversity from noise since the biological systems, being structured organizations, are not exempt of noise[27]. For a definition of biologically significant diversity we could take an operational one: "Any variability upon which evolution does exert modulation". To use this definition experimentally, one has to perform comparative studies, either morphological and/or functional, and these are not always feasible. Nevertheless, if biology is to be understood this kind of approaches are necessary.

Specificity is another term subject to various interpretations. The arguments against the genetic control of the, so called, higher order functions always make the tacit assumption that the genome would have to devote a portion of its information solely to the control of every detail under consideration. Curiously enough, the assumption has always been made by the detractors[30] of the genetic analysis but never by its defendants[28].

Perhaps the problem resides again in the definition of terms. It is conceivable, in fact we will illustrate an example later on, that two investigators one working in membrane permeability and the other working in neural pattern formation would demand different portions of the genome to account for their respective phenomenologies when, in reality, both seemingly different functions are accomplished by the same gene product(s). Thus, it is necessary to reduce our subjective definition of biological functions to the minimum common operation. That is, to learn the genetic language. So far we know the genetic alphabet and a little bit of the syntaxis but nothing of the grammar nor the language proper.

It might be misleading to search for one to one correspondences between the genetic information and biological phenomena, as it is equally misleading to rise arguments against the role of the genome in the control of these phenomena taking for granted such correspondences. Rather, we should dissect the biological phenomena to its components and search for the functional relationships between them.

The unavoidable fact is that the number of different molecular species, either in the form of aminoacids or of nucleotides is the limiting factor and that absolutely all the diversity must be accounted for on these bases. It is obvious that algorithms[30] ought to be found that explain functional relationships at high levels of organization but our point of view is that these algorithms are not as complex as they seem and that they can be found by studying the genome organization directly.

The Shaker gene complex. Its mutant phenotype, molecular basis and functional organization

Shaker (*Sh*;X,16F) was the first behavioral mutant isolated in the genus *Drosophila*[22]. It was detected because of its vigorous shaking of appendages when anesthetized. This abnormal behaviour is in no way caused by the anesthetic agent and it is independent of the state of narcosis[9]. The shaking is continuous during the whole life of the organism although to detect it in larvae, electrophysiological recordings are required[15].

The mutant captured the interest of physiologists and a number of features are known after two decades of work. At the larval neuromuscular junction, the amount of transmitter released is larger than normal[16]. In the intracellular recordings of the cervical giant interneuron of the adult, the action potentials show abnormal repolarization profiles[32]. Finally, voltage clamp studies in adult and larval muscles indicate that a fast, voltage dependent K^+ current of the I_A type is severely affected[29]. In addition, treatment of the cervical giant preparation in normal flies with 4-aminopyridine mimic the *Shaker* abnormality and the blockade of Ca^{++} currents with Co^{++} do not change the abnormal action potentials of *Shaker*[32].

These series of experiments served to establish the working hypothesis that *Shaker* could code for the I_A channel. Driven by the relevance of this hypothesis, a genetic and molecular analyses of the region of the genome where the Shaker mutations reside was initiated.

The first data on the aminoacid sequence of one of the products coded in Shaker have been found recently[2]. They indicate a nervous tissue protein with three membrane spanning domains and a significant accumulation of negatively charged aminoacids in one side, that could possibly act as a selective filter against anions. An important fact for the main theme of this review is that the obtained sequences from the membrane domains show a high homology to similar domains of the Na^+ channel of vertebrates[24]. Although these data correspond to only one of the elements coded in Shaker[9], they are encouraging enough to maintain the expectation that this is the structural gene for the I_A channel or certain of its components.

The genetic analysis shows that Shaker is a gene complex (ShC). A DNA stretch of about 300 Kb (of which 215 have been clonned) can be subdivided in three regions: the maternal effect (ME), the viable (V) and the haplolethal (HL) regions arranged in a distal to proximal order[9]. The product whose sequence is known originates in the V region where the mutations more thoroughly studied in the physiology experiments locate also. However, spread over the entire complex are located other viable and lethal mutations with clear shaking behaviour and abnormal action potentials. Obviously, the physiological phenotype of the lethals is studied in heterozygous flies. It is unfortunate that no voltage clamp data are available for these other genetic variants.

The complex have been saturated with lethal mutations in order to identify the genetically distinct components. The ME region contains three lethal complementation groups and a viable one: the locus minibrain[9,10]. In spite of the existence of lack of function lethal mutations in this region, ME^- males survive if, and only if, their mothers were euploid for this region. Females ME/ME^- can not give rise to ME^- sons[9]. This observation implies two conclusions: a) The products coded in the ME region should be active during oogenesis. Indeed the Northern blot analysis of the transcripts from this region show them to be specially abundant in oocytes. b) Although the lethal mutations can be of the lack of function class, they must interact among themselves, possibly in a multimeric type of structure, in such way that they show an antimorphic behaviour[23]. That is, the lethality of these mutations is not equivalent to the physical deletion of their DNA.

The viable mutations of the V region behave also as antimorphic mutations in gene dosage tests[32]. That is, genetic constructions with the euploid , or even hyperploid, complement of the normal gene show the mutant phenotype as long as a copy of the mutated gene is present.

The HL region also shows a functional organization in which some sort of cooperative effect is required among the products coded in it in order to fulfill the normal activity(ies). In this region, it appears to be difficult to define lethal complementation groups since every lethal mutation shows complementation with all the others. Eight independently induced and recessive lethals have been isolated to date.

In summary, everything that is known about the genetic organization of the ShC is compatible with the working hypothesis of being a gene complex that codes for a family of functionally interrelated elements. At this point it is appropriate to state the following opinions: a) It is unlikely that all the structural products of ShC form only one hypothetical multimere. b) It is also unlikely that only I_A channels are coded in ShC. c) We envision that specific combinations of ShC elements give rise to different multimeres and they perform several biological functions. These functions are likely to be similar to each other and, perhaps coordinated in a particular cellular context. The operations that generate these combinatorial aggregates within the ShC probably require the interaction with other genes or gene products outside of the ShC. It would be important to identify these other regions of the genome that are functionaly related with ShC.

Is there any evidence that ShC codes for other biological functions in addition to the I_A currents?. Certainly yes. In the study of the biology of the mutations and aneuploids of the ShC it was found that HL^- embryos develop a nervous system with total absence of pattern and neurons sprouting in multiple directions. That is, neurons seem to have lost control of growth and polarity. A qualitatively similar phenotype is found in certain neurons of the retina, optic ganglia and mechanoreceptors of mutants from the V region. In these neurons, the terminals are enlarged and frequently multibranched. The mutant <u>minibrain</u>, located in the ME region, shows also a morphological defect (50% reduction of brain size) although, in this case, the abnormality goes in the opposite direction. Finally, the focus of the lethality of all the mutants from the ShC map to the nervous system anlage. It seems that all the functions coded in the ShC play a role in the development of the CNS. However, it should be emphasized that the Sh function best characterized so far, the I_A current, can not be responsible for these developmental abnormalities, since it develops as a terminal differentiation event. Clearly, the components that, when mutated, alter I_A properties of fully differentiated cells, must play a role in some, yet unknown, structures whose biological functions are to lead axons through proper paths, control sprouting, etc.

In classical genetics, these phenomena are called pleiotropy. A concept that has often been used as a synonym of confusion. Our point of view is that pleiotropy is a clear demonstration that our definition of a biological function is all too often wrong. Quite possibly, what we define as the biological function "I_A current" or "axonal sprouting" are meaningless concepts in the genetic language. We should find the elemental operations that the genome handles and that, depending on the level of observation, can be named with one or another subjective term.

Searching for these genetic conversations, we set out to identify other regions in the genome that could have a functional relationship with the ShC. The mutation tetanic (tta;X,14B) was the first result.

The tetanic gene. Its mutant phenotype and the functional relationship with ShC

The gene tta is thought to be functionaly related with ShC on the basis of its behaviour in gene dosage experiments and its own mutant phenotype. The tta phenotype was first characterized by the strong contraction of all appendages, twisting of the neck, lifting of wings, etc. when under anesthesia. When the mutation is accompanied by an extradose of the normal ShC the phenotype becomes more extreme. Finally, tta males carrying two extradoses of the normal ShC are lethal. This dosage sensitivity of tta with respect to ShC makes the former a candidate to be a negative regulator of the later. tta does not show dosage sensisitivity with any other region of the X chromosome.

A more comprehensive study of the biology of tta showed that other biological functions are also affected. Wings are 30% reduced and homozygous females are poorly fertile. However, a surprising result came from an electrophysiological observation. The larval muscles are chronically depolarized and show resting potential values, on the average, 15 mV more positive than normal. The ionic basis of this abnormality is, very likely, a defective K^+ current of the, so called, "leaky channel" responsible for maintainning the resting potential at a proper level in all cells. Changing the extracellular concentrations of several ions, it was found that the abnormality of tta was dependent on the K^+ concentration only. However, the gene tta is, in all probability, not the structural gene for this class of channel since the mutation tta, being X-ray induced, is likely to be an amorph. If tta were the structural gene for this channel the expected phenotype would be total loss of resting potential, surely a cell lethal condition. It is more reasonable to propose, at present, that tta modulates the expression of the structural gene for the K^+ leaky channel. Since tta is sensitive to extradoses of ShC, it is tempting to suggest that the modulated element resides in the ShC, perhaps represented by some of the lethal mutations with which ShC is saturated. In fact, one of the mutations in the ME region is cell lethal[9].

Yet, a more surprising result came from a behavioural experiment. tta flies appear to be deficient in associative learning tests performed in a paradigm with olphactory, negatively reinforced cues.

At this point, it becomes obvious to relate this findings with the data obtained in Aplysia where the conditioning of certain synapses is accomplished by the modulation of the activity of certain K^+ currents[17]. In this system it has been proposed also, that the modulation of the K^+ currents is mediated by the phosphorilation of an element tightly coupled to the channel[6].

The study of tta is still in progress but one result, which is being carefully analyzed due to its relevance, deserves to be mentioned, albeit with due caution. It seems that the tta flies show a significant excess of phosphatase activity. If phosphorilation is a mechanism to close K^+ channel activity, the phosphatase activity is the required counter part to bring the channel under fine modulation. Then, the learning defect in tta could be understood on the basis of K^+ channel modulation defects. Now, the postulated role of tta as a regulator of ShC expression accquires its due relevance.

This is one, among many, of the examples of exquisite coordination of biological activities that seem to bewilder us and for which it seems to be so impossible that the genome could exert a tight control. The history of the analysis of ShC illustrates a case of unwinding a chain of biological functions that were known to act in a precisely coordinated ensemble. If the genetic analysis is a valid procedure for the purpose of understanding the functional relationships in a complex system, then other biological functions related to those of ShC should be identified as well.

Certainly the chain is still unbroken and we continue to pool from it. Although it is too premature to draw further conclusions, some observations point in the right direction. The structural gene for the cAMP dependent phosphodiesterase II: dunce (dnc;X,3D4)[4] shows a synergistic interaction with Sh mutants. The double mutant dnc-Sh is a hypershaking fly that lives only a few days and it is barely capable to stand up. Interestingly, dnc is also a learning defective mutant with subtle morphological consequences in CNS structure (see Heisenberg, this volume). Similarly, the gene rutabaga (rut;X,12E)[21] seems to be a suppressor of certain traits of the Shaker phenotype in double mutants. Once again, rut is a learning defective mutant with morphological consequences (see Dudai, this volume). Finally, the gene no action potential (nap;II,41) whose mutation nap^{ts} causes a temperature sensitive blockade of action potentials at the neuromuscular junction[34], if combined with a Sh mutation, does not show the Sh phenotype[11].

Whether these interactions are physiological or genetical remains to be ellucidated, but every piece of information available to date indicate that, indeed these genomic regions are conversing very actively among them.

A proposition

Following the mathematical definition of this term, we would like to make an statement to be demonstrated. The genome contains informational capacity of a higher level than that of the genetic code of triplets. This information is conveyed in classes of elemental operations between different regions of the genome constituting a genetic language. The genome is considered as the structured organization of all genetic material within a cell.

General remarks

Retaking the introductory comments of this review, we think that a key element to understand the genetic basis of biological complexity resides in the definition of what is a biological function. Our opinion is that different definitions do not necessarely imply different structures and different portions of the genetic material. It might be essential to decipher the genetic language by finding the elemental operations that every subjectively defined biological function is composed of. The elemental operations should not be mistaken by operations at the lowest level of observation. Quite the contrary, even a foremost example of complex organization as it is the architecture of the cortex of high mamalian brains should be analyzed in terms of elemental morphogenetic operations.

To understand the complexity of a system surely requires to know how it is generated. This, in turn, is inseparable from the concept of redundancy. It seems clear that the more complex a biological system becomes, the more redundancy it shows. The best operational test of this redundancy comes from manipulations that result in the reduction of the size of the system (i.e.: ablation of certain parts of the human brain or even 50% reduction of whole brain size in the case of the mutant minibrain from the ME region of the ShC). In general, these manipulations do not cause very severe disfunctions, even they are viable organisms[14].

It is a rather old proposition in genetics that the complexity of the genome organization of eucariotes was attained after duplication of parts of its genetic material[25]. At one point, the duplicated genetic material was redundant but the subsequent specialization rendered it diverse. It is easy to conceive that the lineage of this material might have maintained some sort of ties, we would propose functional ones, among the duplicated and diversified elements. Following this rationale, cellular systems as complex as the human brain cortex could have become so ellaborated after redundant duplication of one or several elemental units.

It would not be unreasonable to propose the existence of a principle of modular organization in every redundant system. Thus, the effort in understanding the complexity of, say, the functional organization of the human cortex[26] or the visual integrative centers that control navigation in flies[13], should be aimed to identify these modules. To define a module one would have to specify the level of observation. Thus, the principle of modular organization might be true when analyzing the morphology or the functionality of the human cortex as well as when studying the structural bases of ionic permeability.

Some functional links might have been kept after the duplication of genetic material as well as the duplication of modules in a neural subsystem. These functional relationships could be the bases for the coordination of biological activities as the modulation of synaptic activity or the sensory integration exemplify. Thus, to understand the apparent complexity of a system one would have to know the elemental operations as well as the lineage of these operations.

Mechanisms that have generated redundancy in the genome are well established. Unequal recombination generates tandem duplications that still maintain a physical contact and, therefore allow the possibility to share the same cis regulatory elements. Transpositions of segments of DNA of varying length can also be generated by especial events of recombination. Movements of small fragments of DNA of the retrovirus type might bring regulatory elements to new positions close to structural elements that will now be controlled in a different way. Although the mechanisms might have been different, the cellular systems might have accquired redundancy first and diversity later by the same principle.

What we know about the molecular bases of many mutations in Drosophila that cause disarrays of pattern formation[1,8,19,33] indicates that, by and large, these are defects in cell proliferation. It seems that there is a substantial amount of genetic energy devoted to control growth. Similarly, the mutations that affect memory and learning, all are cases of enzymes involved in basic metabolism[7]. Obviously, "pattern formation" and "learning" are two biological functions, as we name them, that do not have an immediate correlate in the genetic language and yet, the genome tightly control them as the existence of mutants demonstrates. It is important to realize that all these operations are orchestrated solely by the genome of the organism.

Our point of view is that we should reduce the subjectively defined biological functions into the elemental operations that form the genetic language. This we should do by studying comparatively these functions at several levels of observation as the study of the biology of the ShC has illustrated to us. Then, it is possible that the general principles will emerge and the complexity will not be a synonym of ignorance.

REFERENCES

1. Baker BS, Boyd JB, Carpenter A, Green MM, Nguyen TD, Ripoll P and Smith PD (1976) Genetic controls of meiotic recombination and somatic DNA metabolism in Drosophila melanogaster. Proc. Natl. Acad. Sci. USA 73:4140-4144
2. Baumann A, Krah-Jentgens I, Mueller R, Mueller-Holtkamp F, Seidel R, Kecskemethy N, Casal J, Ferrus A and Pongs,O. (1987) Molecular organization of the maternal effect region of the Shaker complex of Drosophila: Characterization of an I_A-channel transcript with homology to vertebrate Na^+-channel. EMBO J 6:
3. Bullock TH and Horridge GA (1965) Structure and function in the nervous system of invertebrates. Freeman. San Francisco
4. Byers D, Davis RL and Kiger JA (1981) Defective cyclic AMP phosphodiesterase due to the dunce mutation of learning in Drosophila melanogaster. Nature 289: 79-81
5. Cajal SR (1911) Histologie du systéme nerveux de l´homme et des vertebrés. Cons.Sup.Invest.Cient. Madrid
6. Castellucci VF, Kandel ER, Schwartz JH, Wilson FD, Nairn, AC and Greengard P (1980) Intracellular injection of the catalytic subunit of cyclic AMP-dependent protein kinase simulates facilitation of transmitter release underlying behavioral sensitization in Aplysia.Proc. Natl. Acad. Sci. USA. 77: 7492-7496
7. Dudai Y (1985) Genes, enzymes and learning in Drosophila. TINS 8:18-21
8. Ferrús A (1979) Cell functions in morphogenesis. Clonal analysis of new morphogenetic mutations in Drosophila melanogaster. Dev. Biol. 68: 16-28
9. Ferrús A, Tanouye MA, Pongs O and LLamazares S. The genetic organization of the Shaker region of Drosophila (manuscript in prep.)
10. Fischbach KF and Heisenberg M (1984) Neurogenetics and behaviour in insects. J exp. Biol. 112: 65-93
11. Ganetzky B and Wu CF (1982) Indirect suppression involving behavioral mutants with altered nerve excitability in Drosophila melanogaster. Genetics 100: 597-614
12. Goodman CS (1978) Isogenic grasshoppers: genetic variability in the morphology of identified neurons. J Comp. Neurol. 182: 681-706
13. Goetz KG (1980) Visual guidance in Drosophila. In:"Development and neurobiology of Drosophila". (O Siddiqui, P Babu, LM Hall and JC Hall eds.) Plenum Pub. Corp. New York
14. Heitler WJ, Goodman CS and Rowell CHF (1977) The effects of temperature on the threshold of identified neurons in the locust. J Comp. Physiol. Psychol. 117: 163-182.
15. Jan LY and Jan YN (1976) Properties of the larval neuromuscular junction in Drosophila melanogaster. J Physiol. 262: 189-214
16. Jan YN, Jan LY and Dennis MJ (1977) Two mutations of synaptic transmission in Drosophila. Proc. R. Soc. Lond. B. 198:87-108

17. Kandel ER (1981) Calcium and the control of synaptic strength by learning. Nature 293: 697-700
18. Knust E, Dietrich U, Tepass U, Bremer KA, Weigel D, Vassin H and Campos-Ortega JA (1987) EGF homologous sequences encoded in the genome of Drosophila melanogaster and their relation to neurogenic genes. EMBO J 6: 761-766
19. Lehmann R, Jimenez F, Dietrich U and Campos-Ortega JA (1983) On the phenotype and development of mutants of early neurogenesis in Drosophila melanogaster. Roux's Arch. Dev. Biol. 193: 62-74
20. Levinthal F, Macagno E and Levinthal C (1976) Anatomy and development of identified cells in isogenic organisms. CSHQB XL: 321-331
21. Livingstone MS, Sziber PP and Quinn WG (1984) Loss of calcium / calmodulin responsiveness in adenylate cyclase of rutabaga, a Drosophila learning mutant. Cell 37: 205-215
22. Luers H (1936) Shaker, eine erbliche Bewegungsstorung bei Drosophila funebris. Z. induktive Abstam. Vererbungs. 72: 119-126
23. Muller HJ (1932) Further studies on the nature and causes of gene mutations. Proc. Sixth Intl. Cong. Genetics 1: 213-255
24. Noda M, Takahashi H, Tanabe T, Toyosato M, Kikiyotani S, Furutani Y, Hirose T, Takashima H, Inayama S, Miyata T, and Numa S.(1983) Structural homology of Torpedo californica acetylcholine receptor subunits. Nature (Lond.) 302: 528-532
25. Ohno S. (1970) Evolution by gene duplication. Springer.Berlin. Heidelberg. New York
26. Peters A. and Jones EG. (eds.) (1984) Cerebral cortex. Plenum Press. New York and London
27. Prigogine I. (1955) Introduction to thermodynamics of irreversible processes. Wiley. New York
28. Quinn WG and Gould JL (1979) Nerves and genes. Nature (London) 278: 19-23
29. Salkoff L and Wyman R (1981) Genetic modification of potassium channels in Drosophila Shaker mutant. Nature 293: 228-230
30. Stent GS (1980) The genetic approach to developmental neurobiology. TINS 3: 49-51
31. Strausfeld NJ (1976) Atlas of an insect brain. Springer. Berlin. Heidelberg. New York
32. Tanouye MA, Ferrus A and Fujita SC (1981) Abnormal action potentials associated with the Shaker complex locus of Drosophila. Proc. Natl. Acad. Sci. USA 78: 6548-6552
33. Wharton KA, Johansen KM, Xu T and Artavanis-Tsakonas S (1985) Nucleotide sequence from the neurogenetic locus Notch implies a gene product that shares homology with proteins containing EGF-like repeats. Cell 43: 567-581
34. Wu CF, Ganetzky B, Jan LY, Jan YN and Benzer S (1978) A Drosophila mutant with a temperature-sensitive block in nerve conduction. Proc. Natl. Acad. Sci.USA.75: 4047-4051

List of Participants

Mr. Clemens *Allgaier*
Abteilung Pharmakologie I
der Albert-Ludwigs-Universität
Hermann-Herder-Str. 5

7800 Freiburg i. Br.
Fed. Rep. Germany

Dr. Roger *Anwyl*
Department of Physiology
University of Dublin
Trinity College

Dublin 2
Ireland

Dr. Alan *Artola*
Max-Planck-Institut
für Hirnforschung
Deutschordenstraße 46

6000 Frankfurt 71
Fed. Rep. Germany

Dr. Barry *Bank*
Molecular & Cellular Neurobiology
Dept. of Health & Human Services
National Institute of Health

Bethesda, Maryland 20205
U. S. A.

Dr. Lorenzo *Beani*
Istituto di Farmacologia
Universita di Ferrara
Via Fossata di Mortara 23

44100 Ferrara
Italy

Dr. Theodore W. *Berger*
Department of
Behavioral Neuroscience
University of Pittsburgh

Pittsburgh, PA 15260
U. S. A.

Dr. Clementina *Bianchi*
Istituto di Farmacologia
Universita di Ferrara
Via Fossata di Mortara 23

44100 Ferrara
Italy

Dr. Albertino *Bigiani*

Dept. Fisiologia e Biochimica
Via S. Zeno 31

56100 Pisa
Italy

Dr. Tim V. *Bliss*
National Institute
for Medical Research
The Ridgeway, Mill Hill

London NW7 1AA
Great Britain

Dr. David A. *Brown*
Department of Pharmacology
University College London
Gower Street

London WC1E 6BT
Great Britain

Mr. Joachim *Buhmann*
Technische Universität
- Theoretische Physik -
James-Franck-Str.

8046 Garching bei München
Fed. Rep. Germany

Dr. Alberto *Ferrús*
Consejo Sup. de Invest. Científicas
Instituto Cajal
Velázques, 144

28006 Madrid
Spain

Dr. John H. *Byrne*
Dept. of Neurobiology & Anatomy
University of Texas, Medical School
PO Box 20708

Houston, Texas 77225
U. S. A.

Dr. Karl F. *Fischbach*
Institut für Biologie III
der Albert-Ludwigs-Universität
Schänzlestr. 1

7800 Freiburg i. Br.
Fed. Rep. Germany

Mrs. Renate *Czech*
Institut für Biologie III
der Albert-Ludwigs-Universität
Schänzlestr. 1

7800 Freiburg i. Br.
Fed. Rep. Germany

Dr. Elisabeth *Folkers*
Massachusetts Institute
of Technology
Whitaker College E25 - 436

Cambridge, MA 02139
U. S. A.

Dr. Yadin *Dudai*
Dept. of Neurobiology
The Weizmann Institute
of Science

Rehovot 76100
Israel

Dr. Martine *Galissie*
Université Paul Sabatier
Centre de Recherche en Biologie
118, route de Narbonne

31062 Toulouse Cédex
France

Dr. Josef *Dudel*
Physiologisches Institut
der Technischen Universität
Biedersteiner Str. 29

8000 München 40
Fed. Rep. Germany

Dr. H. *Gerschenfeld*
Laboratoire de Neurobiologie
École Normale Supérieure
46, rue d'Ulm

75230 Paris Cédex 05
France

Dr. Peter *Evans*
AFRC Unit of Insect Neurophys.
University of Cambridge
Downing Street

Cambridge CB2 3EJ
Great Britain

Dr. W. H. *Gispen*
Dept. of Molecular Neurobiology
University of Utrecht
Padualaan 8

3508 TB Utrecht
The Netherlands

Dr. Manfred *Göthert*
Institut für Pharmakologie
und Toxikologie der Universität
Reuterstr. 2 b

5300 Bonn 1
Fed. Rep. Germany

Dr. Robert *Hawkins*
College of Physicians & Surgeons
Columbia University
722 West 168th Street

New York, NY 10032
U. S. A.

Dr. Martin *Heisenberg*
Institut für Genetik
der Universität Würzburg
Röntgenring 11

8700 Würzburg
Fed. Rep. Germany

Dr. Georg *Hertting*
Institut für Pharmakologie
der Albert-Ludwigs-Universität
Hermann-Herder-Str. 5

7800 Freiburg i. Br.
Fed. Rep. Germany

Dr. Eamonn *Kelly*
Dept. of Pharmacology & Therap.
University of Leicester
University Road

Leicester LE1 7RH
Great Britain

Dr. Heinz *Kilbinger*
Pharmakologisches Institut
der Universität
Obere Zahlbacher Str. 67

6500 Mainz
Fed. Rep. Germany

Dr. Giulio *Levi*
Laboratorio di Fisiopatologia
Istituto Superiore di Sanitá
Viale Regina Elena 299

00161 Roma
Italy

Dr. Marina *Lynch*
MRC - Division of Neurophysiology
and Neuropharmacology
The Ridgeway, Mill Hill

London NW7 1AA
Great Britain

Dr. Robert *Malenka*
School of Medicine
Dept. of Pharmacology
University of California

San Francisco, CA 94143
U. S. A.

Dr. Jean *Médioni*
Université Paul Sabatier
Centre de Recherche en Biologie
118, route de Narbonne

31062 Toulouse Cédex
France

Dr. Randolf *Menzel*
Institut für Tierphysiologie (WE5)
Freie Universität Berlin
Königin-Luise-Str. 28-30

1000 Berlin - Dahlem 33
Fed. Rep. Germany

Dr. Erich *Muscholl*
Institut für Pharmakologie
der Universität
Obere Zahlbacher Str. 67

6500 Mainz
Fed. Rep. Germany

Dr. R. Alan *North*
Institute for Adv. Biomed. Research
Oregon Health Sciences University
3181 S.W. Sam Jackson Park Road

Portland, Oregon 97201
U. S. A.

Dr. Wilfried *Seifert*
MPI für Biophysikalische Chemie
Abteilung Neurobiologie
Am Faßberg

3400 Göttingen
Fed. Rep. Germany

Dr. Michael *O'Shea*
Laboratoire de Neurobiologie
Département de Biologie Animale
20, Boulevard d'Yvoy

1211 Genève 4
Switzerland

Dr. Hanns-Christof *Spatz*
Institut für Biologie III
der Albert-Ludwigs-Universität
Schänzlestr. 1

7800 Freiburg i. Br.
Fed. Rep. Germany

Dr. Giancarlo *Pepeu*

Dipartimento di Farmacologia
Viale G.B. Morgagni 65

50134 Firenze
Italy

Dr. Klaus *Starke*
Institut für Pharmakologie
der Albert-Ludwigs-Universität
Hermann-Herder-Str. 5

7800 Freiburg i. Br.
Fed. Rep. Germany

Dr. Maurizio *Raiteri*
Universita di Genova
Istituto di Farmacologia
Viale Cembrano 4

16148 Genova
Italy

Dr. Lennart *Stjärne*

Karolinska Institutet
Fysiologiska Institutionen I

10401 Stockholm
Sweden

Dr. Heinrich *Reichert*
Laboratoire de Neurobiologie
Département de Biologie Animale
20, Boulevard d'Yvoy

1211 Genève 4
Switzerland

Ms Michiru *Sugawa*
Institut für Tierphysiologie (WE5)
Freie Universität Berlin
Königin-Luise-Str. 28-30

1000 Berlin - Dahlem 33
Fed. Rep. Germany

Dr. Klaus *Schulten*
Technische Universität
- Theoretische Physik -
James-Franck-Str.

8046 Garching bei München
Fed. Rep. Germany

Mrs. Margrit *Tischendorf*

Georg-Thieme-Verlag
Rüdigerstr. 14

7000 Stuttgart 30
Fed. Rep. Germany

Dr. Tim *Tully*
Department of Biology
Brandeis University
Bassine 235

Waltham, MA 02254
U. S. A.

Mr. Achim *Wachenfeld*

Mülhauser Str. 4

7800 Freiburg i. Br.
Fed. Rep. Germany

Dr. Laura *Vitellaro-Zuccarello*
Dipart. di Fisiologia e Biochimica
Universita degli Studi di Milano
Via Celoria 26

20133 Milano
Italy

Index

D

E

N

O

P

Q

R

S

T

U

V

W

Y / Z

NATO ASI Series H

Vol. 1: **Biology and Molecular Biology of Plant-Pathogen Interactions.** Edited by J.A. Bailey. 415 pages. 1986.

Vol. 2: **Glial-Neuronal Communication in Development and Regeneration.** Edited by H.H. Althaus and W. Seifert. 865 pages. 1987.

Vol. 3: **Nicotinic Acetylcholine Receptor: Structure and Function.** Edited by A. Maelicke. 489 pages. 1986.

Vol. 4: **Recognition in Microbe-Plant Symbiotic and Pathogenic Interactions.** Edited by B. Lugtenberg. 449 pages. 1986.

Vol. 5: **Mesenchymal-Epithelial Interactions in Neural Development.** Edited by J.R. Wolff, J. Sievers, and M. Berry. 428 pages. 1987.

Vol. 6: **Molecular Mechanisms of Desensitization to Signal Molecules.** Edited by T.M. Konijn, P.J.M. Van Haastert, H. Van der Starre, H. Van der Wel, and M.D. Houslay. 336 pages. 1987.

Vol. 7: **Gangliosides and Modulation of Neuronal Functions.** Edited by H. Rahmann. 647 pages. 1987.

Vol. 8: **Molecular and Cellular Aspects of Erythropoietin and Erythropoiesis.** Edited by I.N. Rich. 460 pages. 1987.

Vol. 9: **Modification of Cell to Cell Signals During Normal and Pathological Aging.** Edited by S. Govoni and F. Battaini. 297 pages. 1987.

Vol. 10: **Plant Hormone Receptors.** Edited by D. Klämbt. 319 pages. 1987.

Vol. 11: **Host-Parasite Cellular and Molecular Interactions in Protozoal Infections.** Edited by K.-P. Chang and D. Snary. 425 pages. 1987.

Vol. 12: **The Cell Surface in Signal Transduction.** Edited by E. Wagner, H. Greppin, and B. Millet. 243 pages. 1987.

Vol. 13: **Toxicology of Pesticides: Experimental, Clinical and Regulatory Perspectives.** Edited by L.G. Costa, C.L. Galli, and S.D. Murphy. 320 pages. 1987.

Vol. 14: **Genetics of Translation. New Approaches.** Edited by M.F. Tuite, M. Picard, and M. Bolotin-Fukuhara. 524 pages. 1988.

Vol. 15: **Photosensitisation. Molecular, Cellular and Medical Aspects.** Edited by G. Moreno, R.H. Pottier, and T.G. Truscott. 521 pages. 1988.

Vol. 16: **Membrane Biogenesis.** Edited by J.A.F. Op den Kamp. 477 pages. 1988.

Vol. 17: **Cell to Cell Signals in Plant, Animal and Microbial Symbiosis.** Edited by S. Scannerini, D. Smith, P. Bonfante-Fasolo, and V. Gianinazzi-Pearson. 414 pages. 1988.

Vol. 18: **Plant Cell Biotechnology.** Edited by M.S.S. Pais, F. Mavituna, and J.M. Novais. 540 pages. 1988.

Vol. 19: **Modulation of Synaptic Transmission and Plasticity in Nervous Systems.** Edited by G. Hertting and H.-C. Spatz. 457 pages. 1988.

MIX
Papier aus verantwortungsvollen Quellen
Paper from responsible sources
FSC® C105338

If you have any concerns about our products,
you can contact us on
ProductSafety@springernature.com

In case Publisher is established outside the EU,
the EU authorized representative is:
Springer Nature Customer Service Center GmbH
Europaplatz 3, 69115 Heidelberg, Germany

Printed by Libri Plureos GmbH
in Hamburg, Germany